Evolutionary Conservation Biology

As anthropogenic environmental changes spread and intensify across the planet, conservation biologists have to analyze dynamics at large spatial and temporal scales. Ecological and evolutionary processes are then closely intertwined. In particular, evolutionary responses to anthropogenic environmental change can be so fast and pronounced that conservation biology can no longer afford to ignore them. To tackle this challenge, currently disparate areas of conservation biology ought to be integrated into a unified framework. Bringing together conservation genetics, demography, and ecology, this book introduces evolutionary conservation biology as an integrative approach to managing species in conjunction with ecological interactions and evolutionary processes. Which characteristics of species and which features of environmental change foster or hinder evolutionary responses in ecological systems? How do such responses affect population viability, community dynamics, and ecosystem functioning? Under which conditions will evolutionary responses ameliorate, rather than worsen, the impact of environmental change? This book shows that the grand challenge for evolutionary conservation biology is to identify strategies for managing genetic and ecological conditions such as to ensure the continued operation of favorable evolutionary processes in natural systems embedded in a rapidly changing world.

RÉGIS FERRIÈRE is Professor of Mathematical Ecology in the Department of Ecology at the École Normale Supérieure, Paris, France, and Associate Professor of Evolutionary Ecology in the Department of Ecology and Evolutionary Biology at the University of Arizona, Tucson, USA.

ULF DIECKMANN is Project Leader of the Adaptive Dynamics Network at the International Institute for Applied Systems Analysis (IIASA) in Laxenburg, Austria. He is coeditor of *The Geometry of Ecological Interactions: Simplifying Spatial Complexity*, of *Adaptive Dynamics of Infectious Diseases: In Pursuit of Virulence Management*, and of *Adaptive Speciation.*

DENIS COUVET is Professor at the Muséum National d'Histoire Naturelle, Paris, France, and Associate Professor at the École Polytechnique, Paris, France.

Cambridge Studies in Adaptive Dynamics

Series Editors

ULF DIECKMANN
Adaptive Dynamics Network
International Institute for
Applied Systems Analysis
A-2361 Laxenburg, Austria

JOHAN A.J. METZ
Institute of Biology
Leiden University
NL-2311 GP Leiden
The Netherlands

The modern synthesis of the first half of the twentieth century reconciled Darwinian selection with Mendelian genetics. However, it largely failed to incorporate ecology and hence did not develop into a predictive theory of long-term evolution. It was only in the 1970s that evolutionary game theory put the consequences of frequency-dependent ecological interactions into proper perspective. Adaptive Dynamics extends evolutionary game theory by describing the dynamics of adaptive trait substitutions and by analyzing the evolutionary implications of complex ecological settings.

The *Cambridge Studies in Adaptive Dynamics* highlight these novel concepts and techniques for ecological and evolutionary research. The series is designed to help graduate students and researchers to use the new methods for their own studies. Volumes in the series provide coverage of both empirical observations and theoretical insights, offering natural points of departure for various groups of readers. If you would like to contribute a book to the series, please contact Cambridge University Press or the series editors.

1. *The Geometry of Ecological Interactions: Simplifying Spatial Complexity*
   Edited by Ulf Dieckmann, Richard Law, and Johan A.J. Metz
2. *Adaptive Dynamics of Infectious Diseases: In Pursuit of Virulence Management*
   Edited by Ulf Dieckmann, Johan A.J. Metz, Maurice W. Sabelis, and Karl Sigmund
3. *Adaptive Speciation*
   Edited by Ulf Dieckmann, Michael Doebeli, Johan A.J. Metz, and Diethard Tautz
4. *Evolutionary Conservation Biology*
   Edited by Régis Ferrière, Ulf Dieckmann, and Denis Couvet

In preparation:

Branching Processes: Variation, Growth, and Extinction of Populations
Edited by Patsy Haccou, Peter Jagers, and Vladimir A. Vatutin

Fisheries-induced Adaptive Change
Edited by Ulf Dieckmann, Olav Rune Godø, Mikko Heino, and Jarle Mork

Elements of Adaptive Dynamics
Edited by Ulf Dieckmann and Johan A.J. Metz

Evolutionary Conservation Biology

Edited by

Régis Ferrière, Ulf Dieckmann, and Denis Couvet

CAMBRIDGE UNIVERSITY PRESS
Cambridge, New York, Melbourne, Madrid, Cape Town, Singapore, São Paulo, Delhi

Cambridge University Press
The Edinburgh Building, Cambridge CB2 8RU, UK

Published in the United States of America by Cambridge University Press, New York

www.cambridge.org
Information on this title: www.cambridge.org/9780521116084

First published 2004
This digitally printed version 2009

A catalogue record for this publication is available from the British Library

ISBN 978-0-521-82700-3 hardback
ISBN 978-0-521-11608-4 paperback

Contents

Contributing Authors

Judith Bronstein (judieb@u.arizona.edu) Department of Ecology and Evolutionary Biology, University of Arizona, Tucson, AZ 85712, USA

Reinhard Bürger (Reinhard.Buerger@univie.ac.at) Department of Mathematics, University of Vienna, Strudlhofgasse 4, A-1090 Vienna, Austria

Bruno Colas (bcolas@snv.jussieu.fr) Laboratoire d'Écologie, UMR-CNRS 7625, CC 237, Université de Paris 6, 7 Quai Saint-Bernard, F-75252 Paris Cedex 05, France

Denis Couvet (couvet@mnhn.fr) Muséum National d'Histoire Naturelle, Centre de Recherches en Biologie des Populations d'Oiseaux, 55 rue Buffon, F-75005 Paris, France

Ulf Dieckmann (dieckman@iiasa.ac.at) Adaptive Dynamics Network, International Institute for Applied Systems Analysis, A-2361 Laxenburg, Austria

Régis Ferrière (Regis.Ferriere@biologie.ens.fr) Laboratoire d'Écologie, École Normale Supérieure, CNRS-URA 258, 46 rue d'Ulm, F-75230 Paris Cedex 05, France; Adaptive Dynamics Network, International Institute for Applied Systems Analysis, A-2361 Laxenburg, Austria; and Department of Ecology and Evolutionary Biology, University of Arizona, Tucson, AZ 85721, USA

Richard Frankham (rfrankha@rna.bio.mq.edu.au) Department of Biological Sciences, Macquarie University, New South Wales 2109, Australia

Wilfried Gabriel (wilfried.gabriel@LMU.de) Biologie II, Evolutionsökologie, Zoological Institute of LMU, Karlstraße 23–25, D-80333 Munich, Germany

Oscar E. Gaggiotti (Oscar.Gaggiotti@helsinki.fi) Department of Ecology and Systematics, University of Helsinki, FIN-00014 Helsinki, Finland

Richard Gomulkiewicz (gomulki@wsu.edu) Department of Mathematics, Washington State University, Pullman, WA 99164, USA

Mats Gyllenberg (Mats.Gyllenberg@utu.fi) Department of Mathematics, University of Turku, FIN-20014 Turku, Finland

Ilkka Hanski (ilkka.hanski@helsinki.fi) Metapopulation Research Group, Department of Ecology and Systematics, University of Helsinki, PO Box 17, FIN-00014 Helsinki, Finland

Robert D. Holt (rdholt@zoo.ufl.edu) Department of Zoology, 223 Bartram Hall, PO Box 118525, University of Florida, Gainesville, FL 326ll, USA

Kimberly A. Hughes (kahughes@life.uiuc.edu) Department of Animal Biology, School of Integrative Biology, 505 S. Goodwin Avenue, University of Illinois, Urbana, IL 61801, USA

Alexandra G. Imasheva (imasheva@vigg.ru) N.I. Vavilov Institute of General Genetics, Russian Academy of Sciences, Gubkin Street 3, GSP-1, Moscow-333, 117809, Russia

Joel Kingsolver (jgking@bio.unc.edu) University of North Carolina, CB# 3280, Coker Hall, Chapel Hill, NC 27599, USA

Christoph Krall (christoph.krall@univie.ac.at) Department of Mathematics, University of Vienna, Strudlhofgasse 4, A-1090 Vienna, Austria

Stéphane Legendre (legendre@ens.fr) Laboratoire d'Écologie, École Normale Supérieure, 46 rue d'Ulm, F-75230 Paris Cedex 05, France

Donald A. Levin (dlevin@uts.cc.utexas.edu) Section of Integrative Biology, University of Texas, Austin, TX 78713, USA

Volker Loeschcke (volker.loeschcke@biology.au.dk) Department of Ecology and Genetics, Aarhus University, Ny Munkegade, Building 540, DK-8000 Aarhus C, Denmark

Michel Loreau (loreau@ens.fr) Laboratoire d'Écologie, École Normale Supérieure, 46 rue d'Ulm, F-75230 Paris Cedex 05, France

Claire de Mazancourt (c.mazancourt@ic.ac.uk) Department of Biology, Imperial College at Silwood Park, Ascot, Berkshire, SL5 7PY, United Kingdom

Johan A.J. Metz (metz@rulsfb.leidenuniv.nl) Institute of Biology, Leiden University, Van der

Klaauw Laboratory, P.O.Box 9516, NL-2300 RA Leiden, The Netherlands; and Adaptive Dynamics Network, International Institute for Applied Systems Analysis, A-2361 Laxenburg, Austria

Leonard Nunney (leonard.nunney@ucr.edu) Department of Biology, University of California, Riverside, CA 92521, USA

Kalle Parvinen (kalparvi@utu.fi) Department of Mathematics, University of Turku, FIN-20014 Turku, Finland

David Reznick (david.reznick@ucr.edu) Department of Biology, University of California, Riverside, CA 92521, USA

Helen Rodd (hrodd@zoo.utoronto.ca) Department of Zoology, University of Toronto, Toronto, Ontario M5S 3G5, Canada

Ryan J. Sawby (ryan.sawby@gcmail.maricopa.edu) Department of Biology, Glendale Community College, 6000 W. Olive Ave, Glendale, AZ 85302, USA

Chris D. Thomas (c.d.thomas@leeds.ac.uk) Centre for Biodiversity and Conservation, School of Biology, University of Leeds, Leeds, LS2 9JT, United Kingdom

Michael C. Whitlock (whitlock@zoology.ubc.ca) Department of Zoology, University of British Columbia, Vancouver, BC V6T 1Z4, Canada

Acknowledgments

Development of this book took place at the International Institute of Applied Systems Analysis (IIASA), Laxenburg, Austria, at which IIASA's former directors Gordon J. MacDonald and Arne B. Jernelöv, and current director Leen Hordijk, have provided critical support. Two workshops at IIASA brought together all the authors to discuss their contributions and thus served as an important element in the strategy to achieve as much continuity across the subject areas as possible.

Financial support toward these workshops given by the European Science Foundation's Theoretical Biology of Adaptation Programme is gratefully acknowledged. Régis Ferrière and Ulf Dieckmann received support from the European Research Training Network *ModLife* (Modern Life-History Theory and its Application to the Management of Natural Resources), funded through the Human Potential Programme of the European Commission.

The success of any edited volume aspiring to textbook standards very much depends on the cooperation of the contributors in dealing with the many points the editors are bound to raise. We are indebted to all our authors for their cooperativeness and patience throughout the resultant rounds of revision. The book has benefited greatly from the support of the Publications Department at IIASA; we are especially grateful to Ewa Delpos, Anka James, Martina Jöstl, Eryl Maedel, John Ormiston, and Lieselotte Roggenland for the excellent work they have put into preparing the camera-ready copy of this volume. Any mistakes that remain are, however, our responsibility.

Régis Ferrière
Ulf Dieckmann
Denis Couvet

Notational Standards

To allow for a better focus on the content of chapters and to highlight their interconnections, we have encouraged all the authors of this volume to adhere to the following notational standards:

α	Ecological interaction coefficient
b	Per capita birth rate
d	Per capita death rate
r	Per capita growth rate
R_0	Per capita growth ratio per generation
K	Carrying capacity
N	Population size
N_e	Effective population size
x, y, z	Phenotypic or allelic trait values
G	Genetic contribution to phenotype
E	Environmental contribution to phenotype
P	Phenotype
V_G	Genetic variance(–covariance)
V_E	Environmental variance(–covariance)
V_P	Phenotypic variance(–covariance)
V_A	Additive genetic variance(–covariance)
V_D	Dominance genetic variance(–covariance)
V_I	Epistatic genetic variance(–covariance)
$V_{G\times E}$	Genotype–environment variance(–covariance)
h^2	Heritability
S	Selection coefficient/differential
R	Response to selection
u	Per locus mutation rate
U	Genomic mutation rate
L	Mutation load
F	Inbreeding coefficient
H	Level of heterozygosity
f	Fitness in continuous time ($f = 0$ is neutral)
W	Fitness in discrete time ($W = 1$ is neutral)
t	Time
T	Duration
τ	Delay time

n	Number of entities other than individuals
p, q	Probability or (dimensionless) frequency
i, j, k	Indices
$\mathbb{E}(...)$	Mathematical expectation
$\Delta ...$	Difference
$\bar{...}$	Average
$\hat{...}$	Equilibrium value

1

Introduction

Régis Ferrière, Ulf Dieckmann, and Denis Couvet

Evolution has molded the past and paves the future of biodiversity. As anthropogenic damage to the Earth's biota spans unprecedented temporal and spatial scales, it has become urgent to tear down the traditional scientific barriers between conservation studies of populations, communities, and ecosystems from an evolutionary perspective. Acknowledgment that ecological and evolutionary processes closely interact is now mandatory for the development of management strategies aimed at the long-term conservation of biodiversity. The purpose of this book is to set the stage for an integrative approach to conservation biology that aims to manage *species* as well as ecological and evolutionary *processes*.

Human activities have brought the Earth to the brink of biotic crisis. Over the past decades, habitat destruction and fragmentation has been a major cause of population declines and extinctions. Famous examples include the destruction and serious degradation that have swept away over 75% of primary forests worldwide, about the same proportion of the mangrove forests of southern Asia, 98% or more of the dry forests of western Central America, and native grasslands and savannas across the USA. As human impact spreads and intensifies over the whole planet, conservation concerns evolve. Large-scale climatic changes have begun to endanger entire animal communities (Box 1.1). Amphibian populations, for example, have suffered widespread declines and extinctions in many parts of the world as a result of atmospheric change mediated through complex local ecological interactions. The time scale over which such biological consequences of global change unfolds is measured in decades to centuries. The resultant challenge to conservation biologists is to investigate large spatial and temporal scales over which ecological and evolutionary processes become closely intertwined. To tackle this challenge, it has become urgent to integrate currently disparate areas of conservation biology into a unified framework.

1.1 Demography, Genetics, and Ecology in Conservation Biology

For more than 20 years, conservation biology has developed along three rather disconnected lines of fundamental research and practical applications: conservation demography, conservation genetics, and conservation ecology. *Conservation demography* focuses on the likely fate of threatened populations and on identifying the factors that determine or alter that fate, with the aim of maintaining endangered species in the short term. To this end, stochastic models of population dynamics are combined with field data to predict how long a given population of an endangered

Box 1.1 Global warming and biological responses

Increasing greenhouse gas concentrations are expected to have significant impacts on the world's climate on a time scale of decades to centuries. Evidence from long-term monitoring suggests that climatic conditions over the past few decades have been anomalous compared with past climate variations. Recent climatic and atmospheric trends are already affecting the physiologies, life histories, and abundances of many species and have impacted entire communities (Hughes 2000).

Rapid and sometimes dramatic changes in the composition of communities of marine organisms provide evidence of recent climate-induced transformations. A 20-year (1974 to 1993) survey of a Californian reef fish assemblage shows that the proportion of northern, cold-affinity species declined from approximately 50% to about 33%, and the proportion of warm-affinity southern species increased from about 25% to 35%. These changes in species composition were accompanied by substantial (up to 92%) declines in the abundance of most species (Holbrook *et al.* 1997).

Ocean warming, especially in the tropics, may also affect terrestrial species. Increased evaporation levels generate large amounts of water vapor, which accelerates atmospheric warming through the release of latent heat as the moisture condenses. In tropical regions, such as the cloud forests of Monteverde, Costa Rica, this process results in an elevated cloud base and a decline in the frequency of mist days, a trend that has been associated strongly with synchronous declines in the populations of birds, reptiles, and amphibians (Pounds *et al.* 1999).

Since the mid-1980s, dramatic declines in amphibian populations have occurred in many parts of the world, including a number of apparent extinctions. Kiesecker *et al.* (2001) presented evidence that climate change may be the underlying cause of this global deterioration. In extremely dry years, reductions in the water depth of sites used by amphibians for egg laying increase the exposure of their embryos to damaging ultraviolet B radiation, which allows lethal skin infection by pathogens. Kiesecker *et al.* (2001) link the dry conditions in their study sites in western North America to sea-surface warming in the Pacific, and so identify a chain of events through which large-scale climate change causes wholesale mortality in an amphibian population.

species is likely to persist *under given circumstances*. Conservation demography can advertise some notable achievements, such as devising measures to boost emblematic species like the grizzly bear in Yellowstone National Park, planning the rescue of Californian condors, or recommending legal action to protect tigers in India and China.

A different stance is taken by *conservation genetics*, which focuses on the issue of preserving genetic diversity. Although the practical relevance of population genetics in conservation planning has been heatedly disputed over the past 15 years, empirical studies have lent much weight to the view that the loss of genetic diversity can have short-term effects, like inbreeding depression, that account for a significant fraction of a population's risk of extinction (Saccheri *et al.* 1998). There

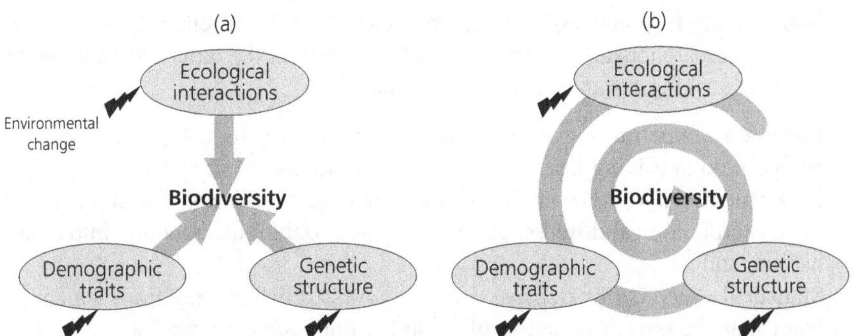

Figure 1.1 The integrative scope of evolutionary conservation biology (b) reconciles the three traditional approaches to the management of biodiversity (a).

is even experimental support for the contention that restoring genetic variation (to reduce inbreeding depression) can reverse population trajectories that would otherwise have headed toward extinction (Madsen *et al.* 1999).

The third branch of conservation biology, *conservation ecology*, relies on utilizing, for ecosystem management, the extensive knowledge developed by community ecologists and ecosystem theorists, in particular of the complicated webs of biotic and abiotic interactions that shape patterns of biodiversity and productivity. All the species in a given ecosystem are linked together, and when disturbances – such as biological invasions, disease outbreaks, or human overexploitation – cause one species to rise or fall in numbers, the effects may cascade throughout these webs. From a conservation perspective, one of the central questions for community and ecosystem ecologists is how the diversity and complexity of ecological interactions influence the resilience of ecosystems to disturbances.

All ecologists and population geneticists agree that evolutionary processes are of paramount importance to understand the genetic composition, community structure, and ecological functioning of natural ecosystems. However, relatively little integration of demographic, genetic, and ecological processes into a unified approach has actually been achieved to enable a better understanding of patterns of biodiversity and their response to environmental change (Figure 1.1). This book demonstrates why such an integrative stance is increasingly necessary, and offers theoretical and empirical avenues for progress in this direction.

1.2 Toward an Evolutionary Conservation Biology

All patterns of biodiversity that we observe in nature reflect a long evolutionary history, molded by a variety of evolutionary processes that have unfolded since life appeared on our planet. In this context, should we be content with safeguarding as much as we can of the current planetary stock of species? Or should we pay equal, if not greater, attention to fostering ecological and evolutionary processes that are responsible for the generation and maintenance of biodiversity?

Evolutionary responses to environmental changes can, indeed, be so fast and so strong that researchers are able to witness them, both in the laboratory and in the wild. Some striking instances (Box 1.2) include:

- Laboratory experiments on fruit flies that illuminate the role of intraspecific competition in driving fast, adaptive responses to pollution;
- Experiments on Caribbean lizards under natural conditions that demonstrate rapid morphological differentiation in response to their introduction into a new habitat; and
- Statistical analysis of extensive data on harvested fish stocks, from which we learn that the overexploitation of these natural resources can induce a rapid life-history evolution that must not be ignored when the status of harvested populations is assessed.

From their review of the studies of microevolutionary rates, Hendry and Kinnison (1999) concluded that rapid microevolution perhaps represents the norm in contemporary populations confronted with environmental change.

Looking much further back, analysis of macroevolutionary patterns suggests further evidence that the interplay of ecological and evolutionary processes is essential in securing the diversity and stability of entire communities challenged by environmental disturbances. Striking patterns of ecological and morphological stability observed in some paleontological records (e.g., from the Paleozoic Appalachian basin) are now explained in terms of "ecological locking": in this view, selection enables populations to respond swiftly to high-frequency disturbances, but is constrained by ecological conditions that change on an altogether slower time scale (Morris *et al.* 1995). Rapid microevolutionary processes driven and constrained by ecological interactions are therefore believed to be critical for the resilience of ecosystems challenged by environmental disturbances on a wide range of temporal and spatial scales.

Such empirical evidence for a close interaction of ecological and evolutionary processes in shaping patterns of biodiversity prompts a series of important questions that should feature prominently on the research agenda of evolutionary conservation biologists:

- How do adaptive responses to environmental threats affect population persistence?
- What are the key demographic, genetic, and ecological determinants of a species' evolutionary potential for adaptation to environmental challenges?
- Which characteristics of environmental change foster or hinder the adaptation of populations?
- How should the evolutionary past of ecological communities influence contemporary decisions about their management?
- How should we prioritize conservation measures to account for the immediate, local effects of anthropogenic threats and for the long-term, large-scale responses of ecosystems?

Box 1.2 Fast evolutionary responses to environmental change

Pollution raises threats that permeate entire food webs. Ecological and evolutionary mechanisms can interact to determine the response of a particular population to the pollution of its environment. This has been shown by Bolnick (2001), who conducted a series of experiments on fruit flies (*Drosophila melanogaster*). By introducing cadmium-intolerant populations to environments that contained both cadmium-free and cadmium-laced resources, he showed that populations experiencing high competition adapted to cadmium more rapidly, in no more than four generations, than low-competition populations. The ecological process of intraspecific competitive interaction can therefore act as a potent evolutionary force to drive rapid niche expansion.

Reintroduction of locally extinct species and reinforcement of threatened populations are important tools for conservation managers. A study by Losos *et al.* (1997) investigated, through a replicated experiment, how the characteristics of isolated habitats and the sizes of founder populations affected the ecological success and evolutionary differentiation of morphological characters. To this end, founder populations of 5–10 lizards (*Anolis sagrei*) from a large island were introduced into 14 much smaller islands that did not contain lizards naturally, probably because of periodic hurricanes. The study indicates that founding populations of lizards, despite their small initial size, can survive and rapidly adapt over a 10–14 year period (about 15 generations) to the new environmental conditions they encounter.

Overexploitation of natural ecosystems is a major concern to conservation biologists. Heavy exploitation can exert strong selective pressures on harvested populations, as in the case of the Northeast Arctic cod (*Gadus morhua*). The exploitation pattern of this stock was changed drastically in the early 20th century with the widespread introduction of motor trawling in the Barents Sea. Over the past 50 years, a period that corresponds to 5–7 generations, the life history of Northeast Arctic cod has exhibited a dramatic evolutionary shift toward earlier maturation (Jørgensen 1990; Godø 2000; Heino *et al.* 2000, 2002). The viability of a fish stock is therefore not just a matter of how many fish are removed each year; to predict the stock's fate, the concomitant evolutionary changes in the fish life-history induced by exploitation must also be accounted for. These adaptive responses are even likely to cascade, both ecologically and evolutionarily, to other species in the food chain and have the potential to impact the whole marine Arctic ecosystem.

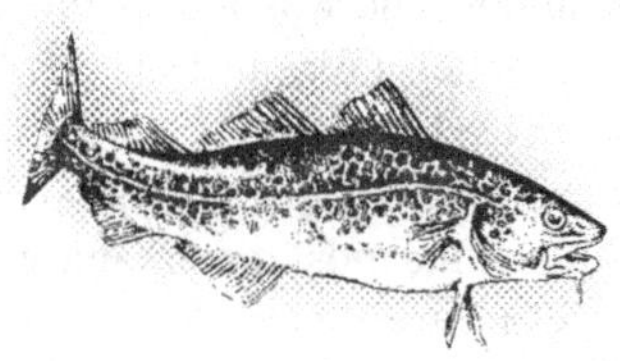

Tackling these questions will require a variety of complementary approaches that are based on a solid theoretical framework. In Box 1.3, we outline the concept of the "environment feedback loop" that has been proposed as a suitable tool to link the joint operation of ecological and evolutionary processes to the dynamics of populations.

1.3 Environmental Challenges and Evolutionary Responses

Complex selective pressures on phenotypic traits arise from the interaction of individuals with their local environment, which consists of abiotic factors as well as conspecifics, preys and predators, mutualists, and parasites. Phenotypic traits respond to these pressures under the constraints imposed by the organism's genetic architecture, and this response in turn affects how individuals shape their environment. This two-way causal relationship – from the environment to the individuals, and back – defines the environment feedback loop that intimately links ecological and evolutionary processes.

The structure of this feedback loop is decisive in determining how ecological and evolutionary processes jointly mediate the effects of biotic and abiotic environmental changes on species' persistence and community structure (Box 1.4). Three kinds of phenomena may ensue:

- Genetic constraints and environmental feedback can result in "evolutionary trapping", a situation in which a population is incapable of escaping to an alternative fitness peak that would ensure its persistence in the face of mounting environmental stress.
- Frequency-dependent selection may sometimes hasten extinction by promoting adaptations that are beneficial from the perspective of individuals and yet detrimental to the population as a whole, leading to processes of "evolutionary suicide".
- By contrast, "evolutionary rescue" may occur when a population's persistence is critically improved by adaptive changes in response to environmental degradation.

The relevance of evolutionary trapping, suicide, and rescue was first pointed out in the realm of verbal or mathematically simplified models (Wright 1931, Haldane 1932, Simpson 1944). Now, however, these concepts help to explain a wide range of evolutionary patterns in realistic models and, even more importantly, have also been documented in natural systems (Box 1.5). Among the most remarkable examples, the study of a narrow endemic plant species, *Centaurea corymbosa*, provides a clear-cut illustration of evolutionary trapping. The collection and analysis of rich demographic and genetic data sets led to the conclusion that *C. corymbosa* is stuck by its limited dispersal strategy in an evolutionary dead-end toward extinction: while variant dispersal strategies could promote persistence of the plant, they turn out to be adaptively unreachable from the population's current phenotypic state. In general, the possibility of evolutionary suicide should not come as a surprise in species that evolve lower basal metabolic rates to cope with the stress imposed

Box 1.3 The environmental feedback loop

Populations alter the environments they inhabit. The environmental feedback loop characterizes these interactions of populations with their environments and thus plays a key role in describing their demographic, ecological, and adaptive dynamics.

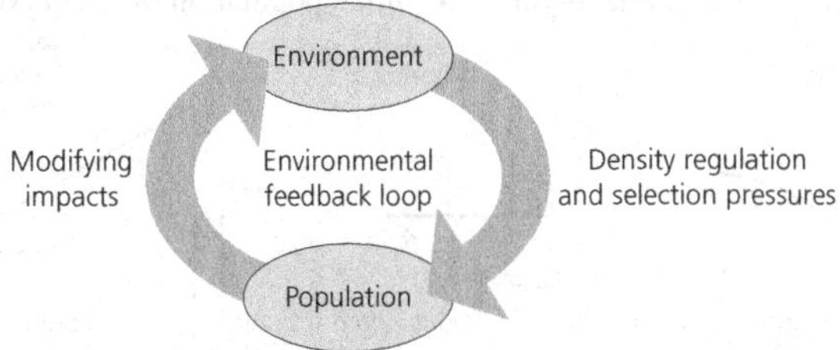

The environmental feedback loop goes beyond the self-evident interaction between a population and its environment. In fact, the concept aims to capture the pathways along which the characteristics of a resident population affect the variables that describe the state of its environment and how these, in turn, influence the demographic properties of resident or variant phenotypes in the population (Metz *et al.* 1996a; Heino *et al.* 1998). Some illustrative examples of variables that belong to these three fundamental sets are given below.

- Population characteristics: mean phenotype, abundance, or biomass, number of newborns, spatial clumping index, sex ratio, temporal variance in population size, etc. All these variables may be measured, either for the population as a whole or for stage- or age-specific subpopulations.
- Environmental variables: resource density, frequency of intraspecific fights, density of predators, helpers, or heterospecific competitors, etc.
- Demographic properties: rate of growth, fecundity, mortality, probability of maturation, dispersal propensity, etc.

The resultant loop structure involves precisely those environmental variables that are both affected by population characteristics and also impact relevant demographic properties. Specifying the environmental feedback loop therefore enables a description of all density- and/or frequency-dependent demographic mechanisms and selection pressures that operate in a considered population.

The minimal number of environmental variables or population characteristics that are sufficient to determine the demographic properties of resident and variant phenotypes is known as the dimension of the environmental feedback loop (Metz *et al.* 1996a; Heino *et al.* 1998; see also Chapter 11). This dimension has two important implications. First, it acts as an upper bound for the number of phenotypes that can stably coexist in the population (Meszéna and Metz 1999). Second, adaptive evolution can operate as an optimizing process and maximize population viability, under the constraints imposed by the underlying genetic system, only if the environmental feedback loop is one-dimensional (Metz *et al.* 1996a).

Box 1.4 Evolutionary rescue, trapping, and suicide

Populations that evolve under frequency-dependent selection have a rich repertoire of responses to environmental change. In general, such change affects, on the one hand, the range of phenotypes for which a population is not viable (gray regions in the panels below) and, on the other hand, the selection pressures (arrows) that, in turn, influence the actual phenotypic state of the population (thick curves).

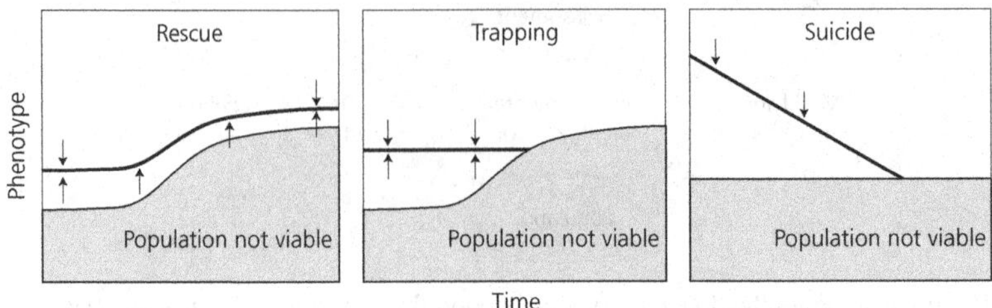

Three prototypical response patterns can be distinguished:

- Evolutionary rescue (left panel) occurs when environmental deterioration reduces the viability range of a population to such an extent that, in the absence of evolution, the population would go extinct, but simultaneously induces directional selection pressures that allow the population to escape extinction through evolutionary adaptation.
- Evolutionary trapping (middle panel) happens when stabilizing selection pressures prevent a population from responding evolutionarily to environmental deterioration. A particularly intriguing case of evolutionary trapping results from the existence of a second evolutionary attractor on which the population could persist: unable to attain this safe haven through gradual evolutionary change, the population maintains its phenotypic state until it ceases to be viable.
- Evolutionary suicide (right panel) amounts to a gradual decline, driven by directional selection, of a population's phenotypic state toward extinction. Such a tendency can be triggered and/or exacerbated by environmental change and is the clearest illustration that evolution cannot always be expected to act in the "interest" of threatened populations.

by an extreme environment, as exemplified by many animals living in deserts. A species that undergoes a reduction in metabolic rates must often divert resources away from growth and reproduction to invest in maintenance and survival. In consequence, reproductive rates fall and population densities decline, while the species' range may shrink. These adaptations confer a selective advantage to particular individuals, but run against the best interest of the species as a whole (Dobson 1996). Evolutionary rescue, on the other hand, is thought to be ubiquitous to maintain the diversity of communities. One example has recently been worked out in detail: the persistence of metapopulations of checkerspot butterflies (*Melitaea cinxia*) in degrading landscapes has been shown to depend critically on the potential for dispersal strategies to respond adaptively to environmental change.

Box 1.5 Evolutionary trapping, suicide, and rescue in the wild

Centaurea corymbosa (Asteraceae) is endemic to a small geographic area (less than 3 km^2) in southeastern France. Combining demographic and genetic analysis, Colas *et al.* (1997) concluded that the scarcity of long-range dispersal events associated with the particular life-history of this species precludes establishment of new populations and thus evolution toward colonization ability, even though nearby unoccupied sites would offer suitable habitats for the species. Thus, *C. corymbosa* seems to be trapped in a life-history pattern that will lead to its ultimate extinction.

Evolution of lower basal metabolic rates in response to environmental stress seems to pave the way for evolutionary suicide. Exposing *Drosophila* to dry conditions in the laboratory for several generations leads to the evolution of a strain of fruit fly with lowered metabolic rates and an increased resistance to dessication; incidentally, this also leads to a greater tolerance to a range of other stresses (starvation, heat shock, organic pollutants). These individuals, however, exhibit a reduction in their average birth rate, and thereby place their whole population at a high risk of extinction.

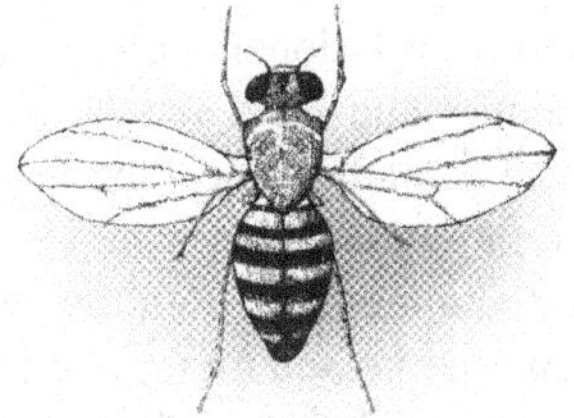

Evolutionary rescue can occur in a realistic metapopulation model of checkerspot butterflies (*Melitaea cinxia*) subject to habitat deterioration (Heino and Hanski 2001). In these simulations, which have been calibrated to an outstanding wealth of field data, habitat quality deteriorates gradually. In the absence of metapopulation evolution, habitat change leads to extinction as habitat occupation falls to zero. By contrast, the adaptive response of migration propensity results in evolutionary rescue.

Evidently, current communities must have gone through a series of environmental challenges throughout their history. Evolutionary trapping and suicide must thus have eliminated many species that lacked the ecological and genetic abilities to adapt successfully, and current species assemblages are expected to comprise those species that are endowed with a relatively high potential for evolutionary rescue (Balmford 1996). This cannot but strengthen the view that to maintain the ecological and genetic conditions required for the operation of evolutionary processes should rank among the top priorities of conservation programs.

1.4 Evolutionary Conservation Biology in Practice

In a few remarkable instances, management actions have already been undertaken with the primary aim of maintaining the potential for evolutionary responses to environmental change.

One such example is provided by the conservation plan devised for the Florida panther (*Felis concolor coryi*). Management of such an apex predator could be critical for the ecological and evolutionary functions of the whole web of interactions to which it is connected. After inbreeding depression was identified as a major threat to the panther population, a conservation scheme was implemented to manage genetic diversity. The aim was to reduce the short-term effects of inbreeding depression, but at the same time preserve those genetic combinations that render the Florida panther adapted to its local environment. Reinforcement with individuals that originated from a different subspecies, the Texas panther *F. concolor stanleyana*, was recognized as the only way to alleviate the deleterious effects of inbreeding in the remnant population of Florida panthers. The two taxa, however, are neither genetically nor ecologically "exchangeable", in the sense of Crandall *et al.* (2000), which implies that they are genetically isolated and adapted to different ecological conditions. A particular challenge for this evolutionary conservation plan was, therefore, to avoid loss of the genetic identity and local adaptation attained by the Florida panther. To address this problem, a mathematical model was constructed to evaluate the proportion of introduced individuals that would eliminate the genes responsible for inbreeding depression and maintain both the genes responsible for local adaptations and the neutral genes expressed by typical characters that distinguish the two subspecies morphologically (Hedrick 1995). Action was then undertaken according to these predictions.

Another characteristic example of a conservation program devised from an evolutionary perspective targets the Cape Floristic Region (CFR), a biodiversity hotspot of global significance located in southwestern Africa. To conserve ecological processes that maintain evolutionary potential, and thus may generate biological diversity, is of central concern to managers of the CFR. Over the past few decades, considerable insights have been gained regarding evolutionary processes in the CFR, especially for those that involve plants. Now the goal has been set to design a conservation system for the CFR that will preserve large numbers of species and their ecological interactions, as well as their evolutionary potential for fast adaptation and lineage turnover (Box 1.6). The currently proposed plan recognizes that extant CFR nature reserves are not located in a manner that will sustain eco-evolutionary processes. The plan also highlights difficult trade-offs between the conservation of either pattern or process, as well as between the requirements for biodiversity conservation and other socioeconomic factors.

The ultimate goal of conservation planning should be to foster systems that enable biodiversity to persist in the face of anthropogenic changes. The two examples mentioned above illustrate the grand challenges that evolutionary conservation biology ought to tackle by identifying ways to preserve or restore genetic and ecological conditions that will ensure the continued operation of favorable

Box 1.6 Evolutionary conservation biology in practice: the Cape Floristic Region

There are very few ecosystems in the world for which an attempt has been made to develop conservation schemes aimed to preserve biodiversity patterns and eco-evolutionary processes in the context of a rapidly changing environment. One such is a conservation scheme suggested for the Cape Floristic Region (CFR) of South Africa, a species-rich region that is recognized as a global priority target for conservation action (Cowling and Pressey 2001). A distinctive evolutionary feature of the CFR is the recent (post-Pliocene) and massive diversification of many plant lineages. Over an area of 90 000 km^2, the CFR includes some 9000 plant species, 69% of which are endemic – one of the highest concentrations of endemic plant species in the world. This diversity is concentrated in relatively few lineages that have radiated spectacularly. There is evidence for a strong ecological component of the diversification processes, which involves meso- and macroscale environmental gradients and coevolutionary dynamics in plant–pollinator systems.

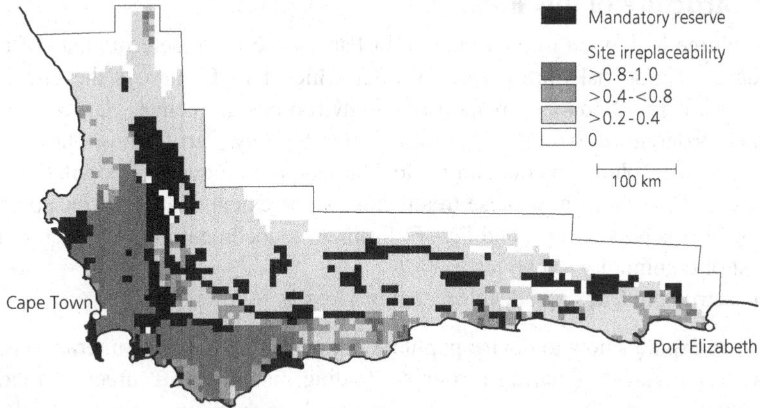

Conservation planning for the CFR aims to identify and conserve key evolutionary processes. For example, gradients from uplands to coastal lowlands and interior basins are assumed to form the ecological substrate for the radiation of plant and animal lineages. Suggested conservation targets amount to preserving at least one instance of a gradient within each of the major climate zones that are represented in the region. In addition, recognized predator–prey coevolutionary processes are motivating recommendations for the strict protection of three "mega wilderness areas". Altogether, seven types of evolutionary processes have been listed for conservation management, and by selecting from areas in which one or more of these seven processes are operating, a system of conservation areas has been designed, based on a map of "irreplaceability" (shown above). Units at the highest irreplaceability level (dark gray) include areas of habitat that are all essential to meet conservation goals, whereas units with lowest irreplaceability (white) comprise patches of habitat in a largely pristine state for which conservation goals can be achieved through the implementation of alternative measures. Black indicates units in which existing reserves cover more than 50% of the area. Each planning unit is sufficiently large to ensure the continual operation of critical ecological and environmental processes (in particular through plant–insect pollinator interactions) and a regular regime of natural fire disturbances.

eco-evolutionary processes in a rapidly changing world. In fact, while protecting species may be hard, there is widespread agreement that the conservation of ecological interactions and evolutionary processes will be more efficient and cost-effective than a species-by-species approach (Noss 1996; Thompson 1998, 1999b; Myers and Knoll 2001). This does not rule out management measures directed at particular species (based on traditional tools such as population viability analysis), but suggests that we reconsider the motivation for doing so. Species-oriented conservation efforts are expected to be more rewarding when they target endangered species that have passed through the extinction sieve of a long history of natural and anthropogenic disturbances, and therefore should possess a higher potential for evolutionary rescue. Management must also prioritize species that are likely to play a crucial role in mediating the effect of global change on the integrity of entire networks of ecological interactions.

1.5 Structure of this Book

This volume is divided into five parts. In Part A, the basic determinants of population extinction risks are reviewed, after which Part B surveys the empirical evidence for rapid adaptive responses to environmental change. Unfolding the research program of evolutionary conservation biology, Part C shows how to integrate demographic, genetic, and ecological factors in models of population viability. Part D explains how these treatments can be extended to describe spatially heterogeneous populations, and Part E discusses embedment into the overarching context of community dynamics.

This structure leads to a development of ideas as follows:

- Part A explains how to devise population models that integrate interactions between individuals (sharing resources, finding mates) with sources of random fluctuations (demographic and environmental stochasticity). Such models are the basis for extinction-risk assessment. Different forms of dependence – which lie at the heart of population regulation and the environmental feedback loop – are shown to differ dramatically in their impact on population viability. In particular, the life cycles and spatial structure of populations must be considered if extinction risks are to be evaluated accurately.
- One motivation behind denial of a role for adaptive evolution in the dynamics of threatened populations might come from a belief that evolutionary change always occurs so slowly (e.g., at the geological time scale of paleontology) that it does not interact significantly with ecological processes and rapid environmental changes. To help overcome this widespread conception, Part B reviews recent observational and experimental studies that provide striking demonstrations of fast adaptive responses of morphological and life-history traits to environmental change. Convincing evidence is available for the existence of substantial genetic variation in life-history traits, and a current exciting line of research investigates whether genetic variability can sometimes even be enhanced by stressful environmental conditions.

- The challenge to assess the quantitative impact of life-history adaptation on extinction risk has nourished new developments in evolutionary theory. Three different stances are presented in Part C. A first option is to capitalize on a well-established modeling tradition in population genetics to investigate how mutations affect the extinction risks of small or declining populations in constant environments. Quantitative genetics offers an elegant alternative approach and allows the study of the conditions under which selection enables a population to track a changing environmental optimum. Integration of all the components of the environmental feedback loop requires the effects of density- and frequency-dependent ecological interactions to be respected, and the framework of adaptive dynamics has been devised to enable this.
- Issues that arise from the spatial dimensions of population dynamics and environmental change are tackled in Part D. Spatial heterogeneity – be it intrinsic to a habitat's structure (given, for instance, by an uneven distribution of resources) or resulting from a population's dynamics (leading to self-organized patterns of abundance) – modifies existing selection pressures and creates new ones. In particular, the option of individual dispersal as an evolutionary alternative to local adaptation exists only in spatially structured settings. In this context, the ecological and evolutionary role of peripheral populations must be analyzed carefully. Empirical studies suggest that processes of evolutionary rescue and evolutionary suicide may have occurred through adaptive responses of dispersal strategies to environmental degradation.
- Today, a scarcity of biological information still tends to confine the scope of viability analyses to single populations. Nevertheless, it is clear that the network of biotic interactions in which endangered species are embedded can strongly affect their viability. Environmental change may impact the focal species directly, or indirectly through its effects on other interacting species. Specific environmental changes that directly act on a single population only may be echoed by feedback responses from interacting species. To elevate our exploration of the adaptive responses to environmental change to the community level provides the motivation for the final Part E.

In addition to pursuing the main agenda of ideas outlined above, this volume also offers coverage of a broad scope of transversal themes. Chapters written in the style of an advanced textbook can be used to access up-to-date and self-contained reviews of key topics in population and conservation biology and evolutionary ecology. Crosscutting topics include:

- Extinction dynamics of unstructured and physiologically structured populations (Chapters 2 and 3);
- Dynamics of metapopulations and evolution of dispersal (Chapters 4, 14, and 15);
- Adaptive responses of natural systems to climate change, pollution, and habitat fragmentation (Chapters 5, 12, and 15);

- Empirical studies of life-history evolution in response to environmental threats (Chapters 6, 7, and 8);
- Population genetics and quantitative genetics of small or declining populations and of metapopulations (Chapters 9, 10, 12, 13, and 15);
- Adaptive dynamics theory and its applications (Chapters 11, 14, 16, and 17);
- Explorations of the demographic and genetic causes and consequences of rarity (Chapters 5, 9, 14, 15, and 18); and
- Community dynamics through evolutionary change in interspecific relations (Chapters 16, 17, and 18).

Merging these approaches will make it possible to acquire new insights into the responses of ecological and evolutionary processes to environmental change, as well as into the implications of these responses for population persistence and ecosystem diversity. The chapters herein are intended to pave the way for such integration.

The aim of this volume is to convince readers of the urgent need for systematic research into eco-evolutionary responses to anthropogenic threats. This research needs to account for, as accurately as is practically feasible, the type of environmental change, the species' life cycle, its habitat structure, and the network of ecological interactions in which it is embedded. This is a call for innovative experimental work on laboratory organisms, for a more integrative assessment of the living conditions of threatened populations in the wild, and for an extension of our theoretical grasp of processes involved in extinction and rescue. We hope that the book will entice students and researchers in ecology, genetics, and evolutionary theory to step into this open arena.

Part A

Theory of Extinction

Introduction to Part A

Local changes in biodiversity happen through migration or speciation and through extinctions. The latter have been at the focus of conservation biology since the field's inception, and the purpose of this opening part is to review the rich theoretical foundations for our understanding of population extinction.

Specifically, we aim to understand how mechanisms that operate at the level of individuals scale up to the dynamics of populations and thus determine extinction risks. In the context of evolutionary conservation biology, this step is necessary to identify potential targets that impact on population viability. Such targets include classic life-history traits (e.g., demographic parameters such as survival probabilities, fecundity, or age at maturity) and behavioral traits that determine the effective interactions between individuals (e.g., propensities to move or migrate, competitive ability, or mate choice).

Connecting individual characteristics to population properties is also necessary to understand the origin of the selective pressures by which populations exert a feedback to individuals. Adaptive evolution usually proceeds by small steps: new phenotypes arise from mutation or recombination, and the individuals thus affected must compete with their conspecifics. Questions of viability and extinction are therefore important to address in assessing whether evolutionary innovations are retained through the persistence of their carriers or, instead, are eliminated through their extinction.

The theoretical material in this part should also be relevant to investigators with a primary interest in population viability analysis (PVA). For more than two decades, PVA has provided a fruitful approach to the quantitative assessment of endangered species; it is used to facilitate the design of management programs and to compare the relative merits of alternative conservation measures prior to their implementation. The species-oriented and short-term perspective of PVAs is not necessarily at odds with the ecosystem-oriented and long-term perspective suggested in this book: there are at least two important reasons for emphasizing the role of PVAs in the context of evolutionary conservation biology.

First, PVAs often target large vertebrates that are the ecological and evolutionary cornerstones of their ecosystems. Major ecological and evolutionary knock-on and ripple effects are expected for smaller species (and, indeed, for biotas as a whole) from the decline or extinction of such keystone species. An example is the current decline of elephants in African savannas. This species and many other large mammals have little hope of innovation in their evolutionary future, but their role in the ecosystem is so central that their extinction could alter the ecological interactions and evolutionary paths of many other species in a disastrous manner. Thus, PVAs are very useful to help maintain keystone species, especially if these are perched on the brink of extinction. This may sometimes win sufficient time to design and implement management measures at the broader level of communities and ecosystems. In a similar vein, the implementation of reserve systems to conserve ecological and evolutionary processes, like the ambitious conservation plan for the Cape Floristic Region, can only be gradual. It is therefore critical that actions be undertaken to minimize the extent to which conservation targets are compromised before measures of evolutionary conservation can take effect.

Second, the endangerment of species targeted by PVAs may often have an evolutionary basis. We now understand that small population size and a resultant high vulnerability to environmental stress can arise as a by-product of behavioral and life-history evolution toward large body size and competitive superiority, both of which have to be traded against low reproductive output. Species that have evolved such attributes are likely to have low abundance; such species must have passed through highly selective extinction sieves during their evolutionary history, and only those endowed with particular demographic and genetic features that enabled them to buffer environmental disturbances have been retained. Thus, rare species still extant today presumably are properly "equipped" by the evolutionary and coevolutionary processes to cope with perturbations. Conservation managers should therefore be aware of how and to what extent current and forthcoming challenges posed by human activities (often unprecedented in their scope and interaction) differ from the evolutionary history and context of a threatened species.

The three chapters in this part introduce the theoretical tools needed to evaluate the risk of extinction for a given population. This issue is addressed, in turn, for unstructured populations (Chapter 2), populations with structured life cycles (Chapter 3), and spatially structured populations (Chapter 4).

How do interactions between individuals influence a population's risk of extinction? In Chapter 2, Gabriel and Ferrière address this question by investigating the properties of unstructured population models in which populations are regulated through density dependence. These models are appropriate for organisms with simple life cycles. Extinction risks, which are inversely proportional to average times to extinction, respond differently to changes in different demographic parameters. Important scaling relationships depend upon the types of stochastic fluctuations to which populations are exposed. Demographic stochasticity originates from the random timing of birth and death events, from individual variation in birth and death rates, and from random fluctuations in the sex ratio. By contrast, external stochastic influences on population dynamics include environmental noise and rare catastrophes. Chapter 2 shows how the type and "color" of stochastic fluctuations interfere with the nonlinear mechanisms of population regulation to shape patterns of population viability and extinction.

As few life-history traits are required to parametrize unstructured population models, these models are particularly amenable to mathematical analysis. Such simplification, however, carries the cost of ignoring those life-history traits that govern transitions in a species' life cycle. This is problematic since developmental transitions, as well as intraspecific interactions that occur in different ways between particular developmental stages, often critically affect population dynamics.

Chapter 3 by Legendre introduces, in a didactical manner, the concepts and tools needed to relate population dynamics to the structure and parameters of life cycles that involve discrete stages. The chapter first focuses on age-dependent stages and transitions. After a review of the basic theory, it is explained how to extend classic models to account for the influence of sexual reproduction on population viability. Traits and interactions involved in mating processes can have a dramatic impact on the extinction risks of populations. As a genetic factor of demographic change induced by sexual reproduction, the consequences of inbreeding depression are discussed.

Space introduces an extra dimension of population structure and presents new challenges for the modeling of extinction dynamics. In Chapter 4, Gyllenberg, Hanski, and Metz describe a general framework for modeling spatially fragmented populations. This enables evaluation of the effects on population viability and persistence of traits that determine spatial population structure (such as offspring dispersal). Although the general treatment is mathematically rather sophisticated, the authors demonstrate the utility of their approach for particular examples, which allows the essentials to be grasped easily. The question of metapopulation growth or decline is addressed by deriving the metapopulation's basic reproduction ratio from life-history traits and environmental characteristics. Relating these parameters to metapopulation viability requires the effects of finite population size to be taken into account, which naturally leads to a discussion of stochastic metapopulation models. The resultant analysis disentangles the relative importance of local resource dynamics, regional habitat structure, and life-history traits on the extinction risk of metapopulations.

2

From Individual Interactions to Population Viability

Wilfried Gabriel and Régis Ferrière

2.1 Introduction

Early life in temporary ponds may be tough for many larval anurans. At extremely high densities, all the tadpoles develop slowly enough, in effect because of food limitation, for them to be driven to extinction. At intermediate tadpole densities, predators like salamanders can have a significant impact on small tadpoles and exert strong selective pressures for faster individual growth. At very low tadpole densities yet another aspect comes into play: predatory salamanders have no appreciable impact because tadpole growth rates are high (resources are plentiful) and encounter rates are low because of both contact probabilities and the availability of refuges.

This classic example of density-dependent selection, demonstrated by Wilbur (1984) and Travis (1984), is instructive in several respects. First, it shows that the risk of extinction of these amphibians depends on their density in a nontrivial way. At high density, regulatory mechanisms become so strong that they may result in population extinction. At very low density, the predation risk is relaxed, which facilitates persistence. At intermediate density, the population undergoes strong selective pressures on those traits for which the adaptive changes feed back onto population density, and thereby influence the risk of extinction. This fascinating case makes it plain that regulatory mechanisms that emanate from individual interactions need to be understood to anticipate the impact of environmental change and evolutionary responses on population persistence. In this chapter we examine how different types of density-dependent mechanisms influence the risk of extinction of unstructured populations subject to three types of chance fluctuations in individual traits: demographic stochasticity, interaction stochasticity, and environmental stochasticity (Box 2.1). Chapters 3 and 4 address the cases of physiologically and spatially structured populations, respectively. These chapters provide the theoretical background necessary to investigate how the risk of extinction is affected by evolutionary processes that impact life-history traits and behavioral interactions.

2.2 From Individual Interactions to Density Dependence

Density dependence is defined as the phenomenon by which the values of vital rates, such as survivorship and fecundity, depend on the density of the population. The underlying mechanisms involve interactions between individuals, which have either negative (e.g., in the case of competition for resources) or positive effects

Box 2.1 Stochastic factors of extinction

Shaffer (1981, 1987) discussed three stochastic demographic factors of extinction:

- *Demographic stochasticity* is caused by chance realizations of individual probabilities of death and reproduction in a finite population. Since independent individuals tend to be averaged out in large populations, demographic stochasticity is most important in small populations.
- *Environmental stochasticity* arises from a nearly continuous series of small or moderate perturbations that similarly affect the birth and death rates of all individuals (within each age or stage class) in a population (May 1974). In contrast to demographic stochasticity, environmental stochasticity is important in both large and small populations.
- *Catastrophes* are large environmental perturbations that act directly upon population size and cause reductions in abundance. Usually seen as rare events, catastrophes in a broader sense may also involve recurrent external perturbations, such as harvesting.

We introduce the notion of *interaction stochasticity* (mating, social interactions) as a further stochastic factor of extinction in closed populations. Interaction stochasticity does not operate at the level of individuals, but at the level of pairs or groups. It involves the stochasticity of encounters between individuals that may arise in the random formation of mating pairs or of social groups.

In spatially structured populations, *migration stochasticity*, that is the chance realization of dispersal probabilities, also influences the local population dynamics, whereas the *stochasticity of extinction–recolonization* processes operate at the regional scale. Extinction–recolonization stochasticity can be regarded as a form of demographic stochasticity that affects patches instead of individuals (see Chapter 4).

(e.g., as in cooperative behavior). Although each individual's vital rates are influenced by local interactions, primarily with neighbors, the aim of a wide range of density-dependent models is to describe mean demographic parameters (i.e., the average over all the individuals present) as functions of total population size or mean population density (the mean being taken across space). Such models are best used for the mathematical exploration of qualitative phenomena. On the empirical side, the unambiguous identification of density dependence in vital rates is notoriously difficult, and the choice and fit of particular density-dependent models turns out to require massive amounts of data and an in-depth understanding of the demographic processes at work in the population (Box 2.2).

The simplest density-dependent models

The notion of population limitation was first reconciled with density-independent models of exponential growth by defining the population carrying capacity as a ceiling at which exponential growth ceases. The population size N has a constant

Box 2.2 The empirical assessment of density dependence

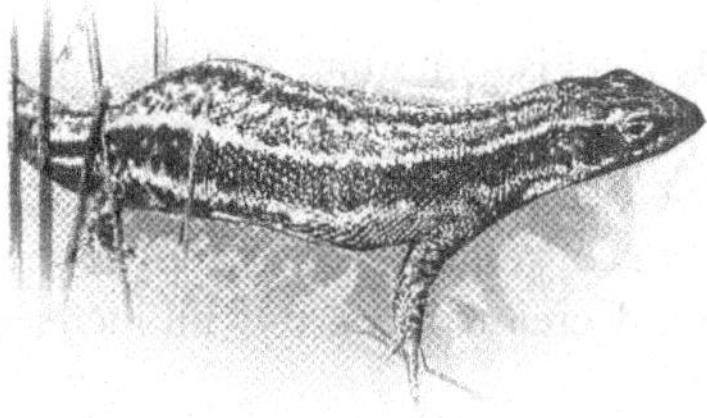

Existing statistical tests developed to analyze trends in population densities often yield conflicting results and in general lack the power to detect even moderate density dependence. In fact, the natural heterogeneity of population parameters that influence density serve to mask the effects of density dependence within a population. Shenk *et al.* (1998) recently reemphasized that to detect density dependence requires investigation of the response of individual life-history traits to changes in population density. This has been achieved in very few studies as yet. By using individual histories of capture–recapture, Lebreton *et al.* (1992) found limited evidence for density dependence of survival probabilities in the roe deer (*Capreolus capreolus*). In contrast, Leirs *et al.* (1997) found that the population dynamics of a murid rodent pest (*Mastomys natalensis*) are driven by both density-independent (stochastic) and density-dependent factors, the latter affecting several demographic traits in different ways. Massot *et al.* (1992) applied the same methodology to data obtained from density manipulation of the common lizard (*Lacerta vivipara*); density was shown to have little effect on the survival parameters, whereas reproductive and dispersal traits responded strongly.

A different approach is to calibrate a structured population model that incorporates hypothesized density-dependent factors to a time series of class-specific population censuses. Dennis *et al.* (1995) used this approach to demonstrate the action of nonlinear density dependence in experimental *Tribolium* populations and to obtain a quantitative assessment of the strength of the density-dependent effects on each parameter of the model.

per capita growth rate r, except at the carrying capacity (ceiling) K where growth stops,

$$\frac{dN}{dt} = \begin{cases} rN & \text{for } 1 < N < K \\ 0 & \text{for } N = K \end{cases} . \tag{2.1}$$

For an initial population size N_0 between 1 and K, the population grows exponentially with time t as $N(t) = N_0 e^{rt}$. If r is positive, population growth continues until K is reached. This simple model of exponential growth to a carrying capacity was analyzed by MacArthur and Wilson (1967), Leigh (1981), and Goodman (1987a, 1987b) in their investigations of demographic and environmental stochasticity. If r is negative, the population declines to extinction, which is defined to occur at a population size of $N = 1$ individual. For a population with an initial population size of $N_0 = K$, the time until extinction $-(\ln K)/r$ then depends on the natural logarithm of the initial size. In the following sections we examine whether a logarithmic dependence of extinction time on the initial population size also holds for stochastic models.

The ceiling-growth model, Equation (2.1), yields important insights into the effect of stochastic factors on extinction risk. Yet it is a very crude representation of population regulation. Instead of piecewise constant growth, that is, at rate r if $N < K$ and at rate 0 if $N = K$, the celebrated Verhulst–Pearl logistic model assumes that the growth rate of the population decreases linearly with increasing population density,

$$\frac{dN}{dt} = rN\left(1 - \frac{N}{K}\right) . \tag{2.2}$$

The logistic model makes several assumptions about the population:

- It has a stable age distribution;
- The response to a change in population density is instantaneous;
- The intrinsic rate of increase is reduced by a constant amount for every individual added to those already present;
- Crowding affects all individuals and life stages of a population equally;
- The environment is constant; stochastic and genetic effects are unimportant.

An interesting feature of the logistic model is that it enables interpretation of the effects of density dependence of birth and death rates at the individual level. For example, density dependence may affect the death rate d linearly while the birth rate b remains constant, which leads to logistic growth if $b = r$ and $d = rN/K$. Notice, however, that the same logistic growth term can be obtained by expressing the birth and death rates in many different ways. This confers a broader scope to the logistic model, but also raises difficulties when defining a stochastic counterpart to Equation (2.2) (Dennis 1989).

The models above approximate birth and death events as processes that are continuous in time. When life-history schedules are markedly seasonal, difference equations formulated in discrete time are more appropriate. The life cycle of many species of plants and animals may often be separated into a few discrete classes with transitions between them over discrete units of time (e.g., a few weeks for beetle cultures, or one year for many birds in temperate regions). By using such units, we give the system time to homogenize, so the critical assumption of a global effect of density on vital rates may be less problematic in this framework.

Density-dependent models in discrete time

Density-dependent models in discrete time take on the generic form $N_{t+1} = \phi(N_t)$, where N_t denotes population size at the time t, and ϕ is a nonlinear function. Beyond the straightforward time-discrete version of the ceiling model, in which

$$\phi(N_t) = \begin{cases} e^r N_t & \text{for } 0 \le e^r N_t < K \\ K & \text{otherwise} \end{cases} , \tag{2.3}$$

there exists a wide variety of unstructured time-discrete, density-dependent models, reviewed in May and Oster (1976), Hassell *et al.* (1976), and Caswell and Cohen (1995). Equation (2.4a) has been used widely ever since it was introduced by Hassell (1975),

$$\phi(N_t) = \frac{e^r N_t}{(1 + aN_t)^\eta} , \tag{2.4a}$$

where $a = (e^{r/\eta} - 1)/K$ and η is a competition parameter. The neat feature of Equation (2.4a) is that it defines a continuum of simple models that range from the so-called Beverton–Holt model, in which $\eta = 1$ (which is equivalent to logistic growth), to the so-called Ricker model, in which η goes to infinity.

Beverton–Holt model. The Beverton–Holt model is relevant when there is a natural limitation to the recruitment of new individuals. If the survival of young is limited by the number of territories or the number of nesting sites, a fairly constant number of young will be recruited, irrespective of the number of offspring produced. This is illustrative of the notion of contest competition that gives rise to compensatory density-dependence: individuals are either fully successful, or they are not successful.

Ricker model. In contrast, the Ricker model, well-known in the form

$$\phi(N_t) = N_t e^{r(1-N_t/K)} , \tag{2.4b}$$

involves an overcompensatory response to population density, which results from scramble competition: all individuals are affected evenly by the competition (Lomnicki 1988). As explained in Box 2.3, the merit of this model is that it relates well-defined properties of individuals that should be accessible to empirical measurement – the size of the home range, the effect of competition per competitor, reproductive success in the absence of competition – to the population behavior. Also, the underlying assumptions (e.g., that of random dispersal) are made explicit in the mathematical derivation of the model.

Other models. Another useful equation was developed by Maynard Smith and Slatkin (1973),

$$\phi(N_t) = \frac{e^r N_t}{1 + (aN_t)^\eta} . \tag{2.4c}$$

It is only superficially similar to Equation (2.4a). Here, a is inversely proportional to the amount of habitat or resource available (approximately $1/K$) and η controls the strength of the dependence of population growth on available resources. A further possibility reads

$$\phi(N_t) = \frac{e^r N_t}{[1 + (aN_t)^\eta]^{1/\eta}} , \tag{2.4d}$$

Box 2.3 Scaling up from individual interactions to population dynamics

Many population dynamic equations in discrete time have been used in the literature. Yet, like the ceiling and logistic continuous-time equations, most of them lack the explicit underpinning of a rigorous derivation that would show them to be mathematically neat derivations from stochastic "first principles" that operate at the level of individuals.

One remarkable exception is provided by the celebrated Ricker model, which assumes discrete generations and a simple life cycle (Royama 1992). Within each time interval (e.g., one year), an offspring may grow to maturity with probability s_0, and then produce offspring and die. Offspring disperse randomly and establish fixed home ranges; this means that individuals are spread across homogeneous space in a Poisson distribution. Population density is measured at the onset of the reproduction period. Mature individuals compete for resources, and the effect of competition is to reduce fecundity. Competition occurs between "neighbors" only, and the effect of competition is captured by reducing the intrinsic (i.e., maximum, in the absence of competition) fecundity b_0 by a constant factor $\kappa < 1$. Any individual is counted as a neighbor to another if their home ranges overlap.

For mathematical tractability, assume that all home ranges are circular with area σ; as a consequence of the random (i.e., Poisson) distribution of individuals in space, given that the population density is N, the probability that an individual has i neighbors ($i \geq 0$) is equal to $(4\sigma N)^i e^{-4\sigma N}/i!$. Hence the expected per capita fecundity is $b_0 e^{-4\sigma(1-\kappa)N}$. The recursion for the expected population density given by Equation (2.4b) readily follows, with $r = \ln(s_0 b_0)$ and $K = \ln(s_0 b_0)/[4\sigma(1-\kappa)]$.

which was used, for instance, by Halley and Iwasa (1998) in their analysis of the effect of environmental and demographic stochasticity on the extinction risk (see Section 2.4).

Allee effects

In the study of the preservation of biodiversity, it seems natural to consider the following question posed by Allee (1938, p. 107): "What minimal numbers are necessary if a species is to maintain itself in nature?" The question arises when the per capita growth rate of a species is initially (i.e., at low population density) an increasing function of population density. A potentially important cause for this phenomenon, commonly termed the "Allee effect" [see recent reviews by Stephens and Sutherland (1999) and by Courchamp *et al.* (1999)], is a shortage of mating encounters in sparse populations (Allee 1931; Haldane 1953; Watt 1968; Wells *et al.* 1998). That to find mates might be difficult to achieve at low density has long been hypothesized, such as for sea urchins (Allee 1931), flour beetles (Park 1933), muskrats (Errington 1940), condors (Mertz 1971), and zooplankton (Gerritsen 1980). Three categories of empirical studies have brought relevant insights into mating rates and Allee effects, as reviewed in Dennis (1989), Stephens and Sutherland (1999), and Courchamp *et al.* (1999):

- Experiments that have detected Allee effects possibly due to mating frequency;
- Experiments (with assorted insects in mating chambers) that have shown positive dependence of mating rates on population density;
- Correlative studies for a few species in the field that have demonstrated positive relationships between mating rates and density.

Few data are available on the mating rates and population growth of rare species. Occasionally, rare organisms proved so adept at finding each other that no effects on the mating rates were detected (Teesdale 1940; Surtees and Wright 1960; Burns 1968). In contrast, a more recent study (Madsen *et al.* 1992) on a small, isolated population of adders (*Vipera berus*) suggests that an important determinant of population growth, litter success, correlates positively with mating frequency. Lande (1988) emphasized that such Allee effects in endangered species could have drastic implications for the theory and practice of conservation biology. Yet mathematical models that relate mating rates to population growth remain scarce. To date one of the most comprehensive studies remains that of Dennis (1989), who developed deterministic and stochastic models to describe the growth, critical density, and extinction probability in sparse populations that experience Allee effects. McCarthy (1997) and Poggiale (1998) have developed more recent advances. The construction of these models involves two steps:

- Starting from behavioral rules that apply to individuals, stochastic models are proposed that predict the probability of mating encounters as a function of population density;
- The mating encounter function is then incorporated into a model of population growth.

In the first step, Dennis (1989) recovered a negative exponential function under the following biological assumptions:

- Constant sex ratio;
- The probability that a female encounters a male after searching a small area is proportional to that area and to the density, and it decreases with the number of previous encounters (i.e., there is a saturation effect);
- The probability of encountering two partners in a small area is negligible.

The negative exponential function is parametrized by the effective mating area of a female, that is, the size of the area over which encounters may occur for any given individual times the proportion of males in the population (sex ratio), and by other parameters related to the presumed aggregation structure of the population. Heterogeneity between individuals in effective mating area can be taken into account, as a consequence of individual differences in, for example, mobility, size of home range, signaling, or attractiveness. From the negative exponential function arises a rectangular hyperbola function, which is mathematically similar to the so-called type II functional response heretofore used in ecological modeling to describe the response of predator feeding rate to prey density.

The mating encounter function may be incorporated *ad hoc* into the logistic Equation (2.2): a term proportional to the probability of not mating is subtracted from the per capita growth rate, to represent the reduction of reproduction because of mating shortage. Let N denote the population density at any given time. Using the rectangular hyperbola function, the probability of mating can be shown to be $N/(\theta + N)$; parameter θ can be seen as a behavioral trait with a value equal to the population density at which the probability of mating is 0.5. Then $1 - N/(\theta + N) = \theta/(\theta + N)$ is the probability of not mating. Thus, the logistic model adjusted for mating encounters is

$$\frac{dN}{dt} = rN\left(1 - \frac{N}{K}\right) - \frac{\delta\theta}{\theta + N}N\,, \tag{2.5a}$$

where the coefficient δ scales the negative effect of not mating. Kostitzin (1940) was the first to publish this growth model, and Jacobs (1984) examined its behavior. Similar equations arose in the context of populations that experienced harvesting or predation (May 1977; Huberman 1978; Ludwig *et al.* 1978; Brauer 1979).

From the individual perspective, one possible interpretation of this phenomenologic model is to assume that r measures the per capita density-independent birth rate, rN/K the per capita density-dependent death rate, and $\delta\theta/(\theta + N)$ the rate at which individuals are removed from the population through not finding a mate. The assumption that not mating leads to permanent removal looks rather artificial. An alternative, and perhaps more natural, way of accounting for the shortage of mating encounters at low density is to condition reproduction upon finding a mate. Assuming negative linear density-dependence of the birth rate and density independence of the death rate yields

$$\frac{dN}{dt} = b\left(1 - \frac{N}{K}\right)\frac{N^2}{\theta + N} - Nd\,, \tag{2.5b}$$

where b denotes the intrinsic (i.e., in the absence of negative density-dependent effects) per capita birth rate, and d the density-independent per capita death rate. Swapping the influence of negative density dependence between birth and death processes leads to the following third model

$$\frac{dN}{dt} = b\frac{N^2}{\theta + N} - N^2 d\,. \tag{2.5c}$$

A feature common to Equations (2.5a) and (2.5b) is that they predict either extinction or bistability, that is, an outcome – extinction versus persistence at an equilibrium density – contingent upon the initial population density. For persistence, the population can reach its viable equilibrium only if the initial density is larger than a critical threshold identified as an unstable equilibrium of the model. This critical threshold is germane to the notion of a minimum viable population (Soulé 1987). As we show later (Section 2.3), the existence of such a critical density has important consequences when the effect of chance factors of extinction on population viability is assessed. In contrast, Equation (2.5c) describes a kind of degenerate

Allee effect: the population growth rate increases with density at low density, but the existence of a viable equilibrium always implies that the extinction equilibrium is unstable.

2.3 Demographic and Interaction Stochasticities

In this section a constant environment is assumed. We concentrate on the most basic extinction risks that result from random fluctuations in the birth and death processes and in the proportion of females in a population (the sex ratio).

Time to extinction under demographic stochasticity

In a finite population, the per capita growth rate r is subject to random variation through the independent chances of individual mortality and reproduction. Thus, for a population of size N, r is a random variable with mean $\overline{r}$ and variance V/N (assuming no autocorrelation). The parameter V is the variance in individual reproduction rate (which comprises birth events and death chance) per unit time (Leigh 1981; Goodman 1987a, 1987b). The growth rate r of a population at a particular time is the mean reproduction rate of individuals in the population, and its variance is equal to the sampling variance of this mean, that is, individual variance divided by population size. The long-run growth rate of a population subject to demographic stochasticity is simply $\overline{r} = r$.

First, we review Lande's (1993) results on the effect of demographic stochasticity on the mean persistence time in the ceiling Equation (2.1). Lande (1993) strongly relies on diffusion theory to approximate the dynamics of stochastic processes; the mathematical basics are introduced in Box 2.4. In Box 2.5, we present the results of Lande's calculations in some detail. These results enable investigation of how the mean extinction time varies with carrying capacity for populations that are initially at the carrying capacity, but that have different mean growth rates. Under the appropriate conditions (made explicit in Box 2.5), there is a nearly exponential scaling of average extinction time with carrying capacity when the mean per capita growth rate $\overline{r}$ is positive (also see Gabriel and Bürger 1992). For $\overline{r} = 0$, a nearly linear dependence is found. For negative $\overline{r}$, the scaling is dominated by a term proportional to the logarithm of the carrying capacity, as in a population undergoing a deterministic decline.

The simplest approach to incorporating demographic stochasticity in the more sophisticated time-discrete density-dependent models described above is to assume that, given the current population size N_t, the number of individuals actually present at time $t+1$ is drawn from a Poisson distribution the mean of which is equal to the deterministic projection $\phi(N_t)$ obtained from the corresponding recursion equation, that is,

$$N_{t+1} = \text{Poisson}[\phi(N_t)] \,. \tag{2.6}$$

Mathematically, this leads to a Markov chain model that is not a branching process (Gabriel and Bürger 1992). (In the following subsection, we describe a modeling alternative based on branching processes.) Monte Carlo simulations can be used

Box 2.4 Diffusion theory for stochastic models

Small or moderate perturbations of the population numbers can be modeled accurately as a diffusion process, provided the mean absolute growth rate per unit time is small (i.e., $|\overline{r}| \ll 1$). Diffusion theory (see Chapter 15 in Karlin and Taylor 1981) can be applied to calculate the mean time of extinction of the population. A diffusion process is described fully by its infinitesimal moments and by the behavior of sample paths at the boundaries. For a population of size N, the infinitesimal mean and variance, denoted by $\mu(N)$ and $v(N)$, give, respectively, the expected change and the variance of the change in population size per unit time. Starting from a given initial size N_0, the mean time to extinction $\overline{T}$ is the solution of the differential equation

$$\tfrac{1}{2}v(N_0)\frac{d^2\overline{T}}{dN_0^2} + \mu(N_0)\frac{d\overline{T}}{dN_0} = -1 \qquad \text{(a)}$$

with the boundary condition $\overline{T}(1) = 0$ and a reflecting boundary at carrying capacity K. The general solution to this equation is (Karlin and Taylor 1981)

$$\overline{T}(N_0) = 2\int_1^{N_0} e^{-G(z)} \int_z^K \frac{e^{G(y)}}{v(y)}\, dy\, dz\,, \qquad \text{(b)}$$

where

$$G(y) = 2\int_1^y \frac{\mu(N)}{v(N)}\, dN\,. \qquad \text{(c)}$$

This formula has been used by Lande (1993) to evaluate the influence of demographic and environmental stochasticity on the persistence of populations that start at carrying capacity.

Typically, catastrophes, which are defined as large and infrequent environmental perturbations, cannot be described by diffusion approximations and need a different mathematical treatment. One approach traces back to Hanson and Tuckwell (1978), who used differential-difference equations for which Lande (1993) found an analytical solution under the assumption of the simple ceiling model. The results confirm the intuitive expectation that, for large populations, catastrophes pose a much greater threat than demographic stochasticity.

to explore the extent to which the risk of extinction depends upon the nature of density dependence. It turns out that the mean extinction time predicted by this model is highly sensitive to assumptions about the mode of density dependence (Figure 2.1).

For undercompensatory density dependence [ϕ given by Equation (2.3), or by Equation (2.4a) with $\eta \leq 1$], the time to extinction increases monotonically with the intrinsic growth rate r. For overcompensatory density dependence [ϕ given by Equation (2.4a) with $\eta > 1$, or by Equation (2.4b)], the mean time to extinction is maximum at intermediate values of r and declines as r increases further: strong oscillations in population numbers are detrimental to long-term persistence. In the Hassell model of density dependence given by Equation (2.4a), it is found

Box 2.5 Scaling laws of time to extinction in the ceiling model

The influence of demographic and environmental stochasticity on the persistence of populations that start at carrying capacity was evaluated by Lande (1993) with the tools of diffusion theory sketched in Box 2.4.

Demographic stochasticity. In a finite population of size N, the per capita growth rate r is a random variable with mean $\bar{r}$ and variance V/N (assuming no autocorrelation). The long-term growth rate of a population subject to demographic stochasticity is simply $\bar{r} = r$. Asymptotic scaling relationships can be derived in cases for which $\bar{r}$ is positive, 0, or negative. To this end, it is useful to set $\varepsilon = 2\bar{r}/V$. With positive $\bar{r}$ and $\varepsilon K \gg 1$,

$$\overline{T}(K) \approx \frac{e^{\varepsilon(K-1)} - 1}{\bar{r}\varepsilon K}\left(1 + \frac{1}{\varepsilon K}\right), \tag{a}$$

where the dominant term is proportional to $e^{\varepsilon K}/K$. With $\bar{r} = 0$,

$$\overline{T}(K) = 2(K - 1 - \ln K)/V. \tag{b}$$

With negative $\bar{r}$ and $-\varepsilon K \gg 1$,

$$\overline{T}(K) \approx \frac{\ln K + \zeta(\varepsilon)}{-\bar{r}}, \tag{c}$$

where $\zeta(\varepsilon)$ is a function of ε only.

Environmental stochasticity. The infinitesimal mean and variance of the diffusion process that approximate the actual dynamics when $\bar{r}$ is small are $\mu(N) = \bar{r}N$ and $v(N) = V_E N^2$, respectively. Transformation of the diffusion process to a logarithmic scale yields the transformed infinitesimal mean and variance as $\bar{r} - V_E/2$ and V_E, respectively, in the domain $0 < \ln N < \ln K$. The quantity $\tilde{r} = \bar{r} - V_E/2$ can be considered as a stochastic analog of r in the deterministic model. Discounting the mean growth rate because of random environmental fluctuations is explained by Lewontin and Cohen (1969) in terms of the finite rate of increase e^r, the arithmetic mean of which determines the expected population size, whereas the smaller geometric mean determines the dynamics of extinction.

With positive $\tilde{r}$ and $c \ln K \gg 1$,

$$\overline{T}(K) \approx 2K^c/(V_E c^2), \tag{d}$$

where $c = 2\bar{r}/V_E - 1$. With $\tilde{r} = 0$,

$$\overline{T}(K) = (\ln K)^2/V_E. \tag{e}$$

With negative $\tilde{r}$ and $-c \ln K \gg 1$,

$$\overline{T}(K) \approx \frac{\ln K + 1/c}{-\bar{r}}. \tag{f}$$

Therefore, under environmental stochasticity the average time to extinction scales faster or slower than linearly with K, depending on whether $\bar{r}/V_E$ is greater than or less than 1.

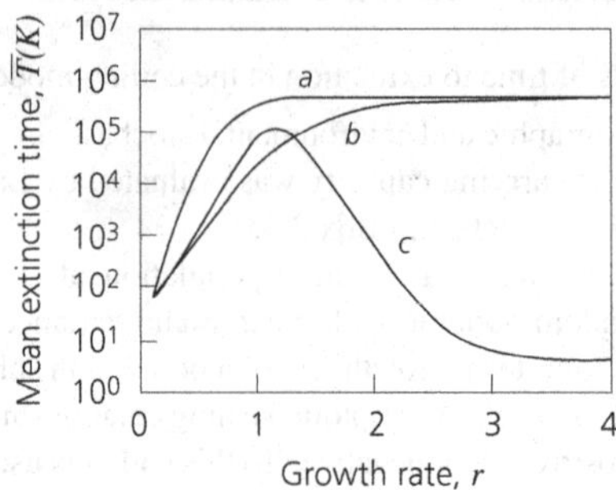

Figure 2.1 Dependence of the mean time to extinction under demographic stochasticity on the intrinsic growth rate r for three modes of density dependence: (a) ceiling model given by Equation (2.3), (b) undercompensatory regulation given by Equation (2.4a) with $\eta = 1$, and (c) overcompensatory regulation given by Equation (2.4b). The carrying capacity is $K = 16$.

that for any positive intrinsic growth rate, the average time to extinction scales approximately exponentially with the carrying capacity K; its maximum value is equal to e^K and is approached as r becomes very large and $\eta = 1$. Thus, the analytical asymptotic scaling of mean extinction time with carrying capacity carries over to more realistic models of density-dependent population dynamics. However, a strong density dependence associated with a high intrinsic rate of increase can cause significant fluctuations in the deterministic system, which translate in the stochastic counterpart into a larger risk of extinction.

Effect of interaction stochasticity

An important form of interaction stochasticity involves the chance formation of mating pairs. The probability of mating may decrease as population size decreases, resulting in the Allee-type of density dependence described in Section 2.2. Dennis (1989) derived extinction probabilities for various discrete birth–death processes that account for demographic stochasticity and integrated birth rates limited by mating encounters. The critical density predicted by the logistic model with mating encounter limits, Equation (2.5a), manifests itself as an inflection point in the probability of extinction plotted as a function of the initial size. The probabilities for a population size at any given time tend to cluster at low values when the initial size is less than this critical size, and they disperse toward higher values when the initial size is larger than the critical threshold.

Dennis' (1989) model assumes a constant sex ratio. However, stochastic variation in the sex ratio is an obvious cause of random fluctuations in the number of mating pairs, which in turn may generate an Allee effect (Lande 1998a). Therefore, sex can influence extinction through chance fluctuations in the sex ratio in small populations. A striking example is provided by the dusky seaside sparrow (*Ammospiza maritima nigrescens*), which was a subspecies once found in a small area of Florida coastal marshes. Habitat destruction, pesticide use, and wildfire reduced its population from a high of perhaps 2000 pairs until, by 1980, only six birds remained. Unfortunately, they were all males. Five of the six were captured, and unsuccessful attempts were made to cross them with females of other subspecies. The last male died in 1987 (Ehrlich *et al.* 1988). The dusky seaside sparrow succumbed to sex ratio stochasticity.

Dusky seaside sparrow
Ammospiza maritima nigrescens

The impact of mating stochasticity (caused by chance variations in the sex ratio) on extinction risk was first demonstrated by Gabriel *et al.* (1991) and Gabriel and Bürger (1992). In the absence of environmental determination or density-dependent regulation of the sex ratio, the distribution of female offspring can be assumed to follow a binomial distribution. Even for an equal probability of male and female production, the expected proportion of females in a population may deviate from 0.5 because of sex-dependent mortality. If both genders are produced and survive with equal probability, the probability that the population consists only of males, or only of females, is $2(1/2)^N$ for a given population size N. Such a population cannot produce offspring and dies out at the next generation. The expected extinction time for this "pure sex-ratio" model is $2^{N-1} + 1$. If r is small, the risk of extinction through sex-ratio stochasticity is negligible compared to the risk induced by demographic stochasticity. In contrast, as r increases beyond 1, the mean extinction time because of sex-ratio stochasticity becomes smaller than the mean extinction time estimated under demographic stochasticity (e.g., with $b = 1$ in the Hassell model, the extinction time approaches e^N with increasing r).

We now address the combined effect of demographic and mating (sex-ratio) stochasticity. A measure of the minimum risk of extinction can be obtained by making the extreme assumption that the total offspring production is independent of the actual number of females in the population; that is, all expected offspring will be produced even if there is only one female in the population. For large intrinsic growth rates r, the mean extinction time is approximately equal to $(1/2)e^{N/2}$ (Gabriel and Bürger 1992). If the maximum fecundity of each female is much smaller than the carrying capacity and if there are only a few females in the population, the total number of offspring produced becomes strongly dependent on the number of females. The resultant time to extinction drops by several orders of magnitude.

Figure 2.2 displays the expected extinction time predicted for the ceiling type of density dependence [Equation (2.3)], under the various scenarios described in the previous two paragraphs, as a function of the intrinsic growth rate r. Even at values of r so high that a population can recover from a stochastic decline within a single time step, the risk of extinction is underestimated by several orders of

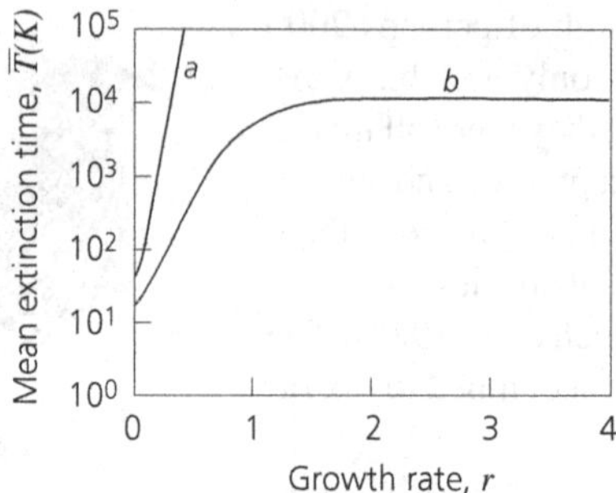

Figure 2.2 Mean time to extinction as a function of the intrinsic growth rate r for the density-dependent relation given by Equation (2.3) under two stochastic scenarios: (a) demographic stochasticity alone, and (b) combined effects of demographic and sex-ratio stochasticities. The carrying capacity is $K = 20$.

magnitude if the stochasticity in sex ratio is neglected. These models suggest that, in sexual populations, the extinction risk at intermediate or high values of r is determined mainly by sex-ratio stochasticity, whereas in species with low r the effect of demographic stochasticity dominates the extinction risk.

Branching processes and quasi-stationarity

As useful as simulations are, they only scratch the surface of the information from demographic and interaction stochasticities in a population model. Approximations based on diffusion methods can yield important analytical results, yet they face serious limitations (see Box 2.4). Further insights into the analysis of demographic and interaction stochasticity can be earned within the framework of stochastic models called *branching processes*. A branching process describes a collection of entities (individuals, in this case) that produce random numbers of new entities according to some probabilistic rule. These offspring produce further offspring in their turn. The descendants of each individual form a family tree; hence the name of the process. Introductions to branching processes can be found in most textbooks on stochastic processes; for detailed treatments, see Harris (1963) and Athreya and Ney (1972). Caswell (2001) pointed out that, in spite of their long mathematical history, their origin in the problem of extinction, and their natural connection with demographic data, ecologists in general, and conservation biologists in particular, have largely ignored branching processes (but see Lebreton 1981; Mode and Pickens 1986; Gosselin and Lebreton 2000).

Hereafter, we give a brief introduction to branching process models for unstructured populations (Chapter 3 introduces multitype processes for age-structured populations). Consider a population that consists of N_t identical individuals. At $t+1$ each individual is replaced by a random number of "offspring", including possibly the individual at time t surviving to time $t + 1$, or new individuals produced by reproduction. We denote the probability distribution of this random variable by p_i, $i = 0, 1, 2...$ (i.e., p_i is the probability of having i offspring), and λ denotes the mean offspring number. Notice that for extinction to be possible $p_0 > 0$ and population growth requires $p_0 + p_1 < 1$. It can be shown that the growth rate of the mean population size is equal to λ, that is, $\mathbb{E}(N_t) = \mathbb{E}(N_0)\,\lambda^t$, where $\mathbb{E}$ denotes mathematical expectation. Populations with $\lambda < 1$, $\lambda = 1$, and $\lambda > 1$ are called *subcritical*, *critical*, and *supercritical*, respectively. Critical and subcritical processes eventually go extinct with probability 1. The probability of extinction for a supercritical process is greater than 0, but less than 1. An important property of subcritical processes is that, although they are doomed to extinction, the size of the population at time t, given that it has yet to go extinct, converges to a fixed, *quasi-stationary distribution* [QSD; see Joffe and Spitzer (1967) for a detailed account on this result, originally derived by Yaglom]. It is possible to calculate the expectation of population size at time t, conditional on nonextinction,

$$\mathbb{E}(N_t \mid N_t > 0) = \frac{\lambda^t}{1 - q_t}, \tag{2.7}$$

where $q_t = \Pr$ (extinction by time t). Both the numerator and the denominator decrease with time, but as t becomes large, their ratio converges to a constant, which is the QSD mean.

Gosselin and Lebreton (2000) have shown that the qualitative properties of density-independent subcritical branching processes carry over to the case of negative density dependence. Assuming that the probability distribution of offspring number at time t depends upon the population size N_t, such that the mean offspring number decreases as density increases, population extinction must eventually occur with a probability of 1. Conditional on nonextinction, the process converges toward a fixed QSD. Remarkably, as t goes to infinity, the probability that a trajectory that is nonextinct at time t becomes extinct at time $t + 1$ converges to a fixed value q, and the distribution of extinction times approaches a geometric distribution with mean $1/q$. How this key parameter q is influenced by the mode of density dependence is currently unknown.

Mating stochasticity can also be approached in the context of branching process modeling. Unfortunately, simple analyses based on probability-generating functions do not work when the model is nonlinear, which ought to be the case here since reproduction depends on the relative abundance of the sexes. The few results available are mainly from Daley *et al.* (1986), but also see Gonzalez and Molina (1996) and Hull (1998). In these models, a "marriage, or mating, function" M of the number of males N_m and females N_f gives the number of mating pairs. The ith pair then gives a number of male offspring and a number of female offspring, independently and identically distributed according to a bivariate offspring distribution. For example, "mating fidelity" may be modeled by taking $M(N_m, N_f) = \min(N_m, N_f)$; "promiscuous mating" with $M(N_m, N_f) = N_f \min(1, N_m)$; and "female dominance" with $M(N_m, N_f) = N_f$. Daley *et al.* (1986) explored how extinction probability scales with the initial number of pairs. Figure 2.3 shows the results: mating fidelity and promiscuous mating both greatly increase the extinction probability compared to mating under female

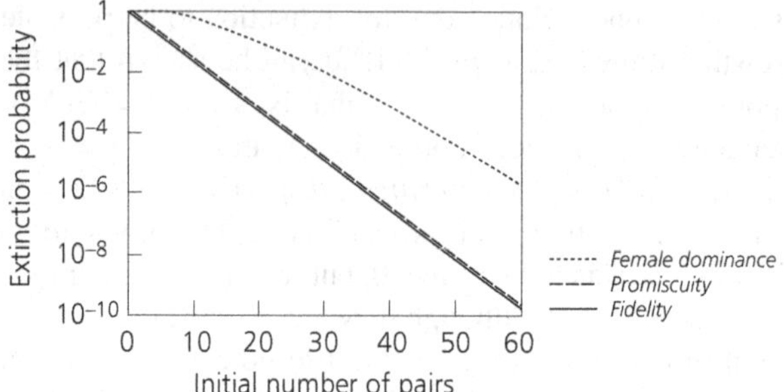

Figure 2.3 Probability of eventual extinction predicted by a two-sex branching process, as a function of the initial number of pairs, for three different mating systems: fidelity (continuous curve), promiscuity (dashed curve), and female-dominance (dotted curve). Male and female offspring numbers are independent Poisson random variables with a mean of 1.2. *Source*: Caswell (2001), after Daley *et al.* (1986).

dominance. Chapter 3 extends this investigation to the case of structured populations, and also addresses the further genetic effects of sexual reproduction in small populations (e.g., inbreeding depression).

2.4 Environmental Stochasticity

In Section 2.3, we assumed a constant environment to study the extinction risks that resulted from demographic and interaction processes only. We now assess the relative effect of environmental stochasticity on the extinction risk. The starting point is again the ceiling Equation (2.1), for the analysis of which we continue to employ the method of diffusion approximations. Environmental stochasticity is modeled by allowing the population growth rate to fluctuate with time as a stationary time series with mean growth rate $\bar{r}$, environmental variance V_e, and no autocorrelation. The scaling of average time to extinction depends qualitatively on the quantity $\tilde{r} = \bar{r} - V_e/2$, which can be considered as a stochastic analog of r in the deterministic model (Lewontin and Cohen 1969). Under the appropriate conditions (see Box 2.5) with a positive $\tilde{r}$, the average time to extinction scales asymptotically and algebraically with the carrying capacity. For $\tilde{r} = 0$, the average time to extinction depends on the square of the logarithm of carrying capacity. If the long-run growth rate $\tilde{r}$ is negative, the dependence of the average time to extinction on carrying capacity is dominated by a logarithmic term.

Accounting for individual interactions and demographic stochasticity

The diffusion approximation is valid only for small intrinsic growth rates r [see the comparisons of diffusion approximations and Monte Carlo simulations in Bürger and Lynch (1997)]. What happens at larger growth rates r? If the carrying capacity K is small, Bürger and Lynch (1997) showed (by means of numerical simulations) that environmental stochasticity poses a weaker threat compared to the risk of extinction caused by sex-ratio stochasticity and demographic stochasticity combined together. Unlike the case for small growth rates dealt with through diffusion approximations, environmental stochasticity adds little to the extinction risk at large values of r. As the carrying capacity K increases, the expected extinction time through environmental stochasticity increases less than exponentially, in contrast with the exponential increase predicted under demographic and sex-ratio stochasticity. Thus, at large carrying capacities, environmental stochasticity appears to be the dominant factor of extinction. These conclusions are underpinned by the ceiling model, which leads to underestimates of the extinction risk. How different modes of interactions between individuals, and hence different forms of density dependence, alter the conclusion remains an open question.

Analyses of the extinction risk in unstructured populations based on the ceiling model are limited in at least three ways. First, it is assumed that environmental stochasticity impacts population dynamics only through the intrinsic rate of increase. Fluctuations of environmental conditions are, indeed, known to have potentially severe effects on birth and death rates. Yet variations often impact the habitat rather than the vital rates directly (e.g., the availability of nesting sites for birds), which might be more appropriately accounted for by incorporating random fluctuations in the carrying capacity. A first important step was taken in this direction by Gyllenberg *et al.* (1994) in the analysis of a stochastic version of the Ricker model, but their study emphasized more the issue of population *regulation* than that of population *extinction*.

Second, the disjoint study of the effects of demographic, sex-ratio, and environmental stochasticities is artificial: demographic stochasticity is inherent to any birth-and-death process, and sex-ratio stochasticity stems unavoidably from sexual reproduction; it is not possible to treat environmental stochasticity in isolation from these other factors. Halley and Iwasa (1998) developed an unstructured Markov chain model for an organism, such as an insect, with two stages (larvae and adults) and discrete generations. The model incorporates individual competitive interactions and environmental stochasticity, which affect reproduction, and demographic stochasticity, which impacts larvae survival to adulthood. Their analysis is based on a decoupling trick: they compute the QSD $Q(N)$ of nonextinct population size N, neglecting demographic stochasticity, which is thus set primarily by environmental stochasticity and density-dependent processes; next, they derive the mean time to extinction through direct transitions to zero caused by demographic stochasticity. More precisely, they show that the asymptotic probability of extinction can be approximated by $\int_0^{+\infty} e^{-x} Q(x)\, dx$. Under the density-dependent Equation (2.4d), the probability distribution Q can be approximated analytically. The results ultimately show that for a given type of individual interaction [fixed η in Equation (2.4d)] and an increase in the intrinsic growth rate r, the asymptotic probability of extinction decreases more if η is larger; furthermore, for a given r, increasing η reduces the probability of extinction, and this effect is more pronounced in species that grow faster (larger r).

Third, and finally, environmental stochasticity is most often handled as "white noise", that is, uncorrelated stationary fluctuations of external variables. How environmental autocorrelations affect the risk of extinction is examined in the next subsection.

The effect of environmental autocorrelation

The limited scope of uncorrelated environmental variations has begun to be widened (Foley 1994; Ripa and Lundberg 1996; Petchey *et al.* 1997; Cuddington and Yodzis 1999). Following the same line of reasoning that led to Equation (2.6), environmental stochasticity and demographic stochasticity are incorporated in an extended version of the Ricker model

$$N_{t+1} = \text{Poisson}\left[N_t \exp(r(1 - N_t/K_{t+1})^{\gamma})\right] . \tag{2.8}$$

The exponent γ is an intraspecific competition parameter that measures the degree of over- or undercompensation ($0 < \gamma < 1$, $\gamma = 1$ indicates overcompensation). Here, environmental noise is included in the carrying capacity, $K_t = K_0 + \phi_t$, where ϕ_t is drawn from a so-called $1/f^{\beta}$ noise function. It has been suggested that $1/f^{\beta}$-type noises, in which variance at each frequency scales according to a power law, describe a wide range of natural phenomena well (Halley 1996). The value of the spectral exponent β indicates the "noise color". Red noise, $\beta = 1$, has been recommended as a better null model of environmental variation than white noise, $\beta = 0$ (Halley 1996). Ecological environmental fluctuations often have a reddened noise spectrum (Pimm and Redfearn 1988; Ariño and Pimm 1995). Geophysical variables are described by a range of spectral exponents from red to brown ($\beta = 2$, such as river height fluctuations) to black noise ($\beta = 3$, such as marine temperature fluctuations; see Schroeder 1991). Cuddington and Yodzis (1999) argue that the $1/f^{\beta}$-paradigm of autocorrelated environmental stochasticity is superior to autoregressive linear models for predicting the effect of noise color on population dynamics.

In Cuddington and Yodzis' (1999) simulations, $1/f^{\beta}$-noises with spectral exponents that ranged from 0 to 3.2 were generated using a so-called spectral synthesis approximation, based on Fourier numerical analysis [see Peitgen and Saupe (1988) for the detailed algorithms]. The results indicate that darker noise increases mean time to extinction (Figure 2.4). This effect interacts significantly with the mode of interaction between individuals around red noise (mean extinction time increases more dramatically with overcompensatory population regulation), whereas there is no effect of under- or overcompensatory regulation in the range of brown-to-black noise. The coefficient of variation (CV) of the extinction time indicates how reliable the mean is as an estimate of extinction risk. The CVs are functions mostly of noise color. They are very high around red noise, which indicates that red environmental variations make population persistence quite unpredictable; this effect is even more pronounced for an overcompensatory regime of density dependence. This can be understood because red noise has a balanced influence of low and high frequencies in the total signal; consequently, long-term trends may reinforce a fast

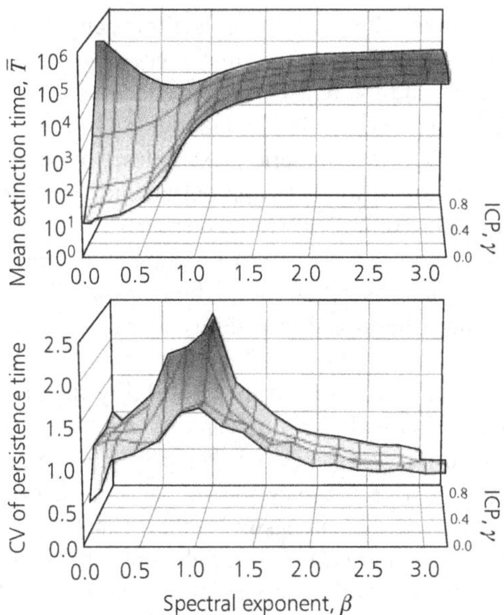

Figure 2.4 The effect of environmental noise color (spectral exponent β) and mode of interaction between individuals (ICP, intraspecific competition parameter γ) on log mean extinction times and the coefficient of variation (CV) of persistence times. In all simulations reported hereafter, the initial population size is set at the initial carrying capacity, with $N_0 = K_0 = 100$. A low intrinsic growth rate ($r = 1.5$) was used to generate deterministic dynamics with a stable point attractor. Population extinction occurs when $N_t < 1$. When a negative value of the carrying capacity was generated, K_t was reset to 5. When extinction did not occur, the persistence time was set to the maximum length of the simulation (200 000 generations). *Source*: Cuddington and Yodzis (1999).

oscillation or ameliorate its effects, with a complex ultimate effect on population dynamics. The conclusion is that individuals who interact in an overcompensatory mode and are exposed to red environmental noise form populations that should be considered much more "at risk" than populations exposed to black noise, and that more conservative management strategies should be adopted. This raises the interesting question of whether ecosystems under conservation concerns could be distinguished by the color of prevailing environmental noise that impacts their major components.

2.5 Density Dependence and the Measure of Extinction Risk

In this chapter, we use the average time to extinction as a relevant measure to assess the influence of the mode of interactions between individuals on the risk of extinction of a population. This is certainly appropriate for declining or regulated populations that have reached their QSD, since the theory of subcritical or density-dependent branching processes shows that the distribution of extinction times approaches a geometric distribution for a population that has survived long enough to "forget" its initial state. However, during the transient phase of convergence to the QSD, the distribution of extinction times may deviate significantly from the geometric distribution, and the density-dependence mode will, in general, influence this discrepancy [see Leigh (1981), Goodman (1987b), Gabriel and Bürger (1992), and Ludwig (1996) for numerical examples]. In other words, whether or not the average time to extinction is an appropriate measure of extinction risk may depend on the very nature of individual interactions.

Gabriel and Bürger (1992) give a verbal argument to explain the observed deviations of extinction time from geometric distribution. For a pure demographic model without sex the conditional probability that a population of size N_t goes extinct within the next generation is e^{-N_t} at any generation number t. Therefore, the extinction probability depends on the probability distribution of population sizes N_t. For a population with carrying capacity K and mean time to extinction $\overline{T}(K)$, the extinction time follows a geometric distribution if the instantaneous extinction probability is $1/\overline{T}(K)$, irrespective of the generation number. For large populations (e.g., $N_t = K$), it can easily be shown that $e^{-K} < 1/\overline{T}(K)$; for N_t sufficiently small, $e^{-N_t} > 1/\overline{T}(K)$. If a population starts with a given population size N_0, the probability distribution of population sizes in the following generation is not independent of N_0 and it may take several generations before the extinction probability becomes nearly independent of the starting condition. How long the initial population size will be reflected in the extinction probability distribution depends dramatically on the kind of population regulation in effect. One important factor can be the time needed to recover from low population sizes, after random fluctuations, to approach the carrying capacity. For example, in the Hassell Equation (2.4a), with $\eta = 1$ this takes much longer for smaller r. Consequently, the observed deviation from the exponential distribution is more pronounced for low r.

2.6 Concluding Comments

As already suggested by Boyce (1992), Burgman *et al.* (1993), Middleton and Nisbet (1997), and Pascual *et al.* (1997), mechanisms of population regulation that result from individual interactions are critical to predictions of the fate of populations subject to chance factors of extinction. Five main conclusions emerge from this review:

- In asexual populations with density-independent growth and an upper limit on population size (carrying capacity), the average time to extinction of a population, starting from the carrying capacity, follows different scaling laws with carrying capacity in response to demographic stochasticity and environmental stochasticity.
- In sexual populations, sex-ratio stochasticity may result in an Allee effect that dramatically alters these relationships. In populations with moderate or large intrinsic growth rates, the mean time to extinction can be reduced by several orders of magnitude compared to the predicted value under demographic stochasticity only. Other forms of interaction stochasticity that are conducive to Allee effects are likely to have similar consequences on population viability.

- Models that explicitly describe the effect of density on vital rates indicate that the scaling of mean extinction time with the population's intrinsic growth rate critically depends on the mode of interaction between individuals. A much analyzed case is that of scramble competition that leads to overcompensatory population regulation; then, for a fixed carrying capacity the mean extinction time may decrease as the intrinsic growth rate increases, as a result of the nonlinear oscillations in population size caused by the overcompensation.
- Whether or not the mean extinction time is an appropriate quantification of the extinction risk of a population also depends on the nature of the density dependence, at least in populations that have not yet reached their QSD.
- Autocorrelations in environmental fluctuations interfere with the mode of operation of density dependence. Reddened environmental noise tends to make the extinction time quite unpredictable, an effect that is even more pronounced with overcompensatory density dependence.

Using brine shrimp (*Artemia franciscana*) as a model organism, Belovsky *et al.* (1999) carried out a set of experimental studies of extinction dynamics that shed some light on the relevance of these conclusions to real populations. Two of their results are of particular interest here. First, they show that population variability through environmental stochasticity is less important to extinction than inherent oscillations through overcompensatory population regulation. Environmental stochasticity can even intensify the density-dependent effects when occasional beneficial conditions increase the population size and temporarily produce overcrowding in subsequent less favorable times. Second, they suggest that the nonlinear dynamics caused by density dependence further explain a substantial departure of the distribution of extinction times from a negative exponential. Overcompensatory dynamics combined with demographic stochasticity and uncorrelated environmental noise yield a narrower probability distribution around the mean extinction time. Altogether, these experimental insights into extinction dynamics emphasize that the inherent nonlinear dynamics caused by individual interactions can be an important influence on the extinction risk. In general, they show the feasibility and fruitfulness of coupling extinction theories with experimental validation.

Brine shrimp
Artemia franciscana

How the risk of extinction of a population is influenced by individual life-history traits very much depends on the density-dependent mechanisms involved in the population's regulation. The pattern of density dependence also determines which selective pressures are exerted on the traits. Behavioral traits involved in the density-dependent mechanisms may themselves change in response to selection that acts on their genetically based variation. Thus, density dependence interlinks individual traits, natural selection, and population viability in a complex interplay. The need to integrate density dependence into models of population viability and extinction has long been acknowledged (Boyce 1992; White 2000). We hope that this chapter will foster this fundamental part of the agenda of evolutionary conservation biology.

3

Age Structure, Mating System, and Population Viability

Stéphane Legendre

3.1 Introduction

The fate of populations depends on the life-history traits of the species and possible adaptive changes in these traits in response to selective pressure. In unstructured population models, life-history traits are compounded into few parameters, like the intrinsic growth rate r and the carrying capacity K (see Chapter 2). Structured population models, based on life-cycle graphs, allow the effects of specific life-history traits (survival rates, fecundities, generation time, age at maturity) on population dynamics to be investigated. For example, sensitivities of the growth rate to changes in life-cycle transitions can be computed. Individual life-history traits are important determinants of a population's extinction risk, and are also both factors in and targets of a population's adaptive response to environmental change.

When population size is small – always a concern in conservation biology – both individual life-history traits and the structure of interactions between individuals and the genetic system are expected to influence viability. The mating system, for example, may be conducive to an Allee effect (see Chapter 2), and inbreeding depression is a potentially important factor of the extinction process of small populations. In this chapter, we study the interplay between population structure, in terms of age and sex, and population persistence. Two-sex structured models that incorporate specific features of the social mating system and possible differences in male and female life cycles are analyzed. Also, attempts to merge genetic factors and demography into integrated models are presented. Size-structured models, more appropriate to plants and some animal species, are not considered here, but lead to similar developments.

3.2 Extinction Risk in Age-structured Populations

A life-cycle graph is a macroscopic description of an average organism within a population, describing the effects of the life-history traits. A population is considered as a set of individuals that share the same life cycle, and is structured in age classes. The life-history trait and resultant demographic parameters (survival, fecundity) quantify the flows of individuals between age classes. Iterating the life cycle in discrete time realizes the dynamics of the average life-history phenotype in a given environment.

Essentials about structured deterministic models

The life cycle translates into matrix form, which enables several demographic quantities to be computed, among which is the long-term growth rate λ (Caswell 1989, 2001; Stearns 1992). A matrix $A = (a_{ij})$ can be assembled, with a_{ij} being the contribution of an individual in age class i to age class j, from one time step to the next. The nonzero entries of the matrix are the demographic parameters. The population trajectory is obtained by iterating over time according to the recursion equation

$$\vec{N}(t+1) = A\vec{N}(t)\,, \tag{3.1a}$$

where $\vec{N}(t) = (N_1(t), ..., N_n(t))^T$ denotes the population vector at time t, and $N_i(t)$ the number of individuals in age class i; there are n age classes. The population size at time t is $N(t)$, the sum of the entries of the population vector. An example of a general female-based matrix A is

$$A = \begin{pmatrix} \sigma\phi_0 b_1 & \cdots & \sigma\phi_0 b_{n-1} & \sigma\phi_0 b_n \\ \phi_1 & \cdots & 0 & 0 \\ \vdots & \ddots & \vdots & \vdots \\ 0 & \cdots & \phi_{n-1} & \phi \end{pmatrix}, \tag{3.1b}$$

with ϕ_0 being the juvenile survival rate, $\phi_1, ..., \phi_{n-1}$ subadult survival rates, ϕ the adult survival rate, and $b_1, ..., b_n$ fecundities. The primary female sex ratio σ (proportion of females at birth) is emphasized here as a parameter, the usual value being $\sigma = 0.5$. When $\phi = 0$, A takes the form of the so-called Leslie matrix. A prebreeding census is assumed, since juvenile survival rate ϕ_0 appears in matrix A as a multiplicative factor of fecundities in the first row.

The main demographic result is that, after transitory damped oscillations, the population enters a stable regime of exponential growth with rate λ, whatever the initial population vector $\vec{N}(0)$, where λ is the dominant eigenvalue of the matrix A. The dynamics of the population depend entirely on the algebraic properties of the matrix. Asymptotically, the population size $N(t)$ is such that $N(t+1) \approx \lambda N(t)$. The celebrated Perron–Frobenius theorem of linear algebra ensures that λ is real and positive. The population either increases exponentially ($\lambda > 1$), which results in demographic explosion, or declines exponentially ($\lambda < 1$), which results in population extinction. The degenerate case $\lambda = 1$ leads to equilibrium. Convergence toward the asymptotic regime is geometric with rate $1/\xi$, where $\xi = \lambda/|\lambda_2|$ is the damping ratio, λ_2 being the second eigenvalue of the matrix. The period of the transient oscillations depends mostly on the angle formed by λ_2 and the real axis in the complex plane.

The population structure at time t,

$$\vec{W}(t) = \left(\frac{N_1(t)}{N(t)}, ..., \frac{N_n(t)}{N(t)}\right)^T, \tag{3.2}$$

is the vector of the proportions of individuals in the various age classes. Except for matrixes A with special structure [see Chapter 4 in Caswell (2001) for more details], this vector converges toward a stable population structure $\vec{W}$, known as the stable age distribution. The vector $\vec{W}$ is the right eigenvector of matrix A corresponding to λ. The left eigenvector $\vec{V}$ of the matrix corresponding to λ gives the series of reproductive values indexed by age. These reproductive values describe which age classes contribute most to population size when the asymptotic regime is reached. More precisely, a population's size is asymptotically given by

$$N(t) \approx \lambda^t \langle \vec{V}, \vec{W}(0) \rangle N(0) , \tag{3.3}$$

where the angular brackets denote the scalar product of vectors. Equation (3.4) yields an estimator of the actual average growth rate until time t

$$\hat{\lambda} = \exp\left(\frac{\ln(N(t)) - \ln(N(0))}{t} \right) . \tag{3.4}$$

Another important quantity defined at the population level from individual life-history traits is the generation time. There are various measures of generation time, one being the mean generation length $\overline{T}$, a weighted sum of the contribution of each age class to offspring once the population has reached its asymptotic regime. For the Leslie matrix, this gives

$$\overline{T} = \sum_{i=1}^{n} i \Phi(i) \lambda^{-i} , \tag{3.5}$$

with $\Phi(i) = \sigma \phi_0 \phi_1 \ldots \phi_{i-1} b_i$.

Factors of population regulation and extinction

The model described above, based on a constant matrix A, can be viewed as a deterministic skeleton upon which density-dependent factors and stochastic processes will operate. Stochastic processes may be endogenous or exogenous: demographic stochasticity and interaction stochasticity pertain to the first kind, whereas environmental stochasticity and catastrophes are of the second kind (see Box 2.1). For small populations, demographic stochasticity, which is unavoidable and strictly dependent on population size, can become the main factor of extinction. The long-term growth rate λ of the deterministic skeleton model can always be defined, but stochastic processes generate variation in the instantaneous growth rate and population size. Predictors of the extinction risk, best described by the distribution of time to extinction, typically involve measures of this variation. Usually, to obtain such measures requires intensive computer simulations, but a good deal of mathematical theory is available to guide the simulations and interpret the results (e.g., Ferrière *et al.* 1996; Mills *et al.* 1996; Fieberg and Ellner 2001).

Density dependence and stochastic factors that affect the population dynamics can be considered, as a first approximation, to be perturbations of the above model, and their respective influences can be assessed from the sensitivities of the growth rate λ to changes in various parameters (Caswell 1989; Tuljapurkar 1990). When a parameter x is varied by an amount ε, the growth rate λ changes by an amount $\varepsilon s_\lambda(x)$, where $s_\lambda(x) = \partial\lambda/\partial x$ is the sensitivity of λ to changes in x. When a parameter x of the model is varied by $\eta\%$, the growth rate λ changes by $\eta e_\lambda(x)\%$, where $e_\lambda(x) = (x/\lambda)s_\lambda(x)$ is the elasticity of λ to changes in x. Thus, elasticity is similar to sensitivity, but it takes the relative magnitude of the parameter change into account. Sensitivities and elasticities enable us to determine those parameters that have the greatest impact on population growth. However, sensitivities only quantify the impact of small independent perturbations (see Mills *et al.* 1999).

With $\mathrm{Var}(a_{ij})$ denoting the variance of temporal fluctuations of the corresponding demographic parameter around its average value a_{ij}, the variance in population growth can be approximated by

$$\mathrm{Var}(\lambda) \approx \sum_{i,j} \left(\frac{\partial\lambda}{\partial a_{ij}}\right)^2 \mathrm{Var}(a_{ij}) \,. \tag{3.6}$$

Thus, the sensitivities to the demographic parameters act as weights to determine the growth-rate variance from these parameters' variance. The sensitivities are therefore important determinants of the extinction risk. A powerful result from Houllier and Lebreton (1986), which has been little appreciated to date, is the following. With c denoting a common parameter that multiplies the fertilities in the first row of A [for example, σ or ϕ_0 in Equation (3.1b)], the elasticity with respect to c is inversely proportional to generation time $\overline{\overline{T}}$, that is

$$e_\lambda(c) = \frac{c}{\lambda}\frac{\partial\lambda}{\partial c} = \frac{1}{\overline{\overline{T}}} \,. \tag{3.7}$$

This implies that short-lived species are more sensitive to fluctuations in fertility parameters than long-lived ones. Conversely, long-lived species are, comparatively speaking, more sensitive to fluctuations in the adult survival rate.

Density dependence arises from resource limitation (exogenous, such as food or space, or endogenous, such as partners for reproduction), and results in demographic parameters being functions of the number of individuals in age classes. As competition may involve different resources (food, territory, breeding sites), density dependence may differentially affect the various stages of a life cycle. Negative density dependence leads to extinction or regulation. In the latter case, complex dynamics (quasi-periodicity, chaos) may occur (May and Oster 1976; Ruelle 1989). Notice that a longstanding common wisdom has been that large and unpredictable fluctuations associated with chaotic dynamics should increase a population's vulnerability to extinction. In fact, chaotic population dynamics can result from the operation of natural selection on life-history traits (Ferrière and Gatto 1993), and chaos may create enough asynchrony between local populations connected by migration to promote long-term persistence at the regional scale (Allen *et al.* 1993). Thus, no simple relationship exists between nonlinear dynamics, the extinction risk, and adaptation (see Chapter 11).

Demographic stochasticity, which stems from the random realization of the life cycle by each individual in the population, is modeled by specifying a multitype branching process based on the above matrix model (see Section 3.3). Under demographic stochasticity, a population either goes extinct or grows at an average rate λ. For $\lambda \leq 1$ extinction occurs with certainty. For $\lambda > 1$ the probability of extinction is strictly larger than 0 and strictly smaller than 1; it depends on the initial population size and structure, and decreases exponentially with initial population size. More precisely, the probability of extinction at time t is

$$q_e(t) = q_1(t)^{N_1(0)} \ldots q_n(t)^{N_n(0)} , \tag{3.8}$$

where $q_i(t)$ is the probability of extinction at time t when the initial population consists of a single individual in age class i. The average population structure is unaffected by demographic stochasticity. For $\lambda < 1$, conditional on being nonextinct, the probability distribution of population size converges toward a constant distribution, known as the population's quasi-stationary distribution. As a consequence, from one time step to the next, the population goes extinct with probability $1 - \lambda$, thus behaving like a single individual with survival probability λ. The quasi-stationary distribution also exists when the population is regulated by density dependence, or is subject to uncorrelated environmental stochasticity: there is a constant parameter $\beta < 1$ such that the population behaves as a single individual with survival rate β (Gosselin and Lebreton 2000).

Under environmental stochasticity, demographic parameters may vary independently of each other, or co-vary (e.g., a decrease in survival co-occurs with a decrease in fecundity), with or without temporal autocorrelations (Shaffer 1987; Tuljapurkar 1990; Lande 1993; Chapter 2). The usual effect is to reduce the expected growth rate, compared to the value that the growth rate would assume if all parameters were fixed at their average value. A stationary population structure is not guaranteed. The ultimate probability of extinction is independent of the initial population size, but the average time to extinction increases with the initial population size.

All of the above factors contribute to various extents to the risk of extinction. For example, in their survey on translocations, Griffith *et al.* (1989) find that the initial population size is a strong predictor of the extinction risk. The probability of extinction decreases as the initial population size increases, and remains constant above some population-size threshold. This is what is expected under the combined effects of demographic and environmental stochasticity. Also, an appreciation of how environmental stochasticity and density dependence combine is crucial to forecasting the dynamics of natural populations accurately (e.g., Leirs *et al.* 1997).

3.3 Effect of Sexual Structure on Population Viability

Demographic models usually describe the dynamics of the female population only. However, for small populations, random fluctuations in sex ratio and pair formation may affect persistence (see Chapter 2). To account for the impact of sexual reproduction on population viability, two-sex life cycles with interactions between sexes must be constructed. The pattern of pair formation (the social mating system), appears to play an important role in extinction processes: for small populations, the shortage of mates can generate an Allee effect because the female population forms a limiting resource (Chapter 2; Courchamp *et al.* 1999; Legendre *et al.* 1999; Stephens and Sutherland 1999; Stephens *et al.* 1999; Møller and Legendre 2001).

Deterministic two-sex models

Models of two-sex life cycles can be constructed by duplicating the life-cycle graph for males and females. Reproductive transitions from the female part of the graph connect to both male and female parts; the relative contribution to each part is measured by the primary female sex ratio σ, which means that female offspring are produced in proportion σ and males in proportion $1-\sigma$.

In two-sex models, reproduction parameters become dependent upon the number of mating pairs that can form. Therefore, the mating system ought to be specified (Box 3.1) in terms of a "marriage function" or "mating function" (Caswell and Weeks 1986; see also Heino *et al.* 1998). This mating function gives the number $M(N_m, N_f)$ of matings as a function of the numbers N_m and N_f of reproductive males and females. Mating models can be designed to account for fidelity, probability of encounter, and asymmetric preferences in males and females depending on social status or age (Gerritsen 1980; Gimelfarb 1988; Castillo-Chavez and Hsu Schmitz 1997). Considering only matings that yield offspring, we assume here that the number of matings is less than the number of sexually mature females, i.e., $M(N_m, N_f) \leq N_f$.

The mathematical properties of the mating function reflect the structure of encounters between sexes (Caswell and Weeks 1986; Martcheva 1999). One important property is homogeneity, which is the fact that, for any $c \geq 0$, one has $M(cN_m, cN_f) = c\, M(N_m, N_f)$. If we define the breeding sex ratio ρ as the proportion of reproductive females in the reproductive population, by virtue of the previous homogeneity property, we have

$$\frac{M(N_m, N_f)}{N_f + N_m} = M\left(\frac{N_m}{N_f + N_m}, \frac{N_f}{N_f + N_m}\right) = M(1-\rho, \rho)\,. \tag{3.9a}$$

Investigating the dynamics of two-sex models is greatly facilitated by considering the following limit function μ_M, which is associated with the mating function M for each value of σ in [0,1], and is given by

$$\mu_M(\sigma) = \lim_{\rho \to \sigma} \frac{M(N_m, N_f)}{N_f + N_m} = M(1-\sigma, \sigma)\,. \tag{3.9b}$$

The limit function μ_M captures the main features of the mating system. It is 0 for $\sigma = 0$ and $\sigma = 1$, and is usually concave with a single maximum (Figure 3.1). Furthermore, for $M(N_m, N_f) \leq N_f$, Equation (3.9b) leads to $\mu_M(\sigma) \leq \sigma$, which entails that the graph of the limit function μ_M lies entirely below the main diagonal. Notice that the main diagonal coincides with the graph of the limit function μ_M recovered when the number of mating paris is merely equal to the number of

Box 3.1 Social mating systems

Several aspects of sexual selection are determined by the social mating system, that is, the way males and females pair for reproduction (Orians 1969; Wade and Arnold 1980; Arnold and Duvall 1994). There are four important mating systems in animals:

Monogamy. A single male and a single female form a strong pair bond, usually involving parental care by both partners. The monogamous mating system is very common in birds (90% of all bird species). However, extra-pair copulations are frequent and can lead to strong sexual selection (Møller and Birkhead 1994).

Polygyny. A single male mates with several females, who mate only with him. Often, males provide nothing but gametes (e.g., lekking species), or they provide little parental care, but rather territory or protection. Sexual dimorphism is often correlated with the degree of polygyny (pinnipeds, ungulates), an extreme example being that of elephant seals, where males outweigh females by more than five times. Harem sizes are typically 50 individuals, and a large proportion of males never reproduce. Nevertheless, several monogamous species are highly dimorphic. There are also examples of species that switch between monogamy and polygyny according to environment (e.g., Höglund 1996). Monogamy and polygyny could coexist as alternative tactics (Pinxten and Eens 1990).

Polyandry. A single female mates with several males, who mate only with her. This rare mating system occurs, for example, in the Dunnock *Prunella modularis*, but polygyny and monogamy are also found in this species, depending on food resources (Davies and Lundberg 1984).

Polygynandry. In this, the most common breeding system, each sex mates with more than one member of the opposite sex. Polygynandry is frequent in mammals. While the offspring require intensive parental care (e.g., female mammals produce milk), the males provide no parental care in 95% of mammal species (Clutton-Brock 1989).

females present in the population, $M(N_m, N_f) = N_f$. In this case, the two-sex model collapses to its one-sex, female counterpart. Figure 3.1 shows the shape of the limit function μ_M for more complex maing systems:

- For the monogamous mating system, males and females pair one-to-one and unpaired individuals do not reproduce. The mating function is $M(N_m, N_f) = \min(N_m, N_f)$, and the corresponding limit function is $\mu_M(\sigma) = \min(1-\sigma, \sigma)$. This function has a tent shape with a single maximum at $\sigma = 0.5$ (Figure 3.1a).
- For the polygynous mating system with harem size θ, a single male mates on average with θ females, giving $M(N_m, N_f) = \min(\theta N_m, N_f)$. The limit function is given by $\mu_M(\sigma) = \min(\theta(1-\sigma), \sigma)$; it has a maximum, as expected at $\sigma = \theta/(\theta+1)$, corresponding to one male mating with θ females (Figure 3.1b).

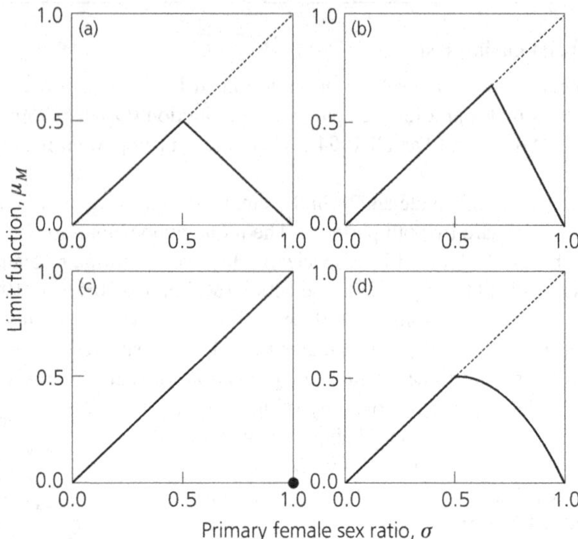

Figure 3.1 Limit functions for alternative mating systems. (a) Monogamy, (b) polygyny with harem size $\theta = 2$, (c) polygyny with unrestricted harem size (the filled circle indicates that μ_M is discontinuous at $\sigma = 1$, with $\mu_M(1) = 0$), (d) harmonic-mean mating function.

- For the polygynous mating system with unrestricted harem size, a single male can mate with as many females as he is willing to. The number of matings is equal to the number of females, except when there are no males, in which case the number of matings is 0. The graph of the limit function coincides with the main diagonal except for $\sigma = 1$, where μ_M is 0 (Figure 3.1c).
- The harmonic-mean mating function $M(N_m, N_f) = \min(2N_f N_m/(N_f + N_m), N_f)$ can be seen as an intermediate pattern between monogamy and polygyny with a harem size of 2. Indeed, each male mates on average with 2ρ females, with ρ being the breeding sex ratio. The graph of the limit function consists of a segment line and half of a parabola (Figure 3.1d).

Under rather general assumptions (the homogeneity property mentioned above being crucial), the two-sex model behaves as the one-sex model, with an asymptotic exponential growth and a stable population structure (Caswell and Weeks 1986; Martcheva 1999). As a result, the realized sex ratio (the proportion of females in the population) stabilizes, as does the breeding sex ratio. As with one-sex dynamics, two-sex dynamics can be decomposed into a transient regime followed by an asymptotic regime of exponential growth. However, convergence toward the equilibrium of the realized sex ratio interferes with convergence toward the stable age distribution, resulting in more complex transient dynamics than in one-sex models. Incorporating density dependence or competition between mates can generate even more complex population dynamics (Caswell and Weeks 1986; Chung 1994; Lindström and Kokko 1998).

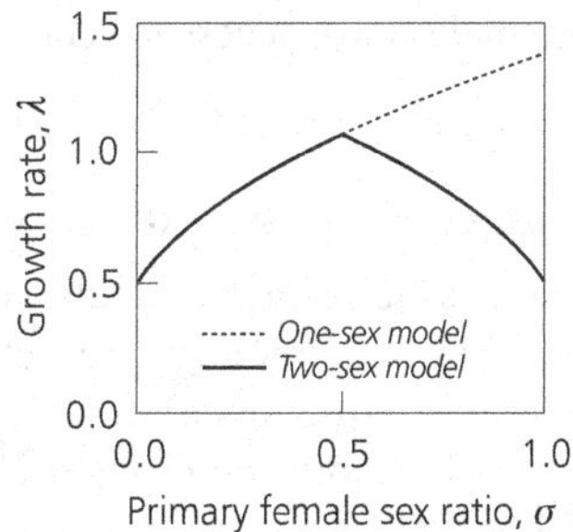

Figure 3.2 Growth rate as a function of the primary female sex ratio for the one-sex model and the two-sex model with monogamous mating function. The two curves coincide for $\sigma \leq 0.5$.

The two-sex expected growth rate, denoted by λ_M, depends on the mating function M and on the male and female parts of the age-specific life cycle. Assuming that:

1. Males and females have identical survival rates,
2. Matings occur with equal probability among reproductive age classes,
3. A limit function μ_M can be associated to the mating function M [Equation (3.10b)],

the growth rate λ_M of the two-sex model with mating function M satisfies (Legendre *et al.* 1999)

$$\lambda_M(\sigma) = \lambda(\mu_M(\sigma)) \, . \tag{3.10a}$$

Thus, the two-sex growth rate is obtained by replacing the primary sex ratio σ in the one-sex model with $\mu_M(\sigma)$. Since the one-sex growth rate $\lambda(\sigma)$ is a continuously increasing function of σ, Equation (3.10a) implies that the two-sex growth rate is always less than, or at most equal to, the one-sex growth rate, that is

$$\lambda_M(\sigma) \leq \lambda(\sigma) \, . \tag{3.10b}$$

Thus, the maximum of the two-sex growth rate corresponds to the maximum of the limit function μ_M (compare Figure 3.1a with Figure 3.2).

For life cycles that are sex-symmetric, i.e., satisfying assumptions 1 and 2 above, the breeding sex ratio ρ is equal to the primary sex ratio σ. However, sex-asymmetric life cycles exist in many species, often in relation to polygamy. In the polygynous mating system with harem size θ, the optimal breeding sex ratio that maximizes the two-sex growth rate is $\rho_{\text{opt}} = \theta/(\theta + 1)$, that is, θ females per male. Assuming a balanced primary sex ratio, $\sigma = 0.5$, the optimal breeding sex ratio can be achieved by reducing the number of reproductive males in several nonexclusive ways:

- Males have lower (adult) survival rates,
- Males have delayed access to reproduction,

- Only a fraction of mature males have access to reproduction.

All three cases are known to occur in polygynous species.

Influence of sexual reproduction on the extinction risk

The effect of demographic stochasticity on the viability of a sexual, age-structured population can be investigated by constructing a branching process model based on the two-sex life cycle and the mating function (Harris 1963; Athreya and Ney 1972; Asmussen and Hering 1983; Gabriel and Bürger 1992; Kokko and Ebenhard 1996; Hull 1998; see also Chapter 2). Numbers of individuals that result from life-cycle transitions are drawn according to integer-valued probability distributions, depending on the number of individuals in age classes. For "all-or-nothing" transitions – such as surviving or not, being born as male or female, being reproductive or not – the number of individuals in the next age class is computed by summing Bernoulli trials (one trial per individual) or, equivalently, by sampling binomial distributions. For fecundity transitions, the number of offspring is computed as a sum of trials according to, for example, a Poisson distribution with a mean equal to the expected fecundity. For example, the two age class matrix

$$A = \begin{pmatrix} 0 & \sigma B \\ \phi_1 & \phi \end{pmatrix} \tag{3.11}$$

leads to recursion Equations (3.12a) and (3.12b) from one time step to the next,

$$N_1(t+1) = \mathrm{Poisson}(\sigma B) * N_2(t) , \tag{3.12a}$$

$$N_2(t+1) = \mathrm{Binom}(N_1(t), \phi_1) + \mathrm{Binom}(N_2(t), \phi) , \tag{3.12b}$$

where $\mathrm{Poisson}(x) * N$ denotes the sum of N samples of the Poisson distribution with mean x, and $\mathrm{Binom}(N, p)$ denotes the sum of N Bernoulli trials of probability p. The corresponding two-sex model is specified by the equations

$$J(t) = \mathrm{Poisson}(B) * M(N_{m2}(t), N_{f2}(t)) , \tag{3.13a}$$

$$N_{f1}(t+1) = \mathrm{Binom}(J(t), \sigma) , \tag{3.13b}$$

$$N_{m1}(t+1) = J(t) - N_{f1}(t+1) , \tag{3.13c}$$

$$N_{f2}(t+1) = \mathrm{Binom}(N_{f1}(t), \phi_{f1}) + \mathrm{Binom}(N_{f2}(t), \phi_f) , \tag{3.13d}$$

$$N_{m2}(t+1) = \mathrm{Binom}(N_{m1}(t), \phi_{m1}) + \mathrm{Binom}(N_{m2}(t), \phi_m) . \tag{3.13e}$$

Equation (3.13a) gives the number J of newborns produced according to the mating function M; Equations (3.13b) and (3.13c) split this number into males and females according to the primary female sex ratio σ; Equations (3.13d) and (3.13e) give the number of adult males and females that result from the survival of either subadult males and subadult females, or adult males and adult females.

Under the assumptions 1 to 3 above, the two-sex population under demographic stochasticity either becomes extinct or grows on average at a rate λ_M, as in the one-sex case. This behavior seems general, but theoretical results are still lacking. Incorporation of both sexes and the mating system complicates the structure of the transient dynamics, and possibly reduces the long-term growth rate as a consequence. Thus, the two-sex branching process has a larger probability of extinction than the corresponding one-sex process, even when the two-sex growth rate is equal to the one-sex growth rate.

Extinction probabilities and the distribution of extinction time turn out to be highly dependent on the mating system. Real data – for passerines introduced to New Zealand (Legendre *et al.* 1999), and bighorn sheep in North America (see Section 3.4) – suggest that demographic stochasticity interacts with the mating system to determine the extinction risk of small populations. Long-lived and short-lived sexual species behave differently with regard to extinction because of the stochasticity of the mating process, one reason being that the elasticity of λ to changes in the primary sex ratio σ is inversely related to the generation time $\overline{T}$ [see Equation (3.5)]. Differentiating Equation (3.10a) with respect to σ by the chain rule,

$$\partial\lambda_M/\partial\sigma = (\partial\lambda/\partial\mu_M)(\partial\mu_M/\partial\sigma)\,, \tag{3.14}$$

shows that the sensitivity $\partial\lambda_M/\partial\sigma$ of the two-sex growth rate to changes in σ, and hence the probability of extinction, depends on the slope $\partial\mu_M/\partial\sigma$ of the limit function. As a result, the same growth rate can lead to different probabilities of extinction depending on the mating system. For example, the monogamous mating function and the harmonic-mean mating function produce the same two-sex growth rate when the primary sex ratio σ is balanced, but the smoothness of the harmonic mean dramatically reduces the probability of extinction, as shown in Figure 3.3a. Furthermore, the value of σ that maximizes the growth rate and the value of σ that minimizes the probability of extinction usually differ. For the polygynous mating system with unrestricted harem size, the growth rate is maximized when the primary sex ratio σ is close to 1 (Figure 3.1c), but the probability of extinction shows a different pattern. If the proportion of females is low, few offspring are produced, and therefore the growth rate is less than 1 and extinction is certain. If the proportion of females is high, then the growth rate is large, but males can go extinct. The extinction risk turns out to be minimum for an intermediate value of σ, as shown in Figure 3.3b. Such contrasting effects of mating structure on a population's growth rate and extinction risk suggest that the adaptive evolution of sex-related life-history traits may have complex and unexpected effects on population viability.

Sexual selection and extinction

Males and females have conflicting interests in reproduction, and as a consequence natural selection operates differentially on each sex. This gives rise to sexual selection (Darwin 1871; Fisher 1958; Andersson 1994). A general pattern is that

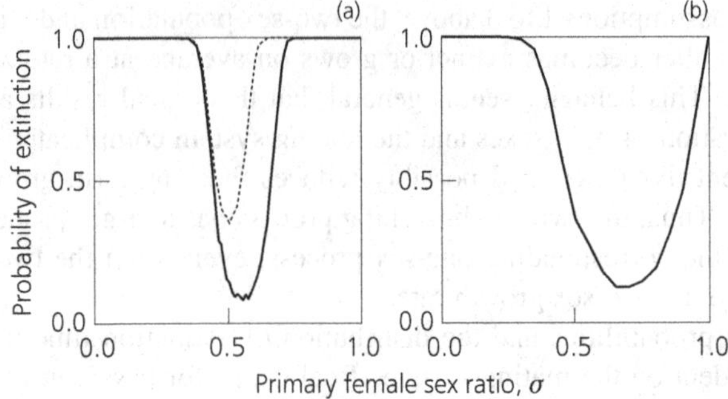

Figure 3.3 Probability of extinction as a function of the primary female sex ratio σ, for the two-sex model as specified in Section 3.3. (a) Monogamous mating system versus harmonic-mean mating function: the smoothness of the harmonic mean reduces the probability of extinction (dotted curve). The initial population comprised 20 adult males and 20 adult females. Monte Carlo simulations involved 1000 trials run over 100 time steps. (b) Polygynous mating system with unrestricted harem size: growth rate is maximal for $\sigma \approx 1$, while the probability of extinction is minimal for $\sigma \approx 0.66$. The initial population comprised five adult males and five adult females. Monte Carlo simulations involved 1000 trials run over 50 time steps. Parameters: $\phi_1 = 0.3$, $\phi = 0.5$, $B = 4.0$; growth rate $\lambda_M(\sigma) = 1.064$ for $\sigma = 0.5$.

females (with a small number of large gametes) are under selection to increase their reproductive success by searching for "good" males, while males can increase their fitness by copulating with many females (as males have a huge number of tiny motile gametes). This generally induces a larger variance in male reproductive success (Bateman 1948; Wade and Arnold 1980), and promotes sexual dimorphism, with the development of exaggerated ornaments, weapons, signals, or behaviors far beyond the expected optimal under the action of individual selection. In effect, sexually selected species seem more prone to extinction (McLain *et al.* 1995; Sorci *et al.* 1998). However, little theory deals with the impact of sexual selection on demography, as most models pertain to the field of population genetics (Lande 1980a; Kirkpatrick 1982; Pomiankowski *et al.* 1991; Iwasa *et al.* 1991) or to game theory (e.g., Maynard Smith 1982; Iwasa and Harada 1998), where, in both cases, demographic structure is usually ignored.

Under sexual selection, the evolution of male and female life-history traits may cause them to diverge, leading to sexual dimorphism. Sexual dimorphism on age at maturity (bimaturism) or survival can result in a strongly skewed breeding sex ratio. For example, in polygynous ungulates, adult males usually have lower survival rates than females; in the bighorn sheep, this difference yields a breeding sex ratio $\rho \approx 0.80$ (see Section 3.4). Results in this section suggest that sexual structure could have a significant impact on persistence. Could the individual behavior that determines the structure of a mating system evolve concomitantly with a reduced risk of extinction for the population? Evolutionary changes in sexual behavior may be tightly constrained, especially in species undergoing strong sexual selection. While the evolution of life histories has been studied in detail, very little is known of the effect of sexual reproduction and its evolution on population dynamics and viability.

3.4 Interfacing Demography and Genetics

In small populations, genetic drift (the genetic equivalent of demographic stochasticity) may beget the fixation of deleterious mutations (see Chapters 5 and 9). Assessing the impact of genetic drift on vital rates and population viability requires the development of integrated models that interface demography and genetics (Lande 1994, 1995; Mills and Smouse 1994; Lynch *et al.* 1995a). To this end, the approach followed by Thévenon and Couvet (2002), aims at tracking the population vector of age-class frequencies and the genetic vector of allelic frequencies simultaneously, while explicitly modeling interactions between the two. The so-called mutation load that results from deleterious mutations (see Chapter 9) affects the demographic parameters and therefore the population size and structure, which in turn modify the genetic composition of the population (Box 3.2). In this approach, the effect of selection against deleterious mutation is accounted for (see Chapter 10). In their seminal work, Mills and Smouse (1994) also describe the combined effects of demography and genetics on the risk of population extinction, but selection was not part of their framework. Mildly deleterious mutations appear to be the most harmful, because, unlike strongly deleterious ones, they are not eliminated. Below, we illustrate these general considerations with a specific example, that of the dynamics of the American bighorn sheep, *Ovis canadensis*.

American bighorn sheep
Ovis canadensis

The American bighorn sheep of the Rocky mountains is a polygynous species that exhibits a strong sexual dimorphism. Females generally start to reproduce when 2–3 years old. Males mature when 3.5 years old on average, but competition between males usually means that they do not participate in reproduction until at an older age. Adult male survival rates are lower than those of females. Senescence starts after 7–8 years, and is more pronounced for males, but animals can live up to 20 years. The bighorn sheep model is summarized in Table 3.1.

Demographic parameters used in the model come from the literature (Geist 1971; Monson and Sumner 1981; Festa-Bianchet *et al.* 1995; Jorgenson *et al.* 1997). Females and males start to reproduce at 3 and 5 years of age, respectively. A polygynous mating system with harem size $\theta = 4$ is used, matings being dispatched evenly between reproductive female age classes. The two-sex matrix is a

Box 3.2 Inbreeding depression in structured population models

In a population, the fate of mutations depends on the initial allelic frequencies, the mutation rate u (and reverse mutation rate u_{rev}), the selective pressures that eliminate deleterious genes, and genetic drift. For a given gene with wild and mutant alleles, we denote by h the dominance of the deleterious allele, and by s_{repr} and s_{surv} the selection coefficients associated with reproduction and annual survival (juvenile survival or adult survival), respectively. For given frequencies p and $q = 1 - p$ of the wild and mutant alleles, the expected frequency q'_{repr} of the mutant allele after reproduction is

$$q'_{\text{repr}} = \frac{pq(1 - hs_{\text{repr}}) + q^2(1 - s_{\text{repr}})}{p^2 + 2pq(1 - hs_{\text{repr}}) + q^2(1 - s_{\text{repr}})} + (up - u_{\text{rev}}q) . \tag{a}$$

After one time step, the expected frequency q'_{surv} of the mutant allele among surviving individuals is

$$q'_{\text{surv}} = \frac{pq(1 - hs_{\text{surv}}) + q^2(1 - s_{\text{surv}})}{p^2 + 2pq(1 - hs_{\text{surv}}) + q^2(1 - s_{\text{surv}})} . \tag{b}$$

The genetic composition of a population of N individuals is described by a genetic vector Γ with $2N + 1$ entries, in which the kth entry gives the probability of a randomly chosen individual to have k mutated genes. Through reproduction N individuals produce N' offspring, among which the number of mutants is drawn from the binomial distribution Binom$(2N', q'_{\text{repr}})$, with q'_{repr} given by Equation (a). The resultant genetic vector $\Gamma' = G_{\text{repr}}(\Gamma, N', s_{\text{repr}})$ has $2N' + 1$ entries. For survival, a hypergeometric distribution is used with expectation q'_{repr}, as given by Equation (b). The mutation load $G_{\text{load}}(\Gamma, s)$ that affects survival and fecundity parameters is computed according to

$$G_{\text{load}}(\Gamma, s) = \prod_{k=0}^{2N} [(1 - hs)^{2pq}(1 - s)^{q^2}]^{n_L \Gamma_k} , \tag{c}$$

where n_L denotes the number of loci, Γ_k is the kth entry of the genetic vector, and $s = s_{\text{repr}}$ or s_{surv} depending on whether the affected trait is a fecundity parameter or a survival parameter. The genetic vectors $\Gamma^{(1)}$ and $\Gamma^{(2)}$ of two sets of individuals can be combined into a genetic vector $\Gamma = G_{\text{comp}}(\Gamma^{(1)}, \Gamma^{(2)})$, where the operator G_{comp} is such that the kth entry of Γ is given by

$$\Gamma_k = \sum_{\substack{k_1, k_2 \\ k_1 + k_2 = k}} \Gamma^{(1)}_{k_1} \Gamma^{(2)}_{k_2} . \tag{d}$$

This associative composition can be extended to any number of genetic vectors.

For each age class i containing N_i individuals, there is a corresponding genetic vector $\Gamma^{(i)}$ with $2N_i + 1$ entries. Likewise, for each life-cycle transition (i, j), there is an associated intermediate population size $N^{(i,j)}$ and an intermediate genetic vector $\Gamma^{(i,j)}$ with $2N^{(i,j)} + 1$ entries. From one time step to the next, the interactions of genetics and demography are incorporated in the one-sex two-age class matrix $\begin{pmatrix} 0 & \sigma B \\ \phi_1 & \phi_2 \end{pmatrix}$ according to the following scheme:

continued

Box 3.2 *continued*

- *Influence of genetics on demography.* Compute the intermediate population sizes $N^{(i,j)}$, taking into account demographic stochasticity and mutational load,

$$N'^{(2,1)} = \text{Poisson}\big(G_{\text{load}}(\Gamma^{(2)}, s_{\text{repr}}) \times \sigma B\big) \times N_2 ,$$
$$N'^{(1,2)} = \text{Binom}\big(N_1, G_{\text{load}}(\Gamma^{(1)}, s_{\text{surv}}) \times \phi_1\big) ,$$
$$N'^{(2,2)} = \text{Binom}\big(N_2, G_{\text{load}}(\Gamma^{(2)}, s_{\text{surv}}) \times \phi_2\big) ,$$
$$N'_1 = N'^{(2,1)} ,$$
$$N'_2 = N'^{(1,2)} + N'^{(2,2)} .$$

- *Influence of demography on genetics.* Update the intermediate genetic vectors $\Gamma^{(i,j)}$, taking the intermediate population sizes $N'^{(i,j)}$ into account, and compute the resultant genetic vectors $\Gamma'^{(j)}$,

$$\Gamma'^{(1,2)} = G_{\text{surv}}(\Gamma^{(1)}, N'^{(1,2)}, s_{\text{surv}})$$
$$\Gamma'^{(2,2)} = G_{\text{surv}}(\Gamma^{(2)}, N'^{(2,2)}, s_{\text{surv}})$$
$$\Gamma'^{(2,1)} = G_{\text{repr}}(\Gamma^{(2)}, N'^{(2,1)}, s_{\text{repr}})$$
$$\Gamma'^{(1)} = \Gamma'^{(2,1)}$$
$$\Gamma'^{(2)} = G_{\text{comp}}(\Gamma'^{(1,2)}, \Gamma'^{(2,2)}) .$$

14 × 14 block matrix with the upper diagonal block standing for the male life cycle, the lower diagonal block for the female life cycle, and the upper nondiagonal block for the production of male offspring by females.

The two-sex growth rate is $\lambda_M = 1.03$. All female age classes have about the same reproductive value. The differences in male and female life cycles mean that the proportion of females in the population is 63%. The breeding sex ratio is $\rho = 0.81$, close to the optimum value $\rho_{\text{opt}} = 4/5 = 0.80$ (see Section 3.3). In fact, harem size has a significant impact on the growth rate, which underscores the importance of the mating system on population persistence. The effective population size, given by

$$N_e = \frac{4N_m N_f}{N_f + N_m} = 4N_m \rho , \tag{3.15}$$

is equal to 36% of the total population size, close to the estimated value of 33% for bighorn populations in Wyoming (Fitzsimmons *et al.* 1997).

Demographic stochasticity is incorporated by treating by randomizing life-cycle transitions and matings as stochastic processes, as explained in Section 3.3. For simplicity, the initial populations are assumed to include individuals that belong to the oldest age class only, with 80% females and 20% males, close to the stable proportion given by the deterministic model. Probabilities of extinction are

Table 3.1 Parameters used in the bighorn sheep population model.

Demographic parameters

		Males		Females
Survival probabilities				
Juvenile	ϕ_{m0}	0.57	ϕ_{f0}	0.57
Yearlings	ϕ_{m1}	0.86	ϕ_{f1}	0.83
Prime age	ϕ_{m2}	0.86	ϕ_{f2}	0.94
	$\phi_{m3}, \phi_{m4}, \phi_{m5}, \phi_{m6}$	0.78	$\phi_{f3}, \phi_{f4}, \phi_{f5}, \phi_{f6}$	0.94
Older	ϕ_{m}	0.63	ϕ_{f}	0.85
Fecundities			b_3, b_4, b_5, b_6, b_7	0.70

Population matrix

$$\left(\begin{array}{ccccccc|ccccccc}
0 & & \cdots & & & & 0 & 0 & 0 & B_{m3} & B_{m4} & B_{m5} & B_{m6} & B_{m7} \\
\phi_{m1} & 0 & \cdots & & & & 0 & 0 & & & \cdots & & & 0 \\
0 & \phi_{m2} & 0 & & & & & & & & & & & \\
\vdots & & \phi_{m3} & 0 & & & \vdots & \vdots & & & & & & \vdots \\
& & & \phi_{m4} & 0 & & & & & & & & & \\
& & & & \phi_{m5} & 0 & 0 & & & & & & & \\
0 & & & \cdots & 0 & \phi_{m6} & \phi_{m} & 0 & & & \cdots & & & 0 \\
\hline
0 & & \cdots & & & & 0 & 0 & 0 & B_{f3} & B_{f4} & B_{f5} & B_{f6} & B_{f7} \\
& & & & & & & \phi_{f1} & 0 & & \cdots & & & 0 \\
& & & & & & & 0 & \phi_{f2} & 0 & & & & \\
\vdots & & & & & & \vdots & \vdots & & \phi_{f3} & 0 & & & \vdots \\
& & & & & & & & & & \phi_{f4} & 0 & & \\
& & & & & & & & & & & \phi_{f5} & 0 & 0 \\
0 & & \cdots & & & & 0 & 0 & & & \cdots & 0 & \phi_{f6} & \phi_{f}
\end{array}\right)$$

Mating system

Number of reproducing females: $N_f = N_{f3} + N_{f4} + N_{f5} + N_{f6} + N_{f7}$

Number of reproducing males: $N_m = N_{m5} + N_{m6} + N_{m7}$

Number of polygynous matings with harem size 4: $M = \min(4N_m, N_f)$

Number of matings involving i-year-old females: $M_i = M\ N_{fi}/N_f$

Primary sex ratio: $\sigma = 0.5$

Number of 1-year-old males produced by i-year-old females: $B_{mi}N_{fi} = (1-\sigma)\phi_{m0}b_i M_i$

Number of 1-year-old females produced by i-year-old females: $B_{fi}N_{fi} = \sigma\phi_{f0}b_i M_i$

Genetic parameters

Dominance	$h = 0.2$
Selection coefficient (reproduction)	$s_{\mathrm{repr}} = 0.01$
Selection coefficient (annual survival)	$s_{\mathrm{surv}} = 0.0017$
Mutation rate	$u = 5\ 10^{-6}$
Reverse mutation rate	$u_{\mathrm{rev}} = 5\ 10^{-7}$
Number of loci	$n_L = 10\,000$

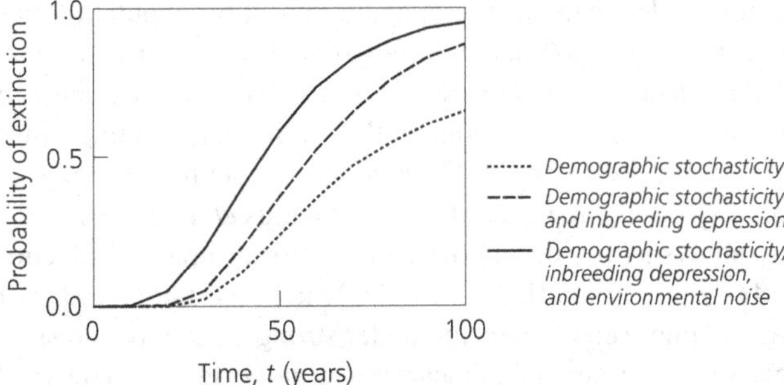

Figure 3.4 Probability of extinction, cumulated over time, as predicted by the bighorn sheep model. The initial population comprised 40 individuals; Monte Carlo simulations involved 1000 trials.

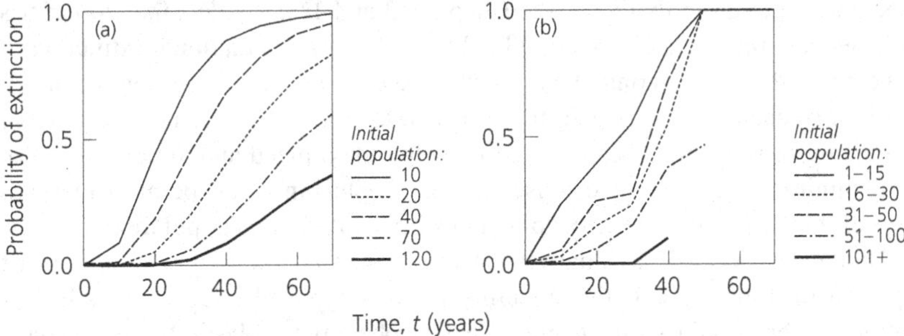

Figure 3.5 Probability of extinction, cumulated over time, for the bighorn sheep population. (a) Model results under the joint effects of demographic stochasticity, inbreeding depression, and environmental noise. Monte Carlo simulations involved 1000 trials. (b) Observed rates of extinction. *Source*: Berger (1990).

computed using Monte Carlo simulations that involve 1000 trials (Figures 3.4 and 3.5). Inbreeding depression was incorporated in the model as described in Box 3.2, with the parameters given in Table 3.1. Accounting for the genetic load induces a marked increase of extinction probability, as shown in Figure 3.4. Finally, environmental noise is incorporated in the model: demographic parameters vary from year to year according to probability beta distributions with means equal to estimated parameter values, and standard deviation fixed to 0.2. This extra noise further increases the probability of extinction (Figure 3.4). Demographic stochasticity, inbreeding depression, and environmental noise all have to be considered for the estimated probabilities of extinction to match the real data shown in Figure 3.5.

3.5 Concluding Comments

Selective pressures that affect phenotypes and the viability of a population both depend upon the structure of the individual life cycle. We have shown in this chapter that both the age structure and the mating system – two important characteristics of a species' life cycle – can have a dramatic impact on population viability. In a conservation perspective, population structure and sexual structure correspond to social bonds that management should take into account. For example, introducing adults is, in many cases, a better strategy than introducing young, immature individuals (Sarrazin and Legendre 2000), and sustainable hunting pressures depend on the breeding system (Greene *et al.* 1998; Wielgus *et al.* 2001).

Sensitivity analysis provides a powerful tool for estimating selection gradients that act on life-history traits (Lande 1982; Benton and Grant 1999). Parameters associated with higher sensitivities are under stronger selective pressures. In populations subject to environmental stochasticity, the resulting adaptive changes are expected to increase the deterministic growth rate and reduce the discount factor that accounts for environmental variation [see Equation (3.7)]. This is expected to reduce the extinction risk (Lande and Orzack 1988). The underlying theory, however, does not account for density-dependent mechanisms that result from interactions between individuals (see Chapters 2 and 11). Models that consider simultaneously the evolution of complex life cycles and population's extinction risk in a density-dependent context still need to be developed. One reason for the current lack of such models is that, traditionally, evolution has been conceived as an optimizing process, with the growth rate being maximized and the extinction risk being minimized as a putative consequence. Chapters in Part C point to alternative ways of thinking about the effect of life-history evolution on population dynamics.

Another aspect of individual responses to environmental change that must be expected to impact population viability is phenotypic plasticity. Very little attention has been paid so far to the effect of phenotypic plasticity on population dynamics. The analysis of an age-structured model of a population of *Drosophila melanogaster* contaminated by the C virus showed that the shortening of developmental time increases the growth rate, but could also increase the extinction risk (Thomas-Orillard and Legendre 1996). This example suggests that the plastic response of life histories to environmental threats may have contrasting and intricate effects on population dynamics and viability. Phenotypic plasticity may itself be adaptive, and study of the combined effects of plasticity and its evolution on population dynamics in changing environments offers a further challenge to eco-evolutionary theorists.

4

Spatial Dimensions of Population Viability

Mats Gyllenberg, Ilkka Hanski, and Johan A.J. Metz

4.1 Introduction

In most parts of the world, habitat loss is the number one threat to endangered species. For instance, in Finland the primary cause of threat is some form of habitat loss or alteration in 73% of the red-listed species (Rassi *et al.* 2001). Typically, a reduced total area of habitat is accompanied by habitat fragmentation, such that the remaining habitat occurs in smaller fragments with reduced connectivity. Many landscapes for many species have become highly fragmented (the habitat fragments are small or relatively small and physically completely isolated), while other landscapes have always been highly fragmented naturally. Species that live in such landscapes necessarily have fragmented populations, which more or less closely approach the metapopulation structure originally envisioned by Levins (1969). Levins' metapopulation is a system of local populations that inhabit individual habitat patches connected, to some extent, by migration. The classic metapopulation concept assumes that local populations may go extinct, and so leave the respective habitat patch temporarily unoccupied, while the metapopulation as a whole may persist in a balance between extinctions and colonizations (Levins 1969; Hanski and Gilpin 1997; Hanski 1999). In a broader sense, any assemblage of local populations connected by migration can be called a metapopulation, regardless of the occurrence of local extinctions (Hanski and Gilpin 1997). What is important is the spatially localized interactions of individuals, which may significantly change the dynamics of the metapopulation as a whole in comparison with a single panmictic population (Hanski 1999).

The metapopulation concept has received much attention from conservation biologists during the past 15 years (Soulé 1987; Western and Pearl 1989; Falk and Holsinger 1991; McCullough 1996; Young and Clarke 2000), ever since it replaced the dynamic theory of island biogeography as the main population ecological paradigm in conservation biology (Hanski and Simberloff 1997). The number of well-studied examples of species with a distinctive metapopulation structure and frequent local extinctions is increasing rapidly; these include butterflies, mammals (like the American pika), plants, and plant–herbivore–parasitoid communities (reviews in Hanski 1999, 2001). For European butterflies, many of which have suffered

Collared pika
Ochotona collaris

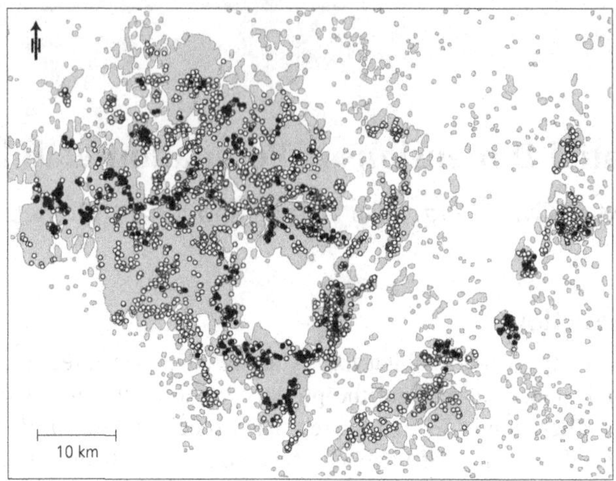

Figure 4.1 Metapopulation of the Glanville fritillary in the Åland Islands (Finland). The map shows occupied (filled circles) and empty (open circles) suitable habitat patches in autumn 2000.

greatly from habitat loss and fragmentation (Pullin 1995), tens of studies have demonstrated the critical role of metapopulation processes in setting the condition for their regional persistence, or extinction (Thomas and Hanski 1997; Hanski 1999). One example is the Glanville fritillary butterfly in the Åland Islands in southwest Finland, which lives in a landscape that is highly fragmented, and probably has been so throughout the period the species has inhabited this area (Hanski 1999). Figure 4.1 illustrates the pattern of habitat patch occupancy in one year, and shows that only some 20% of the suitable habitat is occupied at any one time. Nonetheless, there is much population turnover, extinctions, and colonizations in this metapopulation, and which particular patches are occupied and which are empty changes continuously in time (Hanski 1999). Research on the Glanville fritillary demonstrates conclusively one of the key messages from the metapopulation theory, namely that the currently empty habitat is as important for long-term persistence as the currently occupied habitat.

Glanville fritillary butterfly
Melitaea cinxia

4.2 Deterministic versus Stochastic Metapopulation Models

As in other branches of ecology, mathematical modeling has proved an indispensable tool for understanding the dynamics of metapopulations. The choice of the model depends very much on the real-world situation to be modeled. The number, sizes, and locations of the patches and the sizes of the local populations all influence the choice of model.

Metapopulations with few patches

If both the number of patches and the number of individuals in each patch are small, then the metapopulation must be modeled using stochastic processes (Gyllenberg and Silvestrov 1994, 1999, 2000; Etienne and Heesterbeek 2001; Gyllenberg, in press). In this case the metapopulation becomes extinct on an ecological time scale with the probability of 1. Therefore, if evolution is to be studied at least one of these numbers must be large. Furthermore, in the limiting case, in which one of these numbers is infinite, the long-term viability and persistence of metapopulations can be dealt with in simple qualitative terms. If all the patches have a large carrying capacity, then a deterministic model describes the local dynamics. If the number of patches is small and there are no local extinctions, then the dynamics at the metapopulation level is described by either a system of finitely many ordinary differential equations (in the continuous-time case) or by a set of finitely many coupled maps (in the case of discrete time). In the continuous-time case the analysis of persistence and viability thus reduces to well-known results from the theory of ordinary differential equations (see, e.g., Hofbauer and Sigmund 1988). The discrete time case still presents some technical difficulties (Gyllenberg *et al.* 1993) and even some surprising phenomena; for instance, it is possible that the replacement of a good-quality patch by a poor-quality patch may salvage the metapopulation from extinction (Gyllenberg *et al.* 1996).

Metapopulations with many patches

Another possibility is to assume an infinite number of equally coupled patches. This also allows catastrophes to be incorporated in a relatively simple manner. The price to be paid is that it is no longer possible to model explicitly the spatial configuration of the patches. The deterministic metapopulation models treated in this chapter are therefore based on the assumptions that the local populations are internally homogeneously mixed and that the patches are equally coupled. These assumptions may seem unrealistic, but for many purposes, including the calculations in this chapter, a rule of thumb is that these simplifying assumptions can be used with impunity if, in a spatial configuration, each patch is reached easily from more than 20 neighbors. In Section 4.5 we delve a little deeper into the connection between stochastic reality and our deterministic idealizations of it, and in Section 4.6 we discuss a stochastic metapopulation model in which the simplifying assumptions mentioned above are not made.

A question of utmost importance is, of course, under what conditions will a metapopulation persist. For instance, what is the minimum amount of habitat that will guarantee metapopulation persistence? How does habitat deterioration affect metapopulation persistence? What is the minimum viable metapopulation size? In stochastic models of metapopulations with a finite number of patches with finite local populations, these questions are replaced by questions about the distribution of the extinction time and, in particular, the expected time to extinction.

4.3 Threshold Phenomena and Basic Reproduction Ratios

It is well-known that persistence of metapopulations is linked to threshold phenomena. We illustrate this with the following slightly modified Levins model, which was first used by Lande (1987) to investigate the effect of the amount of habitat upon metapopulation persistence,

$$\frac{dp}{dt} = cp(h - p) - \mu p \,, \tag{4.1}$$

where p is the fraction of occupied patches, c is the colonization parameter, μ is the extinction rate per local population, and h is the fraction of suitable patches. This model is based on the assumption that the colonization rate is proportional to the fraction of occupied patches (from where the potential colonizers come) and the fraction $h - p$ of empty but suitable patches (the patches that can potentially be colonized). A simple calculation shows that the trivial solution that corresponds to metapopulation extinction, that is $p = 0$, is the only steady state and is stable if $ch < \mu$. If $ch > \mu$, the extinction equilibrium is unstable and there exists a unique nontrivial steady state. Thus, the dimensionless parameter R_0, defined by

$$R_0 = \frac{c}{\mu} h \,, \tag{4.2}$$

sets a threshold on metapopulation persistence: the metapopulation persists if and only if $R_0 > 1$.

Basic reproduction ratios and persistence

The quantity R_0 has a clear-cut and important biological interpretation. It is the expected number of new local populations produced by one local population placed in an otherwise virgin environment, that is in an environment with all other patches empty. The parameter R_0 is a direct analog of the R_0 used in epidemic models (Diekmann *et al.* 1990; Diekmann and Heesterbeek 1999); indeed, Equation (4.1) is nothing but the celebrated "susceptible–infected–susceptible" (SIS) model, in which the empty patches are the susceptible individuals and the occupied patches are the infected individuals.

When $R_0 > 1$, the nontrivial steady state is immediately obtained in the usual way by putting $dp/dt = 0$ in Equation (4.1) and solving for p. However, to set the stage for the coming sections we proceed in a slightly more cumbersome, but at the same time more instructive, way. To this end we rewrite the Levins model, Equation (4.1), as

$$\frac{dp}{dt} = [c(h - I) - \mu]p \,, \tag{4.3}$$

where we have simply replaced one p in Equation (4.1) with the symbol I. The point of this seemingly meaningless trick is that, assuming I is given, Equation (4.3) takes the form of a *linear* differential equation. For the nontrivial steady state $I = p$ is a constant different from 0. However, according to Equation (4.3),

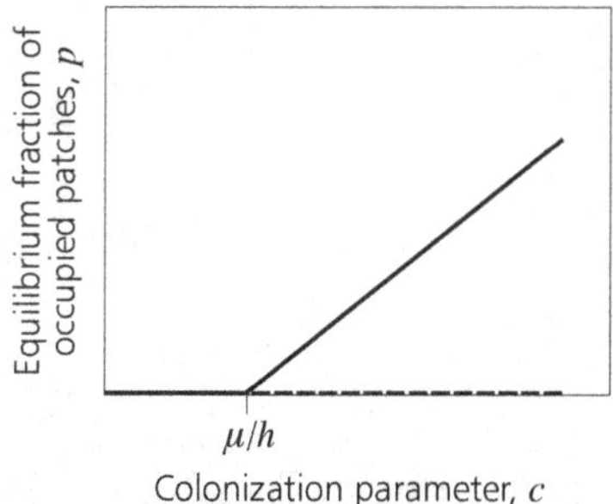

Figure 4.2 Bifurcation diagram of Levins' model [Equations (4.1) and (4.3)], in which the stable and unstable equilibria of $I = p$ are plotted (continous and dashed lines, respectively) against the colonization parameter c, while the other parameters μ and h are kept constant.

this is the case when $c(h - I) - \mu = 0$, or, equivalently, when the quantity defined by

$$R_I = \frac{c}{\mu}(h - I) \tag{4.4}$$

is equal to 1. Then the equilibrium fraction of occupied patches obtained is $p = h - (\mu/c)$.

The introduction of an auxiliary variable (usually called the environmental interaction variable, denoted by I) that cuts the feedback loop and makes the model linear, assuming that the variable is a known function of time, is the leitmotif of modeling structured (meta)populations (Metz and de Roos 1992; Gyllenberg *et al.* 1997; Diekmann *et al.* 2001). An extra advantage is that the steady state condition is formulated in terms of the quantity R_I, which has a biological interpretation similar to that of R_0. It is the expected number of new local populations produced by one local population during its entire life, given that the fraction of occupied patches is I. The steady state criterion $R_I = 1$ thus formalizes the intuitively obvious requirement that a local population on average exactly replaces itself. Note the consistency in the notation: R_I goes to R_0 as I goes to zero.

The results presented above can be summarized and illustrated conveniently by a bifurcation diagram (Figure 4.2). In Figure 4.2 c is (quite arbitrarily) chosen as the bifurcation parameter. If the expected lifetime $1/\mu$ of a local population or the fraction h of habitable patches were chosen as the bifurcation parameter, qualitatively similar diagrams would be obtained.

Persistence and viability

We have introduced the general notion of metapopulation *persistence*. This, by definition, means that the metapopulation extinction equilibrium is unstable. We have shown that a metapopulation governed by the Levins model is persistent if and only if $R_0 > 1$. Also, whenever the Levins metapopulation is not persistent, the metapopulation inevitably becomes extinct. Therefore, in the case of the

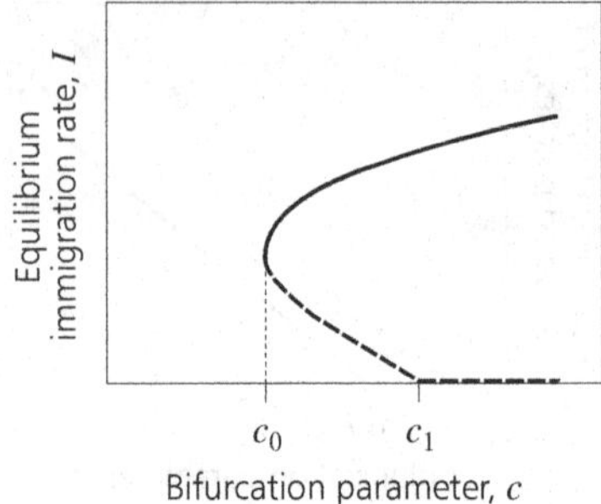

Figure 4.3 Bifurcation diagram of a hypothetical metapopulation model (solid curve, stable equilibria; dashed curve, unstable equilibria). The metapopulation is persistent for all c values larger than c_1, and viable for all c values larger than c_0.

Levins model, persistence coincides with another important notion, namely that of *viability*, which we define in general by the existence of a nontrivial attractor. By "nontrivial" we simply mean "other than the extinction equilibrium"; the "attractor" can be a steady state, a periodic orbit, or even a chaotic attractor, but in this chapter we restrict ourselves to steady states.

We emphasize that the coincidence of persistence and viability is a peculiarity of the Levins model and need not be true for more complicated metapopulation models. To see this, suppose that a model has a bifurcation diagram as that depicted in Figure 4.3; an interval of parameter (e.g., c) values occurs in which the metapopulation is viable, but *not* persistent. In Section 4.5 we give an explicit example of a metapopulation model in which this situation occurs.

The Levins model, Equation (4.1), is simplified in the extreme. In particular, the dynamics is modeled directly at the level of the metapopulation, ignoring local dynamics altogether. This is manifested in the interpretation of the basic ingredients or parameters: μ is the extinction rate *per local population* and c is the colonization rate *per local population and empty patch.* Thus, the persistence criterion $R_0 > 1$, with R_0 given by Equation (4.2), yields a necessary and sufficient condition for persistence in terms of these parameters (and the parameter h) at the metapopulation level. There is no obvious relation between c, μ, and the behavior of individuals. However, one of the main goals of evolutionary conservation biology is to understand how population persistence and viability are influenced by individual traits that may be adaptive. To investigate this question we have to turn to more complicated models that explicitly prescribe local dynamics in terms of parameters that describe individual behavior. Such models are called *structured* metapopulation models and have been treated by, among others, Gyllenberg and Hanski (1992, 1997), Hanski and Gyllenberg (1993), Gyllenberg *et al.* (1997, 2002), Gyllenberg and Metz (2001), Metz and Gyllenberg (2001), and Parvinen (2001a, 2001b). In Section 4.4 the persistence and viability of metapopulations is investigated within the context of structured models. The key technique is to define the basic reproduction ratios R_0 and R_I for these models.

4.4 Modeling Structured Metapopulations

Simple, unstructured metapopulation models face severe limitations. The aim to make predictions that relate to real data raises the need to include an explicit description of migration in terms of the numbers of individuals moving, rather than a description restricted to the colonization rate of empty habitat, as in the Levins model and other patch-occupancy models. It is now widely recognized that both emigration and immigration often have important consequences for the dynamics of especially small local populations (Stacey *et al.* 1997; Thomas and Hanski 1997; Hanski 1998, 1999), and hence also for the dynamics of metapopulations that consist of such small populations. The rescue effect (Brown and Kodric-Brown 1977) – the reduced risk of extinction in a local population because of immigration – is the best-known example of such effects (Hanski 1985), which can be accounted for in a mechanistic manner within the framework of structured models (Gyllenberg and Hanski 1992).

When modeling and analyzing the temporal dynamics of structured populations, the starting point is to describe mechanisms at the individual level, then lift the model to the population level, and finally study phenomena at the population level. As pointed out already by Metz and Diekmann (1986) (see also Diekmann *et al.* 1988, 1989), and later by Gyllenberg and Hanski (1992) (see also Gyllenberg *et al.* 1997), the theory of structured populations can be applied to metapopulations in a rather straightforward manner if an analogy is made between local populations and individuals and between metapopulation and population. We have seen an instance of this already in our brief discussion of the classic Levins model. However, in some cases a more general analog of an individual is needed, examples of which are given later. The entity of a metapopulation that corresponds to an individual in ordinary populations is called the *basic entity*.

As a practical aside, note that below we concentrate on the limiting case of infinite, which in practice means large, local populations. These may become extinct as a consequence of a local catastrophe. The deterministic nature of the model means that a patch which becomes empty as a result of a local disaster is immediately recolonized. Therefore, according to the model there are no empty patches (except in the case of metapopulation extinction). Yet in reality empty patches do exist. When fitting the model to data this dichotomy can be overcome of by introducing a *detection threshold*, by which a patch of local population size less than this threshold is empty.

Defining the environmental interaction variable

A basic entity develops (i.e., its state changes with time) as a consequence of, for instance, patch-quality dynamics, that is the local population growth through births, deaths, and migration. It gives rise to new local entities (e.g., local populations produce dispersers, which colonize empty patches), and some vanish (e.g., when a local population becomes extinct or a disperser dies). To model mechanisms at the local level, two ingredients are needed, one to describe the production of new local entities and one to describe the development and survival of local entities. These ingredients depend on the environmental interaction variable I, which must be chosen such that for a given I the model becomes linear. The value of I, in turn, depends on the metapopulation state and therefore the full model at the level of the metapopulation becomes nonlinear. We refer to articles by Gyllenberg *et al.* (1997), Diekmann *et al.* (2001), Gyllenberg and Metz (2001), and Gyllenberg and Jagers (in press) for accounts of how the modeling task can be carried out in general.

One of the salient features of structured metapopulation models is that they make it possible to model and hence analyze how migration affects local dynamics. Emigration is as easy to model as death is; from the point of view of a local population, it does not make any difference whether an individual is lost through death or emigration. Immigration is more complicated, unless it is assumed that the immigration rate I is known, in which case the modeling task is easy, since I is just an additional contribution to the growth rate of the local population. The immigration rate I depends, of course, on the overall density of dispersers, which in turn depends on the emigration rate and mortality during dispersal. The nonlinear feedback thus takes place through migration, and the immigration rate I qualifies as an environmental interaction variable. Assuming I to be known, a linear problem is obtained, but the true value of I is found by closing the feedback loop.

Defining the basic entity

The goal of this chapter is to understand the determinants of metapopulation persistence and viability. We had seen already, in connection with the Levins model, that this issue can be investigated using the basic reproduction ratios R_0 and R_I. We therefore generalize these quantities to structured metapopulation models. The Levins model, however, is so simple that it is not immediately clear how this should be done. There is a general abstract framework for defining the basic reproduction ratios (Gyllenberg *et al.* 1997; Diekmann *et al.* 1998; Gyllenberg and Metz 2001; Metz and Gyllenberg 2001), but here we are content with a more intuitive approach that takes advantage of biological interpretation. Let us first examine how the basic reproduction ratio is interpreted for "ordinary" structured populations, that is, for populations without the metastructure. For such models the basic reproduction ratio is the expected number of offspring born to a typical individual during its entire life. Here, part of the problem is to define what the word "typical" means.

For a given individual with a known state at birth, the expected lifetime production of offspring can be calculated from the basic model ingredients, and finding R_0 and R_I then amounts to averaging (in the right way!). To translate the ideas of ordinary populations into metapopulations, we must first define the basic entity. The local population may, at first sight, seem the obvious choice. However, in terms of an evolutionary approach to metapopulation viability, it is important to understand the invasion and fixation of rare mutants that have life-history traits that are different from those of a wild type formerly established in the metapopulation. One therefore has to investigate the competition between different types that inhabit the same local population, and therefore the local population itself does not contain sufficient information to qualify as a basic entity.

We restrict our attention to models with two types of local entities: *dispersers* and *resident clans*. A resident clan consists of an individual that arrives at a patch (the "ancestor") and all its descendants (children, grandchildren, great grandchildren, etc.), as long as they stay in the patch. When a resident emigrates we say that the clan gives birth to a disperser. It is convenient to think of the arrival of a disperser to a patch as the simultaneous death of the disperser and the birth of a new resident clan. [Note that "resident", as used in this chapter, means simply an individual who lives in a patch, as opposed to a migrant, and should not be confused with the notion of a resident phenotype as opposed to that of a mutant phenotype in the context of adaptive dynamics theory (see Chapters 11, 14, 16, and 17).]

Note that in the metapopulation context one cannot base the persistence criterion on an individual-based reproduction ratio. An individual may be very prolific locally, but if the dispersal rate is not large enough almost all of its descendants may be lost at a local disaster. It is therefore necessary to take the resident clan as defined above, and not the individual, as the basic unit of the metapopulation.

Defining basic reproduction ratios

To calculate the basic reproduction ratio of a structured metapopulation we first have to find an expression for the expected (cumulative) number and local-state distribution of "offspring" produced by a "newborn" basic entity. At this stage we assume that such an expression is well-defined, given the model. Later we discuss how this expression can be obtained from more detailed models of individual behavior.

We make three important assumptions about individual behavior:

- All dispersers behave in the same way, that is, dispersers are unstructured;
- Dispersers choose their new patch at random;
- The behavior of residents may depend on the state X of their local population.

More complicated models could be treated in the same spirit and formally, using the same abstract method (Gyllenberg *et al.* 1997; Diekmann *et al.* 2001). The following ingredients can now specify the model:

- $E_I(X)$, the expected number of dispersers produced by a clan that was initiated by a disperser immigrating into a local population of state X, given that the immigration rate is I;
- ϕ, the probability that a disperser survives migration and starts a new clan.

With these model specifications, the expected number of new clans produced by a clan initiated by a disperser that arrives at a local population of state X is $\phi E_I(X)$. We assume dispersers choose their new patch at random and therefore the state-at-birth of new clans equals the steady population size distribution p_I that corresponds to the immigration rate I. A "typical" clan is therefore one sampled from the steady population size distribution p_I and we obtain Equation (4.5a) for the reproduction ratio of basic entities,

$$R_I = \phi \int E_I(X) p_I(X)\, dX \,, \tag{4.5a}$$

where the integral is taken over all possible local population states X.

The steady population size distribution p_I depends on the local dynamics, which are not yet specified. It can be shown (Gyllenberg *et al.*, unpublished) that at equilibrium the basic reproduction ratio becomes

$$R_I = \phi \frac{E_I}{I \ell_I} \,, \tag{4.5b}$$

where E_I is the expected number of offspring produced by a local population during its entire life and ℓ_I is the expected lifetime of a local population, given the immigration rate I. This formula has a very intuitive interpretation. Note that the denominator on the right-hand side is the expected number of arrivals at a patch. The whole right-hand side is therefore the expected number of arriving (anywhere in the metapopulation) offspring of a local population divided by the mean number of arrivals at a patch. At equilibrium, this quantity should equal 1, hence R_I. It can also be shown that the equilibrium value of R_I given by Equation (4.5b) tends to R_0 as the immigration rate I tends to zero (Gyllenberg *et al.* 2002).

4.5 Metapopulation Structured by Local Population Density

In this section we illustrate our theory by a simple example which has the population density, denoted as N, as the local state. Our model is specified by the following ingredients:

- h, the fraction of habitable patches;
- $r(N)$, the density-dependent per capita growth rate from local births and deaths;
- $m(N)$, the density-dependent per capita emigration rate;
- ϕ, the probability that a disperser survives migration and establishes a new clan;
- $\mu(N)$, the density-dependent local catastrophe rate.

This enables persistence and viability to be expressed using the expected number E_I of dispersers produced by a local population during its entire life, the expected lifetime ℓ_I of a population, and Equation (4.5b) to obtain R_I at equilibrium, taking the limit as I tends to zero to obtain R_0. This was carried out by Gyllenberg *et al.* (2002). However, to calculate R_0 there is a shortcut, which we shall follow.

Metapopulation persistence

Consider a newly founded resident clan in an otherwise virgin environment. This means that all other patches are empty and that the local population of the clan is size zero; as a consequence, so long as the metapopulation remains small we can neglect the effects of density dependence. Thus, the local per capita growth rate, per capita emigration rate, and catastrophe rate are constant and equal to $r(0)$,

Box 4.1 Deriving a criterion for structured metapopulation persistence

Here we use the model assumptions and notations of Section 4.5 to present a derivation of the persistence criterion based on R_0. We consider a newly founded resident clan in an otherwise virgin environment. The probability that this clan is still extant t time units later is $\exp[-\mu(0)t]$, and if it is extant its size is $\exp(-[r(0) - m(0)]t)$. Each of the individuals in the clan has a probability $m(0)\,dt$ of migrating in the infinitesimal time interval $[t,\ t+dt]$. Summing over all times, we obtain the expected number of migrants produced by a clan, denoted by $E_0(0)$,

$$E_0(0) = \int_0^\infty m(0)\ \exp([r(0)-m(0)-\mu(0)]t)\,dt = \frac{m(0)}{\mu(0)+m(0)-r(0)}\ . \tag{a}$$

Note that the symbol $E_0(0)$ is consistent with the notations used so far: the subscript 0 indicates zero immigration and the argument 0 means that the clan starts from an empty patch, all in accordance with the notion of a virgin environment.

The equality in Equation (a) is, of course, valid if and only if $r(0) < m(0) + \mu(0)$, otherwise the integral is infinite [unless $m(0) = 0$, in which case E_0 assumes the value 0]. We accept the possibility that E_0 will take on infinite values and adopt the usual arithmetic on $[0,\ \infty]$, including the convention $0 \times \infty = 0$.

From the interpretation of the quantities involved it is clear that $R_0 = \phi E_0(0)h$, but we emphasize that this can be rigorously deduced from Equation (4.5a) since the steady population size distribution p_I that corresponds to the population-free case is the point mass of size h concentrated at the origin.

Persistence, then, is found to be determined by the relative values $r(0)$ and $m(0) + \mu(0)$:

- If $r(0) < m(0) + \mu(0)$, the persistence criterion given by Equation (4.6) is obtained.
- If $r(0) > m(0) + \mu(0)$ and $m(0) > 0$, then $R_0 = \infty$ if $h > 0$, and $R_0 = 0$ if $h = 0$. Thus, in this case $R_0 > 1$ if and only if $h > 0$.

$m(0)$, and $\mu(0)$, respectively. Necessary and sufficient conditions of persistence are derived in Box 4.1. Results can be summarized as follows:

- If local growth at zero density is slow enough, namely if $r(0)$ is less than the sum $m(0) + \mu(0)$, then the persistence criterion is

$$R_0 = h\phi \frac{m(0)}{m(0) - [r(0) - \mu(0)]} > 1\ . \tag{4.6}$$

  This persistence criterion is very similar to the corresponding criterion for the Levins model. There is an important difference, though. Whereas the latter was formulated in terms of the parameters c and μ, which measure attributes at the local population level, Equation (4.6) contains the per capita growth and emigration rates $r(0)$ and $m(0)$, which are properties of individuals.
- If local growth at zero density is fast enough, namely if $r(0)$ is larger than $m(0) + \mu(0)$, and if the emigration rate at low density $m(0)$ is not null, then the

Box 4.2 Deriving a criterion for structured metapopulation viability

Here we provide the analysis of metapopulation viability in the context of the structured model introduced in Section 4.5. To this end, we have to calculate R_I at equilibrium by using Equation (4.5a). First, observe that our specification of individual behavior implies that local population growth is governed by the differential equation

$$\frac{dN}{dt} = r(N)\, N - e(N)\, N + I \,. \tag{a}$$

Now, consider a local population at the time of a local disaster and simultaneous recolonization. The probability that it is still extant when it has density N is equal to $\exp\left(-\int_0^N \frac{\mu(x)}{r(x)x-m(x)x+I}\, dx\right)$, and given that it is extant with density N the expected number of dispersers produced in the infinitesimal density interval $[N,\ N+dN]$ is $\frac{m(N)N}{r(N)N-m(N)N+I} dN$. The factor $1/r(N)N - m(N)N + I$ that occurs in these formulas simply reflects a conversion from quantities per unit of time to quantities per unit of population density. Adding up over all sizes gives

$$E_I = \int \frac{m(N)N}{r(N)N - m(N)N + I} \exp\left(-\int_0^N \frac{\mu(x)}{r(x)x - m(x)x + I}\, dx\right) dN \,. \tag{b}$$

Similarly, one obtains

$$\ell_I = \int \frac{1}{r(N)N - m(N)N + I} \exp\left(-\int_0^N \frac{\mu(x)}{r(x)x - m(x)x + I}\, dx\right) dN \,. \tag{c}$$

The expression for R_I at equilibrium follows readily from Equation (4.5a).

persistence criterion $R_0 > 1$ is equivalent to $h > 0$. This entails that fast local growth can compensate for arbitrary loss of habitat and keep the metapopulation alive – any positive amount of suitable habitat is enough to ensure persistence. Notice that in Levins' model, which neglects local dynamics, nothing of this sort is feasible.

Focusing on the specific effect of the emigration rate on metapopulation persistence, two further cases may be distinguished when local growth is slow $[r(0) < m(0) + \mu(0)]$:

- If local growth is too slow the metapopulation is not persistent regardless of the emigration rate, namely, if $r(0) \le \mu(0)$;
- For intermediate rates of local growth, i.e., $\mu(0) \le r(0) < m(0) + \mu(0)$, the metapopulation persistence requires that emigration be less than an upper threshold, denoted by m_1 [obtained by rearranging terms in Equation (4.6)]. This threshold m_1 increases with the difference between the rates of local growth and catastrophe at low density, the probability of survival and establishment for dispersers, and the amount of suitable habitat.

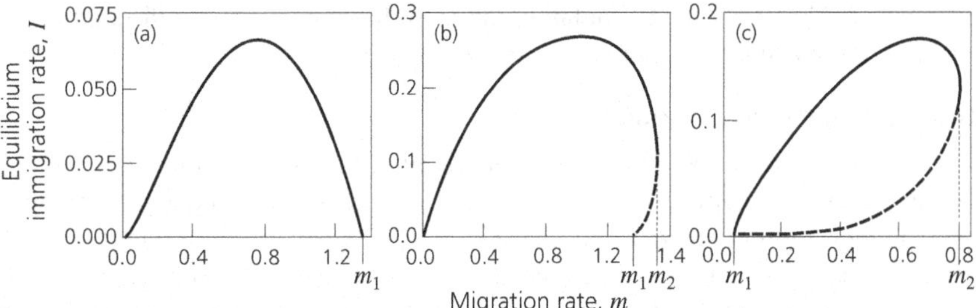

Figure 4.4 Bifurcation diagrams for the structured metapopulation model described in Section 4.5. The stable and unstable equilibria values of I are plotted against the per capita migration rate m, which is density independent and acts as bifurcation parameter. (a) Constant rate of catastrophe: $\mu = 0.4$. (b) The catastrophe rate $\mu(N)$ decreases with increasing local population density N, and $\mu(0) = 0.4$. (c) The catastrophe rate $\mu(N)$ is as in (b), but with $\mu(0) = 1.2$. In all cases the local population growth is logistic, with $r(N) = 1 - N$, $\phi = 0.55$, and $h = 1$.

Metapopulation viability

At the metapopulation steady state the condition $R_I = 1$ holds (see Box 4.2), and Figure 4.4 shows plots of the equilibrium fraction of occupied patches determined by this condition as a function of the emigration rate (from Gyllenberg *et al.* 2002). It may be a surprise that, even for $r(0) \leq \mu(0)$, in which case the metapopulation is not persistent, there might very well exist a range of emigration rates $m(0)$ over which the metapopulation is viable. This is possible even in the case of density-independent migration (i.e., with constant m), and can be seen in Figures 4.4b and 4.4c: in the range $m_1 < m < m_2$ there exists a stable nontrivial equilibrium and hence the metapopulation is viable for migration rates in this range. In Figures 4.4a and 4.4b the condition $\mu(0) < r(0)$ is satisfied, and we know from the previous subsection that there is a range $0 < m < m_1$ above which the metapopulation is persistent. If the catastrophe rate is constant (Figure 4.4a), then persistence and viability coincide. If, on the other hand, the catastrophe rate is a decreasing function of local population density (Figures 4.4b and 4.4c), there is a range $m_1 < m < m_2$ of emigration rates for which the metapopulation is viable but not persistent.

These results illustrate the possibility of alternative locally stable equilibria in metapopulation dynamics (see also Hanski 1985; Hanski and Gyllenberg 1993; Gyllenberg *et al.* 1997). The size of the metapopulation may move from the domain of one stable equilibrium to the domain of the alternative equilibrium following a large environmental perturbation, and at the bifurcation point the metapopulation is predicted to show a deterministic "jump" from the nontrivial equilibrium to metapopulation extinction. The lesson here is that it should not be assumed that slight changes in parameter values will necessarily be reflected in only slight changes in metapopulation size, and it is possible that large long-lasting changes in metapopulation size will occur in response to small environmental perturbations. The feedback between migration and local dynamics, on the one hand, and the dynamics of the entire metapopulation, on the other, may generate discontinuous changes in the abundance and distribution of species.

Toward more realistic models

The structured metapopulation models analyzed in this section omit a description of the spatial population structure in that they assume the habitat patches are identical (though see Gyllenberg and Hanski 1997) and equally connected. This is not a great restriction for systems that consist of large networks of well-connected habitat patches without a strongly aggregated spatial distribution; yet to relate the modeling results to empirical studies it is often useful to account explicitly for the spatial structure in metapopulations. The spatially structured metapopulation models developed by Hanski (1994, 1999) have recently been analyzed mathematically by Hanski and Ovaskainen (2000) and Ovaskainen and Hanski (2001, 2002, unpublished). This line of modeling retains the present or absent description of dynamics in local populations, as in the original Levins model, but allows for finite patch networks with differences in the areas, qualities, and degree of connectivity of the patches. An advantage of these models is that they can be given rigorous parameters for real metapopulations (ter Braak *et al.* 1998; Moilanen 1999, 2000), and therefore establish a firmer link between theory and empirical studies (Hanski 1999).

However, once again, because individuals are not modeled explicitly, the spatially structured models cannot be extended easily to evolutionary studies without resorting to individual-based simulations. [For an example see Heino and Hanski (2001), who combined individual-based simulations with a spatially structured patch occupancy model to constrain the model-predicted long-term dynamics and used a statistical model of individual movement behavior to model migration of individuals among multiple populations.] One challenge for further research is to develop metapopulation models that include both the spatial structure and the local population size structure.

4.6 Persistence of Finite Metapopulations: Stochastic Models

So far we have considered metapopulations that consist of infinitely many patches, which, moreover, in the structured case contain infinitely many individuals. The reason for doing so is that the results for such models can be stated in simple intuitive terms, that we have good tools for studying them, and that we may expect the results to hold to a good approximation when we replace "infinite" with "sufficiently large". Of course, real populations are not infinite. In this section, dedicated to predict relationships between individual traits and metapopulation viability on the basis of stochastic models, we first treat a particular example of a finite metapopulation model and clearly illustrate the concepts involved in a nontrivial manner. We follow this up with a heuristic overview of how the mathematically idealized infinite cases connect to the more realistic finite cases.

Predictions from a spatially explicit stochastic model

To exemplify the main concepts that occur in population models of a more finite kind, we concentrate on a stochastic metapopulation model with a finite number of patches. The spatial arrangements of patches is modeled explicitly. The basic ideas and results are most easily described when time is taken as a discrete variable, and therefore from here onward we switch to discrete-time models. Our model is based on Gyllenberg (in press), but also see Etienne and Heesterbeek (2001). Analogous, but technically more difficult, results for a continuous-time stochastic metapopulation model were derived by Gyllenberg and Silvestrov [1999, 2000; see also Frank and Wissel (1994, 1998, 2002) for practically useful heuristics]. We choose this model structure, which is rather different from those considered previously, because it allows us to show some overarching ideas and to elucidate some of the interpretational problems that occur when comparing models of different origin.

We consider a collection of n patches that can be either occupied or empty at the discrete time instants $t = 0, 1, 2, ...$ Metapopulation extinction corresponds to all patches being empty. The local dynamics are modeled by preassigning an interaction matrix $Q = [q_{ji}]$ in which:

- $q_{ii}(i = 1, 2, ..., n)$ is the probability that, in the absence of migration, the population inhabiting patch i will become extinct in one time-step;
- $q_{ji}(i = 1, 2, ..., n, j = 1, 2, ..., n$, and $j \neq i)$ is the probability that patch i will *not* be colonized in one time-step by a migrant originating from patch j.

Typically, q_{ji} depends on at least the distance between the patches i and j and the area of patch j. This model incorporates the notion of a *rescue effect*, that is, the decreasing extinction rate with increasing fraction of occupied patches. The overall extinction probability of the local population that inhabits patch i may be considerably less than the "internal" extinction probability q_{ii} if there are many large occupied patches in the vicinity (many small q_{ji}).

The analysis of this model (see Box 4.3) requires three conditions:

- No local population is able to colonize another patch in one time-step with probability 1;
- Even in the absence of migration (rescue effect), no local population has extinction probability 1 and no local population is protected from extinction;
- Every local population is able to colonize any other patch either directly or through a chain of patches (stepping-stone dispersal).

These conditions together imply that we are, indeed, dealing with a true metapopulation and not, for instance, with a mainland–island model, in which only migration from the mainland to the islands is allowed, or with a collection of several disconnected metapopulations.

Since there is a finite number of patches, sooner or later all extant populations will simultaneously become extinct and the whole metapopulation will be wiped out. Mathematically speaking, the metapopulation will become extinct with probability 1. In such cases of certain extinction, there is no stationary distribution

Box 4.3 A spatially explicit stochastic model

We consider a collection of n patches that can be either occupied or empty at the discrete time instants $t = 0, 1, 2, \ldots$, and make use of the notations introduced in Section 4.6 to analyze the metapopulation dynamics. The state of patch i at time t is given by the random indicator variable $\eta_i(t)$, which takes on the value 1 if patch i is occupied and 0 if patch i is empty at time t. The state of the metapopulation is described by the vector random process in discrete time $\eta(t) = (\eta_1(t), \eta_2(t), \ldots, \eta_n(t))$, with $t = 0, 1, 2\ldots.$ The state space of the process $\eta(t)$ is $\Xi = \{\xi = (x_1, \ldots, x_n) : x_i \in \{0, 1\}\}$. It has 2^n states. The state $\mathrm{O} = (0, \ldots, 0)$ corresponds to metapopulation extinction.

We assume that the local extinction processes and the colonization attempts from different local populations are all independent. As a consequence of this independence the conditional probabilities $q_i(\xi)$ for patch i to be empty at time $t+1$, given that at time t the metapopulation was in state $\xi = (x_1, \ldots, x_n)$, are given by the product

$$q_i(\xi) = \prod_{j=1}^{n} q_{ji}^{x_j}, \quad i = 1, 2, \ldots, n, \tag{a}$$

where we use the convention $0^0 = 1$.

Having described the local patch dynamics, we can deduce the law that governs the time evolution of the process $\eta(t)$ and gives the state of the metapopulation. This process is a homogeneous Markov chain with state space Ξ and transition probabilities

$$\prod_{i=1}^{n} q_i(\xi)^{1-y_i}[1 - q_i(\xi)]^{y_i}, \quad \xi, \zeta \in \Xi. \tag{b}$$

The process $\eta(t)$ is determined completely by the interaction matrix Q, which is assumed to satisfy the following conditions (stated verbally in the text of Section 4.6):

- $q_{ji} > 0, j \neq i$;
- $0 < q_{ii} < 1, i \in \{1, \ldots, n\}$;
- For each pair (j, i) of patches, $j, i \in \{1, \ldots, n\}$, there exists an integer l and a chain of indices $j = i_0, i_1, \ldots, i_l = i$ such that $\prod_{k=1}^{l}(1 - q_{i_{k-1}i_k}) > 0$.

The process can be described in terms of its quasi-stationary distribution, which is given by the left eigenvector (normalized to a probability distribution) that corresponds to the dominant eigenvalue of the transition matrix Q restricted to the transient class $\Xi \backslash \mathrm{O}$. The dominant eigenvalue measures the probability that a metapopulation *sampled from the quasi-stationary distribution* will not become extinct in one time step.

except the trivial one that corresponds to metapopulation extinction. However, we can define the so-called quasi-stationary distribution (Darroch and Seneta 1965; see also Chapter 2), which is the stationary distribution on the condition that the metapopulation has not become extinct.

If we consider a metapopulation that has been extant for a considerable time, we may use the quasi-stationary distribution as the starting point from which to consider the time to its future extinction; we refer to the corresponding time to extinction as the quasi-stationary extinction time. This solves the problem that, in general, the time to future extinction is highly dependent on the state in which the metapopulation is at present. In particular, we have to distinguish situations such as a reintroduction from situations in which the metapopulation has been present for a long time. Subject to some natural monotonicity conditions, if a metapopulation has known more favorable conditions in the past, the time to extinction under a constant continuation of the current regime will be larger (but usually not much) than the quasi-stationary one. Conversely, if a metapopulation has just been started from a few individuals, the time to extinction will be less than the quasi-stationary extinction time. (For the mathematically inclined reader, here "below" and "above" should be interpreted as inequalities on the corresponding survival functions.)

Conditional on nonextinction, the state distribution will approach the quasi-stationary distribution. It therefore makes sense to view the dynamics of the metapopulation as a two-state Markov process, the states being metapopulation extinction and the quasi-stationary distribution. If q denotes the probability that a metapopulation sampled from the quasi-stationary distribution will become extinct in one time step, the expected extinction time is equal to $1/q$. This is an exact expression for the expected time to extinction, provided that the metapopulation is initially at the quasi-stationary distribution. However, the exact calculation of q (which is obtained as the eigenvalue of a $2^n - 1$ by $2^n - 1$ matrix) becomes computationally prohibitive as the number of patches n grows. For the continuous-time case, a good approximation can be found in Frank and Wissel (2002). We refer to Etienne and Heesterbeek (2001) for examples of how this result can be applied to reach practical conclusions about, for instance, how changing the connectivity of patches, that is changing the q_{ji} values, affects the viability of the metapopulation.

New introductions

As a second consideration, we look at new introductions. In this case the probability of becoming extinct in the next time step is, in general, larger than q, and only decreases to q in the long run. The general rule-of-thumb is that in systems with more or less global coupling, q increases to 1 with increasing system size (see below for a further elaboration), while the time needed for the stepwise extinction probability to converge to q increases much more slowly than $1/q$, the mean extinction time after reaching quasistationarity. This means that for larger system sizes we may consider the probability distribution of the time to extinction as consisting of a discrete mass at zero followed by an exponential tail. In those cases that have global coupling, it is in general possible to calculate the initial mass from a branching process approximation. For example, if each patch may contain at most one individual, then under the assumptions of the Levins model (see Section 4.3) the initial mass equals approximately $[\mu/(ch)]^{\Omega}$ if $\mu < ch$, and

1 (so that there is no tail left) if $\mu > ch$, where Ω is the number of individuals that start the metapopulation. More generally, for the deterministic models considered in Sections 4.3 to 4.5, the initial mass can be written as θ^{Ω}, where θ equals 1 when $R_0 \leq 1$, and $\theta < 1$ when $R_0 > 1$. However, these results have the caveat that it is assumed implicitly that females may always reproduce. Many real populations of conservation interest contain two sexes. This means that, even though there are many individuals, if they all happen to be of one sex a population may have no future. Whether such considerations really matter greatly depends on the detailed reproductive biology of the species (see Chapters 2 and 3 for a discussion of this issue in the context of nonspatial models). Further research into this area should be both a mathematical and a biological priority.

Between stochastic and deterministic models

The main difference between population models with finite total numbers of individuals and those in which these number are thought of as infinite is that in the former the population, in the long run, becomes extinct whatever the value of the parameters. However, the time for this to happen is generally very long when the number of "close-to-independent" entities involved is large. This latter number is referred generally to as the *system size*. In metapopulation models, as considered here, there are actually two system sizes, the number of patches n and the patch size ω (the latter is expressed in units roughly equal to the amount of space needed to support a single individual). We have to consider their interplay to determine what sort of limit is obtained and to establish the scaling relations between the extinction time and system sizes, when either or both of them become large.

By the argument of the above subsection, when system sizes become large, the mean extinction time of a population starting from a nonvanishing population differs from the mean quasi-stationary extinction time only by a relatively small amount. It is these quasistationary mean extinction times that we discuss below. For brevity, we refer to them simply as extinction times.

To establish a feel for the problem, we first consider how the transition to the deterministic model is made for a single local population, without considering immigration from other patches and catastrophes. Let N be the number of individuals in the patch and the rates at which these individuals die, give birth, or emigrate as, respectively, $b(N/\omega)$, $d(N/\omega)$, and $m(N/\omega)$, where the unit of the patch size ω and functions b, d, and m be such that $b(1) - d(1) - m(1) = 0$, that is, the equilibrium density of a deterministic population model based on those functions equals 1. With this scaling we can identify the local system size with ω. We obtain a deterministic limit model for the temporal development of the local population density N/ω by letting ω become large. (The limit holds good over any finite time interval, but not over the full time axis, except for those cases in which the deterministic model predicts certain extinction, since for all finite ω the population becomes extinct if we wait long enough.) In addition, according to accepted wisdom, "in viable systems the extinction time of an established local population is roughly exponential in the system size". More precisely, the extinction time, T_{p}, scales exponentially with ω, written as $T_p \approx e^{\theta\omega}$, to be interpreted as $\frac{1}{\omega} \ln T_p \to \theta$ for $\omega \to \infty$, in which θ is the so-called scaling constant. This result has been shown to hold good in the simplest possible models (MacArthur 1972; see also the figures in Goodman 1987b), is shored up by arguments from statistical physics [see, e.g., Gardiner (1983); also, Schuss (1980), Grasman and HilleRisLambers (1997), Freidlin and Wentzell (1998), and Grasman and Van Herwaarden (1999), who specifically consider metapopulation problems], and is backed up by simulations for some other models. With catastrophes, for large ω the local population size develops deterministically until the first catastrophe, while the extinction time, T_p, is set by the catastrophe rate.

Next, assume that we have n similar patches equally coupled through migration, and that we let the local system size ω become large. We can then make more than one biologically meaningful assumption about the immigration rate.

One such assumption is that the migration parameters are constants, so that at larger patch sizes the number of immigrants grows directly in proportion to ω. Translated into observable quantities, this assumption implies that at least some patches should contain reasonable numbers of individuals and that the average interarrival time of immigrants is small relative to the mean lifetime of individuals. If the local population growth rates decrease with density, and there are no catastrophes, we have a fully deterministic model in which all patches in the long run contain an equal population density; this is positive if the quantity R_0, calculated according to the analysis in the previous section, is larger than 1. If there are catastrophes, the metapopulation may die out even in the limit for large ω, but it may continue forever. The former definitely happens when $R_0 \leq 1$, or the parameter domain that leads to extinction is always larger, with its size shrinking with increasing n, than that given by the deterministic criterion $R_0 \leq 1$ (unpublished results by ourselves).

Another possibility is to change the migration parameters such that the mean interarrival time of immigrants is of the same order of magnitude as the mean lifetime of a local population, T_p. This limit regime applies when the probability of surviving migration and reaching some other patch is of the order of ω^{-1}, or when emigration is relatively rare on a patch basis, or a combination of both these factors. At the same time, we assume that the catastrophes occur at a rate that is slow compared to the speed at which quasistationarity would be reached by immigrationless local populations. Moreover, we consider the metapopulation on the time scale set by T_p. (Despite the somewhat artificial look of the mathematical procedure, the required parameter regime may well be rather common in nature because parameters for community assemblage and selection setting are in a commensurable range.) In this case, patches are either empty or in a quasistationary state almost all of the time, and we have a finite Levins-type model, provided the local populations cut off from immigration have persistent deterministic limits (compare Verboom *et al.* 1991; Drechsler and Wissel 1997). If we now let n become large, while keeping the migration rate into the patches bounded, we recover the deterministic Levins model from Section 4.3, with $\mu = 1/T_p$. The time to metapopulation extinction, T_m, scales linearly with T_p and exponentially with n. In this case we even have available a full asymptotic formula [by applying Stirling's approximation to Equation (6) in Frank and Wissel 2002], applicable for $R_0 = cT_p = c/\mu > 1$,

$$T_m \approx T_p\sqrt{2\pi n}^{-1/2} e^{[(cT_p)^{-1}-1+\ln cT_p](n-1)} . \tag{4.7}$$

For $R_0 < 1$, the average extinction time of a metapopulation starting from any positive fraction of occupied patches increases logarithmically with n.

By taking the Levins-type models as a gauging point, we can now write for persistent metapopulation models

$$T_m = T_p \psi(\omega, n) n^{-1/2} e^{\gamma n} , \tag{4.8}$$

with the function ψ thus defined as measuring the rescue effect. Of course, both γ and ψ depend on all the other system parameters, as well as on how the migration and catastrophe rates are supposed to scale with ω and n. However, we conjecture that for any relevant scaling, ω will be nondecreasing, at least at larger system sizes, because the reliability of the migration stream increases as system size increases.

To show the potential force of the rescue effect, we can compare a Levins model without catastrophes (so that $T_p \approx e^{\theta\omega}$) with a model in which migration increases so fast with ω or n that, effectively, all patches can be considered together as one single population. Combining the relationships found above then gives $\psi(\omega, n) \approx e^{\kappa\omega n} / (e^{\gamma n} e^{\gamma n}) \approx e^{\kappa\omega n}$ for ω and n both large, that is, the rescue effect overwhelms all other contributions to the scaling of T_p. However, the example in which we keep n constant and let ω become large at a constant per capita migration rate makes clear that the limit behavior of the rescue effect can be more complicated than in the example given in this paragraph.

In the above discussion, we implicitly referred to the case where, in the deterministic limit model, the metapopulation is persistent. The case in which the metapopulations are viable, but not persistent, is less clear. However, we also expect an exponential scaling with n (by a rough appeal to the arguments found in Schuss 1980; Gardiner 1983; Freidlin and Wentzell 1998; Grasman and Van Herwaarden 1999). This with the proviso that we expect the scaling constant γ to be roughly proportional to the distance of the equilibrium population state of the deterministic model from the closest point in the state space of that model from where, in the deterministic model, the state would move inexorably toward extinction (where nearness is measured in terms of the "ease of a state transition").

4.7 Concluding Comments

In this chapter the stress is on modeling migration and local dynamics at the individual level rather than at that of local populations. One reason for carefully analyzing the limit relationships between different types of metapopulation models is that this allows us to interpret each of these models from an individual-level perspective. Only in this way can we give a concrete meaning to the model parameters. Achieving such a concreteness is the first step on the arduous path that leads from model results to conservation interventions.

In the chapter we strive to stay within the realm of what is manageable with present-day mathematical methods, while going one step further in the interpretation process than our predecessors. We are well aware that the models we discuss are considerably less concrete than individual- or GIS-based simulation models that purportedly mimic the behavior of specific species. However, even those of our colleagues who believe that we can render a fair fraction of such detailed models right agree that we cannot model all the species we ultimately have to deal with. So we are in dire need of good rules of thumb. It is here that we may hope that the simpler models of this chapter and their future extensions will prove useful.

The second advantage of basing our models on individual-level considerations is the possibility that evolutionary questions, such as the evolution of migration rate, could be addressed (for reviews see Clobert *et al.* 2001; Ferrière *et al.* 2000; Gyllenberg and Metz 2001; Metz and Gyllenberg 2001; Parvinen 2001b; see also Chapter 14). This is simply not possible with patch-occupancy models [see Hanski (1999) for a review], except, perhaps, in the restricted sense of selection that occurs at the level of local populations (group selection). Further merging of ecological and evolutionary dynamics in the context of structured metapopulation models is an exciting prospect for modeling, and one of considerable importance if we are to gain at least a little grip on the potential long-term consequences of human-induced environmental change.

Acknowledgments The work of Mats Gyllenberg has been supported by the Academy of Finland, and the work of Mats Gyllenberg and Johann A.J. Metz by the European Research Training Network *ModLife* (Modern Life-history Theory and its Application to the Management of Natural Resources) Network, funded through the Human Potential Programme of the European Commission (Contract HPRN-CT-2000-00051).

Part B

The Pace of Adaptive Responses to Environmental Change

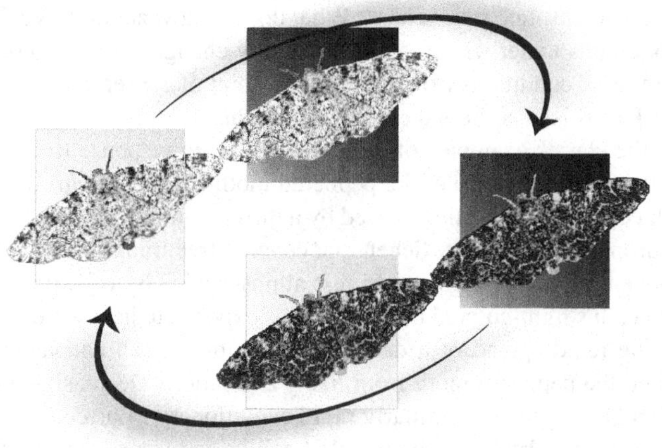

Introduction to Part B

"How fast, as a matter of fact, do animals evolve in nature?" was asked by George Gaylord Simpson (1944) in his renowned book *Tempo and Mode in Evolution.* Ecological and evolutionary processes are often thought to occur on different time scales, so much so that it is common to hear biologists talk about ecological time in contrast to evolutionary time. However, several decades of study in evolutionary ecology and evolutionary genetics have revealed that the time scales of ecological and evolutionary processes can overlap for many crucial questions posed by ecologists. A number of recent studies have reported on the rapid evolution of morphological, physiological, behavioral, and demographic traits over time scales of a few decades, or tens of generations, which coincides with the time horizon of many conservation schemes. How robust is the evidence that leads to the conclusion of commensurate time scales? What do we know about the ecological and genetic conditions under which fast evolutionary change is likely to occur? How relevant are these conditions from the vantage point of conservation biology? The purpose of Part B is to address these important questions.

One of the classic examples of rapid adaptation in response to environmental change is the celebrated case of the peppered moth (*Biston betularia*). At the end of the 19th century, air pollution caused by a thriving industry in the UK and other parts of Europe killed the gray lichen that covered tree trunks on which the moth, with its lichen-mimetic wings, could rest almost invisibly to its predators. As a consequence, this light-colored phenotype was exposed to heavy predation, which promoted the rapid spread of a dark (melanic) mutant, an adaptation that may have rescued the peppered moth from likely extinction. Decreasing air pollution in the 1970s has triggered a similarly rapid evolutionary resurrection of the light-colored phenotype. Whereas the detailed evolutionary mechanisms that underlie this phenomenon are still vigorously debated (Majerus 1998), the case clearly demonstrates how quickly organisms can respond to environmental changes and how ecological predictions that do not account for such adaptations can be in error qualitatively. Not only can adaptations to new environmental conditions occur rapidly, but also they may be amazingly broad in their geographic scope. The rapid evolutionary establishment of a geographic cline in the body size of a fruit fly species (*Drosophila subobscura*), introduced to the New World at the end of the 1970s, shows rates of evolutionary change on a continental scale that exceed almost all previously measured natural rates. Other famous examples of rapid adaptations over large geographic scales are known to involve coevolution of interacting species: rapid and concomitant changes in the virulence of the myxoma virus and the resistance of its host, the European rabbit (*Oryctolagus cuniculus*), were recorded over as few as five years after the disease had spread through Australia in the 1950s.

Evidence for rapid adaptations has been obtained experimentally, too. These experiments fall into four major groups, depending upon whether environmental changes qualify as primarily abiotic (like habitat pollution or climatic change) or biotic (involving, for example, the introduction or removal of predators); and whether these changes correspond to environmental degradation or amelioration. The relevance of such experiments in the context of evolutionary conservation biology can hardly be overestimated: they highlight the important role that common organisms, widely bred in the laboratory, can play in developing an experimental approach to address questions in this emerging field.

Basic empirical evidence for rapid evolutionary responses to environmental threats is reviewed in Chapter 5. After examining the major types of environmental change and their different temporal and spatial scales, Frankham and Kingsolver present a variety of examples for swift responses:

- Recently established latitudinal gradients in fruit flies and codling moths;
- Rapid local adaptation in sockeye salmon;
- Range expansion in admiral butterflies;
- Industrial melanism in peppered moths;
- Acquisition of metal resistance in maple trees; and
- Evolution of heat tolerance in fruit flies and in the bacterium *Escherichia coli*.

The chapter also explains how to assess the evolutionary potential of a threatened population and discusses the special challenges experienced by small populations.

In Chapter 6, Reznick, Rodd, and Nunney present a fascinating review of empirical examples of rapid evolution in natural populations and conclude that fast adaptive changes are not limited to artificially selected organisms. The chapter also examines the sort of ecological and genetic mechanisms that might hamper fast evolution. One important message is that conservation actions that involve environment restoration offer great potential for species recovery to be accelerated by concomitant adaptation.

Genetic variability in life-history traits is necessary for selection to proceed, but selection is also expected to deplete such variability. In Chapter 7, Hughes and Sawby describe the mechanisms that can maintain variability for life-history traits in natural populations. The authors review the empirical evidence, mainly from *Drosophila*, for variability in these traits, and discuss the kinds of mechanisms that could be responsible for the maintenance of variation. Finally, they examine the extent to which these traits are affected by inbreeding depression, and conclude with a review of the evidence for purging inbreeding depression in small populations – a highly controversial topic in conservation biology.

Genetic variation may be depleted or enhanced in the course of adaptations. In Chapter 8, Imasheva and Loeschcke examine the fascinating possibility that external stresses that trigger adaptive responses may also accelerate the production of genetic variation. The chapter offers a review of empirical evidence for such enrichment, and examines the consequences such processes have on the pace and scope of phenotypic evolution.

A review of contemporary rates of evolution (Hendry and Kinnison 1999) concluded that "Claims of rapid microevolution should not necessarily be considered exceptional, and perhaps represent typical rates of microevolution in contemporary populations facing environmental change. [...] Perhaps the greatest contribution that [the study of] evolutionary rates will ultimately make is an awareness of our own role in the present microevolution of life and a cautious consideration of whether populations and species can adapt rapidly enough to forestall the macroevolutionary endpoint of extinction." What is still required, though, is a quantitative assessment of the consequences of rapid evolutionary change on extinction dynamics and population viability. Part C makes theoretical steps forward in this direction, paving the way for the most warranted experimental insights.

5

Responses to Environmental Change: Adaptation or Extinction

Richard Frankham and Joel Kingsolver

5.1 Introduction

All populations are confronted with a plethora of environmental changes and must adapt, shift their range, or face extinction. Adaptation may take two forms:

- The first option involves physiological acclimatization through phenotypic plasticity at the level of individuals.
- Second, the genetic composition of populations may change through natural selection, a change that favors some genotypes at the expense of others.

Whereas plastic adaptation can only cope with environmental change of a limited extent, genetic adaptation allows populations to persist outside their previous tolerance ranges. Therefore, genetic adaptation is of primary concern in conservation biology, in terms of what is required to cope with major or sustained environmental changes. Feasibility and speed of genetic adaptations in response to environmental change depend on a variety of factors, such as a population's genetic diversity, the population size, generation time, and reproduction excess.

This chapter is organized as follows. In Section 5.2 we review the different types of abiotic environmental change that occur in nature, with an emphasis on their characteristic spatial and temporal scales. Section 5.3 explains how changes in local climate affect the physiological and phenological aspects of life histories and shows that evolutionary adaptations to altered climate conditions can be rapid. Sections 5.4 and 5.5 extend this conclusion to responses to thermal stress and pollution, and Section 5.6 highlights the special evolutionary challenges experienced by endangered species.

5.2 Types of Abiotic Environmental Change

Humans are causing widespread changes that involve both physical and biological aspects of the environment. One of the major challenges for conservation biology is that "environmental change" is not a well-defined entity, but rather a heterogeneous assortment of environmental factors that may impact different organisms in diverse ways. These environmental factors operate at various spatial and temporal scales. Evolutionary responses to biotic environmental change are reviewed in Chapter 6; here we focus on adaptations to abiotic environmental change.

In terms of alterations in the physical environment, it is useful to distinguish three general classes of change that typically operate at different spatial scales:

- Biocidal agents and toxins: heavy metals, pesticides, and other environmental contaminants.
- Atmospheric pollutants: acid rain, nitrous oxide, and ozone.
- Climate factors: CO_2 levels, temperature, precipitation, and ultraviolet radiation.

Although, collectively, they have global effects, most biocidal agents (such as heavy metals and pesticides) are applied primarily at local to regional scales. Atmospheric pollutants (such as acid rain) operate at regional and continental scales, as determined by regional atmospheric conditions and weather frontal systems. By contrast, current and future changes in climate occur primarily at the global scale, although the extent of these changes may vary at smaller scales (especially for precipitation). These differences in temporal and spatial scales have important consequences for the potential ecological and evolutionary responses of organisms to environmental changes.

The time scale of environmental changes relative to the generation time of the population in question determines whether the changes (and the potential evolutionary response to change) are experienced as abrupt or gradual. For example, pest insects that have multiple generations each year experience global warming as a gradual environmental change over scores or hundreds of generations, whereas forest trees experience the same climate event within a single generation. This difference in generation time affects both the evolutionary potential (Box 5.1) and the intensity of selection experienced per generation, which can cause differences in the genetic response to selection. For example, laboratory and field studies of insecticide resistance suggest that stronger selection, which results from high pesticide dosages, typically leads to the evolution of resistance that has a monogenic (single locus) basis; by contrast, weaker selection usually leads to the evolution of polygenic resistance (McKenzie and Batterham 1994).

The spatial scale of environmental change is also relevant to potential evolutionary responses. For example, heavy-metal contamination or localized pesticide application may generate intense selection on a small spatial scale; this represents only part of the range of a population or species, and allows spatial refuges in which the population will persist even in the face of strong selection. For pest species that have higher reproduction potentials or excesses, these conditions of intense local selection, together with spatial refuges that prevent population extinction, are likely to be important factors in the rapid evolution of resistance to pesticides and other biocidal agents. By contrast, climate change is occurring on scales much larger than the distribution ranges of most species, which has implications for expected evolutionary responses (Lynch and Lande 1993; Huey and Kingsolver 1993).

Box 5.1 Evolutionary potential and the pace of selection responses

The ability of outbreeding populations to evolve in response to environmental change is called their evolutionary potential. The evolutionary potential of a population is given by the following approximate equation for the annual response R to selection for a quantitative trait

$$R = h_0^2 \frac{1}{tT} \sum_{i=1}^{t} S_i \left(1 - \frac{1}{2N_{e,i}}\right)^{t-1} , \qquad \text{(a)}$$

where t is the number of generations over which selection acts, T is the generation time in years, S_i is the selection differential (difference in the mean of the quantitative traits between selected parents and the mean of all individuals in that generation) in generation i, h_0^2 is the initial heritability for the quantitative trait, and $N_{e,i}$ is the effective population size in generation i. Equation (a) shows that evolutionary potential depends upon genetic diversity (measured by heritability), selection differential, effective population size, and generation time. The maximal selection differential depends upon the fitness of the population, as the higher this is, the greater is its reproductive excess and thus the greater is the selection differential that can be applied.

The maximal total selection response ("the limit to selection") feasible on the basis of the genetic variation that initially exists in a population (i.e., without extra variation being introduced by mutation) is approximately proportional to heritability, selection differential, and effective population size (Robertson 1960),

$$R_{\max} = 2h^2 SN_e . \qquad \text{(b)}$$

This equation is based on the chance loss of favorable alleles through random sampling (genetic drift) in finite populations; Hill and Rasbash (1986) derived a more general expression. The predicted dependence on effective population size has been experimentally validated in *Drosophila* (Jones *et al.* 1968; Weber 1990; Weber and Diggins 1990), mice (Eisen 1975), chickens (Vasquez and Bohren 1982), and plants (Silvela *et al.* 1989). It has been found, however, that the predicted maximal selection responses can be much greater than the observed ones (Weber and Diggins 1990).

Were it not for mutations, populations would lose all their genetic variation through drift and selection. Once the loss through genetic drift is balanced by the gain through mutation, the resultant asymptotic selection response per generation is given by

$$R_{\text{mut}} = 2\frac{V_{G,\text{mut}}}{V_P} SN_e , \qquad \text{(c)}$$

where $V_{G,\text{mut}}$ is the per-generation increment in genetic variation through mutation and V_P is the total phenotypic variation in the population (Hill 1982). Experimental results show that this equation provides approximate predictions of the actual responses (Frankham 1983; Mackay *et al.* 1994). Again, the expected responses are proportional to reproduction excess and effective population size.

5.3 Adaptive Responses to Climate Change

Climate change provides an interesting context in which to explore both the ecological setting and evolutionary potential for adaptation to global change. Temperature and climate impact nearly every physiological and ecological rate process, and physiological and behavioral adaptations to thermal and hydric conditions are ubiquitous (e.g., Johnston and Bennett 1996). Climate is a major factor in determining the geographic distributions of many organisms, and variations in weather and climate directly affect population abundances in many species (Andrewartha and Birch 1954). In addition, climate change on time scales from decades to millennia has occurred repeatedly on earth during the past 200 000 years: climate change is part of the evolutionary history of most present-day species. How will species respond to climate change induced by human activities during the coming century?

The consensus of recent general circulation models indicates that changes in climate will involve not just a simple global increase in mean temperatures, but also will result in rather specific temporal and spatial patterns of change in climate conditions (Houghton 1997). In particular, these climate models predict greater warming at high northern latitudes than at lower latitudes, greater warming in winter than in summer, and greater increases in minimum nighttime temperatures than in maximum daytime temperatures. As a result, anthropogenic climate change will generally increase low temperatures and reduce the degree of diurnal, seasonal, and latitudinal variations in temperature in most regions. In addition, the amount and intensity of precipitation will generally increase, but the effects will vary markedly (and perhaps unpredictably) between different regions.

Physiological and phenological effects of climate change

How will these changes in climate affect the fitness of organisms in ways that might lead to ecological and evolutionary change? It is important to recognize that climate typically influences the distribution and abundance of organisms through its effect on populations at specific stages and times during their life cycle.

For example, the White Admiral Butterfly (*Lagoda camilla*) in Great Britain substantially expanded its geographic range to the north between 1920 and 1940 (Figure 5.1a; Dennis 1993). This was associated with unusually warm conditions during the month of June between 1930 and 1942 (Figure 5.1b). Detailed life-history studies of this species showed that warmer weather in June typically leads to increased larval and pupal survival, probably as a result of reduced bird predation because of the shorter developmental times. In addition, population abundances of *L. camilla* and many other temperate-zone butterflies are strongly influenced by the realized fecundities (the number of eggs laid) of females in the field; in turn, realized fecundities correlate strongly with warmer and sunnier weather during

White Admiral Butterfly
Lagoda camilla

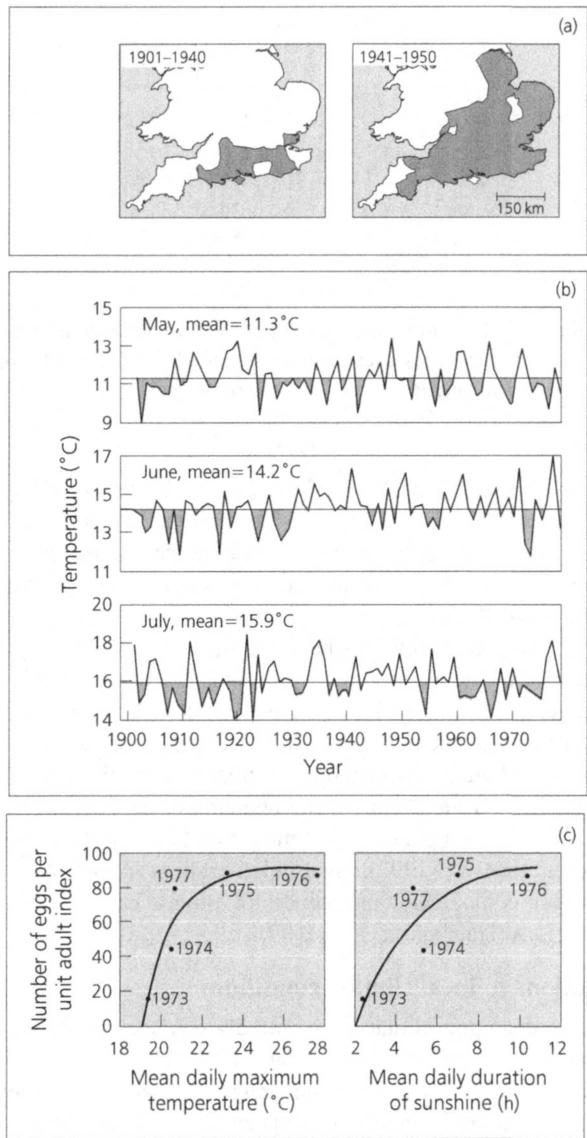

Figure 5.1 Expansion of the geographic range of the White Admiral Butterfly (*L. camilla*) in Britain in relation to changes in temperature. (a) Distribution of *L. camilla* in southern England, 1900 to 1950. (b) Mean monthly temperatures in central England for May, June, and July, 1900 to 1977; horizontal lines show temperature means over the whole time period. (c) Index of egg numbers laid per adult butterfly (counts on transect) as a function of mean daily maximum temperature (left panel) and of mean daily duration of sunshine (right panel) for the period 1973 to 1977. *Source*: Dennis (1993), after Pollard (1979).

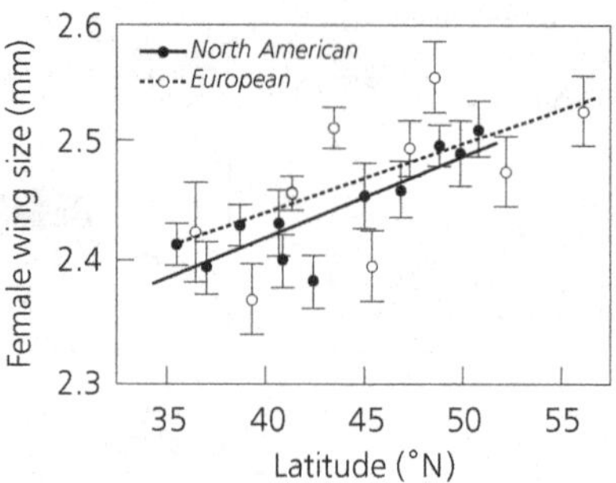

Figure 5.2 Latitudinal clines in female wing size of *D. subobscura* in Europe and North America. During the two decades after the introduction of an ancestral European form to North America, the North American cline has had time to build up and converge toward its European counterpart. *Source*: Huey *et al.* (2000).

flight seasons in June and July (Figure 5.1c; Pollard 1988). Similarly, the ranges of the northern winter boundaries of many North American bird species correlate strongly with minimum January temperatures; in at least some species this range limit may result from the costs of maintaining a positive energy balance during cold conditions (Root 1988).

These examples illustrate how weather and climate conditions may affect fitness at specific stages in many species. A corollary of this observation is that climate changes may alter the timing of life-cycle events of organisms relative to that of their resources or natural enemies. For instance, experimental studies indicate that the effect of increased winter temperatures is greater on the timing of egg hatching of winter moths than on the timing of the budburst of the Sitka spruce on which they feed. This results in a phenological mismatch between the herbivore and its host plant (Dewar and Watt 1992). Such relative shifts in timing may be among the most important ecological consequences of climate change in temperate and high-latitude regions (Harrington *et al.* 1999).

Rapid adaptations to local climate conditions

As a response to changing climate, how rapidly can species evolve adaptations in their thermal physiology or in the timing of phenological events? Studies of organisms that have recently colonized new regions provide some useful clues.

Drosophila subobscura is a widespread species throughout Europe and northern Africa. The cline in body size among European populations is long-established, with increasing size in the more northerly populations. *D. subobscura* colonized the west coast of North America in the 1970s, probably first in the region near Vancouver, British Columbia (Prevosti *et al.* 1988). During the past 25 years it has expanded throughout the west coast from California to British Columbia. Recent studies of *D. subobscura* along this latitudinal and climate gradient indicate the evolution of differences of size: flies from lower latitudes have smaller wing size than those from higher latitudes (Figure 5.2; Huey *et al.* 2000).

Studies of colonizing agricultural pests provide similar insights on the evolution of diapause traits that control the timing of life-cycle events (Tauber *et al.* 1986). For example, the codling moth, a major pest on fruit trees, colonized North America from its native India in the late 18th century, and subsequently expanded its distribution over a wide latitudinal range. Like many insect species, codling moths overwinter in a diapause stage, the initiation of which is triggered by day length. Genetic analyses demonstrate that different critical day lengths initiate diapause in different geographic populations, with earlier dates of diapause in more northern populations, a pattern seen in many temperate insects (Riedl and Croft 1978).

Codling moth
Cydia pomonella

Rapid adaptation to novel environmental conditions has also been demonstrated for sockeye salmon (*Oncorhynchus nerka*) introduced to a new range of environments (Hendry *et al.* 1998). After its introduction to Lake Washington in the 1930s and 1940s, this salmon species adapted to the encountered diverse local climate conditions within only nine to 14 generations: survival rates and body sizes at emergence became adjusted to local temperatures. Genetic divergence has been identified as the most likely cause of this pattern of local adaptation.

These examples clearly indicate that at least some species have the capacity to evolve physiological and phenological adaptations rapidly in response to climate. As expected (Box 5.1), these examples involve species with large population sizes and high reproduction potential.

5.4 Adaptive Responses to Thermal Stress

Evolutionary responses to extremes of temperature and desiccation have been explored in the laboratory for a number of model study organisms (Huey and Kingsolver 1993). One insight to emerge from these studies is that evolutionary responses to low temperatures may be quite different to those that result from high temperatures.

The thermal niche of a population is defined as the temperature range over which a population can maintain a mean population fitness above its replacement rate. Bennett and Lenski (1993) examined the evolution of the thermal niche in the bacteria *Echerichia coli*, by allowing populations to evolve at different temperatures in the laboratory. Evolution for 2000 generations at 20°C (near the ancestral population's lower thermal limit) caused modest increases (9%) in mean population fitness (Figure 5.3a), and decreased both the lower and upper thermal limits by 1–2°C (Figure 5.3b). By contrast, evolution for 2000 generations at 42°C (near the ancestral population's upper thermal limit) caused substantial increases (33%) in mean population fitness at 42°C (Figure 5.3a), but did not alter the lower or upper thermal limits of the population (Figure 5.3b).

Artificial selection experiments with the parasitoid wasp *Aphytes* showed a different pattern of asymmetry (Huey and Kingsolver 1989). In this system, selection

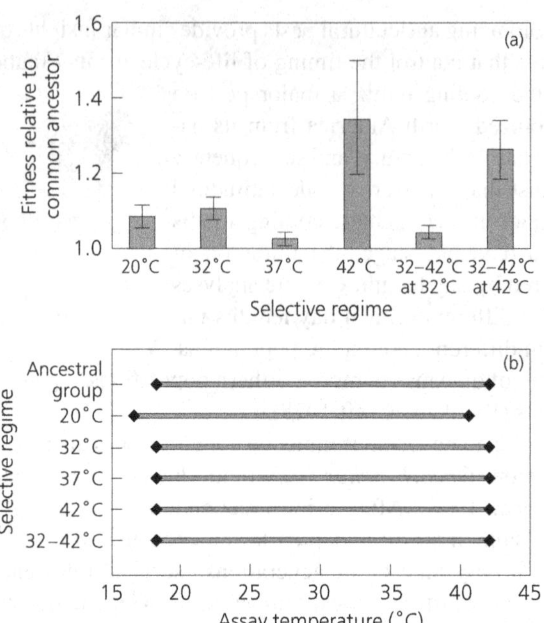

Figure 5.3 Evolution of the thermal niche of *E. coli* after 2000 generations of experimental evolution at different temperatures. (a) Mean fitness of groups of *E. coli* populations relative to the common ancestor. Relative fitness was assayed at each group's selective temperature. Error bars represent 95% confidence intervals. (b) Each bar indicates the range of temperatures over which mean fitness exceeded zero for groups of *E. coli* evolved at different temperatures and for the ancestral group. Only the 20°C group showed a significant evolutionary shift in its thermal niche relative to the ancestral group. *Source*: Mongold *et al.* (1996).

for increased tolerance to cold causes evolutionary increases in tolerance to both cold and heat, whereas selection for increased tolerance to heat only increased tolerance to heat (White *et al.* 1970).

Many laboratory studies of evolution in response to temperature changes with *D. melanogaster* also demonstrate asymmetric evolutionary responses at different temperatures. For example, Partridge *et al.* (1995) tested the thermal adaptation of *D. melanogaster* lines maintained at 16.5°C or 25°C for four years. When exposed to different experimental temperatures, females from lines evolved at 25°C exhibited a relatively small change in lifetime progeny production in response to experimental temperature (Figure 5.4). In contrast, females from lines evolved at 16.5°C showed a high lifetime progeny production at 16.5°C, but a low production at 25°C (Figure 5.4). Similarly, selection for increased tolerance to desiccation with *D. melanogaster* typically increases tolerance to desiccation, starvation, and heat, but has no effect on tolerance to cold (Hoffmann and Parsons 1989).

These different evolutionary responses at low and high temperatures imply that different physiological processes may limit key components of fitness at different

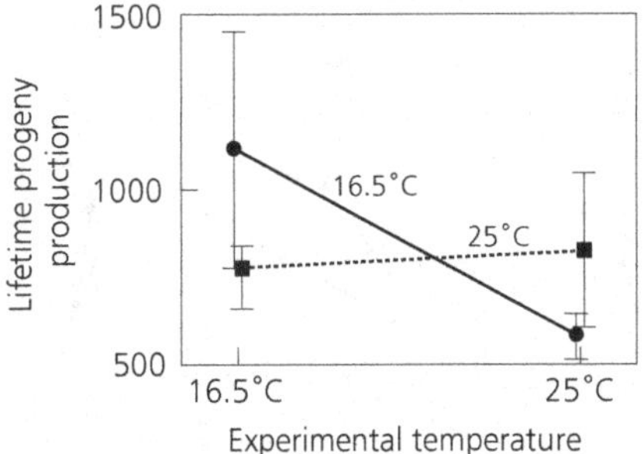

Figure 5.4 Mean and 95% confidence intervals of lifetime progeny production by females of *D. melanogaster* from populations evolved at 16.5°C (circles, continuous line) or at 25°C (squares, dashed line) and tested at experimental temperatures corresponding to these two selective regimes. *Source*: Partridge *et al.* (1995).

temperatures. Studies on *Pieris rapae* and *Manduca sexta* caterpillars corroborate this, as they show that short-term growth rates increase with increasing temperatures between 10°C and 30°C, are maximal at around 34°C to 36°C, and decline rapidly at higher temperatures (Figure 5.5a). Physiological analyses with *M. sexta* suggest that the growth rate at high temperatures is primarily limited by the effects of temperature on the rate of food intake, whereas growth at low temperatures is limited by the effects of temperature on both food intake and the uptake of amino acid across the gut wall (Figure 5.5b; Kingsolver and Woods 1997). Similarly, studies with the herbivorous grasshopper *Melanoplus bivittatus* show that the growth rate is strongly limited by digestive throughput at low temperatures, but not at intermediate or high temperatures (Harrison and Fewell 1995).

Results of these artificial selection experiments highlight that the physiological mechanisms and genetic architectures that underlie traits relating to stress tolerance may be complex, and may vary even with respect to single environmental factors such as temperature. This complexity will likely lead to surprises in the ways that organisms evolve in response to global warming and to other aspects of environmental change.

5.5 Adaptive Responses to Pollution

Evolutionary changes in response to environmental pollution have been documented in nature and in the laboratory. Examples of adaptive evolutionary change in response to anthropogenic change in nature include industrial melanism, heavy-metal tolerance, and insecticide resistance.

Industrial melanism is defined as a correlation between the heritable coloration of insects and the effects of industrial pollutants on the visual characteristics of their environments. Evolution of industrial melanism has been observed in over 200 species of moths in polluted areas (Kettlewell 1973; Majerus 1998). Whereas the particular ecological mechanisms by which abiotic pollutants alter selection pressures that arise from biotic interactions are species specific and still debated

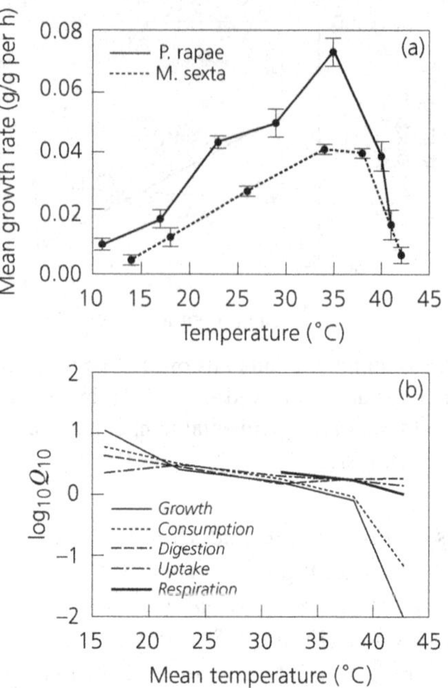

Figure 5.5 Physiological responses to temperature in *M. sexta* and *P. rapae* caterpillars. (a) Mean mass-specific growth rates (grams per grams per hour) as functions of temperature for early fifth instar *P. rapae* (continuous curve) and *M. sexta* (dashed curve). Error bars represent plus/minus one standard error. (b) Q_{10} values of growth, consumption, protein digestion, methionine uptake, and respiration rates as functions of temperature for fifth instar *M. sexta* caterpillars. [Q_{10} is a standard physiological term that refers to the multiple by which a physiological rate process (e.g., metabolic or growth rate) increases due to a 10°C increase in body temperature. For example, a Q_{10} of 2 indicates that the rate increases two-fold from a 10°C temperature increase.] *Sources*: Kingsolver and Woods (1997) and Kingsolver (2000).

(Majerus 1998), industrial melanism remains a prime example of evolution in action. Throughout the second half of the 20th century, levels of air pollution in industrial countries have been reduced significantly and the resultant selection responses in the relative frequency of melanic (dark) morphs have been well documented. The time series in Figure 5.6a depicts the rapid evolutionary response of peppered moths (*Biston betularia*) in northwestern England. A period of 25 years was long enough to bring the frequency of the dark-colored *B. betularia carbonaria* down from 90% to 20% (Majerus 1998). Figure 5.6b shows how the relative frequencies of five differently colored *B. betularia* morphs have changed in southern Holland between 1969 and 1999 (Brakefield and Liebert 2000).

The evolution of heavy-metal tolerance has occurred in many plants. Examples include tolerance to copper, zinc, mercury, and cadmium in grasses that colonize polluted mine wastes (Jones and Wilkins 1971; Briggs and Walters 1997). Also,

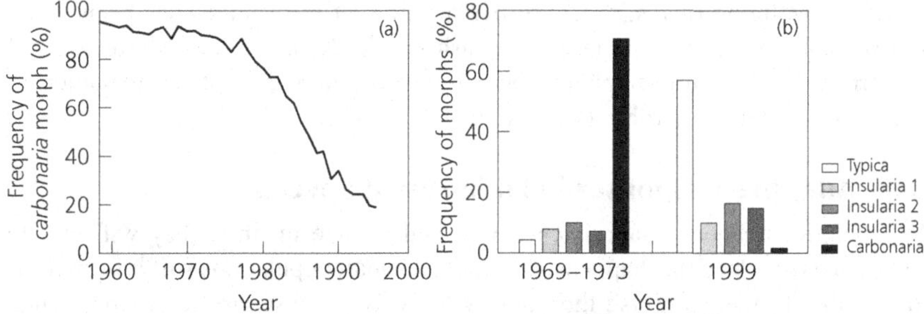

Figure 5.6 Rapid decline of industrial melanism in Europe. (a) Changes in the relative frequency of the melanic peppered moth *B. betularia carbonaria* in northwestern England between 1959 and 1995. *Source*: Majerus (1998). (b) Changes in the relative frequency of five morphs of *B. betularia* in southern Holland between 1969 and 1999. *Source*: Brakefield and Liebert (2000).

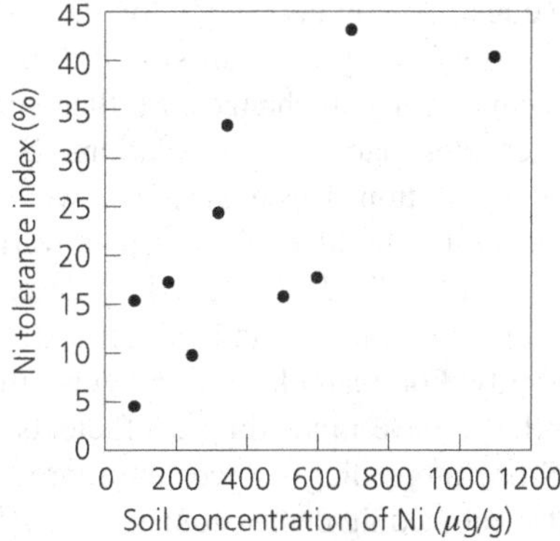

Figure 5.7 Nickel (Ni) resistance of red maple trees (*A. rubrum*) on mining sites in Eastern Canada. A Ni tolerance index shows a strong positive correlation (0.74) with Ni concentrations in the soil that surrounds the trees used to establish the cell lines. (The tolerance index used here is defined as the ratio between the increase in fresh weight of the callus of cell lines grown on a medium with a metal concentration of 10 mg Ni per liter and the corresponding increase on a metal-free medium). *Source*: Watmough and Hutchinson (1997).

trees have been shown to evolve metal tolerance rapidly. For instance, 100 years after mining and smelting operations began on sites in Ontario and Quebec, Canada, cell lines of red maple trees (*Acer rubrum*) show tolerance levels against nickel concentrations that correlate positively with nickel concentration in the soil that surrounds the sampled trees (Figure 5.7; Watmough and Hutchinson 1997). Evolutionary changes in response to pollution have also been reported for a range of pollutants in laboratory experiments. For example, *D. melanogaster* and *D. willistoni* have both responded to selection for resistance to sodium chloride (NaCl) and copper sulfate ($CuSO_4$) pollutants (Tabachnick and Powell 1977; Ehrman *et al.* 1991).

Rapid evolution of resistance to several different insecticides has been found in over 200 species of pest insects (Roush and McKenzie 1987; McKenzie and Batterham 1994). A related phenomenon is the evolution of resistance to numerous antibiotics in many microbes (Garrett 1994).

5.6 Adaptive Responses in Endangered Species

If populations cannot adapt to environmental change in time, they will greatly reduce in size. Once this happens, the resultant small populations suffer from three effects that further diminish their ability to evolve in response to environmental change:

- Loss of genetic variation at higher rates than in larger populations;
- Inbreeding is more rapid and thus reductions in fitness occur; and
- The risk of accumulating deleterious mutations by chance effects is higher.

Clear relationships occur between population size and genetic diversity, such as smaller populations have less genetic diversity (Box 5.2). Consequently, populations and species with a prior history of small size will, on average, be less able to evolve in response to environmental changes. Further, smaller populations lose genetic diversity at greater rates than larger populations (Box 5.2). Consequently, populations with smaller population sizes evolve at slower rates than larger populations, even if they have similar initial levels of genetic variation.

Inbreeding is the mating of individuals related by descent. It is an inevitable consequence of a limited population size: individuals come to share common ancestors with high probability. For example, after 50 generations, populations with effective sizes of 10 and 100 have inbreeding coefficients of 92% and 22%, respectively (Box 5.2). Since inbreeding reduces the reproduction fitness in naturally outbreeding species (inbreeding depression; Wright 1977; Charlesworth and Charlesworth 1987; Ralls *et al.* 1988; Thornhill 1993; Frankham 1995a), it reduces reproduction excess and thus the ability of affected populations to evolve. Further, inbreeding directly increases the risk of extinction through adverse effects on demographic rates: in this way, it may directly lead to a negative population growth and eventual extinction (Frankham and Ralls 1998; Saccheri *et al.* 1998). Simple single-locus models suggest that reproduction fitness decreases linearly with increasing inbreeding depression; experiments agree with this expectation or show an even steeper decline in fitness (Charlesworth and Charlesworth 1987; Falconer and Mackay 1996).

The third adverse effect of a small population size is the higher risk of a chance fixation of deleterious mutations (Lande 1995; Lynch *et al.* 1995a, 1995b; see Chapter 9). Since this effect leads to a lower reproduction fitness, it may eventually cause population decline and extinction ("mutational meltdown"). Lynch *et al.* (1995b) suggested that this may be an issue of concern in conservation biology over time scales of 50 or more generations. An experimental evaluation of the importance of mutational accumulation has been carried out in *Drosophila* by testing for the effects of deleterious mutations in populations maintained at effective

Box 5.2 Evolution of genetic diversity

Evolutionary change does not occur in the absence of genetic diversity. Such diversity may either be pre-existing or result from new mutations. Since mutation rates are low in eukaryotes of conservation concern, evolutionary change arises from pre-existing genetic diversity. Outbreeding species typically contain high levels of quantitative genetic variation (Lewontin 1974) and of genetic diversity at the level of deoxyribonucleic acid (DNA), as detected by allozyme electrophoresis (Ward *et al.* 1992). For example, the average species is polymorphic at 28% of allozyme loci, and 7.3% of loci are heterozygous in an average individual. More variability is found for microsatellites; average heterozygosities for polymorphic loci are 60% to 80% in species with large population sizes, and loci have an average of 6 to 11 alleles (Frankham *et al.* 2002). Compared with outbreeding species, natural inbreeding species, such as selfing plants, contain less genetic diversity and harbor this diversity more between (rather than within) individuals and between (rather than within) populations (Hamrick and Godt 1989).

Levels of genetic diversity in populations and species are related to long-term average abundance (Soulé 1976; Frankham 1996). The rate of loss of neutral genetic diversity is greater in populations with smaller effective population sizes, and such losses continue in each generation,

$$\frac{H_t}{H_0} = \prod_{i=0}^{t} \left(1 - \frac{1}{2N_{e,i}}\right), \tag{a}$$

where H_t is the population's heterozygosity in generation t in the absence of mutation (Crow and Kimura 1970). The coefficient of inbreeding in generation t is defined as $F_t = 1 - H_t/H_0$; it rapidly increases from 0 to 1 if effective population sizes $N_{e,t}$ become small.

Over long periods of time, mutation cannot be ignored as a source of genetic diversity. Eventually, a balance between loss through drift and gain through mutation is reached. For neutral loci (those that are not subject to natural selection), the probabilities of the chance loss of alleles through genetic drift decrease with increasing effective population size. For a per-locus mutation probability u the expected asymptotic level of heterozygosity is given by

$$H = \frac{4N_e u}{4N_e u + 1}. \tag{b}$$

Although observations do not follow the shape of this relationship for cases in which the assumptions used in its derivation are not strictly fulfilled (Frankham 1996), the equation correctly predicts that heterozygosities are higher in abundant populations and species than in smaller ones.

sizes of 25, 50, 100, 250, and 500 for 45 to 50 generations: no indications of mutational accumulations were detected in three experiments (Gilligan *et al.* 1997). However, all replicates of the treatments with effective population sizes of 25 and 50 showed inbreeding depression (Woodworth 1996). It therefore appears that inbreeding depression must be regarded as a much greater conservation threat than

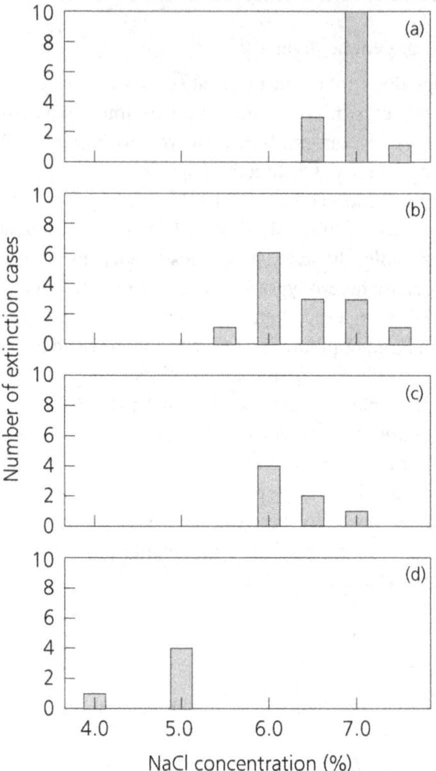

Figure 5.8 Effects of population-size bottlenecks on evolutionary potential. Four types of populations were considered: (a) outbred populations, (b) populations with a bottleneck caused by one generation of single-pair mating, (c) populations with bottlenecks caused by three generations of single-pair mating, and (d) highly inbred populations. All the populations were initiated with 500 parents and subjected to increasing concentrations of NaCl until extinction occurred. The four panels show the distribution of the resultant extinction cases across NaCl concentrations. *Source*: Frankham *et al.* (1999).

mutational accumulation; naturally outbreeding populations are likely to be driven to extinction by inbreeding depression before mutational accumulation becomes a serious threat.

Laboratory experiments have confirmed the expected effects of population size restrictions on the evolutionary potential of *D. melanogaster*. Frankham *et al.* (1999) showed that a single-generation bottleneck of one pair reduced evolutionary potential in response to NaCl concentrations (Figure 5.8). Here, evolutionary potential was measured as the NaCl concentration at which cage populations were driven to extinction (after having been subjected to a regime of successively increasing NaCl concentrations; this measure encompasses all the factors that contribute to evolutionary potential, including genetic diversity and reproduction excess). A significant positive correlation between evolutionary potential and effective population size has also been found in populations of *D. melanogaster* maintained for 50 generations at effective population sizes of 25, 50, 100, 250, and 500 (Frankham *et al.*, unpublished data).

In summary, the ability of species that are already endangered to respond to environmental change may be greatly compromised:

- Since endangered species, by definition, have smaller effective population sizes than nonendangered species (IUCN 1994), they evolve at slower rates than nonendangered species.
- Endangered species typically have lower genetic diversity than their nonendangered relatives and presumably exhibit lower heritabilities (Frankham 1995a).
- Since species with long generation times are more prone to become endangered, endangered species are more likely to have long generation times.
- Endangered species often have low reproduction rates. Consequently, potential reproduction excesses and selection differentials are small.
- At small populations sizes, reproduction rates are also more likely to be reduced by inbreeding depression. Such inbreeding, whether imposed deliberately (Frankham 1995b) or occurring naturally in small populations (Latter *et al.* 1995), can in itself be so severe as to cause extinction.

The effects of all these factors combine and lead to the conclusion that endangered species generally have lower evolutionary potential than nonendangered species.

5.7 Concluding Comments

In response to environmental change, populations either adapt or are driven to extinction if they cannot adapt rapidly enough. Such environmental change is ubiquitous and can involve changes in both the abiotic and biotic environments. The ability to adapt depends on current genetic diversity, population size, breeding system, reproduction excess, the speed of the environment change, and the nature of the change (whether gradual or catastrophic). Endangered species have compromised abilities to adapt as they typically have less genetic diversity, lower reproduction rates, and smaller population sizes than nonendangered species. Adaptation in threatened species relies overwhelmingly on pre-existing genetic variation, rather than new mutations.

The alleviation of extinction risks involves:

- Minimization of anthropogenic environmental change;
- Remediation of altered environments;
- *Ex situ* conservation and reintroductions;
- Genetic management to minimize inbreeding and retain genetic diversity.

Management of wild populations to recover genetic diversity and reproduction fitness by exchanging individuals among fragmented populations is a much-needed procedure, but in very few cases has it been, or is it being, carried out. The red-cockaded woodpecker is, as far as we are aware, the only case for which such a procedure is part of the management of a wild outbreeding species (Haig *et al.* 1993; Kulhavy *et al.* 1995).

Red-cockaded woodpecker *Picoides borealis*

The brief survey of empirical literature in this chapter highlights an important dichotomy about the potential role of adaptive evolution in response to environmental change. For endangered species and other taxa of immediate conservation concern, adaptive evolution is unlikely to be an important component of successful management: at best we may be able to reduce the rate at which genetic variation is lost in such cases. By contrast, many examples from field and laboratory studies demonstrate that organisms with high evolutionary potential readily evolve adaptations in response to a variety of environmental changes – including climate factors, thermal stress, pesticides, and pollutants – on time scales of years to decades.

Between these two extremes, we have very little quantitative information about the relative likelihood of adaptation or extinction, of where the balance is most critical between selection that leads toward adaptive evolution and reduction in mean population fitness that leads toward extinction. Yet, populations and species that occupy this middle ground today may become the endangered taxa 50 years from now. A better understanding of the possible role of adaptation for the long-term survival of species of currently moderate evolutionary potential, where the interplay of evolutionary and ecological factors is most complex, should be one important contribution of evolutionary conservation biologists in the coming decades (Kareiva *et al.* 1993).

6

Empirical Evidence for Rapid Evolution

David Reznick, Helen Rodd, and Leonard Nunney

6.1 Introduction

All organisms have the capacity to evolve in the face of a changing environment. Our general goal is to learn about the limitations to this process of evolution. How quickly can organisms evolve? How much change is possible? Can this capacity for change be predicted from factors like the genetic variation present within the population or the structure and demography of the population? Such issues are clearly of importance in the context of conservation biology, because species can be endangered by a changing environment and their limited capacity to evolve in response to that change.

Our first goal is to describe the process of adaptation in natural populations of guppies, and then generalize this approach to other organisms. We begin with guppies because it has been possible to study their evolution from an experimental perspective in natural populations. They therefore provide an empirical example of the process of evolution by natural selection and, in this single special case, characterize the limitations to this process. The first perspective is to describe how guppy life histories differ between localities in terms of the predators that they co-occur with. The second perspective is to use duplicated experimental episodes of directional selection to characterize the process of natural selection in real populations. We then use the results of these and other studies of rapid evolution to describe some of the limitations to evolution by natural selection and to characterize the circumstances under which rapid evolution has been seen to occur.

6.2 Guppy Life-history Evolution

"Life history" refers to the timing of development and allocation of resources to reproduction. For guppies, the most important traits are the age at maturity, frequency of reproduction, number and size of offspring, and "reproductive effort", or the proportion of consumed resources that is devoted to reproduction as opposed to other functions such as growth or maintenance. The life history is thus a complex phenotype determined by a large number of genes.

The study system

Guppies (*Poecilia reticulata*) are small, live-bearing fish found in coastal regions of northeastern South America and some near-shore Caribbean islands (Rosen and Bailey 1963). They can attain sexual maturity in less than three months. Females produce litters of young at 3–4 week intervals thereafter. Guppies are well

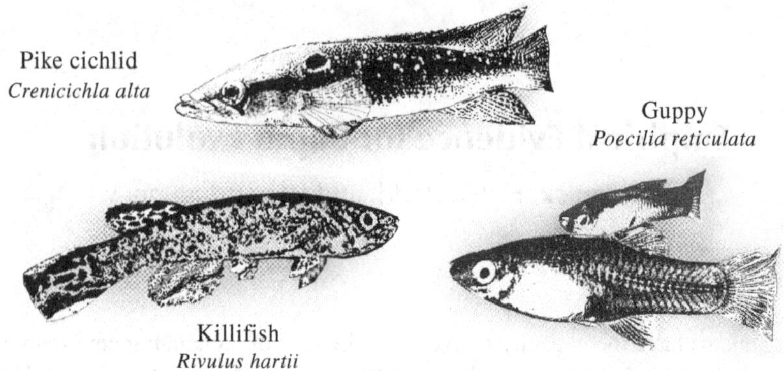

known for their pronounced sexual dimorphism; males are smaller than females, are brightly colored, and their color patterns are remarkably polymorphic. Guppies are abundant and easy to collect in their native environment. They are easy to maintain and breed in the laboratory.

We study guppies in the seasonal rainforests on the Northern Range Mountains of Trinidad. These mountains receive 3 m or more of rain per year and have a dry season that extends from January through May. Their numerous streams flow throughout the year, but since the terrain is steep, these streams are often punctuated by waterfalls. Waterfalls are frequently barriers to the upstream migration of some species, and thus divide the streams into discrete types of communities that live in very similar physical environments. The fish communities found in these streams contain relatively few species, as is typical of islands. We concentrated on a contrast between "high"- and "low"-predation communities. High-predation communities are those in which guppies co-occur with species of fish, like the pike cichlid (*Crenicichla alta*), that frequently prey on guppies. In low-predation communities guppies are found with only one potential predator, the killifish *Rivulus hartii*. These communities are often found in the same streams as high-predation communities, but above barrier waterfalls. High- and low-predation environments are replicated in a large number of river systems (Figure 6.1). These comparisons can therefore be made across a number of localities, which may differ in environment, but share the same fish communities.

The association between predation and life histories

Mark–recapture studies (Box 6.1) demonstrate that guppies from high-predation sites experience much higher mortality rates than their counterparts from low-predation localities (Reznick *et al.* 1996). Life-history theory predicts that the higher mortality rates in high-predation localities will select for individuals that attain maturity at an earlier age and have higher reproductive efforts, that is, devote more of their consumed resources to reproduction (Charlesworth 1980; Gadgil and Bossert 1970; Law 1979; Michod 1979). Our goal was to test this hypothesis in natural populations of guppies.

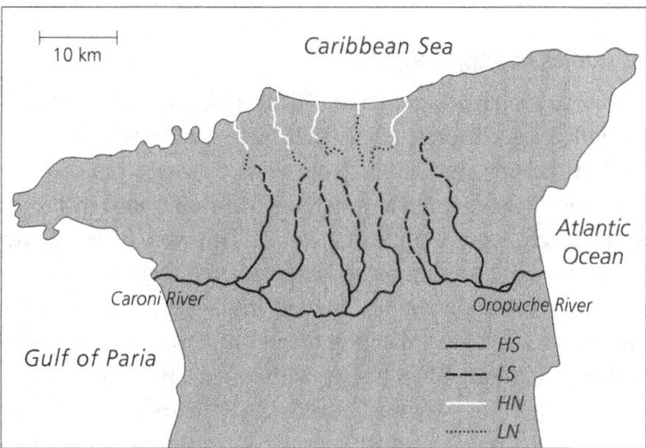

Figure 6.1 Distribution of high- and low-predation environments in Trinidad. Many of the smaller tributaries are not included, and neither are the precise locations of the divisions between high- and low-predation sites, since the distribution of predators has not been accurately mapped in most rivers. Nevertheless, the repeated occurrence of high-predation (continuous lines) and low-predation (dashed lines) environments in different drainages is depicted. The Aripo and El Cedro rivers are part of the Caroni drainage. (HS, high predation, southern slope; LS, low predation, southern slope; HN, high predation, northern slope; LN, low predation, northern slope.)

We first used two methodologies to compare the life histories of guppies from high- and low-predation environments. First, we characterized life-history phenotypes, which involved collecting and preserving guppies from natural populations. We then analyzed these guppies and estimated a series of variables that characterize the life history, including the size distribution of mature individuals, litter size, and offspring size (Reznick and Endler 1982; Reznick *et al.* 1996). Second, we characterized the genetic basis of differences in life histories of guppies from different environments. To do so, we began with 20 to 25 wild-caught adult females from each locality. Adult female guppies reproduce continuously and store sperm. Each female was isolated and gave birth to a series of litters, which were then reared in a common laboratory environment. The laboratory-born offspring were mated, using a design that prevented inbreeding and preserved the genetic diversity present in the original sample, to produce a second laboratory generation. These isolated, second-generation guppies were reared on controlled levels of food availability. We quantified the age at maturity, fecundity, offspring size, time interval between broods of young, and the proportion of consumed resources that were used for reproduction, growth, and maintenance for these fish (Reznick 1982; Reznick and Bryga 1996). Average differences in life-history traits between localities that persist after two generations in a common environment are assumed to have a genetic basis. As evaluating life-history phenotypes is relatively easy, it has been possible to characterize the life histories of guppies from 60 localities in Venezuela, Trinidad, and Tobago; however, these data do not allow a precise estimate of critical variables, such as age at maturity or reproductive effort.

Box 6.1 Field studies of guppy populations

Many guppy streams are small and have a riffle-pool structure, which means that the stream is divided into discrete pools that have a relatively low current. Each pool is bounded at the up- and downstream ends by more rapidly flowing water. Guppies congregate in pools and rarely swim from one pool to the next, so it is relatively easy to catch every guppy in a pool. This sort of population structure lends itself to a series of observations and experiments that treat pools as sampling units and guppy populations as being divided into these discrete sampling units. Mark–recapture studies involve collecting the entire population in a pool, and measuring and marking each individual guppy with a mark that indicates its size class, to the nearest millimeter. We previously used acrylic latex paints (Liquitex) diluted in physiologic saline and now use elastomer marking agents developed specifically for marking aquatic organisms. The mark is a small dot of pigment injected subcutaneously with a very fine needle. Guppies that are longer than 12 mm are marked in this fashion. Guppies shorter than 12 mm are marked by dipping them in a buffered solution of calcein, which is a substance that binds to calcium-bearing tissues (Wilson *et al.* 1987), and so fish that are too small to mark with paints can be followed. The fish are then released and recaught after intervals that range from 12 days to up to six months, depending on the study. Reznick *et al.* (1996) present these methods in more detail.

Such studies yield a wealth of information about guppy populations, especially when they are applied to a series of high- and low-predation localities. For example, Reznick *et al.* (1996) use them to characterize the size-specific mortality rates of guppies from these two types of habitats. Such data also make it possible to characterize size-specific growth rates, the age and size at maturity, recruitment of new young into the population, and the movement of individuals among pools. Fecundity is estimated from the dissection of adult females. These data in turn make it possible to model population growth and to use life-table methods to estimate generation time, both of which are part of studies described in Boxes 6.2 to 6.4 and the text. We now also use variations on this theme to study metapopulation dynamics in natural communities, density regulation, and aging.

Laboratory estimates of life-history traits take nine months to one year to complete, which limits the number of localities that can be investigated (20 thus far), but they allow precise estimates of the variables that characterize the life history. Our conclusions are based on the combination of these two approaches.

These two types of data yield the same answer, namely that guppies from high-predation sites do mature at an earlier age and smaller size than their counterparts from low-predation sites. The reproductive efforts of the guppies from high-predation environments are also higher because they begin to reproduce at an earlier age, have shorter time intervals between successive broods, and devote more resources to each brood (reproductive allotment; Figure 6.2). In summary, the differences in life histories of guppies from high- versus those from low-predation environments are consistent with the predictions of life-history theory.

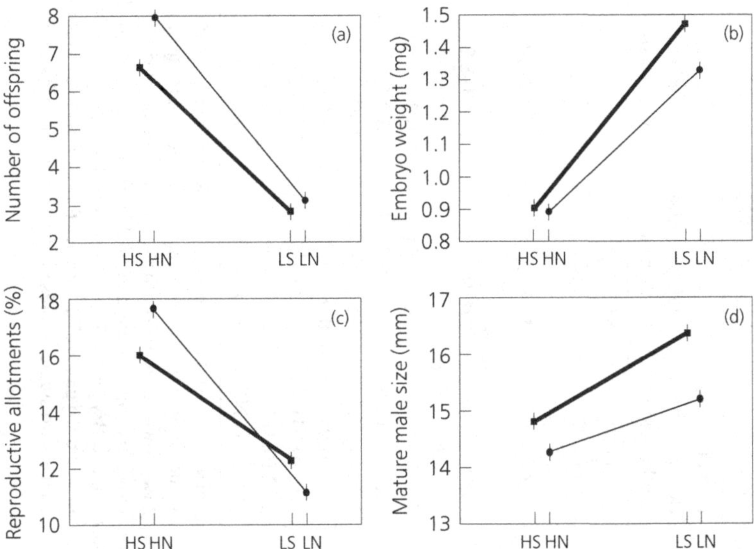

Figure 6.2 Least-square means (± standard errors) of life-history traits for guppies from high- and low-predation sites on the north slope from analyses of the life-history phenotypes of wild-caught guppies [thin lines, from Reznick *et al.* (1996)]. For comparison, included are the results for the same analyses of south slope data [thick lines, from Reznick and Endler (1982)]. (a) Number of developing offspring per female, adjusted for the female's somatic dry weight; (b) dry weight of developing offspring, adjusted for their stage of development; (c) reproductive allotment of females with developing offspring, adjusted for their stage of development, measured as the percentage of total dry weight that consists of developing embryos; and (d) average size of mature males. (HS, high predation, southern slope; LS, low predation, southern slope; HN, high predation, northern slope; LN, low predation, northern slope.)

6.3 Selection Experiments

The comparisons establish correlations between guppy life histories and predation. Our goal was to establish causation. We thus executed experiments that mimicked an episode of directional selection so that we could test directly the hypothesis that predation selects for the evolution of life histories. In the process, we were also able to characterize the process of evolution by natural selection.

Methods

Barrier waterfalls often impede the upstream dispersal of all fish species except *R. hartii*. In two such streams, we collected guppies from the high-predation sites below the waterfalls and introduced them into the guppy-free, low-predation sites above the barrier waterfalls. We predicted that natural selection would favor those individuals that delayed maturity and devoted fewer resources to reproduction, as is typical of low-predation localities. Guppies from below the barrier waterfall

Table 6.1 Results of the Aripo tributary selection experiment[a].

	Reznick *et al.* (1990)		Reznick and Bryga (1987)		Reznick (1982)	
Life-history trait[b]	Control (high predation, southern slope)	Introduction (low predation, southern slope)	Control (high predation, southern slope)	Introduction (low predation, southern slope)	High predation, southern slope	Low predation, southern slope
Male age at maturity (days)	60.6 (1.8)	72.7 (1.8)[c]	48.5 (1.2)	58.2 (1.4)[c]	51.8 (1.1)	58.8 (1.0)[c]
Male size at maturity (mg-wet)	56.0 (1.4)	62.4 (1.5)[c]	67.5 (1.2)	76.1 (1.9)[c]	87.7 (2.8)	99.7 (2.5)[c]
Female age at first parturition (days)	94.1 (1.8)	95.5 (1.8) NS	85.7 (2.2)	92.3 (2.6)[d]	71.5 (2.0)	81.9 (1.9)[c]
Female size at first parturition (mg-wet)	116.5 (3.7)	118.9 (3.7) NS	161.5 (6.4)	185.6 (7.5)[c]	218.0 (8.4)	270.0 (8.2)[c]
Brood size, litter 1	2.5 (0.2)	3.0 (0.2) NS	4.5 (0.4)	3.3 (0.4)[d]	5.2 (0.4)	3.2 (0.5)[c]
Brood size, litter 2	6.3 (0.3)	7.0 (0.3)[e]	8.1 (0.6)	7.5 (0.7) NS	10.9 (0.6)	10.2 (0.8) NS
Brood size, litter 3[f]	–	–	11.4 (0.8)	11.5 (0.9) NS	16.1 (0.9)	16.0 (1.1) NS
Offspring size (mg-dry), litter 1	0.91 (0.02)	0.87 (0.02) NS	0.87 (0.02)	0.95 (0.02)[e]	0.84 (0.02)	0.99 (0.03)[c]
Offspring size, litter 2	0.93 (0.02)	0.86 (0.02)[d]	0.90 (0.03)	1.02 (0.04)[d]	0.95 (0.02)	1.05 (0.03)[d]
Offspring size, litter 3[f]	–	–	1.10 (0.03)	1.17 (0.04) NS	1.03 (0.03)	1.17 (0.04)[c]
Interbrood interval (days)	24.9 (0.4)	24.89 (0.4) NS	24.5 (0.3)	25.2 (0.3) NS	22.8 (0.3)	25.0 (0.03)[c]
Reproductive effort (%)[g]	4.0 (0.1)[d]	3.9 (0.1) NS	22.0 (1.8)	18.5 (2.1) NS	25.1 (1.6)	19.2 (1.5)[d]

[a] Values are means (s.e.) and represent data from Reznick (1982), Reznick and Bryga (1987), or Reznick *et al.* (1990). The data from Reznick *et al.* (1990) represent the Aripo Introduction Experiment 11 years after the introduction. The data from Reznick and Bryga (1987) represent the El Cedro Introduction Experiment four years after the introduction. The data from Reznick (1982) represent two high- and two low-predation localities. All data are derived from the second generation of laboratory offspring. NS = not significant. [b] Difference in mean values among experiments are attributable to differences in food availability. Reznick (1982) had the highest levels, Reznick *et al.* (1990) was intermediate, and Reznick and Bryga (1987) had the lowest levels. [c] $p < 0.01$. [d] $p < 0.05$. [e] $0.05 < p < 0.10$. [f] Fish were only kept until they produced two litters of young in Reznick and Bryga (1987). [g] Values for reproductive effort in Reznick and Bryga (1987) represent a single estimate made at the end of the experiment; those for the other two studies represent the sum of four consecutive estimates. See Reznick (1982) for details on the latter analysis.

served as controls to evaluate these predictions, which were tested with our laboratory life-history assay. One such experiment was initiated by John Endler in 1976 on a tributary to the Aripo River (Endler 1980); the response to selection was evaluated after 11 years (Reznick *et al.* 1990). A second was initiated by one of us (DR) on a tributary to the El Cedro River in 1981; the response to selection was evaluated after four and 7.5 years (Reznick and Bryga 1987; Reznick *et al.* 1990, 1997).

Results

In both experiments, the life histories of the introduced guppies evolved as predicted (Table 6.1). In the El Cedro River, after four years experimental males were significantly older and larger at maturity than the controls, but there were no other differences in life histories. After seven years, females from the introduction site also displayed delayed maturity. In the Aripo River, after 11 years we observed significant increases in the age and size at maturity in males and females, plus a reduction in the number of offspring in the first litter, an increase in offspring size, and a decrease in reproductive effort early in life. The males showed more change in age and size at maturity than did the females. All of these differences are consistent with the theoretical predictions and with the comparative studies. They suggest that the differences among populations in life-history traits represent an adaptation to the prevailing mortality rate within that population and that evolution can be fast.

Intensity of natural selection on different traits

Next, how natural selection worked in this system was assessed. The demography of natural populations of guppies was characterized (Box 6.1), and these results combined with laboratory data in a selection gradient analysis (Lande 1979). This type of analysis is traditionally applied to an episode of selection, such as a comparison of individuals that survived for an interval of time versus those that did not. Instead, we applied it to an episode of evolution by comparing guppies from the control and introduction sites (see Box 6.2 for details).

We confined our analysis to the age and size at maturity in males and females because this response was common to both introduction experiments. We found generally strong direct selection on the age at maturity (Table 6.2). Size at maturity tended to change because it was genetically correlated with age. We also found that males evolved more rapidly than females, largely because they had more genetic variation upon which natural selection could act. This result seemed strange, since one might assume that the genetic basis for age and size at maturity would be the same in males and females. It suggests some genes have a major effect on male age and size at maturity on the Y chromosome, providing males with a pool of genetic variation that is not shared by females. Such Y-linked variation has been reported for other species in the guppy family (e.g., Kallman 1989). Theory predicts that the rate at which a trait evolves is a function of the amount of genetic variation present in the population (Falconer 1960). This difference between males and females in both rate and genetic variation is consistent with this prediction.

Box 6.2 Selection gradient analysis

This analysis is a modified version of that originally presented by Lande (1979). It requires three types of information [see Reznick *et al.* (1997) for additional details]:

- The response to selection. In this case, the response is estimated by the genetic differences between guppies from the control and experimental populations, as estimated from second-generation laboratory-reared descendants of wild-caught fish (e.g., the Control and Introduction values in Table 6.1).
- The genetic variance–covariance matrix. Laboratory-reared fish were crossed in a breeding design that allowed us to estimate the percentage of variation in life-history traits among individuals within a population that could be attributed to genetic, as opposed to environmental, causes. We also estimated the degree to which different life-history traits shared a common genetic basis (genetic correlations). For example, a very high correlation occurred between the age and size at maturity, indicating that the two traits are largely controlled by the same genes.
- Estimation of generation time. Evolution is often scaled as change per generation, rather than as change per unit time. Generation time was estimated with a formal life-table analysis, derived from the mark–recapture studies on natural populations. More details on this approach are given in Box 6.1.

Responses to selection are estimated initially as the difference in the mean values of life-history traits in the control and experimental populations. For example, if the average ages at maturity for guppies from the control and introduction populations are 70 and 80 days, respectively, the response is 10 days. This response is converted to standard deviation units and divided by the number of generations between the beginning of the experiment and the time when fish were collected for the laboratory study. This process converts the response to an estimate of the rate of change in life-history traits in units of standard deviation per generation (R_{std} in Table 6.2). The direct effect of selection on a given trait, or selection gradient, is estimated as

$$B = G^{-1}R\,, \tag{a}$$

where B is the selection gradient (or the change in relative fitness as a function of the change in a given trait as other traits are held constant), G^{-1} is the inverse of the genetic variance–covariance matrix, and R is the response to selection. The coefficients of selection, or the covariance between a trait and fitness, can then be estimated as

$$S = V_P B\,, \tag{b}$$

where V_P is the phenotypic variance–covariance matrix. The coefficients of selection estimate the total influence of selection on a trait, which includes both the direct effects of selection and the indirect effects, or the degree to which a trait evolves because it is correlated genetically with some other trait that is influenced by selection.

Table 6.2 Summary of the selection gradient analysis from the Aripo River introduction experiment (R, R_{std}, B, and S are defined in Box 6.2). The results for the El Cedro introduction experiment are qualitatively similar.

		R	R_{std}	No. of generations	B	S
Males	Age	9.6 days	0.062	18	0.193[a]	0.210[a]
	Mass	8.6 mg	0.038	18	–0.127[a]	0.023[a]
Females	Age	7.8 days	0.031	18	0.290[b]	0.375[b]
	Mass	26.8 mg	0.035	18	0.013[b]	0.191[b]

[a] These are coefficients derived from a bivariate analysis. It was not possible to evaluate their significance. The corresponding coefficients from univariate analyses are all significantly greater than zero.

[b] These coefficients are significantly greater than zero, based upon a parametric bootstrap.

We can show (Box 6.3) that, the rates of evolution that cause these subtle changes in life-history traits are actually 10 000 to 10 000 000 times faster than rates of evolution observed in the fossil record. Similarly, rapid rates of evolution have been seen in most other contemporary studies of adaptation (Box 6.3).

6.4 Limits to Adaptation

If organisms have the capacity for such rapid evolution, why do they go extinct? This issue is addressed from a theoretical perspective in Parts C and D. For example, Holt and Gomulkiewicz (Chapter 13) evaluate the balance between the rate of population decline caused by a change in the environment and the rate of evolutionary response to that change. They predict a critical population size below which extinction becomes very likely because of demographic stochasticity and conclude that "... only mildly affected populations at high natural densities can reasonably be expected to be rescued by evolution in novel environments." Empirical studies bring life to this theory.

The complete history of introduction experiments in Trinidad includes five introductions of guppies from high- to low-predation environments and two from low- to high-predation environments. All five high-to-low introductions established thriving populations of guppies on the first try. Both low-to-high introductions failed. Consideration of the characteristics of the guppies and environment in each type of introduction indicates why this happened. Guppies from high-predation environments mature at an early age and have more offspring early in life. When they are introduced into a low-predation environment, they experience a reduction in adult mortality rate. This combination of life-history traits and reduced mortality rates should yield rapid population growth. Guppies from low-predation environments mature at a later age and produce fewer young early in life. When they are introduced into a high-predation environment, they experience an increase in adult mortality rate. In the latter case, it seems much more likely that the introduced population will dip below the critical population size before they can be rescued by evolution. We illustrate this point in two ways. The

Box 6.3 Rates of evolution

Guppy life histories can evolve quickly, but how fast are they? To answer this question requires the rate of evolution to be quantified, and this rate to be compared with similar rates from other types of studies. One convenient rate metric is the Darwin (Haldane 1949a). The Darwin equals the difference between the log-transformed value of a trait at the beginning and end of a time interval divided by the duration of the time interval, multiplied by 10^6. For example, if the control population matures in 70 days while the experimental population matures in 80 days, then the difference is 10 days. Log-transformation results in an estimate of proportional rates of change. This means that changes from 7 to 8 days or from 700 to 800 days yield the same rate since both represent the same proportional change. The Darwin can be applied to any types of data that provide a description of the organism at the beginning and end of a time interval and an estimate of the interval's duration. It is thus possible to compare rates of evolution from data as diverse as the fossil record and artificial selection.

Gingerich (1983) compiled summaries of rates of evolution that included three types of fossil data and artificial selection. The fossil data revealed average rates of change from less than 0.1 Darwins to 3 Darwins. Artificial selection yields rates that average 60 000 Darwins, or ten thousand to ten million times greater than those of the fossil record. Most people think of the fossil record as our best indicator of how quickly organisms can evolve, so this difference between artificial selection and the fossil record seems extraordinary. Artificial selection is generally discounted as an unrealistic representation of an organism's capacity for evolutionary change because investigators can make the coefficient of selection arbitrarily large. Stearns (1992) amended Gingerich's compilation with rate estimates from natural populations, including the guppy introduction experiments. The average rate of evolution in guppy body size and life-history traits was 12 000 Darwins. Thus, the subtle changes in our experiments yield rates of evolution that are of the same order of magnitude as those of artificial selection and ten thousand to ten million times greater than those inferred from the fossil record.

The likely source of the remarkable differences between fossil and contemporary studies is statistical bias. The fossil record averages patterns of change over long intervals of time. Such intervals will include periods when organisms do not change and periods when the direction of change is reversed. Estimating rates of change from fossils thus compares the beginning and end of long intervals, with no record of the irregularities between those endpoints, and yields a single average rate for the entire period. Nevertheless, it is the source of our concept of the rates and patterns of evolutionary change.

How representative are our results for guppies? While the number of studies that quantify rates of evolution during episodes of directional selection are few, they imply that our results may not be unusual. For example, the studies of Galapagos finches by Peter and Rosemary Grant and their colleagues recorded even higher rates of change in body size and bill morphology in response to the climatic changes associated with El Niño events (e.g., Gibbs and Grant 1987). Scott Carroll studied the colonization of newly introduced host plants by soapberry bugs and was able

continued

Box 6.3 *continued*

to infer similar rates of change for beak length and a diversity of other features of morphology (e.g., Carroll and Boyd 1992). Similar observations for butterflies are reviewed in Chapter 15, Section 15.2. Finally, there are a number of other cases of rapid evolution, as inferred from historical records, including industrial melanism in many species of moths (Berry 1990), insecticide resistance in a diversity of insects (Mallet 1989), or heavy-metal tolerance in plants (e.g., Antonovics and Bradshaw 1970). The available evidence documents the capacity for rapid evolution in a large number of organisms for a diversity of traits. These observations suggest that the capacity for rapid evolution, at least in the short term, may be quite common.

first is a modeling of population growth, based upon our knowledge of the demographics of natural populations of guppies. The second is a modeling of evolution that combines our demographic model with a simple genetic model of life-history evolution.

Modeling population dynamics

Our simulation model (Box 6.4), was based on a series of mark–recapture experiments in natural populations (Box 6.2). The simulation includes the probability of an individual of a given size surviving, its expected amount of growth, and its expected production of offspring during a given time interval. These "vital rates" are then used to parametrize models of population growth, and to evaluate the consequences of changes in the nature of the environment. A series of simulations were run to characterize the population growth of guppies from high- and low-predation environments in their own habitats. Think of these as experimental controls. We then simulated the consequences of moving guppies from a high-predation environment to a low-predation environment, and vice versa (Box 6.4).

These simulations (Table 6.3) predict that guppies tend to persist in their native environment (as they well should). All of the low-predation populations in low-predation environments persisted for all three years of simulation. On average, their population sizes more than doubled. Out of 20 high-predation populations in high-predation environments, 16 persisted for three years, but these populations tended to decline in size. We assume that if our simulations perfectly characterized population growth there would be persistence and stable population sizes in all of our simulations. These "controls" do not perfectly match these expectations, which suggests that the underlying assumptions of the simulations are also not perfect in some way. Possible sources of imperfection might include temporal or spatial variation in the properties of the environment. For example, our mark–recapture studies in high-predation environments were all executed in pools that we knew contained predators, but some pools lack predators during some portions of the year. It is possible that a "metapopulation" model of a high-predation site that included a mixture of pools with and without predators would result in more stable

Box 6.4 Individual-based models of guppy population biology and evolution

Population dynamics. An individual-based simulation model was used to predict the dynamics of populations of guppies if individuals from one type of predator locality were introduced to the other type of locality. Full details of the model are provided in Rodd and Reznick (1997). Each run of the simulation was started with 150 individuals (a population of moderate size for guppies in a single pool). After each 12-day interval, whether an individual lived or died, how much it grew, and whether or not it reproduced were determined. An individual's fate was based on size- and sex-specific data collected from several high- and low-predation populations. Size- and sex-specific mortality and growth rates, as well as size-specific fecundity and sex-specific sizes at maturity, were all determined in data collected from the field using mark–recapture studies (Box 6.1). The initial composition of the population was also based on data for natural populations. Interbrood interval and offspring size at birth were measured on wild stocks held in the laboratory. A 12 day interval was chosen because that was the duration of the mark–recapture studies. After each iteration, animals that died were removed from the population and ones that were born were added. Further iterations were carried out until either the population size declined to zero or to a maximum of 90 iterations (three years). To predict the response of guppies moved from one locality to another, we used life-history data (fecundity, offspring size, size at maturity, interbrood interval) of guppies from the original predator locality and mortality and growth rate data for guppies from the new predator locality.

The model assumes that growth rates and mortality rates are a property of the environment, so both variables were based on data from the new predator locality. The schedule of development and reproduction was assumed to be a property of the genotype, so these variables were held constant as a property of the locality of origin on the guppies, rather than as a property of their environment. These are simplifying assumptions, since we know that there is an interaction between life-history variables and environmental variables. For example, we have shown in laboratory studies that when guppies receive less food, they grow more slowly, mature at a later age, and have fewer young (e.g., Reznick and Bryga 1996). We also know from mark–recapture studies that guppies collected from the low-predation communities in Trinidad tend to grow more slowly than those from high-predation communities and that this difference is, in part, because they have lower levels of resource availability. The simulation described here thus simplifies nature and exaggerates the consequences of these introductions, because moving a guppy from a high- to a low-predation community tends to increase the age at maturity and to reduce fecundity, while moving a guppy from a low- to a high-predation community tends to decrease the age at maturity and to increase fecundity. A simple version of the simulation is presented here for the sake of illustration.

Genetic dynamics. The original simulation model was modified to incorporate the possibility of a genetic response to a new environment. Genetic differences were incorporated by assuming that the difference in maturation times between high- and low-predation guppies was controlled by one or more additive gene loci (the

continued

Box 6.4 *continued*

specific number was varied between simulations). Other minor modifications included survival being determined daily, and the initial introduction of guppies being staggered over a period of 24 days. This last modification was designed to avoid any dynamic artifacts caused by the pulsed introduction of a single cohort on a single day. A total of 200 fish were introduced, but the effects of daily mortality meant that by the end of the introduction period fewer than 200 adult fish were actually present in the population. Receptive females chose their mate at random from all the adult males present, and the genotype of each offspring was determined by Mendel's rules, assuming that all of the loci were unlinked.

Table 6.3 Results of simulated introductions. 20 populations were initiated, each with 150 individuals, for all combinations except the introduction of high-predation guppies into a low-predation environment. See Box 6.4 for details on the simulations. Recorded here is the probability of extinction one, two, and three years after the initiation of the population, and the mean population size after three years. Again, results are recorded differently for the high-to-low combination because all nine populations exceeded 1000 individuals within 1 year.

From	to	1 year	2 years	3 years	No. of survivors
High	High	0/20	1/20	4/40	97
High	Low	0/9	[a]	[a]	>1000
Low	Low	0/20	0/20	0/20	320
Low	High	0/20	4/20	15/20	11

[a] Simulation discontinued.

populations, with some pools producing temporary surpluses and some temporary deficits of guppies.

When we simulate the introduction of guppies from a high- to a low-predation environment, the populations invariably explode. This simulation was run only nine times because every time the population exceeded 1000 individuals in less than one year the simulation had to be discontinued. When low-predation guppies are introduced to high-predation environments, 75% of the populations became extinct within three years. The remaining populations were reduced, on average, to only 11 individuals, which means that they are also destined for extinction in the near future. The results of these simulations are thus quite consistent with the results of our natural introductions. Furthermore, they tell us something important about the circumstances under which guppies are capable of rapid evolution. The introduction of guppies from a high- to a low-predation environment likely results in at least a temporary increase in population growth rate and population size, making it more likely that the populations would persist for the time required to adapt to the new circumstances. The simulations also suggest that the way they

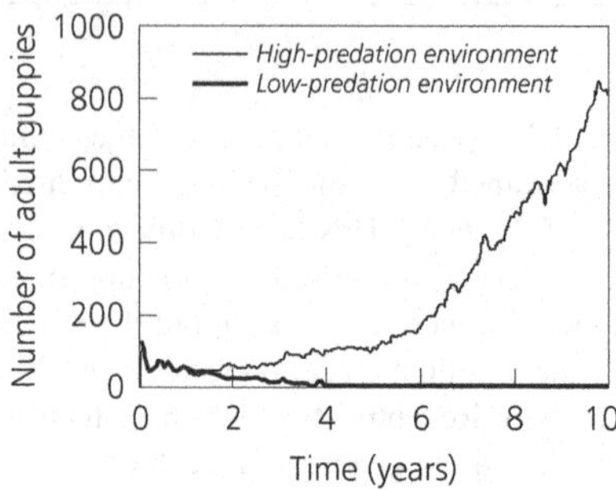

Figure 6.3 Simulation results for adult population size in a high-predation environment over a 10-year period. The initial introduction was of 200 adult guppies (either all from a high-predation environment or all from a low-predation environment) spread over 24 days.

adapt may not be entirely attributable to a change in mortality rate, as we originally predicted; guppy life histories may also have evolved in response to the high population densities that they experienced in their new environment. Furthermore, these simulations suggest that guppies are likely to fall below the critical population size when they are moved from a low- to a high-predation environment, making it far less likely that they will be able to adapt to their new circumstances before extinction.

Modeling genetic dynamics

We investigated the balance between extinction and adaptation by simulating the joint introduction of high- and low-predation guppies into a high-predation environment. These simulations incorporated a model of the genetic differences in age at maturity between the two types of guppy (Box 6.4); age at maturity was controlled by either one or several loci. We implicitly assume that the genetic differences among high- and low-predation guppies are not fixed, but are rather a function of differences in allele frequencies, so that not all low-to-high introductions begin with the same genetic composition.

Our genetic simulations showed that a population of guppies transferred from a low-predation site to a high-predation site was ultimately doomed to extinction. However, even for this relatively short-lived fish, the process of extinction often took several years. Indeed, after a year the difference between low-predation and high-predation populations introduced to a high-predation environment could appear quite minor, even though one population was destined for extinction and the other for success (Figure 6.3).

In this first set of simulations, we assumed that the difference in the maturation time of guppies from high- and low-predation sites is controlled by a single locus with two alternative alleles. As expected, the probability of a population persisting increased with the percentage of "high-predation" alleles included in the initial

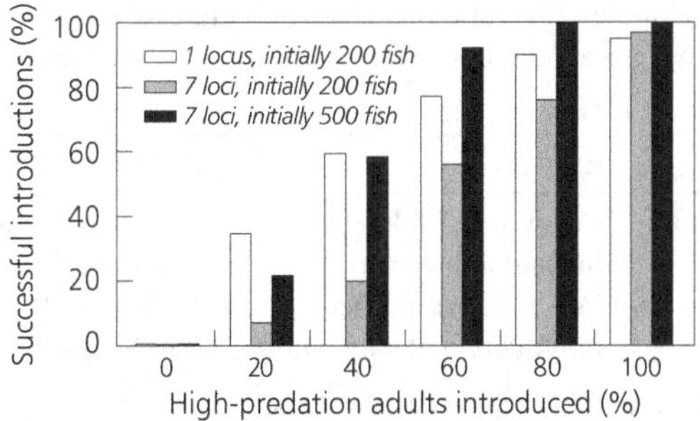

Figure 6.4 The success rate of mixed introductions into a high-predation environment. The results reflect 200 simulations of each mix of fish of high- and low-predation origin and success was defined as persistence for 10 years.

introduction. However, this was not a simple linear relationship. We found that the presence of just a few "high-predation" alleles among those introduced significantly improved the chances that the population would survive (Figure 6.4). For example, increasing the percentage of "high-predation" alleles from 0% to 20% increased the chance of the population surviving for 10 years from 0% to 35%. This suggests a rapid adaptation occurring on a time scale fast enough to substantially alter the extinction risk.

To test this conclusion, we simulated a genetic system that leads to slower adaptation. Our prediction was that this would require a higher percentage of fish from a high-predation environment in the initial population to produce a similar lowering of extinction risk. We did this using a seven-locus system. By shifting the genetic determination of maturation time from a single major gene to seven genes, each of small effect, natural selection is much weaker per gene. This slows the response to selection and increases the influence of genetic drift (Figure 6.4). Unlike the one-locus case, increasing the percentage of "high-predation" alleles from 0% to 20% in the seven-locus system only slightly reduced the extinction risk, since population success only increased from 0% to 7%. Thus, the slower time scale of adaptation of the seven-locus system dampened the influence of high-predation alleles.

As a final test, we increased the number of guppies initially introduced. Our expectation was that the success of introductions containing a small proportion of high-predation fish would be substantially improved. This was the pattern seen (Figure 6.4). The success of introductions of only low-predation guppies remained unchanged, but the success of introductions with 20% or 40% high-predation guppies more than doubled. In conclusion, the results of our genetic simulations largely parallel those of the population simulations. They also demonstrate that the size and genetic composition of the population and the genetic basis of fitness-related traits all influence the probability of extinction.

Table 6.4 Summary of the different circumstances, with examples from the literature, in which investigators have seen rapid evolution. The different categories are defined in the text.

Scenario	Examples	References
Parapatric colonization	Heavy-metal tolerance	Antonovics and Bradshaw (1970)
	Industrial melanism	Berry (1990)
	Insecticide resistance	Mallet (1989)
	Introduction of exotic host	Carroll and Boyd (1992)
Allopatric colonization	Introduction experiments:	
	Guppies	This chapter
	Anolis lizards	Losos *et al.* (1997)
Allopatric environmental change	Climate change associated to El Niño events:	
	Galapagos finches	Gibbs and Grant (1987)

6.5 Conditions that Favor Rapid Evolution

What are the circumstances under which organisms appear to have rapidly adapted to a change in their environment? We summarize some of the most famous examples of rapid evolution under three different scenarios that characterize the underlying population processes (Table 6.4). Our criteria for inclusion in this table are that we know the history of the introductions and that the evolved response has been shown to have a genetic basis. The one exception with regard to genetics is the *Anolis* introduction experiments.

- Our first category, parapatric colonization, represents the largest number of examples of rapid evolution. Here we envision a novel environment that appears within the pre-existing range of a species and that is initially inhospitable. One such example is found in mine tailings, which are a new soil type that is toxic to all plant species because of the high metal content. Another is the advent of the use of insecticides, which increase mortality rates in the islands of habitat in which they are applied. A third is industrial melanism, in which some aspect of industrialization causes an environmental change centered on industrial regions. A fourth is the introduction of an exotic species of plant that is a potential host. In all cases, this new patch of environment has a reduced flora and fauna, creating an opportunity for colonization by surrounding populations.
- Allopatric colonization refers to the introduction of a species into a new locality from which it had been excluded previously by some form of geographic barrier. This description applies well to our introduction experiments and to the introduction of *Anolis* lizards to Caribbean islands (Losos *et al.* 1997).
- Allopatric environmental change refers to an isolated population that experiences a global change in its environment. The work on Galapagos finches by the Grants and their colleagues (e.g., Gibbs and Grant 1987) represents the only well-studied example of evolution under these circumstances that we are aware of.

These three categories differ in how well they represent the circumstances of species in danger of extinction. In the case of parapatric colonizations, successful adaptation can be thought of as a trial-and-error process that involves sifting through the potentially large number of colonists and repeated colonization attempts, which either succeed or fail. Our records of rapid evolution represent the successes among an unknown number of failures. There always remain viable source populations that are presumably not endangered. These examples do not impress us as good models for an endangered species, but may serve well as a model for potential invaders. The other two types of category represent models of an organism in a natural population, often a population that has been fragmented into small isolates, that is exposed to some change in its environment and is threatened with local extinction. In both cases, we have examples of organisms that have persisted and evolved in the face of environmental change, although in the case of guppies we have argued, as did Gomulkiewicz and Holt (1995), that such persistence is most likely in specific circumstances (i.e., the population increase associated with the introduction of high-predation guppies into a low-predation environment, but not vice versa). The important point is that, if we want to consider cases most relevant to evolutionary conservation biology, then the number of such examples is restricted. Both theoretical and empirical studies argue that successful adaptation requires special circumstances. We must also realize that our examples represent cases of positive evidence. It is not so easy to evaluate the converse, or studies in which adaptation failed and extinction resulted, since there is a consistent bias against observing these in the first place and in the reporting of negative evidence in the literature.

6.6 Concluding Comments

Experimental and observational studies demonstrate that organisms are capable of very rapid evolution by natural selection. Observable rates of adaptive evolution are four to seven orders of magnitude faster than that revealed by the fossil record. Such observations might create a false sense of security with regard to the capacity of species to persist in the face of a changing environment. In guppies, such rapid evolution appears most likely in circumstances that promote population growth and persistence. This is because they are associated with moving fish from an environment with high mortality rates to one with low mortality rates. This introduction likely results in rapid population growth, which is far more permissive to evolution by natural selection than an initial interval of population decline. Other literature reports of rapid evolution are also often associated with special circumstances that are unlikely to represent effectively circumstances encountered by populations under the threat of local extinction.

There are three consequences from a management perspective:

- Act pre-emptively, in such a way that the adaptive response of populations may develop before the environmental conditions become so adverse that population sizes are compromised too severely;
- Take life-history adaptations into account when considering reintroductions in restored habitats (adaptive evolution may contribute significantly to reducing the extinction risk of restored populations if the size of founding populations is sufficiently large);
- Seek genetic mechanisms that may enhance or hamper the sort of fast adaptive responses demonstrated in this chapter.

These agenda items should provide strong motivation for future work on identifying conservation actions that could ameliorate or remedy such genetic predicaments.

7

Genetic Variability and Life-history Evolution

Kimberly A. Hughes and Ryan Sawby

7.1 Introduction

The persistence of populations in the face of environmental change depends upon their ability to adapt to changing conditions. Since genetic variation (specifically additive genetic variation) is a prerequisite for adaptation, a critical concern for the conservation biologist is that threatened and endangered species should retain both genetic variation and adaptive potential. Loss of genetic variation in small populations can also have other deleterious consequences, such as inbreeding depression. One of the primary goals of the conservation geneticist is to understand how genetic variation can be maintained within small and/or captive populations.

To understand maintenance of genetic variation is also a major concern for evolutionary biologists. Adaptation both requires additive genetic variation and erodes it at a rapid rate (Fisher 1930). Richard Lewontin also described the apparent ubiquity of genetic variation, even for traits that correlate highly with fitness, as a "paradox" (Lewontin 1974, p. 189) that evolutionary genetics had yet to satisfactorily explain. Although since 1974 progress has been made in understanding the maintenance of genetic variation, many investigators still consider it the central problem of evolutionary genetics. In this chapter, we describe possible solutions to Lewontin's paradox, empirical evidence that relates to these solutions, and some of the conservation implications of these results.

7.2 Genetic Variation and Life Histories

The paradox of genetic variation is particularly relevant to traits known as *fitness components*. These are traits (such as developmental rate, mating success, and fecundity) that directly determine Darwinian fitness. In general they are quantitative traits that are affected by a few or many genes and by the environment. Formally, a fitness component is a trait for which, all else being equal, an increase in the trait leads to an increase in total fitness. Fitness itself is measured by the intrinsic population growth rate or by the net reproduction rate associated with the demographic parameters that characterize a particular genotype in a population (Charlesworth 1987, 1994; Charlesworth and Hughes 2000). Juvenile survival, development rate, fecundity, mating success, and longevity are all examples of fitness components. Such traits are sometimes called *life-history traits*, because they relate to the timing and success of development, reproduction, and senescence. In contrast, morphological traits are not fitness components because they do not have a predictable monotonic relationship with fitness.

Box 7.1 Parameters for genetic and environmental influences on quantitative traits

Genetic and environmental influences on quantitative traits are characterized by several measures of variation. The *heritability* of a trait is the proportion of all phenotypic variation (V_P) accounted for by genetic variation. The *narrow-sense heritability* h^2 is the proportion of phenotypic variation accounted for by a particular kind of genetic variance, the *additive variance* V_A, where $h^2 = V_A/V_P$. V_A is the variance of *breeding values*. The breeding value is the sum of the average effects of all alleles that affect a trait (see Falconer and Mackay 1996 for a more complete description and definitions of these terms). V_A is the variance that is strictly heritable in the sense that it is responsible for the resemblance between parents and offspring.

Narrow-sense heritability h^2 does not include nonadditive components of genetic variance, such as dominance variance (V_D) and epistatic variance (V_I). V_D and V_I are variance terms that arise from dominance interactions between alleles at a locus and epistatic interactions between alleles at different loci, respectively. However, these nonadditive terms do contribute to the *broad-sense heritability* H^2, which is the proportion of variation accounted for by total genetic variance.

Since life-history traits are subject to very strong natural selection, it might be predicted that they should show less genetic variation than do other quantitative traits, like morphological characters. However, formal comparisons have shown that life-history traits have more, not less, genetic variation than morphological traits (Houle 1992; Hughes 1995a), even though they generally have lower heritabilities (Roff and Mousseau 1987). As a result of their importance for population persistence and the unexpected magnitude of variation, this chapter specifically focuses on the maintenance of variation in life-history traits.

7.3 Forces that Maintain Genetic Variation in Life-history Traits

Potentially, several phenomena could account for Lewontin's paradox, and details of how each can maintain variation are fairly well understood. However, the relative importance of the different forces is often debated. Several forces that potentially contribute to within-population variation are described below, along with empirical evidence that relates to the ubiquity of each mechanism. [Note: between-population migration as a force for maintaining variation is treated in several other chapters in this volume (see Chapter 12), so it is not described here.] Before proceeding, readers may wish to consult Box 7.1 for definitions of heritability, genetic variance, and additive and nonadditive components of variance.

Mutation

Mutation is the ultimate source of all genetic variation and the most ubiquitous of the phenomena that maintain variance for fitness and its components. Nevertheless, the debate continues among evolutionary geneticists as to how much of the variation within a population results from a balance between new mutational variation and the elimination of this (mostly deleterious) variation by natural selection (see Chapter 10 for an extended discussion of mutational variation and adaptation.)

The information required to answer this question (average effects and average dominance of new mutations, genomic mutation rate for fitness) is difficult to obtain and requires very large and laborious laboratory experiments. It will probably not be possible to establish this kind of information for threatened and endangered species; we therefore have to rely on data from model experimental organisms such as mice, *Drosophila, Daphnia*, and *Arabidopsis*.

In *Drosophila melanogaster*, life-history traits have been the object of mutation experiments and quantitative genetic studies for over 40 years. A recent analysis of this literature supports the view that mutation can account for between one-third and two-thirds of the genetic variation seen in *Drosophila* life-history traits (Figure 7.1). Mutation does not seem able to account for all the variation, however, and the excess seems to arise from some form of balancing selection such as genetic and environmental interaction, frequency-dependent selection, or sexual antagonism. The pattern of variation is not consistent with the maintenance of variation by heterozygote advantage (overdominance), which generates dominance, but not additive variation.

Other recent studies of mutational and standing genetic variation in *Drosophila* (Houle *et al.* 1997) and *Daphnia pulex* (Lynch *et al.* 1998) also support the idea that the majority of standing genetic variance for fitness components results from recurrent deleterious mutation. The *Daphnia* study is unique in using measures of standing variation taken from natural rather than laboratory-adapted populations.

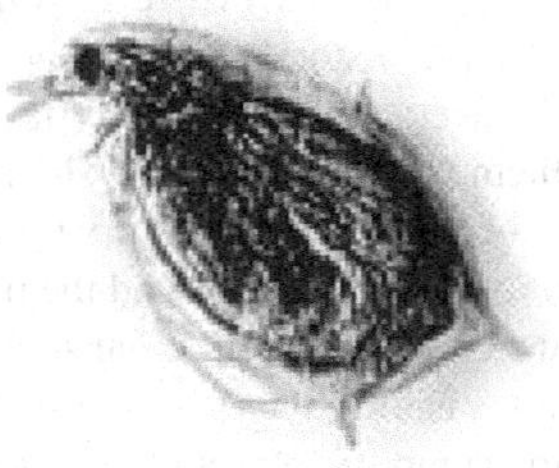

Daphnie
Daphnia pulex

A disturbing conclusion arises from the view that a large fraction of standing variation is caused by recurrent mutation. If most mutational variation is unconditionally deleterious, it will not be available for adaptive evolution. Models that incorporate unconditionally deleterious mutations predict "mutational meltdown" in small populations and suggest that populations must have an effective population size $N_e > 10\,000$ to retain potentially adaptive genetic variation over the long term (Lynch and Gabriel 1990; Lande 1995; Lynch *et al.* 1995b; Chapters 9 and 10). Long-term maintenance of large populations is obviously impossible in most captive breeding programs. *In situ* conservation or reclamation of natural habitat is the only method that will maintain such large populations over long time periods.

Frequency-dependent selection

Frequency-dependent selection occurs when the fitness of a genotype depends upon its frequency in the population (Crow and Kimura 1970, pp. 256–257). This type of selection is theoretically capable of maintaining large amounts

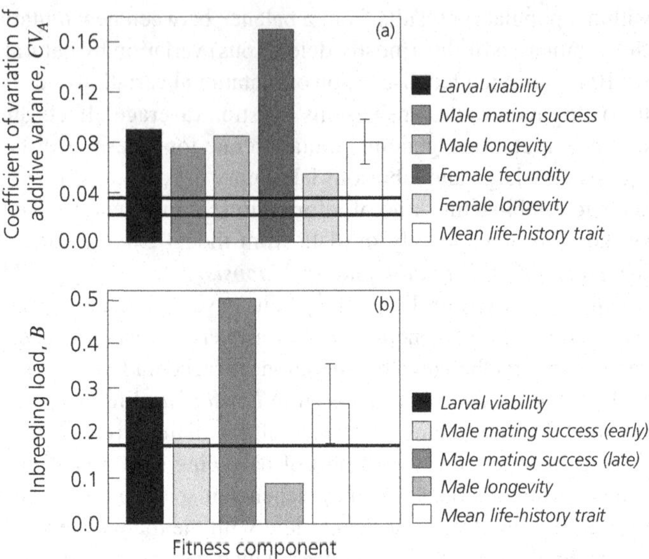

Figure 7.1 Comparison of observed values of genetic variation and inbreeding depression with values expected under mutation–selection balance. (a) Range of values for expected V_A (expressed as its coefficient of variation of additive variance, $CV_A = \sqrt{V_A}/\overline{x}$), where $\overline{x}$ is the trait mean) is shown by the horizontal lines. Expected V_A was calculated as $V_A = Uz^2\overline{hs}$, where U is the genomic mutation rate for mutations that affect a particular life-history trait, z is the mean effect of new mutations on the life-history trait, $\overline{h}$ is the average dominance of these mutations, and $\overline{s}$ is the mean selection coefficient against them (Charlesworth and Hughes 2000). The upper and lower limits were determined by using maximum and minimum estimates of z and $\overline{hs}$. Observed values of CV_A for individual life-history traits and the mean value over all these traits are shown. Since these values were taken from different studies and the statistical significance was calculated in different ways in different studies, error bars for individual traits are not shown; the standard error for the composite "mean life-history trait" was calculated from the values for individual traits. (b) The expected value for the inbreeding load B, $\frac{1}{2}Uz(\overline{h} - 2)$, is shown by the horizontal line. Observed values of B for individual life-history traits and the mean value over all these traits are shown (no estimate is available for female fecundity). The error bar represents the standard error of the individual life-history traits. *Source*: Modified from Charlesworth and Hughes (2000).

of genetic variation within populations, but very few examples in animals are well-documented. Perhaps the best-known case is that of the land snail *Cepea nemoralis*. Multiple alleles that affect color and banding pattern are maintained within populations, and the polymorphism seems to result partly from frequency-dependent predation by birds (Cain and Sheppard 1954). Some other cases of morphological and behavioral polymorphism may be maintained by frequency-dependent selection. Hughes *et al.* (1999) describe experiments in which male guppies with novel color patterns have higher reproductive success than do males with common color patterns. Frequency-dependent mating success may also help maintain a genetic polymorphism for breeding plumage and reproductive behavior in ruffs, *Philomachus pugnax* (Lank *et al.* 1995).

Dominance variance and heterozygote advantage

In principle, heterozygote advantage (overdominance) can maintain genetic variation for fitness and its components. If the heterozygote at a locus has higher fitness than any homozygote, then at equilibrium all genetic variation contributed by the locus is nonadditive variation. In fact, the genetic variance in fitness is equal to V_D, the dominance variance (Haldane 1949b; Schnell and Cockerham 1992). So heterozygote advantage produces no additive variance V_A at equilibrium, and does not contribute to h^2. Therefore, a testable prediction of this model is that large amounts of V_D should occur relative to V_A if heterozygote advantage maintains substantial genetic variance. Reviews of the *Drosophila* data showed that (with few exceptions) life-history traits typically have high V_A and very little V_D, consistent with the hypothesis that overdominance is generally not an important contributor to the maintenance of variation for fitness components (Charlesworth and Charlesworth 1987; Charlesworth and Hughes 2000).

Spatial and temporal variation in fitness

It seems intuitively obvious that spatial or temporal variation in environmental conditions should maintain genetic variation. However, mathematical models show that variable environments maintain variation only under fairly restrictive conditions (Hoekstra *et al.* 1985; Hedrick 1986). Maintenance of variance requires either strong selection on individual loci or a form of heterozygote advantage such that (averaged over all environments) the heterozygote has a higher fitness than any homozygote (Maynard Smith and Hoekstra 1980; Hoekstra *et al.* 1985). Hedrick (Hedrick *et al.* 1976; Hedrick 1986) concluded that spatial differences in the environment are more likely to maintain genetic variation than are temporal differences, although conditions for both cases are stringent. Habitat selection (Hoekstra *et al.* 1985; Hedrick 1990) and limited gene flow (Christiansen 1975) among different environments make the conditions somewhat less restrictive.

Theory concerning variable selection and the conditions under which variation is maintained has been continuously developed and refined (Gillespie and Turelli 1989; Gillespie 1991). Yet few empirical tests relate quantitative variation to environmental variation. Mackay (1980, 1981) observed increased V_A and increased h^2 for three morphological traits in *D. melanogaster* when she varied the concentration of environmental alcohol, both spatially and temporally. Hedrick (1986) criticized these results on methodological grounds, citing a lack of appropriate controls, and highlighting several results of the experiment that are inconsistent with the predictions of the model. Equivocal results were also reported in several experiments conducted by R.A. Riddle and co-workers. They were unable to document a consistent association between genetic variation and variation in culture conditions in two species of flour beetles (Dawson and Riddle 1983; Zirkle and Riddle 1983; Riddle *et al.* 1986). Therefore, despite the intuitive appeal of environmental variation as an important factor in maintaining genetic variation, its general importance is uncertain. (See Chapter 8 for further discussion of this topic.)

Sexual antagonism

If an allele increases fitness in one sex, but causes a decrease in fitness in the other, it is said to display sexual antagonism. Sexual antagonism can maintain genetic variation. Haldane (1962) and Livingstone (1992) described conditions for the simple case of one locus and two alleles. In general, sexual antagonism can lead to the maintenance of variance in two ways:

- By causing fitness of the heterozygote to be higher than either homozygote when averaged over sexes;
- By generating rather large fitness differences between the sexes.

This phenomenon has received little attention compared to other mechanisms that maintain variance. However, recent *Drosophila* studies indicate that populations are polymorphic for sexually antagonistic alleles and that new mutations can have different effects in males and females (Mackay *et al.* 1992a, 1994, 1995; Rice 1992, 1996; Mackay and Fry 1996). These studies have generated new interest in this phenomenon and its potential to maintain variation.

Genetic correlations

Remember that one of the sources of Lewontin's paradox is that genetic variation for fitness itself should be eroded by natural selection. High additive variances for fitness components can be consistent with low additive variance for total fitness, however. Negative genetic correlations (genetic "trade-offs") between different traits can lead to substantial genetic variance for fitness components, but little variation for fitness itself. Box 7.2 examines this concept in more detail.

The hypothesis that life-history traits are genetically correlated negatively has been tested in two ways:

- By direct measures of the correlation;
- By selection experiments.

Direct measures of genetic correlations using quantitative genetic breeding designs sometimes reveal negative genetic correlations, but often do not (Barton and Turelli 1989; Roff 1992). However, these direct estimates of genetic correlation have very large standard errors, and are extremely sensitive to changes in both allele frequencies and environmental conditions. In contrast, directional selection to increase late-life survival or reproduction typically results in decreased reproductive performance early in life (reviewed in Roff 1992; Charlesworth 1994). This pattern suggests that negative genetic correlations between life-history traits may be quite common. In particular, negative correlations between reproduction and survival, between early- and late-life reproduction, and between reproductive effort and survival have been demonstrated repeatedly in laboratory populations of

Box 7.2 Trade-offs, genetic correlations, and genetic variance for life-history traits

A genetic trade-off is defined as a negative genetic covariance (or correlation) between different traits. These trade-offs can be caused by individual alleles that have effects on more than one fitness component (pleiotropy) or by linkage disequilibrium among alleles that affect different traits. Linkage disequilibrium occurs when alleles at different loci are inherited nonindependently (i.e., particular alleles are inherited together more often than expected under Mendel's law of independent assortment, usually because they are located very close together on the same chromosome).

Negative genetic correlations between fitness components allow a population to have low V_A for fitness, even if some of the traits have high V_A. This concept is best explained algebraically. If fitness is controlled by many loci with small effects on different fitness traits, the additive genetic variance in total fitness is given approximately by the sum of the genetic variances for each trait, plus the additive genetic covariances between each pair of traits,

$$V_A = \sum_{ij} g_{ij} \frac{\partial w}{\partial z_i} \frac{\partial w}{\partial z_j},$$

where the summation is carried out over the different components of fitness. The i and j subscripts refer to specific traits (e.g., fecundity, longevity, and male mating success), z_i represents the different values that a specific trait can assume (e.g., one, two, or three offspring for the fecundity trait), and g_{ij} is the additive genetic covariance between traits (Charlesworth 1984). Since the additive genetic variance of a trait is the same as the additive covariance of the trait with itself, these variances enter the summation when $i = j$. The other terms in the equation describe the change in total fitness caused by a small change in a particular fitness component. These are the partial derivatives of fitness w with respect to its components and are denoted as $\partial w / \partial z_i$ and $\partial w / \partial z_j$. The summation in the formula is carried out over all the fitness components. The partial derivatives are positive by definition (see the definition of a fitness component given in Section 7.2) and the genetic variances also must be positive. There is thus only one way that V_A for total fitness can be small when additive variances for individual components (g_{ii}) are large: when at least some of the genetic covariances (g_{ij}) are negative (Dickerson 1955; Rose 1982). This leads to the main prediction of the model: negative genetic correlations should occur between some fitness components. However, when many different life-history traits contribute to fitness, only one correlation must be negative to generate additive variance for fitness components. So positive correlations between some pairs of traits are still consistent with the notion of trade-offs.

Drosophila, *Tribolium*, and *Daphnia* (reviewed in Roff 1992; Charlesworth 1994) and in natural populations of guppies (Reznick *et al.* 1990) and other fish (Law and Grey 1989).

Even if negative correlations are common, they do not necessarily contribute to the maintenance of genetic variation. Maintenance of variation requires a pattern of gene action known as "reversal of dominance", in which the advantageous allele must always be dominant. For example, if a diallelic locus has one allele A_e that is beneficial with respect to early fecundity, and an alternate allele A_l that is beneficial with respect to late fecundity, A_e must be dominant in its effects on early fecundity, while A_l must be dominant in its effects on late fecundity. There is no direct evidence that reversal of dominance is a common feature of loci that affect life-history traits.

In summary, most recent reviews (Roff 1992; Partridge and Barton 1993; Charlesworth 1994; Roff 2002) agree that genetic trade-offs between fitness components are well established, at least in some cases. However, the importance of this mechanism in maintaining genetic variation is uncertain.

7.4 How Much Variation is There?

In this section, we describe methods that measure quantitative variation, and how these measures may be adapted to studies of threatened and endangered species. We also summarize empirical studies of the effects of restricted population size on quantitative variation.

The short-term response to selection is directly proportional to the narrow-sense heritability (Falconer and Mackay 1996 and Chapter 5). However, the long-term response to selection is probably determined more by the additive variance than by the heritability of a trait (Houle 1992; Lande 1995). Conservation biologists are therefore particularly interested in the maintenance of additive variance in small and/or isolated populations.

Measuring variation in quantitative traits

Methods to measure genetic and environmental influences (see Box 7.1) and other variance components are described in Box 7.3. Methods with various levels of sophistication are available to estimate quantitative genetic parameters (Box 7.3). Unfortunately, these methods are much more practical in captive or laboratory populations than in natural populations, because:

- Relatedness is difficult to estimate in natural populations;
- The methods rely on assumptions such as independence of families and environments, and constancy of environments over time.

The first difficulty has recently become less of a hurdle. Molecular methods have become available to determine relatedness among individuals in natural populations (Altmann *et al.* 1996; Ritland 1996) and the pedigrees derived from natural populations can be analyzed using maximum-likelihood estimation procedures.

However, the second limitation is more difficult to overcome. Nonindependence of families and of environmental sources of similarity can be alleviated if captive populations of threatened and endangered species are used. However, sample sizes of captive populations are usually too small to produce meaningful estimates of quantitative genetic parameters. Also, if *in situ* preservation of natural populations is planned, we are interested in knowing how much quantitative

Box 7.3 Estimating heritabilities and variance components

An excellent introductory treatment of these methods is given in Falconer and Mackay (1996) and more advanced methods are described in Lynch and Walsh (1998). Here, we give a brief overview of some commonly used approaches. We focus on problems associated with the measurement of these parameters in natural populations or in small captive populations. We describe sources of bias and some common misuses of these techniques that are particularly relevant to conservation biology.

Quantitative geneticists typically use one of a few standard methods to estimate variance components and heritabilities in laboratory or agricultural populations. Unfortunately, these methods are difficult to apply to natural populations because they require large numbers of individuals of known relatedness. Even if this kind of information is available, critical assumptions are often violated. Even so, the estimates can be informative and useful if the constraints of the analysis are kept in mind. Below we outline some of the methods available for use in natural populations and point out potential sources of bias and inaccuracy.

A few long-term studies of natural populations have estimated heritabilities by using parent–offspring regression (reviewed in Lynch and Walsh 1998). This is a straightforward analysis in which the heritability is estimated by the regression coefficient that relates a trait in parents (one or the mean of both parents) to the trait mean in their offspring:

- $\beta = \frac{1}{2}h^2$, regression of offspring on one parent;
- $\beta = h^2$, regression of offspring on the mean of both parents.

where β is the linear regression coefficient. These equations are strictly true only if there is no epistatic variance and no environmental source of similarity between parents and offspring. In natural populations, environmental similarity between relatives is probably the rule rather than the exception. If moderate-to-large numbers of sibships are available, the covariance among sibs can be estimated by variance-partitioning methods, such as analysis of variance (Falconer and Mackay 1996) and maximum likelihood (Shaw 1987). The covariance among half sibs estimates $\frac{1}{4}V_A$ and the covariance among full sibs estimates $\frac{1}{2}V_A + \frac{1}{4}V_D + V_{Ec}$, where V_{Ec} is the variance through the sharing of a common environment among full sibs. V_{Ec} includes variance from maternal effects. Thus, if only full sibs are available, an upper limit to V_A can be measured. However, this estimate may be inflated substantially if dominant variance contributes, or if a common environment among the sibs contributes to similarity. If paternal half-sibships are available, a nested analysis of variance with full-sib families (same mother and father) nested within half-sib families (e.g., same father, two or more different mothers) gives estimates of both V_A and the combined effects of V_D and V_{Ec} (including maternal effects). This kind of analysis is valid when fathers do not contribute to the care of the offspring and when we can assume that females mated to the same male are no more similar than females chosen at random from the population. To the extent that these assumptions are violated, the estimates of heritability are biased.

variation exists in nature, rathert than in a zoo- or greenhouse-adapted population. Unfortunately, it is in natural settings that biases are most likely to occur:

- Offspring often experience an environment similar to that of their parents;
- Different mates of the same male may inhabit adjacent home ranges and may themselves be related;
- Polygamous males may contribute to the care of offspring by providing food or by territory and nest defense.

A hybrid approach can sometimes alleviate problems of nonindependence and environmental sources of family resemblance. In this method, one measures a trait in wild individuals, collects them, and mates them randomly under controlled laboratory conditions. The trait is then measured in offspring of the random matings and regressions of laboratory-reared offspring are used on wild parents to calculate a lower bound to the heritability in the field, h^2_{min}. The expected value of this regression coefficient is $\gamma^2 h^2_{\text{nat}}$, where γ^2 is the additive genetic correlation between the field and laboratory environments and h^2_{nat} is the heritability of the natural population (Riska *et al.* 1989; Lynch and Walsh 1998). Since γ^2 is less than or equal to one, h^2_{min} is an underestimate of h^2_{nat}, unless the genetic correlation across environments is perfect.

Another problem is that unequal sample sizes among families can make the calculation of variance components quite complicated (Searle 1971). Of course, unequal family sizes is the norm in studies of natural populations. However, if many different types of relatives are available, maximum-likelihood statistical methods allow all the relatives to be used in a single analysis and can accommodate unequal sample sizes among families.

Can we estimate quantitative variation from molecular variation?

Since the widespread use of quantitative genetic analyses in natural populations seems impractical, many biologists have sought a surrogate measure of genetic variation. Both protein- and DNA-based measures of variation have been used to make inferences about the ability of natural populations to evolve. Unfortunately, the evidence for a direct relationship between molecular variation and quantitative variation is weak (Karhu *et al.* 1996; Butlin and Tregenza 1998).

Although relationships between allozyme heterozygosity and fitness traits have often been suggested, experimental results are mixed. Some investigators have found that allozyme heterozygosity correlates positively with fitness traits, while others have failed to find an association (for recent reviews, see Houle 1989a; Pogson and Zouros 1994; Savolainen and Hedrick 1995). For example, bivalve mollusks consistently show allozyme heterosis. Curiously, the same organisms also consistently show less heterozygosity than expected under Hardy–Weinberg equilibrium (Zouros *et al.* 1988). However, aside from mollusks, organisms with large mobile populations and panmictic population structure generally do not show an association between allozyme heterozygosity and fitness (Houle 1989a; Savolainen and Hedrick 1995). Houle (1989a) concludes "there is little evidence for allozyme heterosis in any population which is not likely to be partially inbred, have very small N_e, or have recently undergone strong directional selection." Use of allozyme diversity measures to make conclusions regarding quantitative variation is therefore unwarranted, unless a positive association between fitness traits and molecular markers has been demonstrated previously for a particular population. This conclusion is quite disappointing, because allozyme studies are much easier and less expensive than quantitative genetic studies, and they can be conducted on a wider variety of organisms.

There is also little evidence for associations between quantitative variation and other forms of molecular variation. For example, Karhu *et al.* (1996) reported that populations of Scots pine (*Pinus sylvestris*) highly differentiated for an adaptive quantitative trait (bud-set date) were indistinguishable with respect to molecular markers [allozymes, restriction fragment length polymorphism (RFLP), random amplified polymorphic DNA (RAPD), and microsatellites]. Savolainen (1994) reviews allozyme and DNA data, transplant experiments, and common-garden experiments in coniferous trees. She concludes, "marker studies have so far not been a substitute for direct measurements of quantitative variation."

Although disappointing, the lack of congruence between molecular and quantitative variation should not be surprising. Marker alleles are generally assumed to be approximately neutral and to have frequencies near mutation–migration equilibrium, while alleles that affect quantitative traits are subject to selection and so their frequencies are near the selection–migration balance. Equilibrium frequencies for the two types of alleles are expected to differ under this scenario (Savolainen 1994).

General patterns of genetic variability

It is often stated that traits subject to strong directional selection should have less genetic variation than traits subject to weak directional or stabilizing selection. Yet one of the most important conclusions to emerge from recent studies is that, compared with morphological traits, fitness components have higher levels of genetic variance (Houle 1992; Figure 7.2). It is true that fitness components tend to have lower heritabilities (Mousseau and Roff 1987; Roff and Mousseau 1987) than other kinds of traits. Low heritabilities seem to result from high levels of nongenetic sources of variation (environmental variation, sampling error, and measurement error). The paradoxic pattern of higher genetic variance for traits subject to strong selection may result because more loci affect life-history traits than affect morphological traits (Houle 1992).

As described in Section 7.3 (subsection Genetic correlation), negative genetic correlations between different life-history traits are often observed in natural and laboratory populations. This suggests that genetic trade-offs may be quite common. What is not clear is whether these trade-offs play an important role in maintaining genetic variation. However, genetic correlations can affect the results of natural and artificial selection, and are therefore important to conservation biologists even if they do not substantially contribute to the maintenance of variation.

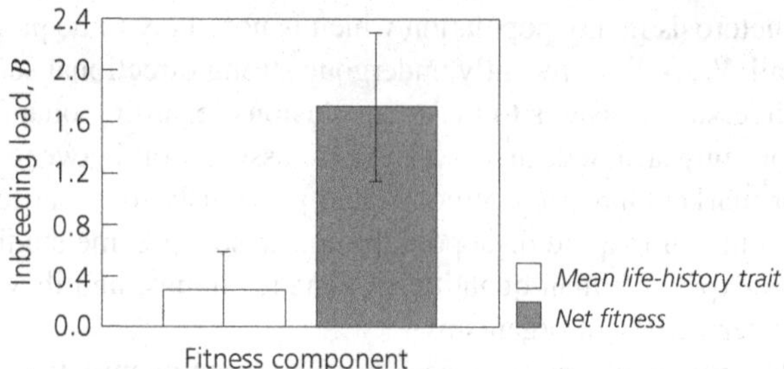

Figure 7.2 Comparison of the average inbreeding load for individual components of fitness in *D. melanogaster* (Hughes 1995b) with that for net fitness (Simmons and Crow 1977). The error bar for the "mean life-history trait" gives the standard error of inbreeding load calculations for the traits shown in Figure 7.1b. The estimate for total fitness comes from only two studies, so the error bars in this case represent the range. *Sources*: Hughes (1995b) and Simmons and Crow (1977).

In the context of conservation issues, we emphasize that not all life-history traits are created equal. Within species, different life-history traits seem to have different levels of genetic variation and qualitatively different genetic architectures (see Hughes 1997). Life-history traits may also vary greatly in their effects on fitness. The sensitivity of fitness and population viability to particular life-history traits depends upon birth and death rates in all age classes, on stochastic variation in the environment, on mechanisms of density regulation, and on trade-offs between different life-history traits (Charlesworth 1994). For threatened and endangered species, usually we will not have most of the information needed to calculate sensitivities analytically. In these cases, simulation models that investigate a range of demographic, environmental, and genetic conditions are probably the best tool (Miller and Lacy 1999). As a result of variation among populations in demographic regimes, ecological conditions, and the genetic architecture of life-history traits, we should expect fitness and population viability sensitivities to vary between species and between populations of the same species.

Effects of small population size and inbreeding

The expected effects on the quantitative variation of small population size have been reviewed elsewhere (see Chapters 5 and 10). In general, we expect inbreeding and genetic drift to decrease the additive genetic variance of quantitative traits. Here, we examine experimental studies that attempted to measure these effects.

Many experiments have been carried out to measure the effects of restricted population size on allele frequencies, heterozygosity, and/or polymorphism (reviewed in Frankham 1996). However, only a handful of studies have measured directly the effects of changing population size on quantitative variation. Fewer still have investigated population-size effects on genetic variation in life-history traits. In one of these rare cases, Lopez-Fanjul and Villaverde (1989) found that the response to selection on egg viability was several times higher in inbred lines of *D. melanogaster* (inbreeding coefficient $F = 0.25$) compared to lines that were not inbred. Nevertheless, the response to selection was only half the inbreeding depression experienced by these lines. Thus, the increase in heritability and increase in selection response were not enough to offset the deleterious effects of inbreeding.

A few more studies such as these have involved morphological variation. Somewhat surprisingly, Bryant *et al.* (1986) observed increases in additive variance for morphological traits after single-generation bottlenecks in populations of houseflies. Genetic drift can cause additive variance to increase if rare recessive alleles increase in frequency (Willis and Orr 1993). Still, if drift continues (rather than being restricted to a single bottleneck event), any increase in genetic variance is temporary. As alleles that affect quantitative traits drift to fixation, both additive and nonadditive variances should decline toward zero. Supporting this view, a study of the long-term effects of a restricted population size showed that all the measures of variation (V_A, h^2 for bristle number, and allozyme heterozygosity H) decreased with time in a recently captured population of *D. melanogaster* (Briscoe *et al.* 1992). In contrast to these experimental studies, a review of natural populations revealed little support for associations between population size and quantitative genetic variation (Houle 1989b).

7.5 Inbreeding Depression in Life-history Traits

Mathematical descriptions of inbreeding depression can be found elsewhere (Morton *et al.* 1956; Crow and Kimura 1970; Charlesworth and Charlesworth 1987). Discussions of the effects of inbreeding on demography and population persistence are presented in Chapters 3 and 9. In this section we briefly describe how inbreeding depression is measured and whether inbreeding depression can be reduced by selection in captive or natural populations.

Measuring inbreeding depression

The inbreeding load B is the most commonly used measure of inbreeding depression in the conservation literature. B is a measure of the rate of decrease in fitness as the inbreeding coefficient increases, assuming the loci that affect fitness have independent multiplicative effects. This is expressed as $Z = e^{-(A+BF)}$, where Z is a measure of survival or some other component of fitness, A represents genetic causes of death not associated with inbreeding, B represents genetic causes of death as a result of inbreeding, and F is the inbreeding coefficient (Morton *et al.* 1956). Usually, B is estimated as the coefficient of linear regression of a (log-transformed) fitness component on the inbreeding coefficient F. B is also known as the number of *lethal equivalents* associated with a particular population. Several other measures of inbreeding depression are used in evolutionary and conservation biology (reviewed in Charlesworth and Charlesworth 1987; Hedrick and Kalinowski 2000), including a recently introduced maximum-likelihood approach to measuring inbreeding depression from pedigrees (Kalinowski and Hedrick 1998).

General patterns of inbreeding depression

Substantial deleterious effects of inbreeding have been found in invertebrates and vertebrates (Charlesworth and Charlesworth 1987; Ralls *et al.* 1988; Lacy 1993). These effects are not limited to any particular life stage, but are manifest in most life-history traits that have been measured (Miller and Hedrick 1993; Hughes 1995b; Margulis 1998a, 1998b). A survey of inbreeding depression in captive mammals reported a median *B* of 3.1 for juvenile survival (Ralls *et al.* 1988). This is very likely to be an underestimate of the inbred load in mammals, since only a single fitness component was measured under very benign conditions. Total fitness probably has a much higher load than any single component, as for *D. melanogaster*, for which the mean *B* for male life-history traits is 1.4 (Hughes 1995b), but that for net fitness is nearly three times higher (3.9; Simmons and Crow 1977).

Few estimates are available for the effects of inbreeding in natural populations of animals. However, one recent study showed substantial inbreeding depression in song sparrows after a natural population bottleneck (Keller *et al.* 1994) and another demonstrated high inbreeding depression in a re-introduced *Peromyscus* population (Jimenez *et al.* 1994). Also, in a few cases introductions from other populations have seemingly increased the fitness of inbred populations (reviewed in Hedrick and Kalinowski 2000).

Population size and purging of inbreeding depression in animals

Under some conditions, inbreeding depression can be reduced by selection in inbred populations (Hedrick 1994; Fu *et al.* 1998; Fu 1999; Wang *et al.* 1999). Inbreeding increases the frequency of homozygotes, so the effects of recessive deleterious alleles are exposed to selection more often in inbred than in noninbred populations. Intentional inbreeding of a population, with the consequent selection for inbred individuals of high fitness, is known as "purging" of inbreeding depression.

Some biologists have argued that purging can be used as a tool to manage threatened and endangered species (Templeton and Read 1983, 1984). If the deleterious effects of inbreeding can be removed or substantially reduced, then captive populations can be maintained at a small size without the risk of inbreeding depression. However, the expected effects of selection on the reduction of inbreeding depression depend on characteristics of the genes that cause inbreeding depression. Purging is most effective when inbreeding depression is caused by a few highly deleterious recessive alleles, is less effective when inbreeding depression is caused by many mildly deleterious alleles, and is completely ineffective if inbreeding depression results from overdominant alleles (Charlesworth and Charlesworth 1987; Charlesworth *et al.* 1992; Hedrick *et al.* 1995). Partial dominance of deleterious alleles and strong synergism among deleterious alleles also increase the effectiveness of purging (Fu *et al.* 1998; Fu 1999). Even when purging effectively reduces inbreeding depression, it may often result in a decrease in population fitness and an increase in extinction risk, because mildly detrimental alleles can become fixed in the population (Hedrick *et al.* 1995; Lynch *et al.* 1995a; Wang *et al.* 1999).

In animals, the results of experimental studies have been mixed, with some workers reporting evidence of purging (Templeton and Read 1983, 1984; Latter *et al.* 1995; Ballou 1997) and some finding no evidence (Lynch 1977; Sharp 1984; Brewer *et al.* 1990; Lacy and Horner 1996). In a large experimental study, Lacy and Ballou (1998) described the effects on multiple life-history traits of ten generations of inbreeding in three subspecies of oldfield mice (*P. polionotus*). Each subspecies responded somewhat differently to inbreeding:

- One showed evidence of purging in four of seven life-history traits;
- One showed increased inbreeding depression as the inbreeding coefficient increased;
- Another showed no effect of increased inbreeding.

Increased inbreeding depression with inbreeding may seem surprising, but overdominant alleles contribute to this pattern, as does directional dominance if the relationship between fitness and the number of homozygous recessives is concave downward (a pattern of gene action called *synergistic epistasis*). Another explanation for this variability in the results of experimental tests of purging is that inbreeding depression itself has a variance. Depending upon the particular individuals chosen for an experiment, highly variable levels of inbreeding depression will result in their offspring, for purely stochastic reasons (Schultz and Willis 1995).

A case study of purging in an endangered mammal

The most widely discussed case of potential purging is that of a captive population of Speke's gazelle (*Gazella spekei*). All the individuals in the data set were descended from one wild-caught male and three wild-caught females, which were transferred to the St Louis Zoo in 1972. Beginning in 1980, a breeding program was established by choosing breeders that were moderately inbred, but healthy (an additional criterion was that breeders were chosen to equalize founder representation in the gene pool).

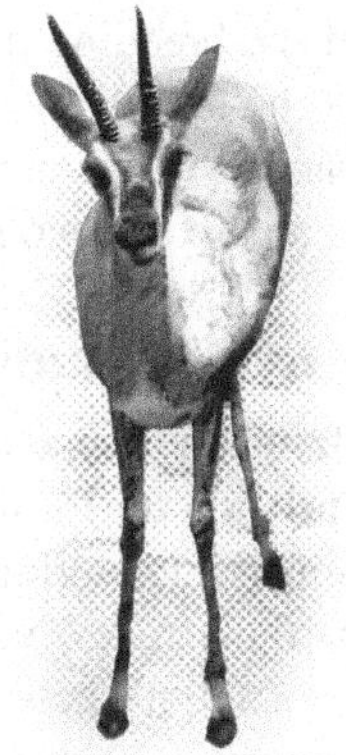
Speke's gazelle
Gazella spekei

After following this breeding program, Templeton and Read (1984) reported that juvenile survival increased and inbreeding depression decreased after one or more generations of inbreeding. Other workers suggested that the results potentially arose from statistical artifacts (Lacy 1997; Willis and Wiese 1997) or from the improvement in environmental conditions (Frankham 1995a). A recent reanalysis using a maximum-likelihood model (Kalinowski *et al.* 2000) confirms lowered inbreeding depression in the offspring of inbred parents. However, the analysis showed that the reduction in inbreeding depression occurred during the middle of the first generation of inbreeding, not after the first generation of inbreeding.

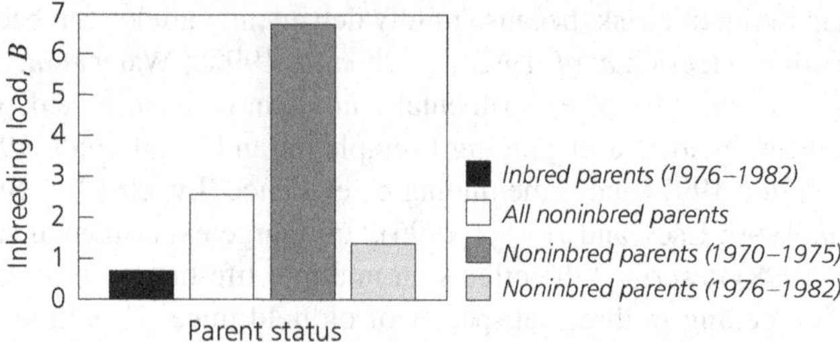

Figure 7.3 Estimates of inbreeding load B (measured as lethal equivalents) from the reanalysis of the Speke's gazelle purging experiment by Kalinowski *et al.* (2000). When all animals with inbred parents are compared to all animals with noninbred parents, those with inbred parents have much less inbreeding depression, as predicted by the purging hypothesis. However, all offspring of inbred parents were born after 1975. When B for offspring of noninbred parents is estimated separately for 1970–1975 and 1976–1982, most of the inbreeding load disappears during the later years. The authors attribute the decrease in load to improved animal husbandry during the later years, rather than to purging of inbreeding depression in the offspring of inbred parents. All the values shown were significantly different from zero. Confidence intervals were complex functions of the model parameters and are therefore not shown here. *Source*: Kalinowski *et al.* (2000).

This result was puzzling, because if purging caused the reduced inbreeding depression, its effects would be apparent only in the second and subsequent generations, not in the middle of the first generation. Further, a highly efficient model of selection was unable to account for the observed large improvement in inbred viability (Kalinowski 1999). After eliminating selection as a cause of the reduction in inbreeding depression, two hypotheses remained that were consistent with the data: improvement in zoo husbandry techniques during the early stages of the breeding program and epistatic interactions associated with recombination of the founder's genomes. Kalinowski *et al.* (2000) concluded that improved husbandry was a more parsimonious explanation of the results, but was unable to compare the two hypotheses rigorously (Figure 7.3).

7.6 Concluding Comments

Based on several recent analyses, it seems safe to conclude that much of the genetic variation for life-history traits that exists within populations results from unconditionally deleterious mutations, and that inbreeding depression is caused largely by partially recessive deleterious alleles. However, at least one study suggests that a substantial fraction (one-third to one-half) of the standing variation within populations may be maintained selectively. From a conservation perspective, we should be much more concerned about the loss of selectively maintained variation than about the loss of neutral or unconditionally deleterious variation. Realistically, however, to distinguish the three classes of variation is likely to be impossible for the foreseeable future. Furthermore, in small populations genetic drift and inbreeding will eliminate neutral, deleterious, and beneficial variants at nearly the same rate (Crow and Kimura 1970). It therefore seems likely that conservation geneticists will need to continue to focus on methods that preserve as much quantitative genetic variation as possible, even if this means the bad is preserved as well as the good.

Many other important problems face the conservation-oriented quantitative geneticist. A few of the most pressing are:

- Our inability to use molecular markers to measure quantitative variation or to assess local adaptation;
- The difficulty of preventing adaptation to captive conditions in captive-propagation programs;
- The problem of identifying those traits that have the greatest effects on fitness and on population viability;
- The theoretical predictions that indicate very large population sizes are necessary to prevent loss of adaptive variation and "mutational meltdown" over the long-term.

This list of problems may be much more intractable than the relatively straightforward task of preserving genetic variation. Thus far, these difficult topics have generated much discussion in the conservation community. However, very few practical solutions have been proposed (much less evaluated) for any of them.

8

Environmental Stress and Quantitative Genetic Variation

Alexandra G. Imasheva and Volker Loeschcke

8.1 Introduction

Exposure to environmental stress is a common feature in the life of most natural populations of plants and animals that live in the wild. Primarily, this is because natural populations inhabit environments that are heterogeneous in space and constantly change in time; some of these changes are bound to be hostile. Numerous factors impose stress on living organisms. Not only catastrophic events, such as earthquakes, floods, and fires, but also many less extreme environmental shifts experienced by organisms may be stressful for them. The nature of these stresses can be abiotic (e.g., climatic changes, reduction in food resources) or biotic (e.g., overpopulation, invasion of novel parasites or predators). In addition to environmental stressors that are part of their "natural" environment, in the twentieth century plant and animal populations faced human-related challenges that had previously never been encountered by them throughout their evolutionary history. Loss and deterioration of natural habitats because of land-clearing for agricultural and urbanization purposes, global warming, and environmental pollution are the main threatening ecological hazards created by human activity. It appears that organisms live in a world full of stress and the ability to cope with it is of crucial significance to their persistence.

The consequences of stress exposure are obviously relevant to conservation biology. By definition, stress has a detrimental effect on population fitness, that is the ability of a population to survive and reproduce. Stress often leads to a reduction in population size, which increases the risk of extinction, especially in populations that are already small and isolated or patchily distributed (see Chapter 4). Environmental stresses are also extremely hazardous for marginal populations (i.e., populations that live at the boundaries of the species range), as such populations are generally less abundant and frequently subject to adverse environmental factors.

The concept of stress was first introduced and developed by Hans Selye in relation to human physiology. Selye (1955) defined stress as a nonspecific physiological response of an individual to an adverse external impact. Later, the stress concept was generalized and extended from the individual to the population. Stress can be caused not only by external factors (environmental stress), but also by internal factors (genomic stress). Internal factors that result in genomic stress are, for instance, inbreeding and hybridization. Numerous definitions of environmental stress, some of which are listed in Forbes and Calow (1997), have been suggested by different authors. Hoffmann and Parsons (1991) define environmental stress as an "environmental factor causing a change in a biological system, which is potentially injurious." This definition includes both individual and population aspects of stress. Regarding stress in the evolutionary context, many authors emphasize that environmental stress impairs the fitness of a population (see, e.g., Sibly and Calow 1989).

A slightly confusing side to stress definitions is that, as noted by Hoffmann and Parsons (1991), they generally refer both to the adverse factor that affects organisms and to their response to this factor. These two aspects of stress are closely related in that the degree of stress can be evaluated only from the response of the individual or population subjected to it. Thus, stress includes the environmental and biological component as integral parts. To avoid confusion, some authors refer to the environmental component as "stressor".

Environmental stress plays an important role in ecological and evolutionary processes. It can act as a selective force that eliminates less fit individuals and promoting selection for a specific stress resistance (i.e., resistance to a particular stress factor). By drastically reducing population size, it can also deplete genetic variation in a population and cause inbreeding depression. These two mechanisms are evident and generally recognized. However, there is a third possibility: stress conditions may themselves generate variation by some mechanism (perhaps more than one) that starts to operate only in critical situations in which biological functions of the organism are seriously impaired.

The view that extreme environmental conditions may increase quantitative genetic variation has a long-standing history (Mather 1943; Belyaev and Borodin 1982; Parsons 1987; Hoffmann and Parsons 1991). Based more on general considerations than on experimental evidence, this concept is very appealing from the standpoint of evolution and adaptation. If environmental stress does enhance the level of genetic variation, then it provides more material for selection to act upon, which increases the rate of evolutionary change, and thereby promotes adaptation. Thus, environmental stress may act both as a selective force that eliminates less fit individuals and, simultaneously, as a source of variation that is a prerequisite for natural selection. Both of these processes could lead to faster adaptation, so that a population threatened by extinction because of dramatic deterioration of the environment may be able to escape this fate by adapting to the new environmental conditions.

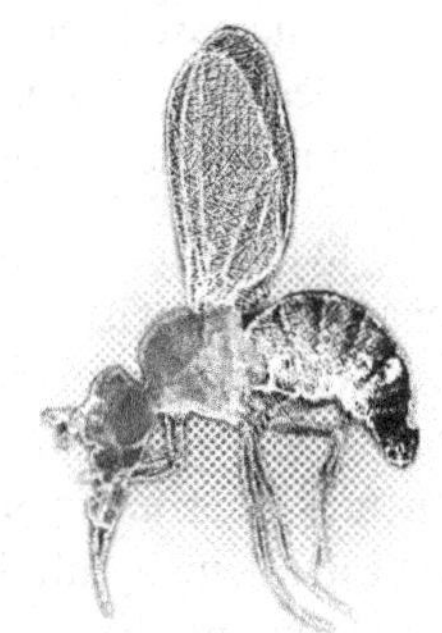

Fruit fly
Drosophila melanogaster

Until recently, the problem of the impact of environmental stress on quantitative traits was specifically addressed in very few studies (see, e.g., Parsons 1961; Westerman and Parsons 1973). Moreover, extreme environments were very often excluded from analyses of phenotypic plasticity on the grounds that they were "too drastic" and "untypical for the species". The 1990s, however, were witness to a renewal of interest in the relationship between stress and quantitative variation. This relationship was explored in numerous experimental works, most of which used *Drosophila* as a model organism. These efforts have shown that the actual situation is far more complicated than the analysis outlined above. In this chapter, we consider the effects of stressful environments on variation, on the basis of results mostly obtained for *Drosophila*.

8.2 Hypotheses and Predictions

Stress can enhance the level of variation in two ways:

- By producing new variation via mutation and recombination;
- By revealing the variation that already exists in the population, but that is not expressed and is thus hidden from natural selection.

Some evidence indicates that the first mechanism occurs in *Drosophila*, although its generality and evolutionary significance are unclear (Box 8.1). In what follows, we focus on the second possibility.

Hypotheses on the effect of stress on quantitative genetic variation

Several hypotheses that relate environmental stress and the level of genetic variation have been advanced. Hoffmann and Merilä (1999) provided a comprehensive review in which they listed eight different hypotheses, which shows that our understanding is not very clear. These authors classified these hypotheses according to their predictions:

- An increase in genetic variation under stress;
- A decrease in genetic variation under stress;
- An unpredictable outcome.

The hypotheses are not mutually exclusive: different patterns of stress-induced change may exist in different organisms. The view that stress reduces genetic variation is based on the assumption that differences between genotypes would not be implemented fully under unfavorable conditions because of limited resources, such as poor nutrition. Also, genetic differences under stress may be obscured (but not actually diminished) by an increase in the environmental component of variation, and hence a decrease in heritability (for definitions of heritability and variation components, see Chapter 7 and below). An unpredictable (with respect to organism or trait) effect of stress on variation is expected if genotype–environment interactions under stress enhance genetic differences in some traits (or organisms) and reduce them in others.

However, as noted above, the most popular and long-standing view is that genetic variation increases under adverse environmental conditions. This is explained by the so-called "selection history" hypothesis, which is based on the consequences of natural selection in the population in the past. The rationale that

Box 8.1 Mutation and recombination under stress

If environmental stress induces higher recombination rates and mutation rates, it would directly generate variation for selection to act upon. For recombination rates, several studies in *Drosophila* show an increase at extreme high temperatures [see Hoffmann and Parsons (1991) for a review]. That the spontaneous mutation rate can be enhanced by exposure to harsh environmental conditions was first pointed out by Schmalhausen (1949). However, although this was shown in some cases, the results obtained in *Drosophila* are not fully convincing. The mutation rate can be increased by stress-induced mobilization of transposable elements, as reported by some authors (Junakovic *et al.* 1986; Ratner *et al.* 1992).

Specifically, it was shown that various stressors, such as heat shock, chemical pollutants, and some types of genomic stress, led to transpositions of *copia*-like elements. These and other findings were interpreted as evidence for the adaptive role of stress-induced transpositions (McDonald 1995). However, other authors failed to detect any stress induction of transpositions of mobile genetic elements (e.g., Arnault and Biemont 1989; Arnault *et al.* 1991). In any case, still unclear are:

- How general the phenomena of stress-induced mutation and recombination are;
- Whether or not these phenomena can appreciably change the level of variation in populations such that the rate of evolution is affected.

If the mutation rate does increase under stress, the consequences for fitness may be important. Houle *et al.* (1996) hypothesized that, since fitness traits are controlled by a larger part of the genome than morphological traits, the former should have greater mutational variances (V_M, the amount of genetic variation produced in each generation by mutation). These authors conducted a review of literature data, which revealed that the mutational coefficients of variation (V_M standardized by trait means) are, indeed, higher in fitness traits than in morphological ones. This "mutational target" hypothesis predicts that mutations (including stress-induced ones) would affect fitness traits more than morphological traits do.

underlies this hypothesis is as follows. Under the conditions it normally encounters for long periods of time the species experiences selection that decreases the expression of genetic variation. This decrease can be achieved via selection that removes variation under normal conditions and so leads to the so-called erosion of variance, which is supposed to be especially pronounced in adaptive traits. For instance, deleterious mutations that are expressed only in "bad" environments can be selected against. Another case of selection that reduces variation is selection for canalization of developmental processes: traits with highly canalized development would have lower phenotypic (and, maybe, genetic) variation (Waddington 1961). Under stressful conditions, which are presumably encountered more rarely by the species, hidden variation is expressed, since there was no history of selection under these conditions. As stabilizing selection is thought to operate frequently in natural populations, this scenario is expected to be quite common.

Box 8.2 Heat-shock proteins

Besides assisting the folding (i.e., chaperoning of nascent proteins and refolding of denatured proteins), HSPs serve a multitude of functions within the cell, being involved in protein targeting, stabilization, degradation, signal transduction, and disease tolerance. These functions are performed by members of the many size-class families of HSPs. An extensive overview of the nomenclature of the HSPs is given in Gething (1997). Their functional diversity is reviewed, for instance, by Feder and Hofmann (1999), while their biophysical actions are reviewed in Lorimer (1997).

The inducible family members of the HSPs are also significant for evolution and thermal adaptation, as they are induced by heat and involved in the acquisition of thermal tolerance of cell cultures, embryos, larvae, and adults of *Drosophila* and other organisms. HSP70, one of the best-known members of the family, acts as a molecular chaperone, promoting correct refolding and preventing aggregation of denatured proteins (e.g., Parsell and Lindquist 1994).

Acquisition of thermal resistance and HSP expression both follow after a conditioning heat treatment. Such heat hardening or short-term acclimation increases not only heat-stress resistance (Loeschcke *et al.* 1994), but probably also resistance to other stress types, since HSP induction occurs after many stress forms, including heat, cold, heavy metals, ethanol fumes, and various chemicals (Hoffmann and Parsons 1991). Expression of HSP70 in an unstressed state, however, carries some costs, including retarded growth and cell division (Feder *et al.* 1992) and decreased fecundity (Krebs and Loeschcke 1994). Lines that express relatively high amounts of HSP70 after induction showed a reduced larva-to-adult survival in the absence of stress (Feder and Krebs 1997). These effects may explain why HSP70 is removed in the absence of stress and why the quickest removal occurs in young stages with active cell divisions, where excess HSP70 might be most harmful (Parsell and Lindquist 1994). This tight regulation suggests that a strong trade-off exists between the expression of HSP70, heat-stress resistance, development, and fertility or fecundity.

There is evidence that natural variation in HSP expression exists and can respond to selection (Krebs and Feder 1997). In *Drosophila*, HSP70 variation was related to experimental regimes and survival (e.g., Feder and Krebs 1997; Dahlgaard *et al.* 1998), while selection for heat resistance was found to have caused changes in allele frequencies in HSP68 and HSR-*omega* (McColl *et al.* 1996).

A specific mechanism that reveals cryptic genetic variation that pertains to the developmental pathways in *Drosophila melanogaster* was suggested recently by Rutherford and Lindquist (1998). These authors argue that extensive variation affecting the development of morphological structures exists in laboratory and natural populations of this species. Under normal conditions this variation is hidden, since its expression is buffered by a heat-shock protein 90 (HSP 90; Box 8.2), which is involved in the regulation of signaling pathways. The range of variation presumably suppressed by HSP90 is surprisingly wide and includes developmental abnormalities of the wings, eyes, legs, bristles, etc. According to Rutherford and

Lindquist (1998), this hidden variation may be revealed and exposed to selection under stressful conditions, when the concentration of HSP90 in cells is reduced because of its binding to stress-damaged proteins.

Predictions from models of genetic variation under stress

The view of stress as an enhancer of quantitative genetic variation is supported by mathematical models. In reaction norm models of the evolution of phenotypic plasticity, under the assumption of linearity of the reaction norm, the additive genetic variance becomes a quadratic function of the environment, and the lowest values correspond to the most optimal environment (de Jong 1989, 1990). A more explicit analysis of the evolution of quantitative traits in spatially heterogeneous environments, including nonadditive effects, showed that the genetic variance of adaptive traits was higher in niches that were rare and stressful (in the sense that individuals in these niches had low fitness), than in common unstressful (high fitness) environments (Zhivotovsky *et al.* 1996). For temporally changing environments Zhivotovsky (1997) showed that a population that continuously evolved in one environment under deleterious recurrent mutation exhibits larger genetic variation in fitness when exposed to a novel and stressful environment.

8.3 Stress and Phenotypic Variation

Phenotypic variation of a quantitative trait in a population is usually estimated by the variance. The variance measures the scatter of individual trait values around the mean and is calculated from individual measurements made in the population. A related measure is the coefficient of variation, which is the square root of the variance divided by the trait mean. The use of the coefficient of variation makes comparisons across traits possible. However, in stress studies, this measure should be used with caution: it is inversely related to the mean trait value, which typically decreases under stress, and thus inflates the coefficient of variation.

Increased phenotypic variation under stress

In many studies with various organisms, it has been shown that phenotypic variation tends to increase under adverse conditions. In *Drosophila*, this has already been demonstrated in early studies and confirmed by various later experiments (Table 8.1; Figure 8.1). The experiments using *Drosophila* mostly follow a general scheme: animals are subjected to stress at the pre-adult developmental stages (from eggs or first-instar larvae to emergence of adult flies) and the variation is measured at the adult stage and compared to that in flies reared in nonstressful conditions. A variety of stressors was commonly tested, such as extreme temperatures, nutritional stress (depleted medium), high larval density, high concentrations of ethanol or other chemical compounds, and combined stresses (i.e., when two or more stress factors are applied simultaneously or in succession).

The effect of stressful conditions on phenotypic variation appears to be rather general, as it has been shown for various traits and stress types. However, in *D. melanogaster* there seems to be a difference in phenotypic stress response

Table 8.1 Increase in phenotypic variation in *Drosophila* induced by various stress factors.

Trait	Stress	Reference
Thorax length	Temperature, food	Barker and Krebs 1995; Imasheva *et al.* 1997, 1998; Loeschcke *et al.* 1999
Wing length	Temperature, food, combined stress	Tantawy and Mallah 1961; David *et al.* 1994; Imasheva *et al.* 1999; Loeschcke *et al.* 1999; Woods *et al.* 1999
Sternopleural bristle number	Temperature, food	Parsons 1961; Imasheva *et al.* 1999
Abdominal bristle number	Food	Imasheva *et al.* 1999
Arista branch number	Temperature, food	Imasheva *et al.* 1998, 1999
Ovariole number	Temperature	Delpuech *et al.* 1995
Developmental time	Temperature	Parsons 1961; Loeschcke *et al.* 1999
Fecundity	Temperature	Sgrò and Hoffmann 1998b

between meristic and size-related traits. In size-related traits (for instance, thorax length and wing length), significantly higher values of phenotypic variances and/or coefficients of variation have been observed under stress in many experiments, whereas in meristic traits, such as bristle number, the upward changes are recorded as a trend and may be absent or reversed (Woods *et al.* 1999; Bubli *et al.* 2000).

Fluctuating asymmetry does not reliably show stress exposure

The phenotypic variance and the coefficient of variation provide information on phenotypic variation at the population level. Another quantity related to phenotypic variation, which estimates variance at the individual level, is fluctuating asymmetry (FA), the nondirectional deviation from bilateral symmetry, which is measured as an unsigned difference between trait values of the right and left sides of an individual (Palmer and Strobeck 1986). Developmental noise, that is slight random developmental deviations, is thought to be caused by FA (Palmer 1996), and is thus used to assess any departure from developmental stability. Several indices have been proposed to estimate the average FA level in a population [see Palmer (1994) for an excellent summary]. As it is easy to measure and straightforward to interpret, recently FA has been employed extensively for various purposes in ecological and population studies.

Since stress is supposed to enhance developmental instability, the magnitude of FA in a population is expected to be associated positively with the stress level. For this reason, FA has been suggested as an indicator of environmental stress in natural populations (Leary and Allendorf 1989; Graham *et al.* 1993a). However, Palmer (1996) cautions that two important points must not be neglected when employing FA as a tool to biomonitor environmental stress:

- The level of FA in a population that is presumably subjected to stress should significantly differ from that in unstressed populations;

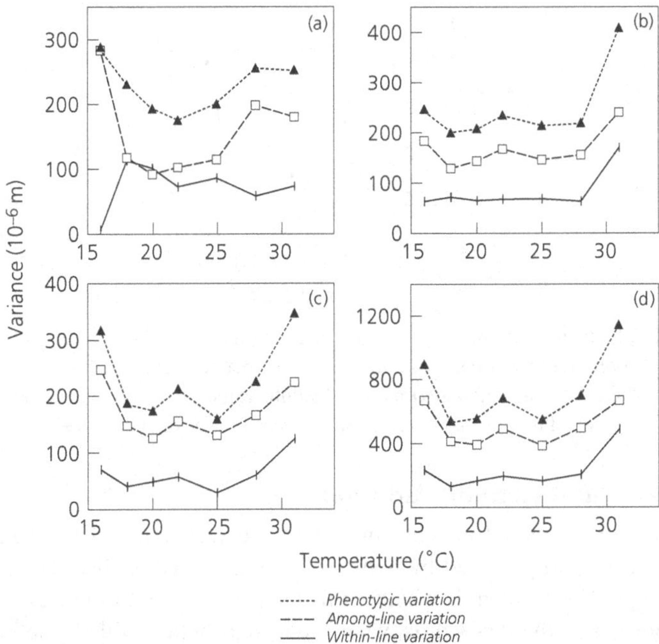

Figure 8.1 Phenotypic and genetic variation of ten isofemale lines of *D. aldrichi* at different temperatures: (a) thorax length, (b) wing length (proximal length of third longitudinal vein), (c) wing width, and (d) wing area. The upper and lower temperatures shown here are stressful for this species. *Source*: Loeschcke *et al.* (1999).

- To serve as an early-warning system to detect potential population damage, FA should increase before a substantial stress-induced decline in fitness rather than after it.

For various species, an ample body of data documents higher FA levels in adverse environments (Møller and Swaddle 1992). Surprisingly, little experimental evidence comes from *Drosophila* studies, but the results obtained for this organism are rather discouraging. In some works, a significant increase in FA was recorded in *D. melanogaster* exposed to chemical substances (Graham *et al.* 1993b), stressfully high and low temperatures (Imasheva *et al.* 1997; see Figure 8.2), or reared on depleted medium (Imasheva *et al.* 1999). However, in other studies, adverse environmental factors were not found to have a clear-cut effect on FA in *Drosophila* (Hurtado *et al.* 1997; Loeschcke *et al.* 1999; Woods *et al.* 1999; Bubli *et al.* 2000). Moreover, there is virtually no evidence of FA changes under stress for natural *Drosophila* populations. Thus, for *Drosophila*, whether or not FA can be used to monitor environmental stress in nature remains unresolved. It may well be that some trait or a combination of traits will prove useful in detecting the impact of stressors in this organism, but the indiscriminant use of one randomly chosen trait for this purpose seems unwarranted.

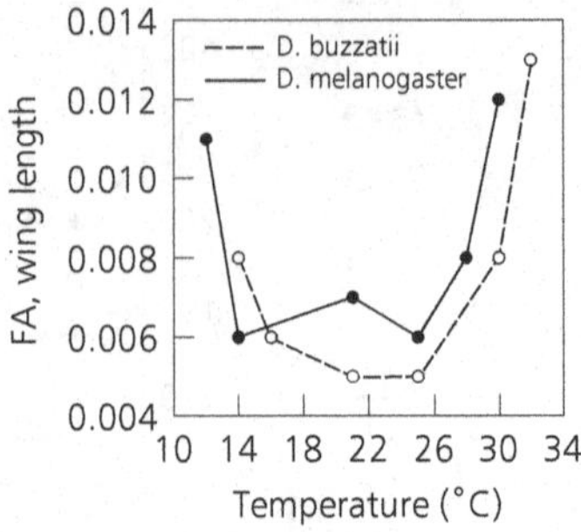

Figure 8.2 Fluctuating asymmetry (FA) of wing length in *D. melanogaster* and *D. buzzatii* reared at different temperatures. Note the difference between species: in *D. buzzatii* FA is generally lower. FA was calculated as the absolute value of the difference between right and left, standardized by the character mean. *Source*: Imasheva *et al.* (1997).

8.4 Stress and Genetic Variation

Apparently, the observed increase in phenotypic variation under adverse conditions is at least partly caused by nongenetic factors such as disturbed individual development. The problem is whether this increase is also, to some extent, genetically determined. In other words, what is the contribution of the genetic variance to the observed increase in the phenotypic variance? To assess this, the proportion that genetic variance contributes to the total phenotypic variance must be estimated and compared under normal and stressful conditions.

Estimating genetic components of variation

Partitioning the total phenotypic variation into its components and the various measures of genetic variation that are pertinent to selection response are considered explicitly in Chapter 7. Here, we give some basic definitions that are needed to understand further discussion.

Assuming no correlation between genotypes and the environment, the total phenotypic variance V_P can be represented (Falconer and Mackay 1996) as

$$V_P = V_A + V_D + V_I + V_E + V_{G\times E} \,, \tag{8.1}$$

where V_A is the additive genetic variance, which measures the variation that is passed from parent to offspring, V_D and V_I are, respectively, the dominance and epistatic variance components, and $V_{G\times E}$ is the variance attributable to genotype–environment interactions. The sum $V_A + V_D + V_I$ constitutes the genetic variance V_G.

The genetic variance and its additive components are absolute measures of genetic variation. By standardizing the genetic variance with different parameters, relative measures of genetic variation present in the population can be obtained. A widely used relative measure of genetic variation is heritability (h^2), which estimates the proportion of genetic variation in the total phenotypic variation;

broad-sense and narrow-sense heritabilities refer, respectively, to the fractions of the additive and the total genetic variation (V_G / V_P and V_A / V_P). Another proposed relative measure is evolvability (Houle 1992), which in the case of directional selection is determined as V_A/M, where M is the trait mean.

The measurement of genetic variance is not so simple in practice, because, in contrast to phenotypic variance, it cannot be estimated directly from observations in a single population. Numerous methods based on measuring the degree of resemblance between relatives have been devised to this end; some of them are briefly outlined in Box 8.3. However, an exact estimation of additive genetic variation, and especially of its comparison in different environments, still presents a difficult task.

Collecting experimental results

To test the hypothesis of a stress-induced increase in the expression of variation, many experiments on *Drosophila* have been conducted. These experiments yielded very heterogeneous results. It is unclear at the moment whether this heterogeneity stems from methodological inconsistency or from the actual complexity of the underlying processes, or, which is more likely, from both. Systematization of the results is complicated by the scarcity and biased character of the available data. The overwhelming majority of *Drosophila* studies used *D. melanogaster* as an experimental model and the isofemale line technique as the method to estimate genetic variation. The emphasis has been put on the effect of high (and, more recently, low) temperature as a key ecological factor for this organism, but other stress types were almost overlooked. Another complication arises from the lack of consensus on the definition of stress. Severity of stress is rarely monitored in any way, and some authors consider as stressful mildly deleterious conditions that apparently do not substantially impair fitness and thus can hardly be expected to have a substantial effect on variation. All this makes comparisons across studies difficult. However, some generalizations, albeit tentative, can still be made.

Trait-specific effects of stress on genetic variation

Genetic variation of a number of traits that belong to different categories was studied with respect to stress effects. Here we differentiate between meristic (e.g., bristle counts), morphometric (mostly traits related to body size, such as thorax length and wing length), and life-history traits (e.g., developmental time, viability, fecundity). Of these, meristic traits bear the least relation to fitness; size-related traits in *Drosophila* are positively correlated with fitness as larger flies have higher fecundity and mating success, and life-history traits are, by definition, related to fitness, and sometimes are called fitness components (problems related to the genetic variation of fitness components are discussed in Chapter 7). Experimental evidence indicates that the behavior of variation under stress in *D. melanogaster* depends on the trait type:

Box 8.3 Methods for estimating genetic variation

Numerous approaches, based on different crossing designs, to the estimation of additive genetic variation have been developed (see Chapter 7 for an extensive discussion and comparisons). The most straightforward method is to estimate heritability as the mean offspring on mid-parent regression, which quantifies the resemblance between parents and offspring and thus gives the closest approximation to narrow-sense heritability. Another method is the full-sib design, in which heritability is calculated from the covariance between full siblings. In this design, gene interactions are included in the variation, so this estimate is closer to broad-sense heritability. Another well-known method of estimating heritability is the diallel cross design. It involves all possible matings between several genotypes that can be represented by, for example, clones or homozygous or inbred lines. The advantage of the diallel cross design is the possibility that variation can be partitioned into additive and nonadditive components.

A modification of the full-sib design is the isofemale line technique (Hoffmann and Parsons 1988), which has been widely used to assess variation in *Drosophila*. An isofemale line is a line derived from a single wild-caught female. In this method, isofemale lines are treated as families and variation among them is assumed to have a genetic basis. To estimate genetic variation, the intraclass correlation, or the proportion of the total variance attributable to among-family components, is measured, from which a heritability estimate can be derived easily.

None of these methods can give an exact estimate of the additive component of variance; in all of them, a nonadditive or interaction fraction is present:

Method	Nonadditive variation component	$G \times E$	Inbreeding
Diallel cross	–	–	–
Parent–offspring regression	$V_{AA}/4$	–	–
Full sibling	$V_D/4 + V_{AA}/4 + V_{AD}/8 + V_{DD}/16$	+	–
Isofemale line	$V_D/4 + V_{AA}/4 + V_{AD}/8 + V_{DD}/16$	+	+

$G \times E$: genotype–environment interaction; V_D, V_{AA}, V_{AD}, V_{DD}: variance components attributable to dominance (V_D) and epistatic interactions (V_{AA}, V_{AD}, V_{DD}).

The isofemale line technique has additional disadvantages. As isofemale lines do not, in reality, consist of full siblings, all the measures of genetic variation obtained with this technique are approximations, being only proportional to the real values; moreover, these lines are often subject to inbreeding and, because of the small size, possibly also to genetic drift, which can change their genetic composition.

- *Meristic traits.* In all the experiments conducted so far, the genetic variation of meristic traits (numbers of sternopleural, abdominal, and orbital bristles; number of arista branches) has not shown a consistent pattern of change (if any) with any method used (Imasheva *et al.* 1998, 1999; Woods *et al.* 1999).
- *Life-history traits.* In life-history traits, additive variance and/or heritability have been shown to increase under some types of stress, using both isofemale line analysis (Westerman and Parsons 1973; Imasheva *et al.* 1998) and more precise methods of heritability estimation (Gebhardt and Stearns 1988, 1992;

Sgrò and Hoffmann 1998b). For instance, this was demonstrated convincingly for fecundity under combined ethanol–nutrition–cold–shock stress (Sgrò and Hoffmann 1998b). However, there are exceptions to this [e.g., developmental time at high temperature (Imasheva *et al.* 1998)] and the number of experiments is still too small to allow broad generalizations.

- *Size-related traits.* Although the number of experiments conducted on size-related morphometric traits is much greater, the situation with this trait category seems to be even more complicated. Essentially, the results obtained with this trait type depend on the method used. In numerous studies using the isofemale line technique, an increase of genetic variance and heritability of thorax length and, to a lesser extent, of wing length is recorded under stress, although for heritabilities these differences are usually reported as trends, probably because of the notoriously high standard errors in heritability estimates (David *et al.* 1994; Barker and Krebs 1995; Morin *et al.* 1996; Noach *et al.* 1996; de Moed *et al.* 1997; Imasheva *et al.* 1998, 1999; Loeschcke *et al.* 1999). By contrast, a few studies undertaken using more exact methods of heritability estimation, such as parent–offspring regression, have not revealed any increase in the additive variation under stress (Sgrò and Hoffmann 1998a; Hoffmann and Schiffer 1998; Woods *et al.* 1999). In fact, in one case, wing length heritability decreased under stressful conditions, because of an increase in the environmental variance (Sgrò and Hoffmann 1998a).

Causes of higher genetic variance in size-related traits under stress

The reasons why these discrepancies are maintained, even with experimental designs, are unclear. On the basis of the difference between the methods used, three nonexclusive explanations can be given:

- A stress-induced increase in *dominance and epistatic gene interactions.* In the full-sib design, on which the isofemale line technique is based, a rather large proportion of the heritability estimate is attributable to dominance and epistasis. Indeed, some *Drosophila* experiments provide evidence of an increase in gene interactions in adverse environments (Blows and Sokolowski 1995), although this result was obtained for a life-history trait.
- *Genotype–environment interactions*, which occur if different genotypes respond differently to a change of environmental conditions. Heritability estimates obtained using the full-sib method contain a fraction attributable to this interaction. If differences between genotypes are increased in unfavorable environments, estimates of the additive genetic variance and/or heritability obtained using this method and the isofemale line technique may appear inflated under stress conditions.
- *Inbreeding* can also enhance the effect of stress on variation in isofemale lines, since these are usually maintained in the laboratory in comparatively small numbers over many generations. This may lead to the differential fixation of recessive deleterious alleles in isofemale lines. Among-line variation may thus partly be the result of a different load because of the differential accumulation of these recessives in different isofemale lines.

On the basis of the available evidence, it is not possible to make a definite choice between these possibilities. However, their evolutionary implications are different. If an increase in the genetic variance is based mainly on an increase in the additive variation, it would, as noted above, result in an increase of the evolutionary rate and the rate of adaptation. Generally, the same outcome may be expected in the case of an increase in the dominance and epistatic variance. It has been shown that under some conditions (such as population bottlenecks, which may well occur as a result of stress exposure), nonadditive variance components can be partly converted into additive variance (Goodnight 1988; Willis and Orr 1993). Both under stress and in a temporally fluctuating (stressful-to-nonstressful) environment, these effects may increase the selection response. The situation with genotype–environment interactions is less clear. The possible effects of stress on variation in the presence of genotype–environment interactions are schematically shown in Figure 8.3. When stress does not affect the additive variance component, genotype–environment interactions do not result in higher variation (Figure 8.3b). If the additive variation is enhanced in a stressful environment, genotype–environment interactions may significantly increase the selection response by augmenting phenotypic differences between genotypes (Figures 8.3c and 8.3d). However, if the rank order of the genotypes is changed (Figures 8.3b and 8.3d), different genotypes will be selected for under stressful and nonstressful conditions. In the long-term perspective, if stress conditions disappear and the environment changes back to "normal", this could promote the preservation of genetic variation (Gillespie and Turelli 1989). In a fluctuating stressful-to-nonstressful environment, genotype–environment interactions may decrease the selection response.

8.5 Experimental Selection under Stress

Artificial selection experiments could provide a definitive solution to the problem of the impact of environmental stress on the rate of evolution. An increase in the additive genetic variation in a population should lead to an increased selection response. Hence, if stress enhances the expression of additive genetic variation, selection must be more successful under stressful than under normal conditions. Unfortunately, the number of such experiments is to small to enable any general conclusions. However, the available evidence partly supports (albeit not very conclusively) the hypothesis of an increase in the additive variation of life-history traits under stress. Selection for longevity in *D. melanogaster* was shown to be critically dependent on larval density: larval crowding is generally required to produce a selection response (Luckinbill and Clare 1985). Neifakh and Hartl (1993) obtained a selection response for faster embryonic development in *D. melanogaster* when eggs were exposed to an extremely high temperature, while at normal temperatures this response did not occur. However, the selection response for knockdown resistance in this species was the same under low-density conditions and under larval crowding (Bubli *et al.* 1998). Negative results were also obtained by the same group when they compared the selection response on sternopleural bristle number under extremely high versus normal temperatures (Bubli *et al.* 2000).

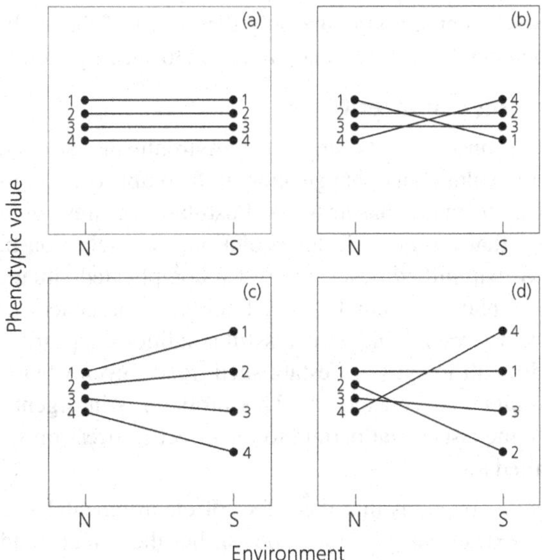

Figure 8.3 A diagrammatic representation of some possible outcomes of stress exposure. The hypothetical population consists of four genotypes designated by numbers 1 to 4; their phenotypic values are plotted on the vertical axis. N and S refer to nonstressful and stressful environments, respectively. The variation is given by the range between the uppermost and lowermost genotypes. (a) Stress has no effect on any variation component; (b) stress produces genotype–environment interaction, but the additive variance remains unchanged; (c, d) stress increases additive variation and genotype–environment interaction; in (d), the rank order of the genotypes is changed.

Two preliminary conclusions can be drawn from the results of this limited and heterogeneous set of experiments:

- Exposure to stressful conditions may have an effect on evolutionary rate in life-history traits, but not on other trait categories. This seems to be in general agreement with results of the estimation of genetic variation under stress for the various trait types considered above. One can speculate that, apart from the implications of erosion of variance in fitness traits, this stress effect may be related to the fact that fitness traits are always controlled by numerous genes, which augments the influence of gene interactions. This is in line with the finding that stress has no effect on the selection response on sternopleural bristles (Bubli *et al.* 2000) – a trait with much simpler genetic architecture controlled by a relatively small number of genes.
- It is noteworthy that in no selection experiment conducted so far has stress accelerated the rate of change in trait values in response to selection. In all the above examples, stress produced a selection response in traits that did not respond to selection under more beneficial conditions.

As selection experiments provide the only direct test of the evolutionary significance of environmental stress, more are required to shed light on this question.

8.6 Concluding Comments

The crucial issue concerning the impact of environmental stress on genetic variance is the rate of evolutionary change under unfavorable conditions. This problem poses more questions than it has answers. Postulating a stress-induced increase in genetic variation that promotes faster evolution, the "selection history" hypothesis offers an oversimplified description of a complicated situation, although in some cases this explanation may be true. Differences in genetic variances and/or heritabilities under stressful and nonstressful conditions appear to be general, but currently it is difficult to speak of established trends, even when only one model organism, *Drosophila*, is considered. This problem is in urgent need of further investigation. Some issues that pertain to the effect of stress on variation and that need to be clarified are:

- *Trait specificity.* It seems that the effect of environmental stress on variation depends on the trait category (see above), but the actual trends are yet to be determined.
- *Stress specificity.* There are indications that different stressors might not act alike on variation. For instance, the effect of cold stress on phenotypic (genetic?) variation in *D. melanogaster* appears to be stronger than that of heat stress (Delpuech *et al.* 1995; Imasheva *et al.* 1997, 1998; Loeschcke *et al.* 1999). It is conceivable that differences between stresses could even be qualitative. Of particular interest in this respect are combined stresses, which are probably very common in nature: the effects of individual stresses that constitute a combined stress might not be simply additive (Loeschcke *et al.* 1997). If proved, stress specificity would imply that results obtained for one stress type cannot be extrapolated to other stresses.
- *Species specificity.* As mentioned above, most experiments that test for the effect of stress on variation have been carried out using *D. melanogaster*. Helpful as this species is as a model system, examination of other species, even within the genus *Drosophila*, might show quite a different picture (e.g., see Imasheva *et al.* 1997; Loeschcke *et al.* 1999).
- Finally, it should be emphasized that virtually all the results currently available on stress-mediated changes in genetic variation stress have been obtained in model experiments for experimental laboratory populations. What the impact is of unfavorable conditions on variation in natural populations remains an open question. Only future research can tell whether environmental stress plays an appreciable role in determining rates of evolutionary change.

Acknowledgments This work was supported by the NATO Linkage grant GRG.LG.972856 and grants from the Danish Natural Science Foundation. Volker Loeschcke is grateful to the Institute for Advanced Study and the Center for Environmental Stress and Adaptation Research at La Trobe University for their hospitality during his stay while the final revision of the manuscript was made.

Part C

Genetic and Ecological Bases of Adaptive Responses

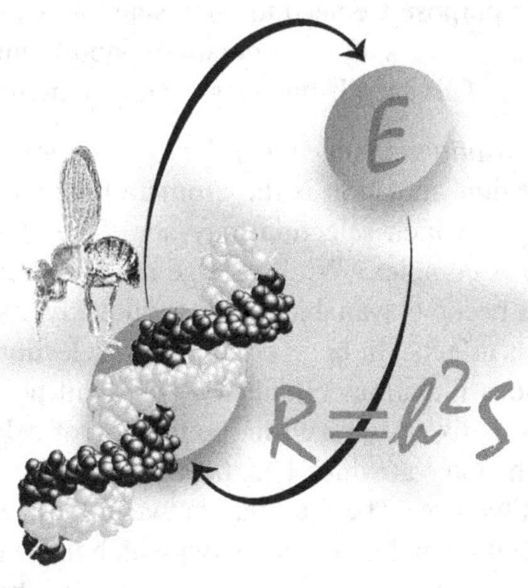

Introduction to Part C

As Part B shows, there is now strong empirical evidence that a rapid adaptive evolution of life-history traits may occur in response to environmental change in the wild. However, very few empirical studies document the consequences such adaptations have for population dynamics and viability.

Continuing the establishment of a theoretical platform for evolutionary conservation biology, begun in Part A for populations with a static genetic composition, Part C introduces models to assess the extent to which contemporary evolutionary change can contribute to, or hamper, population persistence in the face of environmental threats. For this purpose we need to understand how the different components of the evolutionary process affect population viability under different types of environmental change. This agenda involves a series of more specific questions:

- *How does genetic variation influence population viability?* In populations unaffected by immigration, mutation is the ultimate long-term source of all the genetic variation upon which selection may act. The effect of mutation on population viability is complex. While a large mutation rate implies that new favorable mutations become available relatively quickly, most mutations have primarily deleterious effects. In large populations, selection is expected to remove such deleterious mutations efficiently. In small populations, however, random genetic drift is likely to result in the fixation of deleterious mutations. At the same time, the rates at which beneficial mutations are incorporated decrease with population size. The interplay between these effects suggests the existence of a threshold population size below which the net effect of mutation on population viability is negative. Estimation of this threshold and understanding how it depends on the genetic system under consideration is important in conservation.
- *When is adaptive change expected to be fast enough to promote evolutionary rescue from a degrading environment?* In simple models of phenotypic evolution, a population possesses an optimal phenotype that depends on the given environment. As environmental conditions vary, a crucial question is whether the population has any opportunity to track the optimal phenotype. The risk of extinction increases and adaptation may fail to rescue the population if the lag between its phenotypic composition and the optimal phenotype becomes too large. How closely a population can track its evolutionary optimum depends on the available genetic variability, on the strength of selection, and on the type and pace of environmental change that causes the evolutionary optimum to vary.
- *Under which conditions does the selective process promote phenotypes that enhance population viability?* The metaphor of evolutionary optimization is suitable only if the fitness (i.e., the rate of increase from low density) of phenotypes is independent of the population's frequency distribution and the total

density of individuals that express different phenotypes. Only under these conditions does the phenotype that maximizes fitness, given the genetic constraints (e.g., linkage disequilibria), also minimize a population's risk of extinction. In contrast, as has already been pointed out by Haldane (1932), a phenotypic adaptation that confers an advantage to an individual may be detrimental to the population as a whole. Under frequency- and/or density-dependent selection, conditions for population viability are not related to conditions for evolutionary stability, and the actual direction of selection may therefore not always coincide with the one that leads to the most efficient reduction in extinction risk. To understand the interplay between population viability and evolutionary dynamics requires complete knowledge of the environment feedback loop that describes the dependence of selection pressures on the adaptive status of populations.

The following chapters address these three important questions for closed single-species populations. The challenges of how to extend these approaches to spatially structured habitats and to multispecific interactions are tackled in Parts D and E, respectively.

In Chapter 9, Whitlock and Bürger investigate, within the framework of population genetics models, the effects of reduced or declining population sizes on population viability. The authors review the current understanding of a process known as "mutational meltdown" and discuss several of the genetic and demographic features of species that can substantially affect this process. Elaborating on the classic results of Malécot and Kimura concerning the rate of fixation of alleles, the process of fixation of beneficial and deleterious mutations in small or declining populations is explored. The review concludes with several strong qualitative statements on how genetic components of evolutionary processes are likely to influence the extinction risks of populations that experience reductions in their population sizes.

The dynamics of environmental change are considered explicitly in the quantitative genetics models that Bürger and Krall present in Chapter 10. The authors investigate how effective the adaptive response to environmental degradation can be given the amount of genetic variation, the strength of selection, and the pace of environmental change. Three types of environmental variation are envisioned: stationary environmental stochasticity (temporally correlated or not), directional change (like gradual habitat loss or increased pollution), and a single catastrophic change. The outcome is that the speed and predictability of environmental deterioration are critical in determining the effect of adaptation on population viability. The authors discuss the conditions under which evolutionary rescue is expected to occur and the conservation measures that back up such natural processes. The chapter also draws attention to situations in which genetic variability detrimentally affects population persistence.

The dichotomy between rescuing and suicidal adaptive responses to environmental change is probed further in Dieckmann and Ferrière's Chapter 11. In contrast with quantitative genetics models (in which ecological interactions are usually subsumed in a few compound parameters that describe environmental change and selection), this chapter focuses on the ecological factors that determine the evolutionary responses of threatened populations. After highlighting the importance of eco-evolutionary feedback between populations and their environment, the authors review the theory of adaptive dynamics as a versatile tool for examining the evolutionary consequences of environmental change. The chapter then describes how evolutionary processes directly influence biodiversity. First, the diversity of communities is enhanced through speciation, which is recognized increasingly as being driven by frequency-dependent selection that results from ecological interactions. Second, diversity is reduced evolutionarily through selection-induced extinction, a process during which a species' adaptive dynamics jeopardize its viability. Based on various examples, the chapter delineates the ecological conditions that favor the evolutionary enhancement or reduction of biodiversity.

The material presented in this part has two basic implications for the improvement of conservation strategies. First, conservation planning is most efficient when it is proactive, such as by implementing measures that prevent the decline of populations and thus prevent the harmful genetic effects associated with small population size. Yet, it must be kept in mind that the restoration of abundances or habitat may result in conflicting selection pressures, either by introducing new phenotypes that have undergone different selection histories, or by opening up new habitats that induce different selection pressures. Second, understanding interactions between individuals turns out to be essential for the long-term success of conservation efforts. This applies not only to predicting the short-term fate of populations in a static world (see Chapter 2), but also to assessing the direction and relative strength of selection pressures that act on life-history traits in response to environmental change, and to the eventual impact of such adaptive responses on population viability.

9

Fixation of New Mutations in Small Populations

Michael C. Whitlock and Reinhard Bürger

9.1 Introduction

Evolution proceeds as the result of a balance between a few basic processes: mutation, selection, migration, genetic drift, and recombination. Mutation is the ultimate source of all the genetic variation on which selection may act; it is therefore essential to evolution. Mutations carry a large cost, though; almost all are deleterious, reducing the fitness of the organisms in which they occur (see Chapter 7). Mutation is therefore both a source of good and ill for a population (Lande 1995).

The overall effect of mutation on a population is strongly dependent on the population size. A large population has many new mutations in each generation, and therefore the probability is high that it will obtain new favorable mutations. This large population also has effective selection against the bad mutations that occur; deleterious mutations in a large population are kept at a low frequency within a balance between the forces of selection and those of mutation. A population with relatively fewer individuals, however, will have lower fitness on average, not only because fewer beneficial mutations arise, but also because deleterious mutations are more likely to reach high frequencies through random genetic drift. This shift in the balance between fixation of beneficial and deleterious mutations can result in a decline in the fitness of individuals in a small population and, ultimately, may lead to the extinction of that population. As such, a change in population size may determine the ultimate fate of a species affected by anthropogenic change.

This chapter reviews the genetic changes that occur as a result of a decrease in population size. We particularly focus on the fate of new mutations in a small population and the conditions under which a species may be at risk of extinction from harmful changes in fixation rates. We confine most of our comments to sexual, randomly mating populations.

9.2 Purging and Fitness Changes in Declining Populations

One of the unfortunate results of human activity in recent years is that the number of individuals in many species has declined rapidly and drastically, and this change seems likely to continue. This sudden change in population size can have several effects on the average fitness of a population.

Even a large population is expected to carry many copies of deleterious mutations, segregating at low frequencies at most of the loci in the genome. These rare alleles cause the mean fitness of the population to be lower than it would otherwise be; this reduction in fitness is called the *mutation load* (see Box 9.1). If the

Box 9.1 Genetic load

Genetic load is the term used to describe the reduction in fitness of a population, relative to some ideal population, due to the actions and interactions of various population genetic processes. The term *load* entered population genetics as a result of H.J. Muller's (1950) article *Our load of mutations*. Muller was concerned with the selective deaths required to remove new deleterious mutations from a population; we now refer to this type of load as *mutation load*, the reduction in fitness caused by segregating alleles brought into the population by mutation. An important result of Haldane (1937) is that if an infinitely large population is in a mutation–selection balance, then the mutation load L is independent of the fitness effects of the mutations. More precisely, if μ denotes the total mutation rate from the wildtype (the allele with the highest fitness, normalized to 1) to all other possible alleles, then the mean fitness of the population $\overline{W}$ is

$$\overline{W} = 1 - L = 1 - \mu \,, \tag{a}$$

independent of the fitness values of the mutants (except for completely recessive alleles). This principle was extended to very general fitness and mutation patterns (Bürger and Hofbauer 1994; Bürger 2000, Chapter IV.5). The following theory assumes that each mutation occurs at a new locus, that loci are statistically independent (linkage equilibrium and no epistasis), and that every mutation reduces the fitness of its homozygous carrier by a factor $1 - 2s$, where s is the *selection coefficient*. The assumption that each mutation occurs at a new locus is adequate because mutation rates per locus are usually very low, so a mutation is lost or fixed before the next occurs. This, however, does not preclude the possibility of a high genome mutation rate U, because there are tens of thousands of genetic loci in genomes of larger organisms. Under the assumption that the actual number of mutations per individual is Poisson distributed with a mean U, one obtains

$$\overline{W} = 1 - L \approx e^{-U} \,. \tag{b}$$

In a finite population, the load L_S from segregating mutations approaches a stationary value with time and is close to zero in very small populations (less than 10–20 individuals) because of the reduced heterozygosity in such small populations. It increases to a maximum value at an intermediate population size [for genes with additive effects, this population size is approximately $1/(2s)$] and thereafter decreases to the infinite-population expectation, as $N_e s > 5$, that is, $L_S \approx 1 - e^{-U}$ (see Kimura *et al.* 1963; Lynch *et al.* 1995a, 1995b). Here, N_e denotes the so-called effective population size; it is explained in Section 9.4.

Alleles also sometimes fix through genetic drift. The resultant progressive reduction in mean fitness is sometimes referred to as *drift load*. Since, in a population of N individuals, UN mutations occur per generation, the expected fitness reduction caused by fixation in each generation is

$$\Delta \overline{W} = \frac{\overline{W}'}{\overline{W}} = (1 - 2s)^{UNu_f} \,, \tag{c}$$

continued

Box 9.1 *continued*

where $\overline{W}'$ is the population's mean fitness in the next generation, u_f is the fixation probability of a deleterious mutant (see Box 9.2). Mutation load is common to all populations, since all species have nonzero mutation rates for deleterious alleles. The drift load, on the other hand, is only important in relatively small populations.

Other types of load have been considered, such as the *recombination load* (a reduction in fitness that results from the breakup of fit gene combinations), the *lag* or *evolutionary load* (the deficit in fitness because of incomplete adaptation to a changing environment), and so on [see Crow and Kimura (1970) for a list]. We are not concerned with these kinds of load here (but see Chapter 10). Finally, the concept of a load depends on the assumption of the existence of an optimal genotype or phenotype and, as such, is an idealization that has to be applied with caution.

population's size becomes much lower than it was previously, there are a number of consequences. Many of the deleterious mutations already segregating in the population are lost immediately (or within a few generations) through sampling, but some may be fixed. Most of the deleterious mutations that persist in the population are, to some extent, recessive to the wildtype alleles, and a smaller population is more likely to express these alleles as homozygotes. Therefore, the strength of selection is increased proportionally, and the frequency of deleterious alleles becomes lower. This process is referred to as *purging*. While purging can reduce the mutation load for populations that are temporarily small [although not by much, Kirkpatrick and Jarne (2000)], in populations that are small for an extended period new mutations soon return the mutation load to a similar level as before the population size changed (see Box 9.1; Charlesworth *et al.* 1993a; Lynch *et al.* 1995a). Purging is not likely to increase the fitness of permanently small populations.

In addition, and more importantly over a longer time scale, some of the mutations that occur at relatively low frequency in the large population drift to a higher frequency after the population has become smaller. If the new population is small enough, then some of these mutations fix, and the average fitness of the population is reduced as a result. The fitness reduction caused by the fixation of deleterious mutations through genetic drift is called *drift load*. If the population remains small, this process will continue with new mutations until the mean fitness of the population is sufficiently low that it cannot sustain itself. This decrease in fitness through the drift of deleterious mutations is thought to be the major *genetic* factor in determining the probability of extinction of a small population (Lande 1994; Lynch *et al.* 1995a, 1995b). The next sections deal with this issue in more detail.

9.3 Fixation of Deleterious Mutations: Mutational Meltdown

A species reduced to a small population size continues to have the same rate of mutation to inferior alleles, but if it is small enough some of these mutations increase in frequency through genetic drift until they replace the more fit allele. If

Box 9.2 Fixation of beneficial and deleterious alleles by genetic drift

When an allele appears initially in a population as a result of a mutation event, its frequency is $1/(2N)$, where N is the census size of the population. For a new allele with fitness $1 - s$ as a heterozygote and $1 - 2s$ in the homozygous state, the probability u_f that the allele will ultimately fix in the population was derived by Malecót (1952) and Kimura (1957), on the basis of a diffusion approximation, to be

$$u_f = \frac{e^{2sN_e/N} - 1}{e^{4sN_e} - 1}\,, \tag{a}$$

which for $2sN_e/N \ll 1$ is approximated well by

$$u_f = \frac{2sN_e/N}{e^{4sN_e} - 1}\,. \tag{b}$$

The probability of fixation is therefore a function of the effective size of the population, the census size, and the strength of selection that acts on the allele. Bürger and Ewens (1995) showed that a better approximation is obtained by replacing s in Equations (a) and (b) by $s/(1 - s)$. Similar, but more complicated, equations that include the effects of dominance are available (see Crow and Kimura 1970).

Looking more carefully at Equation (a), we see that for beneficial alleles (in this notation given by $s < 0$), the probability of fixation is approximately $2|s|N_e/N$, if $|s| > 1/(2N_e)$. Thus, even an allele with a strong favorable effect has a low probability of fixation. If the strength of selection is such that $|s| < 1/(2N_e)$, the allele is said to be nearly neutral and the probability of fixation becomes almost independent of selection and approaches the neutral value, $u_f = 1/(2N)$.

The probability of fixation of deleterious alleles decays nearly exponentially as the effective population size increases and becomes negligible if $s \gg 1/(2N_e)$. If, however, $s < 1/(2N_e)$ the probability of fixation increases rapidly and approaches the neutral value of $1/(2N)$. In contrast to that of advantageous alleles, the fixation probability of a detrimental allele increases with decreasing effective population size (see Figure 9.1).

The decline of a population toward extinction begins when the mean absolute fitness drops below 1. Once this point is reached, population extinction occurs almost deterministically and very quickly. The mean extinction time of the population is determined mainly by the phase during which the mutations accumulate, mean fitness decreases, but the population size remains constant. The mutation load in the small population is approximately the same as that in the original larger population, so the mean fitness of the smaller population can be expressed as a function of its drift load and of its initial mean absolute fitness at low density, $\overline{W}_0$. If T is the number of generations over which mutations accumulate by drift before the mean fitness becomes less than 1, then

$$\left(\Delta\,\overline{W}\right)^T \overline{W}_0 = 1\,, \tag{c}$$

where $\Delta\,\overline{W}$ is the change in mean fitness per generation and $\overline{W}_0$ is the average mean absolute fitness at low density.

continued

Box 9.2 *continued*

Taking logarithms and substituting Equation (c) from Box 9.1 and Equation (b) from this box, the mean time in generations to extinction is given approximately by (Lande 1994; Lynch *et al.* 1995a)

$$T \approx \frac{(e^{4N_e s} - 1)\ln \overline{W}_0}{4UN_e s^2} . \qquad \text{(d)}$$

For mutations of variable effect, the expected decrease in fitness per generation at equilibrium is approximately

$$\Delta\overline{W} \approx \int_0^{1/2} NU\ u_f(s) 2s\Psi(s)\ ds\ , \qquad \text{(e)}$$

where $\Psi[s]$ is the probability that a new mutation has effect s as a heterozygote. If we assume that the distribution of mutational effects is exponential with mean λ, the fixation flux as a result of drift can be found

$$\Delta\overline{W} \approx \frac{U\ \zeta(3, 1 + \frac{1}{4N_e\lambda})}{8N_e^2\lambda} , \qquad \text{(f)}$$

where ζ is the generalized Riemann zeta function (Lande 1994). For cases in which the effective size is large enough such that the average mutant is not nearly neutral, the value of this ζ function is within 20% of 1, and thus the decline in fitness with each generation in this model is approximately

$$\Delta\overline{W} \approx \frac{U}{8N_e^2\lambda} . \qquad \text{(g)}$$

From this it is obvious that as the population size increases, the rate of decline in fitness due to the fixation of deleterious mutations rapidly becomes small.

a population has a reproductive rate R_0 (that is, each individual can produce R_0 offspring), then on average $1/R_0$ of these offspring must survive to reproduce and keep the population from decreasing in size. As a population accumulates deleterious mutations, its intrinsic rate of increase becomes closer and closer to zero, until the mean fitness is below the point at which the population can survive. As the population size declines, deleterious alleles become more likely to fix and further reduce population size. This process has been called a mutational meltdown (Lynch and Gabriel 1990). This path to extinction will be fastest when:

- Deleterious mutations fix at a high rate;
- Fixed alleles have large effect; and
- The reproductive excess of the population is small.

The factors that accentuate these three terms are sometimes contradictory. As we show later, mutations of large effect are both rarer and less likely to fix. We consider each of these factors in turn.

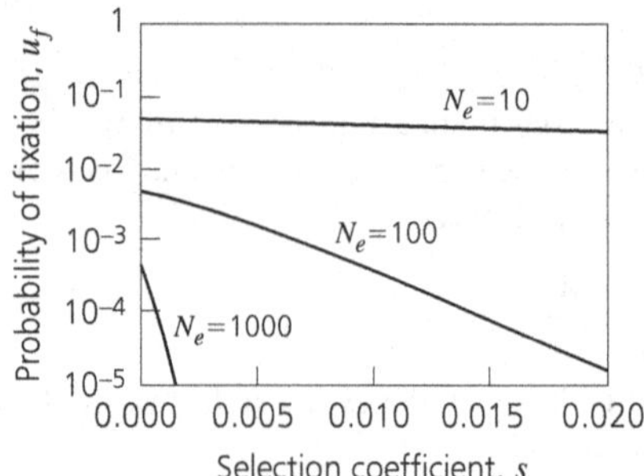

Figure 9.1 The probability of fixation of deleterious alleles depends on the strength of selection against them and the effective population size. The allele is initially present at frequency $1/(2N_e)$.

The rate at which new harmful mutations are fixed depends on both the rate at which new mutants appear in the population and the probability that these new alleles fix. Empirical estimates suggest that the rate of deleterious mutations could be as high as one new mutation per gamete per generation, but this may be strongly species dependent (Crow and Simmons 1983; Eyre-Walker and Keightley 1999; Lynch *et al.* 1999; but see García-Dorado *et al.* 1999; also see Chapter 7). Few of these mutations have very large effects on fitness; the vast majority have very small effects. On average, the homozygous effect of new mutations each generation is a few percent of fitness, maybe 2–15% (Crow and Simmons 1983; Caballero and Keightley 1998; Lynch and Walsh 1998; Fry *et al.* 1999; Lynch *et al.* 1999; see also Chapter 7). It is clear that without the counterbalance of effective selection, mutation would quickly erode the fitness of a population.

The ability of selection to keep deleterious alleles at a low frequency is diminished in small populations, because random genetic drift allows even a harmful allele to fix. The probability that a deleterious allele will fix is therefore a function of both its fitness effects and the effective size of the population. Box 9.2 summarizes the basic theory. Two main results are important. First, the probability of fixation is an exponentially decreasing function of the population size. For mutations of large effect, the effective population size must be very small to allow fixation (see Figure 9.1). Second, alleles with a selective effect that is less than about $1/(2N_e)$ are most likely to fix in the population. These alleles are called *nearly neutral.*

The effect on mean fitness of these fixed alleles is the product of the rate at which the alleles are fixed and their effects on fitness when homozygous. Alleles that have a very small effect are more likely to fix, but do not affect mean fitness much when they do. In contrast, alleles of large effect, if fixed, cause large changes in fitness, but this fixation is unlikely. It turns out that the largest effect on fitness results from mutations with mildly deleterious effects, with s near $0.4/N_e$ (Gabriel and Bürger 1994; Lande 1994); their fixation rate is relatively high but the effects on fitness are not negligible (see Figure 9.2).

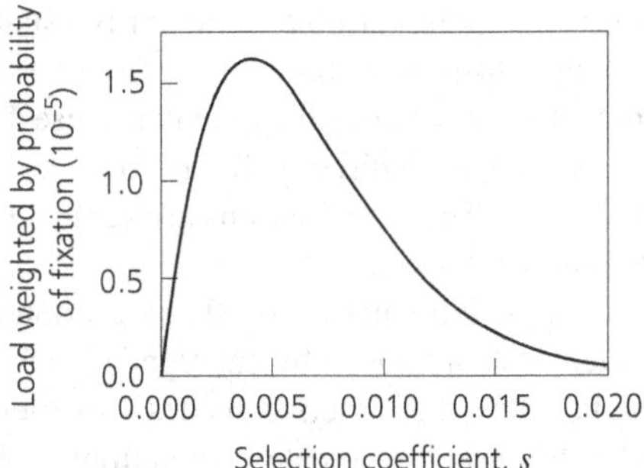

Figure 9.2 The total drift load attributable to an average new mutation is the product of the effect of that mutation when homozygous and the probability of fixation of that mutation. This load is maximized near a value of $s = 0.4/N_e$. Here, N_e is set at 100.

If the population size is small enough that deleterious alleles accumulate, the mean fitness of the population slowly drops. Initially, this drop in fitness is relatively unimportant to the persistence of the population, because most species have reproductive excess, that is, much more reproductive capacity than required to replace dying members of the population. Eventually, though, as the less fit alleles accumulate, this excess decays and the population becomes incapable of sustaining itself, unless it is large enough to generate sufficient beneficial mutations. Therefore, whether a population eventually drops below this minimum fitness is a function of the mutation rate and population size, but how quickly this process leads to extinction is a function of reproductive excess as well. Populations with lower reproductive excess to begin with, as is often the case for much of the macrofauna beloved of conservation posters and polemics, are likely to be much more sensitive to genetic extinction than other species.

9.4 Factors Affecting Fixation of Deleterious Mutations

In this section we discuss several of the genetic and demographic properties of species that can substantially affect the basic process of mutation meltdown.

Effective population size and the Hill–Robertson effect

The *effective population size*, N_e, of a population is the size of an idealized population that has the same properties with respect to loss of variation by random drift as the population in question. As a rule, N_e is smaller than the census size N of the population, sometimes much smaller, and therefore genetic drift occurs at a faster rate than would be expected for N. The effective size is reduced by variance in reproductive success among individuals, including variance caused by unequal sex ratios, selection, and environmental effects. Populations are typically extremely variable in the reproductive success of their members; therefore N_e is usually lower than N. The effective size is also decreased by variation over time in population size; the best description of the effects of drift in a population over time is given by the harmonic mean of the effective size of each generation. The harmonic mean of a set of highly variable numbers is usually much lower than the arithmetic mean, so this form of averaging over time results in N_e being much smaller than N. A recent review showed that when these factors are taken into account, or when N_e is estimated indirectly, the ratio of N_e/N is often as low as 0.1 or less (Frankham 1995c). Thus, populations are subject to much more genetic drift than their numbers alone may indicate.

One additional factor that influences the effective population size is genetic variance for fitness, such as that caused by segregating deleterious alleles. This not only results in variance in reproductive success, as mentioned in the previous paragraph, but also these alleles create correlations over the generations in which genotypes are successful. As a result, the variability in reproductive success is compounded over generations, and some alleles, independent of their own effects, may rise to high frequencies. This *background selection* (Charlesworth *et al.* 1993b) and *hitchhiking* (Maynard Smith and Haigh 1974) can result in a much smaller effective population size and so change the probability of fixation of other alleles (Hill and Robertson 1966). Simulations of the mutational meltdown process that include multiple loci showed that deleterious mutations accumulate much faster than expected by theory that does not account for the effects of background selection, particularly if the mutation rate is high and in relatively large populations (Lynch *et al.* 1995a).

Distribution of mutational effects

The probability of fixation of a deleterious mutation is a function of its selective effect. As mentioned above, the probability of fixation is greatest for small values of s, but the effect on mean fitness is maximized when s is around $0.4/N_e$ (see Figure 9.2). Thus, populations with many mutations that have selection coefficients around this value will decline in fitness rapidly relative to populations with the same number of mutations, but with larger or smaller effects.

For this reason, variance in mutational effects can make a large difference in the time to extinction. If the mean selection coefficient of a new mutation is much higher than $0.4/N_e$, then variance in selective effect results in a faster decline of the population mean fitness (Lande 1994). This variance can result in orders of magnitude differences in the time to extinction. In contrast, if the mean mutation effect is close to this maximum effect size, then variance in mutational effect can only decrease the rate of loss of fitness (Lande 1994; Schultz and Lynch 1997). We have little direct evidence about the variance in mutational effects or, indeed, anything else about the shape of the distribution of the mutations, but it is clear that not all mutations have the same effect on fitness. Analysis by Keightley (1994) of Mukai's (1964) mutation accumulation data suggests that the distribution of mutational effects is extremely variable. Mackay *et al.* (1992b) showed that the distribution of the effects of the mutations caused by transposable element insertions is approximately exponential. Davies *et al.* (1999) show that many deleterious mutations have undetectably small effects.

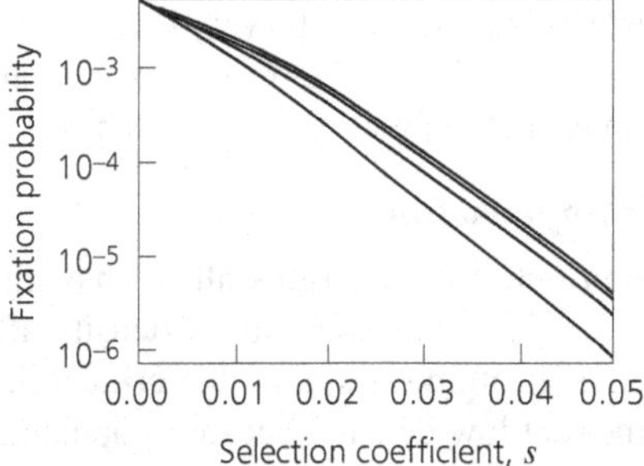

Figure 9.3 Deleterious recessive alleles are more likely to fix through drift than are additive alleles with the same homozygous effect, but this difference is not great. The five curves correspond (from top to bottom) to values of h equal to 0, 0.01, 0.1, 0.5, and 1, respectively. The curves for $h = 0$, 0.01, and 0.1 overlap almost completely on this graph. The fitnesses of the three genotypes are 1, $1 - 2hs$, and $1 - 2s$, respectively. The effective population size is 100.

Dominance

The quantitative conclusions above are based on alleles that interact additively with other alleles at the same locus. In fact, many deleterious alleles are recessive to their more fit counterparts. This recessivity increases the rate of fixation of deleterious mutations through drift (Kimura 1957; Crow and Kimura 1970; see Figure 9.3). The difference in the rates, though, is less than an order of magnitude. The reason for this increase in fixation rate is that the longer an allele segregates in a population at some frequency, the more chances there are that drift will fix that allele. Recessive alleles remain in populations longer than additive alleles with the same homozygous effect. Moreover, the probability of fixation is an increasing function of the allele frequency; recessive alleles are expected to have a much higher allele frequency than are additive alleles because of the mutation–selection balance. As a result of these two facts, the rate of fixation by drift of recessive alleles is somewhat greater than that of additively interacting ones. Indeed, some empirical analyses suggest that many deleterious mutations are nearly recessive (García-Dorado and Caballero 2001).

Epistasis

Not all alleles interact with alleles at other loci in an independent way, as assumed so far. If there is epistasis between loci, in particular if the deleterious effects of two loci combine to make an individual with a fitness worse than that predicted by the product of the fitness effects of the two loci, then the rate of fixation of deleterious mutations can be lower than predicted from single locus theory (Lande 1994; Schultz and Lynch 1997). This *synergistic epistasis* has been observed in *Drosophila* (Mukai 1964), but the evidence for an average level of epistasis being synergistic is weak. Other studies have found synergistic interaction in other species, but only among some pairs of loci and not others (Elena and Lenski 1997; de Visser *et al.* 1997; Whitlock and Bourguet 2000, reviewed in Phillips *et al.*

2000). Synergistic epistasis is unlikely to be very large on average, but even if it is large it will be unable to much change the time to extinction due to the fixation of deleterious mutations in sexual populations (see Butcher 1995).

Nongenetic fitness compensation

Several factors can contribute to deleterious alleles having more of an effect in large populations than in small. For example, if density affects the reproductive capacity of the species, as it often does, then the critical fitness to determine extinction rates is the fitness at low density, because populations must pass through a period of low density before extinction.

A more subtle factor is that the fixation rate of a deleterious allele is a function of the fitness of the allele relative to alternative alleles, while the drift load that results depends only on the absolute reduction in the number of surviving offspring per parent. Therefore, any mechanism that increases the differences in relative fitness of the alleles without increasing their effects on the absolute mean fitness will reduce the rate of the mutation meltdown. For example, an allele that reduces the competitive ability of males to attract mates can greatly reduce the relative fitness of an individual in a variable population and hence is much more unlikely to fix (Whitlock 2000). However, in a population fixed for this allele, the mean male mating success may be unaffected, as long as all females still mate. This allele would therefore be selected against and be less likely to fix through drift than if it did not affect male mating success, but once fixed this reduction in the relative mating success would have no effect on the productivity of the population as a whole. This effect is potentially large; there is a strong positive correlation between the effects of deleterious alleles on the productivity of a population and their effects on competitive mating success (Whitlock and Bourguet 2000). Consequently, the strength of selection against these alleles is much greater than would have been predicted from their effects on productivity alone, and they fix less often than would otherwise be expected. As a result of the nonlinear relationship between the strength of selection and probability of fixation, a small change in s that does not result in a proportional change in mean fitness decline when fixed can reduce the effects of drift load by several orders of magnitude (Whitlock 2000).

Sex and selfing

So far only the effects of deleterious mutations in sexual, randomly mating populations have been discussed. The amount of drift load, and therefore the pace at which the mutational meltdown can occur, is expected to be much greater in either asexual populations or in populations that have a high degree of selfing. While these topics are too broad to give the detailed treatment they deserve in this chapter, we give the basic ideas of some of the results here.

The mutational meltdown process was first defined in terms of asexual populations (Lynch and Gabriel 1990). Without recombination between competing genotypes, deleterious alleles accumulate in asexual populations as a result of the stochastic loss of the fittest class with the fewest deleterious mutations. Without recombination, alleles are not required to fix in the population to reduce the evolutionary potential. If the class of genotypes with the fewest deleterious alleles is lost then it cannot be recreated by recombination, even if the fit versions of those alleles still exist in the population. This process is referred to as *Muller's Ratchet* because, as Muller described it, the population continually loses its fittest class, in which case the mean fitness can only decrease (Muller 1964; Felsenstein 1974; Charlesworth and Charlesworth 1997). The pace at which fitness is lost in an asexual population through new deleterious alleles is expected to be much faster than that in a sexual population of the same size.

Selfing increases the pace of meltdown for similar reasons. Deleterious alleles are more likely to fix in selfing populations, because the effective population size tends to be smaller and because alleles that fix within a selfing lineage are less likely to recombine with genotypes from other lineages. As a result, the effectiveness of selection is diminished, and the population declines more rapidly in mean fitness (Pamilo *et al.* 1987; Charlesworth *et al.* 1993a; Schultz and Lynch 1997).

9.5 Fixation of Beneficial Mutations

The theory of mutational meltdown presented thus far assumes that all mutations are deleterious and that no mutations have a beneficial effect for the organisms that carry them. Of course, some mutations must be beneficial, or else evolution could not be a positive process and all species would become extinct. This section explores the effects of a small population size on the rate of incorporation of these beneficial mutations.

Rate of back, beneficial, and compensatory mutations

The justification for ignoring beneficial mutations is that beneficial mutations seem to be much rarer than deleterious ones, a fact borne out by mutation accumulation experiments (see Chapter 7). However, many lines of evidence indicate that mutations with selective benefit do occur at nontrivial frequencies. Before this evidence is discussed, a few definitions are required. After a population has accumulated one or more deleterious mutations, it is possible for a new mutation to be beneficial even if it was not favored before. A new mutation can re-create a fit allele at a locus that had previously been fixed for a less fit allele; this is referred to as a *back*, or *reverse*, mutation. Furthermore it is possible that an allele at another locus, previously not favored by selection, may become selectively advantageous, perhaps because its action replaces or compensates for some of the effect of a deleterious mutant. These alleles are referred to as *compensatory* mutations. Finally, it is possible for a new allele to arise that is favored selectively in both the contexts of the deleterious allele and its absence; these mutations we include in the general group of beneficial mutants.

Back mutations for simple single-substitution mutations occur at a very low rate, and for some types of forward mutations, such as large deletions, back mutation is nearly impossible [see the references in Lande (1998b)]. However, back

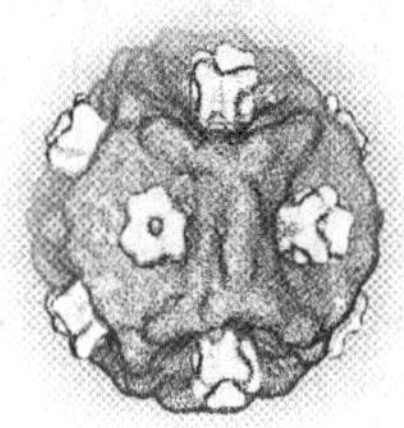

Phage $\phi 6$

mutations probably represent only a small fraction of the mutations that are able to ameliorate the fitness loss of deleterious mutations. Biological systems are typically highly redundant, and changes in one part of that system can be compensated for by changes elsewhere. A particularly striking example of this was provided by an experiment with the RNA virus $\phi 6$ by Burch and Chao (1999). In this experiment, replicate populations of RNA viruses were fixed for a deleterious allele that arose spontaneously in their cultures. These populations were then kept at different population sizes, ranging from $N_e = 60$ to 60 000, and the mean fitness of the populations was measured over time. In all but the smallest populations the fitness recovered to nearly the level of the nonmutant original strain. Moreover, this recovery took place in a series of discrete steps, which implies that the evolution subsequent to the original deleterious mutation took place as a result of the fixation of new alleles at other loci, *not* merely by the fixation of a back mutation at this locus. As RNA viruses have much higher mutation rates than large eukaryotes, the rate of their recovery may be different to that expected for the class of organisms for which there is the most conservation concern; however, this experiment strongly indicates the potential for compensatory mutations that, if prevalent, can dramatically change the dynamics of meltdown to extinction, as shown below.

Finally, a good geometric argument can be made that the number of beneficial mutations must increase as a population becomes more maladapted. Fisher (1930) argued that as a population neared its optimum, only mutations with a small effect could be beneficial, but if that same population were farther away from the optimum a much larger range of mutational effects would be adaptive. This argument could be advanced equally from the assumptions of quantitative genetics: if fixations of some alleles moved critical traits away from their fitness maxima, then any mutation that brings those traits back toward the optimum would be favored. A trait close to its best value would give a lower fitness if changed in value in either direction, while a trait away from its optimal value would be improved by that half of the possible changes that moved it back toward the optimum. By this line of argument, the possibility for compensatory mutations is large. The evolutionary consequences of genetic variability caused by such mutations with conditional fitness effects are explored in Chapter 10.

Rate of fixation of beneficial mutations in small populations

The relative rarity of beneficial or compensatory mutations has caused some authors to ignore their effects when determining the rate of loss of fitness of a small population. In contrast, Lande (1998b) showed that the time to extinction can be changed by an order of magnitude or more if the effects of back mutations alone are included. Compensatory mutations would effectively increase the rate of back mutations and further slow or even halt the loss of fitness through drift [Poon and

Otto 2000; Whitlock 2000; see also Wagner and Gabriel (1990) for a study of asexual populations]. While beneficial mutations may occur at a much lower rate than that for deleterious mutants, the selection process amplifies the fitness effects of these mutations because they have a relatively high probability of fixation.

There is, however, a critical population size at which the loss of fitness through the fixation of bad alleles cannot be compensated for by the fixation of favored alleles (Schultz and Lynch 1997). As the population size becomes small, the number of new favorable mutations that enters a population becomes low, because it is proportional to the population size. Simultaneously, the probability of fixation of beneficial mutations decreases as the effective population size decreases, while that of deleterious alleles increases. For small values of N_e, both classes have nearly the same fixation probability, namely that of neutral mutations (see Box 9.2). As a result there is a critical effective size at which the effects of the fixation of beneficial mutations cannot balance the fixation of deleterious mutations, and the population begins to lose mean fitness. This balance point is likely reached with effective population sizes in the low hundreds and depends on the mean effect of mutations and the proportion of mutations that are beneficial (Whitlock 2000).

Rate of fixation of mutations in declining populations

The paragraphs above deal with the effects of a constant small population size on the probability of fixation of new mutations. In reality, threatened populations may have a population size that is declining over time because of external factors. In such cases, the probability of fixation of new alleles with selective advantage $|s|$ becomes $2(|s| + r)N_e/N$, where r is the growth rate of the population size per generation and $r < 0$ in declining populations (Otto and Whitlock 1997). For values of r just below zero, this can nearly negate the probability of fixation of beneficial alleles. For negative values of $(|s|+r)$, there is essentially no possibility that the beneficial allele will be fixed.

Similarly, the probability of fixation of deleterious alleles in a declining population is greatly increased. In populations that are steadily reducing in size, the probability of fixation of an allele is given by the standard equation [Equation (a) in Box 9.2], but with the fixation effective population size given by $N_e \approx N(1+r/|s|)$ (Otto and Whitlock 1997). If the rate of population decline is a substantial fraction of the selection coefficient, then the effective size can be drastically reduced by population decline.

9.6 Time Scales for Extinction, Evolution, and Conservation

The most important question about this topic from a conservation biology perspective is, "How fast will a population of this size go extinct?" This question has been addressed by a number of authors, and the answer is, "It depends." We can, however, make some strong qualitative statements.

When a population is reduced in size artificially, many of the deleterious mutations carried by it before the population decline are lost through genetic drift or purging, but some become fixed through genetic drift. The population continues

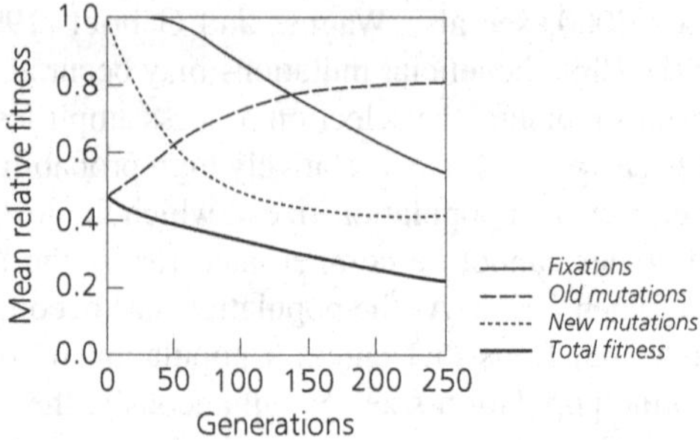

Figure 9.4 The evolution of mean fitness in a finite monoecious population of effective size $N_e = 32$, the ancestral population of which was infinite and in mutation–selection balance. The genomic mutation rate is $U = 0.75$, and the heterozygous effect of each deleterious mutation is $s = 0.025$, which gives an initial mean fitness of $e^{-U} = 0.472$. The fitness from "old" mutations includes purging and the fixation of deleterious alleles in the founder individuals, whereas the curve for "new" mutations refers only to segregating genes. The curve "Fixations" represents the fitness that results from the fixation of new and old deleterious mutations. *Source*: Lynch *et al.* (1995a).

to have new mutations, most of which are quickly lost, some of which persist for long enough to affect the population by mutation load, and a few of which are eventually fixed in the population through drift. It is this last phase of the process that presents the most danger to the population; the continued fixation of deleterious mutations may eventually result in a loss of fitness of the population such that it cannot sustain itself and it declines to extinction (see Figure 9.4). The critical effective population size at which this occurs depends on mutation rates and on assumptions about the possible effects the mutations might have.

The earliest estimates of the potential for drift load to cause extinction of a sexual population suggested that this was only likely to occur with a population size less than 100 (Charlesworth *et al.* 1993a). This study used a fixed mutational effect with only deleterious alleles, but more importantly it assumed that each individual was capable of producing an unlimited number of offspring. Hence, the population was unlikely to ever go extinct. Subsequent studies that relaxed this last assumption and set a limit on the reproductive capacity of individuals found that populations with effective size in the hundreds could be expected to become extinct within a few hundred or thousand generations (Lande 1994; Lynch *et al.* 1995a, 1995b; see Figure 9.5). These population sizes are effective sizes and so are likely to be an order of magnitude lower than the census size, which implies that for populations to be sustainable over more than a few hundred generations they must number in the thousands or tens of thousands.

Several factors make these predictions approximate. We have few data on the distribution of mutational effects; there is even great uncertainty about the mutation rate itself. It may be that mutational effects on the order of $0.4/N_e$ are uncommon, which could substantially prolong the expected lifespan of a species (Lande 1994). Furthermore, mutations of beneficial effect are likely to be non-negligible

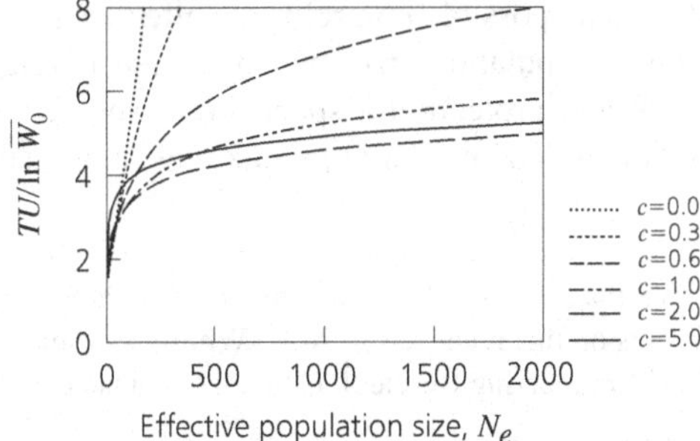

Figure 9.5 Length T of the phase of mutation accumulation until mean fitness is less than 1 [Equation (c) in Box 9.2] as a function of the effective population size N_e; T is scaled by the genomic mutation rate U and the logarithm of the initial mean fitness $\overline{W}_0$. It is assumed that the distribution of mutational effects $\Psi(s)$ is a gamma distribution with a mean of 0.025 and a coefficient of variation of c. The choice $c = 1$ yields an exponential distribution. For large values of c, the distribution is highly kurtotic (e.g., with $c = 5$, the kurtosis is 150). *Source*: Lande (1998b).

and reduce the risk of extinction substantially (Lande 1998b; Whitlock 2000). On the other hand, these models do not account for demographic stochastic effects, which could substantially increase the risk of extinction of an already mutation-weakened population. The calculations typically assume additive allelic effects, which somewhat underestimates the probability of fixation and therefore the rate of fitness loss. Finally, a small population size may substantially impair the ability to respond to environmental change (see Chapter 10). With this uncertainty, prudence argues for policies that err on the side of caution.

Unfortunately, these ideas are difficult to test experimentally. The declines in mean fitness expected through the meltdown are likely to take tens or hundreds of generations and should not show much effect on population productivity in the short term. It is essential that we learn more about the distribution of mutational effects and that careful long-term experiments be carried out to test the outcomes of these models (see Chapter 5).

9.7 Concluding Comments

If a species' population size was reduced to the hundreds, would it persist long enough for these genetic considerations to matter? To some extent this is not known, although current theory suggests that for populations of this size the risk presented by mutational meltdown is greater than the threat of demographic stochasticity and on a par with the risks of environmental stochasticity (Lande 1994). Of course, the risk of extinction caused by these longer term factors is irrelevant if sufficient habitat and protection is not secured for the short term.

The key parameters in determining whether a population will go extinct for genetic reasons and how long this will take are the effective population size, the mutation rate, and the distribution of deleterious and beneficial mutational effects.

The parameter that we may possibly control is the effective population size, which is related to the census population size, which in turn is related to the amount of undisturbed habitat and protection a species receives. It behooves us to use that control to maintain sufficiently large populations to prevent any meltdown to extinction.

Acknowledgments We thank Lance Barrett-Lennard, Art Poon, Sally Otto, and Mike Lynch for their comments on this manuscript. M.C. Whitlock is supported by a grant from the Natural Science and Engineering Research Council of Canada.

10

Quantitative-Genetic Models and Changing Environments

Reinhard Bürger and Christoph Krall

10.1 Introduction

Mutation is the ultimate source of genetic variability. However, a large fraction of mutations reduce the fitness of the individuals in which they occur (Chapter 7). The evolutionary consequences of mutations with an unconditionally deleterious effect are manifold and have been the subject of intense investigation (Charlesworth and Charlesworth 1998; Chapter 9). Since many mutations affect several traits and the developmental pathways are complex, their fitness effects may also depend on the genetic background in which they occur, and on the kind of selective pressure to which the population is exposed. For instance, if for a given trait, say birth weight, there is an optimal phenotype, a mutation that increases birth weight will be detrimental if it arises in a genotype that, otherwise, would have the optimal or a higher birth weight, but will be beneficial in other genotypes. In a changing environment, the selective value of an allele will change with time if different values of a trait affected by the allele are favored at different times. The fitness effect of a given mutation, therefore, depends on the effect it has on this trait, and on the current environment. Hence, a newly introduced allele may, in spite of its immediate adversary effect on fitness, prove to be beneficial at a later stage of the population history. Mutation itself thereby gains an additional role as a provider of the genetic variation that allows adaptation to occur. In this chapter, we evaluate the role of genetic variation, as caused by mutation, for population persistence if environmental change induces selection on one or more quantitative traits.

For the present purpose, environmental changes may be grouped roughly according to the mode in which they occur in time:

- Stochastic fluctuations of a certain parameter around a constant mean (e.g., temperature in tropical regions);
- Periodic fluctuations around a constant mean (e.g., seasonal fluctuations, oscillations in the life cycles of prey, predators, or parasites) that are at least partially predictable;
- Directional changes, such as global climatic changes, increasing concentration of certain substances because of increasing pollution, or gradual loss of habitat through human settlement or spread of a predator or pest (see Chapter 5);

- Single abrupt changes in the local environment, as caused by the sudden introduction of pests or pesticides; an abrupt change may also be faced by a founding colony in a novel habitat.

The different types of changes can be superimposed in arbitrary combination, and thereby pose different challenges upon the population and prompt different responses. These may range from immediate extinction to evolution sustained over long periods, possibly resulting in speciation (Chapter 7).

The response to environmental change will be influenced in various ways by ecological, demographic, and genetic factors. Ecological considerations take into account that environmental changes often influence a population not only by direct effects, but also via their effects on the focal population's preys, predators, mutualistic partners, or competitors. The interplay between direct and indirect effects determines the selective forces. So what may be experienced as a changing environment for a single species may, at a higher level of observation, be described as the intrinsic dynamics of the ecosystem under constant environmental conditions (see Chapters 16 and 17).

Demographic factors include the reproductive system, population size, intrinsic population growth rate, migration patterns, and so on. For instance, small populations are more affected by stochastic influences, and a high growth rate enhances population recovery after a bottleneck caused by a catastrophic event (see Chapters 2 to 4).

Consideration of the relevant properties of the genetic system of the population must include at least the degree of ploidy, the number of loci, and the way these interact, and must determine the phenotype (e.g., pleiotropy, if one gene has effects on several traits), recombination, and mutation. Mutation is crucial in the long run because it generates new genetic variability. Recombination breaks associations between alleles at different loci, and thus allows beneficial alleles to spread through the population and accelerates the elimination of deleterious alleles. Pleiotropy may impede adaptation by coupling the selectively advantageous change of one character with the maladaptation of pleiotropically connected characters.

The combined action of all these factors mentioned ultimately determines the genetic composition of the population and, via the amount of genetic variation, the rate of response to environmental challenges. For various scenarios of environmental change, we elucidate the:

- Role of genetic variation for adaptation;
- Rate of adaptive response; and
- Extinction risk of a population, as measured by the expected extinction time.

Demographic and genetic parameters are incorporated explicitly into the models, whereas ecological interactions are subsumed in the parameters that describe environmental change and selection.

10.2 Quantitative Genetics and Response to Selection

Quantitative characters are traits that exhibit continuous or almost continuous variation and can be measured on a metric scale. Typical examples are weight, height, various morphological measurements, yield, or fitness. Usually, such traits are controlled by a large number of gene loci, often with small effects. Since individuals in a population differ in their trait values, the state of a population is best described by the probability distribution of the trait. Quite often, the mean value and the variance are sufficient to predict the evolutionary response of a population to selection. In contrast to the frequencies of the genes that determine the character, the values of the mean, the variance, and (sometimes) the higher moments can be estimated accurately from real data.

Pioneering analyses to elucidate the genetic basis of inheritance and the response to selection of the mean of a quantitative character were made by Galton (1889), Pearson (1903), Fisher (1918), and Wright (1921), and their students Smith (1936) and Lush (1937). While the analyses of Galton (1889) and Pearson (1903) were of a purely statistical nature and based on regression theory, Fisher reconciled their biometric description with Mendelian genetics by assuming that a large number of unlinked loci with small additive effects determine the character. The work of Fisher (1918) and Wright (1921) forms the basis of classic quantitative genetic theory and its applications to animal and plant breeding (see, e.g., Bulmer 1980; Mayo 1987; Falconer and Mackay 1996; Lynch and Walsh 1998; Bürger 2000). Box 10.1 summarizes the basic aspects of the additive model of quantitative genetics.

The so-called breeder's equation, Equation (c) in Box 10.1, allows prediction of the change between generations on the basis of the selection differential. For many questions of evolutionary interest, however, selection is conveniently modeled by a fitness function $W(P)$, which assigns a fitness value to each phenotypic value P. In the simplest case, it measures the probability that an individual survives viability selection. Lande (1976, 1979, and later articles) extended the classic approach and derived dynamic equations for the change of mean phenotype of a set of quantitative characters in terms of the additive genetic covariance matrix and the so-called selection gradient. The fundamentals of his theory are summarized in Box 10.2 and form the basis for the subsequent analysis herein.

We now set up the general model on which the present results are based. A finite, sexually reproducing population of diploid individuals is assumed; it mates at random and, with respect to the traits considered, it has equivalent sexes. For simplicity, fitness is determined by a single quantitative character under Gaussian stabilizing selection on viability, with the optimum phenotype $\hat{P}_t$ exhibiting temporal change (but see subsection Pleiotropy and changing optima in Section 10.3 for some multivariate results). Thus, the more a phenotype deviates from the optimum the lower is its fitness. In mathematical terms, the fitness (viability) of an

Box 10.1 The classic additive genetic model

Quantitative traits are influenced by genes at many (ℓ) loci and by the environment. Genetically identical individuals may have different phenotypes because of external conditions (e.g., nutrition), developmental "noise", and cytoplasmatic effects. All these are lumped together into the so-called environmental contribution E, which (in the simple model discussed here) is assumed to be independent of the genotype and normally distributed with a mean of zero and a variance of V_E. Alleles at each locus have an effect on the character, measured by a real number. Let $\overline{x}_i$ and x_i denote the allelic effect of the maternally and paternally inherited alleles, respectively, at locus i. The fundamental assumption is that the alleles between the loci interact additively and, in the diploid case (as assumed here), show no dominance. Thus, the phenotypic value P of an individual is assumed to be

$$P = G + E = \sum_{i=1}^{\ell} (\overline{x}_i + x_i) + E \, , \tag{a}$$

where G is the genotypic value. Since the environmental contribution is scaled to have zero mean, the mean phenotypic value $\overline{P}$ equals the mean genotypic value $\overline{G}$. A consequence of the assumption of no genotype–environment interaction is that the mean phenotypic variance V_P can be decomposed into

$$V_P = V_G + V_E \, , \tag{b}$$

where V_G is the (additive) genetic variance (i.e., the variance of genotypic values). The assumption of additivity of allelic effects rests on the fundamental concept of the average or additive effect (Fisher 1930, 1941) and may be viewed as a least-squares approximation. Thus, the additive effects are found by an analysis of variance; indeed, Fisher invented the analysis of variance for this purpose. Often, an appropriate scale of measurement can be chosen so that the additivity assumption is a close approximation (Falconer and Mackay 1996). The variance of additive effects, in this case equal to V_G, is the main determinant of the response to selection.

Let S denote the selection differential, that is the within-generation difference between the mean phenotypes $\overline{P}$ (before selection) and $\overline{P}_S$ (after selection but before reproduction). The expected change in the mean phenotype across generations is then equal to

$$\Delta\overline{P} = h^2 S \, , \tag{c}$$

where $h^2 = V_G / V_P$, the ratio of additive genetic to phenotypic variance, and is called the heritability. It measures the fraction of variance that is heritable. Equation (c) is called the breeder's equation and is of fundamental importance because it allows prediction of the expected selection response from measurable quantities (see Mayo 1987; Falconer and Mackay 1996).

individual with phenotypic value P is conveniently described as

$$W(P,t) = \exp\Big(-\frac{(P-\hat{P}_t)^2}{2\omega^2}\Big) , \tag{10.1}$$

where ω^2 is inversely proportional to the strength of stabilizing selection and independent of the generation number t. Selection acts only through viability selection, and each individual produces b offspring. Initial populations are assumed to be in a stationary state with respect to stabilizing selection and genetic mechanisms when environmental change commences.

The following types of environmental change are modeled here:

- A phenotypic optimum that moves at a constant rate κ per generation,

$$\hat{P}_t = \kappa t ; \tag{10.2a}$$

- A periodically fluctuating optimum,

$$\hat{P}_t = A \sin(2\pi t/T) , \tag{10.2b}$$

  where A and T measure amplitude and period of the fluctuations, respectively;
- An optimum fluctuating randomly about its average position;
- A single abrupt shift of the optimum phenotype.

Under each of these models, the population experiences a mixture of directional and stabilizing selection. Such models of selection have been investigated previously by Lynch *et al.* (1991), Charlesworth (1993a, 1993b), Lynch and Lande (1993), Bürger and Lynch (1995, 1997), Kondrashov and Yampolsky (1996a, 1996b), Lande and Shannon (1996), Bürger (1999), and Bürger and Gimelfarb (2002).

The quantitative character under consideration is assumed to be determined by n mutationally equivalent, recombining loci. The additive genetic model of Box 10.1 is assumed and, as usual, the scale of measurement is normalized such that $V_E = 1$. Neither the genetic nor the phenotypic variance is assumed to be constant, which is indicated by a subscript t. The parameter $V_S = \omega^2 + V_E = \omega^2 + 1$ is used to describe the strength of stabilizing selection on the genotypic values G (Lande 1975). We then have $V_S + V_G = \omega^2 + V_P$.

Since this chapter is concerned with finite populations of effective size N_e, theoretical predictions are needed for the distribution of the mean phenotype, because it will fluctuate around its deterministically expected position. Let

$$s_t = \frac{V_{G,t}}{V_{G,t} + V_S} \tag{10.3}$$

be a measure for the strength of selection. Under the assumption of a Gaussian distribution of phenotypic values and a constant genetic variance, the distribution of the mean phenotype $\overline{P}_{t+1}$ in generation $t+1$, conditional on $\overline{P}_t$ and $\hat{P}_t$, is Gaussian. Its expectation is given by

$$\mathbb{E}(\overline{P}_{t+1} | \overline{P}_t, \hat{P}_t) = \overline{P}_t + s_t(\hat{P}_t - \overline{P}_t) . \tag{10.4}$$

This is a consequence of Equation (d) in Box 10.2 because mean fitness is calculated to be

$$\overline{W}_t = \frac{\omega}{v_t} \exp(-\tfrac{1}{2}(\overline{P}_t - \hat{P}_t)^2/v_t^2) , \tag{10.5}$$

where $v_t^2 \approx V_S(1 + 1/2N_e) + V_{G,t}$ (Latter 1970; Lande 1976; Bürger and Lynch 1995). These equations are very general and hold for arbitrary fitness functions of the form in Equation (10.1), so long as the phenotypic values remain approximately Gaussian.

Under prolonged environmental change, mean fitness may become very low. Since it is assumed that individuals can only produce a finite number b of offspring, a constant population size cannot necessarily be maintained because the (multiplicative) growth rate

$$R_t = b\overline{W}_t \tag{10.6}$$

may fall below 1. Therefore, a simple kind of density-dependent population regulation is imposed to ensure that the population size is close to the carrying capacity K, as long as the growth rate R_t is larger than 1, but allows extinction otherwise (Box 10.3).

The above theory and several of the consequences derived below are not based on a detailed genetic model, but assume a Gaussian distribution of phenotypes. In particular, the theory does not specify the mechanism by which genetic variation is generated and maintained. Therefore, computer simulations have been performed that use an explicit genetic model and enable the analytic approximations to be tested and the consequences of various assumptions about genetic parameters to be explored. This requires the mechanism by which genetic variability is maintained to be specified. It is assumed that this mechanism is mutation (see Box 10.3).

10.3 Adaptation and Extinction in Changing Environments

Classic quantitative genetics predicts that a population which experiences selection according to Equation (10.1) responds by shifting its mean phenotype according to Equation (10.4). If the optimum phenotype changes continuously, the mean phenotype will lag behind the fitness optimum. If this deviation is too large, the mean fitness may decrease to such an extent that the population cannot replace itself and declines, possibly to extinction. We are primarily concerned here with the role of genetic variation for the extinction risk that results from various forms of temporally varying environments.

Sustained directional change

For a model in which the optimum moves at a constant rate κ per generation, as in Equation (10.2a), a critical rate of environmental change κ_c has been identified beyond which extinction is certain because the lag increases from generation to generation, thus decreasing the mean fitness of the population below $\overline{W} < 1/b$, the level at which the population starts to decline. With a smaller population size, genetic drift reduces the genetic variance, which leads to an even larger lag, a

Box 10.2 Lande's phenotypic model of selection

Consider a randomly mating, large population such that random genetic drift can be ignored and assume that the sexes are equivalent with respect to the characters considered. The phenotype P of an individual is characterized by measurements of n traits, that is $P = (P_1, ..., P_n)^T$, where ^{T} denotes vector transposition. In analogy with the theory in Box 10.1, a decomposition $P = G + E$ is assumed with independent distributions of G and E that are multivariate normal with mean vectors $\overline{G}$ and 0, and covariance matrices V_G and V_E. Thus, the phenotypic covariance matrix is $V_P = V_G + V_E$. The normality assumption is justified if, as is often the case in practice, a scale can be found on which phenotypic values are approximately normally distributed (Falconer and Mackay 1996). This is also expected from the central limit theorem, because many loci, as well as environmental effects, contribute to quantitative traits.

If $W(P)$ denotes the fitness of an individual with phenotype P, then the mean fitness of the population is

$$\overline{W} = \int \phi(P) W(P)\, dP \,, \tag{a}$$

where $\phi(P)$ is the (multivariate) probability density of P. After selection, the mean vector is

$$\overline{P}_S = \frac{1}{\overline{W}} \int P \phi(P) W(P)\, dP \,, \tag{b}$$

so that the vector of selection differentials is $S = \overline{P}_S - \overline{P}$. In a generalization of Equation (c) in Box 10.1, the change between generations is calculated to be

$$\Delta \overline{P} = V_G V_P^{-1} S \,. \tag{c}$$

Let ∇ denote the gradient operator, that is $\nabla g = (\partial g / \partial x_1, ..., \partial g / \partial x_n)^T$ is the vector of partial derivatives of the real-valued function g depending on the n arguments $x_1, ..., x_n$. Then, the mean fitness, expressed as a function of the vector of mean phenotypes $\overline{W}(\overline{P}_1, ..., \overline{P}_n)$, can be viewed as an adaptive topography that determines the response of the mean phenotype to selection. Indeed, Lande (1976, 1979) derived the fundamental relation

$$\Delta \overline{P} = V_G \, \nabla \ln \overline{W} \,, \tag{d}$$

where the derivatives in the selection gradient $\nabla \ln \overline{W}$ are taken with respect to $\overline{P}_1, ..., \overline{P}_n$. For a single trait and with the notation from Box 10.1, Equation (c) reduces to

$$\Delta \overline{P} = V_G \frac{1}{\overline{W}} \frac{d\overline{W}}{d\overline{P}} \,. \tag{e}$$

Lande and others have applied this theory (and several generalizations) to numerous problems of evolutionary biology. The application to long-term evolution, however, requires knowledge of the genetic variances and covariances. Many of these analyses relied on the assumption that the phenotypic and genetic covariance matrices change on a much slower time scale than the mean values. These assumptions have been the subject of intense discussion and analysis (see Turelli 1984, 1988; Barton and Turelli 1989; Turelli and Barton 1994; Bürger 2000).

Box 10.3 A numerical model for adaptation in changing environments

The simulation model has been adapted from that used in Bürger *et al.* (1989) and Bürger and Lynch (1995). It uses direct Monte Carlo simulation to represent each individual and each gene. The genotypic value of the character is determined by ℓ additive loci with no dominance or epistasis. In the present investigation $\ell = 50$ was chosen.

Following Crow and Kimura's (1964) continuum of alleles model, at each locus an effectively continuous distribution of possible effects for mutants is assumed. Thus, provided an allele with effect x gives rise to a mutation, the effect of the mutant is $x + \xi$, where ξ is drawn from a distribution with mean zero, variance γ^2, and no skewness. Hence, the number of possible segregating alleles per locus is limited only by population size. The mutation rate per haploid locus is denoted by u, the genomic mutation rate by U, and the variance introduced by the mutation per generation per zygote by $V_m = U\gamma^2$. Unless otherwise stated, a Gaussian distribution of mutational effects with a mean of zero and variance $\gamma^2 = 0.05$ and a (diploid) genomic mutation rate of $U = 0.02$ per individual and generation are assumed. This implies that the variance introduced through mutation in each generation is $V_m = 0.001$. These values have been suggested as gross averages by reviews of empirical data (see Lande 1975; Turelli 1984; Lynch and Walsh 1998).

The phenotypic value of an individual is obtained from the genotypic value by adding a random number drawn from a normal distribution with a mean of zero and variance $V_E = 1$.

The generations are discrete, and the life cycle consists of three stages:

- Random sampling without replacement of a maximum of K reproducing adults from the surviving offspring of the preceding generation;
- Production of offspring, including mutation, segregation, and recombination;
- Viability selection according to Equation (10.1).

Modification of this model to allow for two pleiotropically related traits is straightforward. Each allele is now written as a vector (x, y), in which the two entries represent the effect of the allele on each of the traits it influences. Mutation is modeled by the addition of a vector (ξ, ζ), the components being drawn independently from a Gaussian distribution as in the single-character case. Viability selection acts on both characters and is modeled by a bivariate extension of the fitness function given by Equation (10.1). It is assumed that the optimum moves in the direction of the first trait, thus leading to directional selection on this trait, while the second trait remains under stabilizing selection. The width of the fitness functions that acts on the first and second traits are denoted by ω_1 and ω_2, respectively.

The maximum number K of reproducing adults may be called the carrying capacity. The $N_t(\leq K)$ adults in generation t produce bN_t offspring, an expected $R_t N_t$ of which will survive viability selection. In this way, demographic stochasticity is induced. The sex ratio among parents is always 1:1 and $N/2$ breeding pairs are formed, each of which produces exactly $2b$ offspring. If the actual number of

continued

Box 10.3 *continued*

surviving offspring is larger than K, then K individuals are chosen randomly to constitute the next generation of parents. Otherwise, all the surviving offspring serve as parents for the next generation. The effective population size is $N_e = 4N/(V_f + 2)$, where the variance in family size is $V_f = 2(1 - 1/b)[1 - (2b - 1)/(bN - 1)]$.

For each parameter combination, a certain number of replicate runs with stochastically independent initial populations were carried out. Each run was over 10^5 generations, unless population extinction had occurred previously. The initial populations were obtained from a preceding initial phase of several hundreds or thousands of generations (depending on N) during which mutation–selection balance had been reached. The number of replicate runs per parameter combination was chosen such that the standard errors were less than 5% (and often on the order of 1%).

further decrease of mean fitness, and rapid extinction (Lynch and Lande 1993; Bürger and Lynch 1995, 1997).

If the rate of environmental change is sufficiently low, then the mean phenotype lags behind the optimum, but after several generations evolves parallel with it. From Equation (10.4) it can be derived easily that the asymptotic average lag is given by κ/s, with $s \approx s_t$ as in Equation (10.3) and $V_{G,t} = V_{G,\text{move}}$ the asymptotic genetic variance (Lynch *et al.* 1991; Lynch and Lande 1993). The genetic load induced by this lag has been called the evolutionary load (Lande and Shannon 1996; see Chapter 9 for genetic loads) and can be calculated. Indeed, from Equation (10.5) and the fact that the lag converges to κ/s, the asymptotic mean fitness is readily calculated to be

$$\mathbb{E}(\overline{W}_{\text{move}}) \approx \frac{\omega}{v} \exp\left(-\tfrac{1}{2}\kappa^2/(s^2v^2)\right) , \tag{10.7}$$

where $v^2 = V_S(1 + 1/2N_e) + V_{G,\text{move}}$. Differentiation shows that $\mathbb{E}(\overline{W}_{\text{move}})$ is an increasing function of the genetic variance if $V_{G,\text{move}}^3 > 2V_S^2\kappa^2$, that is unless the genetic variance is very large or κ is very small (Figure 10.1a; see Charlesworth 1993b; Lande and Shannon 1996). This, however, does not imply that in a real genetic system the variance evolves such as to maximize mean fitness (see below).

The critical rate of environmental change κ_c is defined as the value of κ at which the population can just replace itself, so that $b\mathbb{E}(\overline{W}_{\text{move}}) = 1$. Unless the population size is very small (less than, say, two dozen) or the stabilizing selection component is extremely weak, κ_c can be approximated by

$$\kappa_c \approx V_{G,\text{move}}\sqrt{2(\ln b)/V_S} , \tag{10.8}$$

see Lynch and Lande (1993) and Bürger and Lynch (1995). Equations (10.7) and (10.8) are deceptively simple because the determinants of the genetic variance have not yet been elucidated. However, it is obvious from Equation (10.8) that the

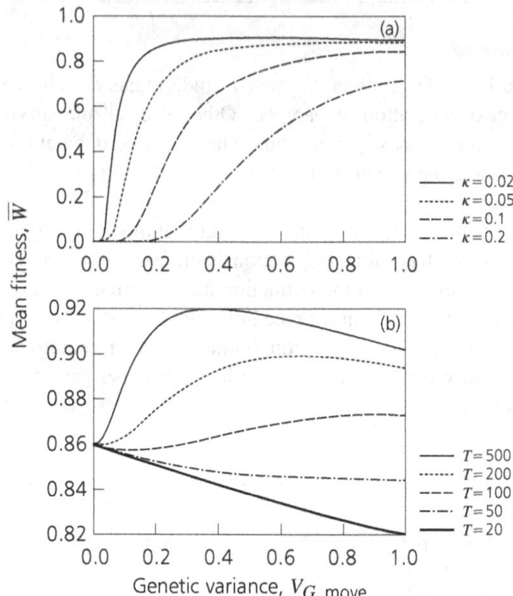

Figure 10.1 Dependence of mean fitness on genetic variance. (a) displays the mean fitness of a population subject to environmental change according to the moving optimum model, Equations (10.1) and (10.2a), as function of the genetic variance $V_{G,\text{move}}$ for four different rates of environmental change κ. The curves are calculated from Equation (10.5). The population is assumed to be infinitely large and the width of the phenotypic fitness function is $\omega = 3$ ($V_S = 10$). For large $V_{G,\text{move}}$ the mean fitness decreases because of the stabilizing component of selection. (b) Analagous to (a), but for a periodically changing environment according to Equations (10.1) and (10.2b). The amplitude is $A = 2\omega$ and the curves, calculated from Equation (10.11), are for five different periods, as indicated. Obviously, more genetic variance is beneficial only for long periods T. At $V_{G,\text{move}} = 0$ the derivative of mean fitness is always negative.

genetic variance is the major limiting factor for the rate of environmental change that can be tolerated by a population.

This theory can be extended to derive an approximate expression for the mean time to extinction by recognizing that for $\kappa > \kappa_c$, the extinction process consists of two phases (Bürger and Lynch 1995). During phase 1, the multiplicative growth rate R_t [Equation (10.6)] decreases to 1, but the population size remains at the carrying capacity. The length t_1 of this phase is easily estimated by substituting Equation (10.7) into Equation (10.6) and solving the equation $R_t = 1$ for t. This produces

$$t_1 \approx -\frac{1}{s} \ln\left(1 - \frac{\kappa}{\kappa_c}\right) . \tag{10.9}$$

The length t_2 of the second phase can be obtained by numerical iteration of the recursion $N_{t+1} = R_t N_t$ until the population size reaches 1. The second phase

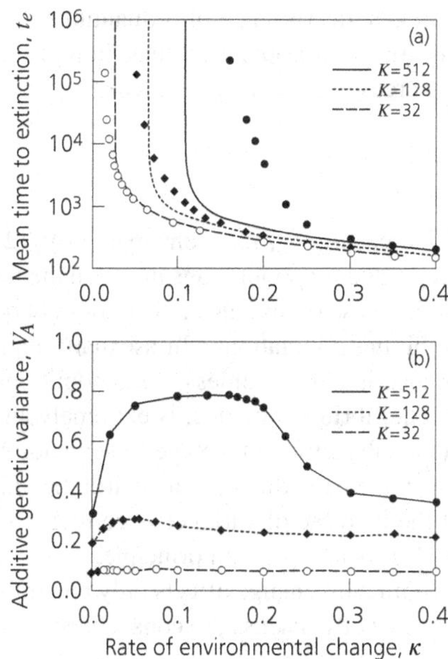

Figure 10.2 Evolution and extinction in a directionally changing environment. (a) The mean time to extinction as a function of the rate of environmental change κ for three different population sizes K. Data points are from Monte Carlo simulation, whereas lines are based on the quasi-deterministic theory for t_e, as described in Equation (10.9) and the text below. This approximation assumes that the genetic variance does not change after the onset of environmental change, an assumption that is valid only in small populations. Since all genetic variances are less than 1 (see Figure 10.2b) and the environmental variance is normalized to 1, all phenotypic variances are between 1 and 2. Hence, the value $\kappa = 0.1$ corresponds to less than 10% of a phenotypic standard deviation. (b) The observed genetic variance in the simulations of Figure 10.2a. The mutational parameters are as in Box 10.3, and the other parameters are $b = 2$ and $\omega = 3$. Most data points at which extinction occurred are averages over 100 replicate runs. *Source*: Numerical data mostly from Bürger and Lynch (1995).

is typically much shorter than the first one. This theory for the mean time to extinction, $t_e = t_1 + t_2$, produces good approximations if $\kappa \ll \kappa_c$ (Figure 10.2a), although it neglects several sources of stochasticity (fluctuations of R_t about its mean, demographic stochasticity, stochasticity and autocorrelation resulting from genetic events like mutation and recombination). Most importantly, it requires knowledge of the actual genetic variance of the population.

An important observation is that the genetic variance actually increases in response to the moving optimum, but only if the population size is sufficiently large (Figure 10.2b). Therefore, if the population size is higher than a few hundred individuals, the mean time to extinction is longer than predicted by the above theory if this uses the initial genetic variance (see the case $K = 512$ in Figure 10.2a). Recently, the determinants of this increase were investigated in some detail (Bürger

1999). It was shown that the genetic variance of a quantitative trait determined by many freely recombining loci in a population that, initially, is in mutation–selection–drift equilibrium, increases at least by the factor

$$\frac{\gamma}{2\sqrt{uV_S}}\,\frac{N_e + V_S/\gamma^2}{N_e + \frac{1}{2}\sqrt{V_S/(u\gamma^2)}}\,, \tag{10.10}$$

unless κ is very small. For increasing N_e, this converges to $\gamma/(2\sqrt{uV_S})$. It was also shown that in sexually reproducing populations in which the trait is controlled by completely linked loci, an increase of variance either does not occur or is much smaller than in freely recombining populations. In asexually reproducing populations, this increase of variance is absent, unless the genomic mutation rate for the trait, and thus the initial equilibrium variance, is extremely small. This flexibility of the genome confers a substantial advantage to sex and recombination if the population is subject to sustained and directional environmental change. Equation (10.10) also shows that the increase of variance is constrained by the genetic system and not (directly) guided by an optimum principle.

The assumption that environmental change affects only one trait is a gross simplification. In the following we briefly discuss the consequences of pleiotropy.

Pleiotropy and changing optima

Many genes have effects on several characters, and thereby cause a statistical association of heritable variation among different phenotypic traits. This pleiotropic connection of characters may have important consequences for their responses to selection. We employed a simplified version of a model of Lande (1980b), in which two traits are determined by the same set of loci, as described in Box 10.3. Mutation modifies an allele's contribution to the two characters independently, and there is no selectional correlation between the characters. Figure 10.3 displays the results of Monte Carlo simulations that evaluate the role of pleiotropy on the evolutionary capacity of the population, in which the first trait is under sustained directional selection while the second is under stabilizing selection, as described in Box 10.3. The curves in the figure correspond to different strengths of stabilizing selection on the second trait. The top curve represents the case of a single trait, subject to the moving optimum model described in the previous section (Sustained directional change), since this is equivalent to a neutral second trait ($\omega_2 = \infty$).

Figure 10.3 shows that in the single-character case the dependence of the mean time to extinction on the strength of stabilizing selection is bell shaped and has a maximum near $\omega_1 = 2$, which corresponds to strong selection. This is so because very strong selection focuses the population mean close to the actual optimum, but destroys most of the genetic variance needed to respond to further changes, whereas weak stabilizing selection admits more genetic variance, but leads to a very large lag (see Huey and Kingsolver 1993; Bürger and Lynch 1995). This bell shape of the curve persists if the second character is exposed to increasingly strong selection. However, increasing stabilizing selection on the second character always accelerates population extinction, for three reasons:

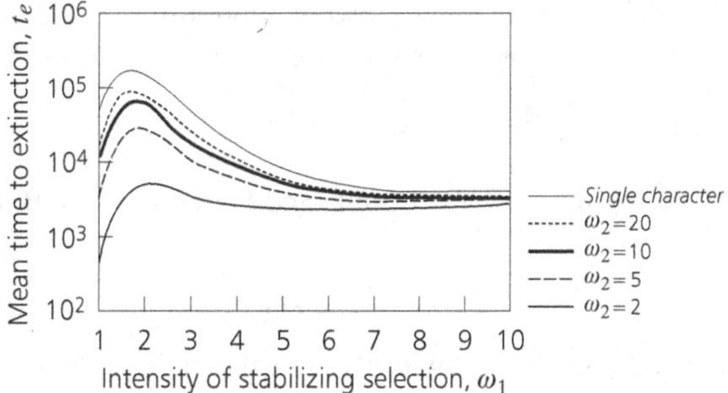

Figure 10.3 Dependence of mean time to extinction in a directionally changing environment with pleiotropic gene action. Mean extinction time is displayed as a function of stabilizing selection strength ω_1 on a character with a moving optimum for various intensities of stabilizing selection ω_2 on a second, pleiotropically coupled character with a constant optimum. The rate of environmental change in the direction of the first trait is $\kappa = 0.055$ per generation, and for both characters the mutational parameters are as in Box 10.3. Parameters ω_1 and ω_2 are the widths of the phenotypic fitness function for each of the two characters. The case of single character (no pleiotropy) is equivalent to $\omega_2 = \infty$. Other parameters: $K = 64, b = 2$.

- Stabilizing selection on a genetically variable second trait reduces the mean fitness of the population by introducing an additional load.
- As a result of pleiotropy, stabilizing selection on the second character reduces the equilibrium variance of the first one (Lande 1980b; Turelli 1985; Wagner 1989). Therefore, when the optimum starts to move (remember that we allow the population to reach mutation–selection equilibrium in a constant environment before environmental change commences), the selection response is reduced; see Equations (c) and (d) in Box 10.2, and Equation (10.4).
- A moving optimum increases genetic variance by favoring mutations with positive effects. Since these mutations also affect the second trait, stabilizing selection tries to eliminate them.

In the present simulations, no correlation between the mutational effects on the two characters was assumed. Therefore, on average, the traits are uncorrelated. Hence, the lag of the wandering character is affected only by the decrease of variation caused by selection on the pleiotropically connected trait. With correlation, however, mutations in a favorable direction for the moving trait have a tendency to push the mean value of the second character away from its optimum. Therefore, selection on the second, now correlated, trait impedes adaptation of the first trait even more, and thereby increases the lag and extinction risk even further (results not shown).

Periodic change

In a periodically varying environment, Equation (10.2b), more genetic variance is not necessarily beneficial for population persistence. This can be seen from

Figure 10.1b, which is based on the following approximation for the mean fitness averaged over one full cycle, after a sufficiently long initial phase has elapsed,

$$\overline{W}_{\text{per}} \approx \frac{\omega}{v} \exp\left(-\lambda + \tfrac{1}{4}\lambda^2\right) , \tag{10.11a}$$

where the expected log mean fitness (Lande and Shannon 1996; Bürger 1999) is

$$\lambda = \tfrac{1}{2}\mathbb{E}(\Delta^2)/(V_S + V_G) \approx \frac{A^2\pi^2}{V_S(s^2T^2 + 4\pi^2)} . \tag{10.11b}$$

Here, s is given by Equation (10.3) with $V_{G,t}$ equal to the genetic variance averaged over a full cycle. The above equations assume a large population size but, as shown by comparison with Monte Carlo simulations, yield close approximations for populations above 100–200 individuals.

The detailed dynamics of evolution and extinction for both finite sexual and asexual populations were investigated by Monte Carlo simulations, as described in Box 10.3. No assumptions are imposed on the distribution of phenotypic (or genotypic) values. Some of the results are summarized in Figure 10.4. Further results are found in Bürger (1999, 2000). Figure 10.4 displays the mean time to extinction and the average genetic variance of a freely recombining sexual population, of a nonrecombining sexual population, and of an asexual population as a function of $\kappa = 4A/T$. These populations differ substantially in their initial genetic variance (with recombination and sex more genetic variance is maintained at mutation–selection balance than without; see Bürger 1999), as well as in their ability to adapt their genetic variance to a higher, selectively favorable, level (Figure 10.4b). Figure 10.4a shows that for intermediate periods of T, the freely recombining population persists for much longer than the two other populations. The reason is that in this case the populations have to adapt to the changing optimum, which requires much genetic variance. The freely recombining population not only has a higher initial variance, but is also able to increase its level of variance, and thus obtains a substantial advantage over the two other populations. For very long periods, all three types of population harbor enough genetic variance to track the optimum. If the optimum changes rapidly, so that it returns to its initial state every few generations, more genetic variance is not beneficial. In this case, it makes sense for a population to stay where it is and wait until the environmental optimum returns. Clearly, this requires that the population be able to maintain a minimum viable population size during periods of low fitness.

Stochastic fluctuations

Several types of models have been investigated to evaluate population persistence in stochastically fluctuating environments. A large body of literature on environmental stochasticity neglects the genetic structure of populations and studies extinction of monomorphic populations under a variety of assumptions about the demography and ecology of the population [see Chapter 2, and Bürger and Lynch (1997) for reviews].

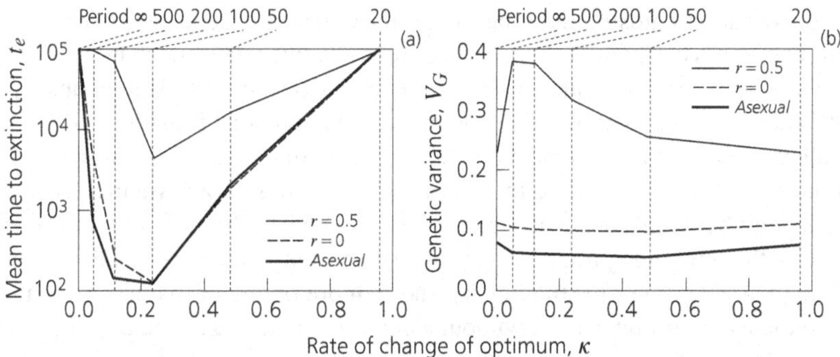

Figure 10.4 Evolution and extinction in a periodic environment. (a) The mean time to extinction as a function of $\kappa = 4A/T$ for a freely recombining sexual population, a non-recombining sexual population, and a diploid asexual population. Here, $\kappa = 4A/T$ can be interpreted as the rate of change of the optimum, averaged over one full cycle (during which the optimum moves $4A$ units, measured in multiples of V_E). Dynamically, an infinitely long period T is equivalent to a resting optimum. The amplitude A of the periodic optimum is chosen to be 2ω, which implies that at the most extreme position of the optimum (A units from the origin) the originally optimal phenotype (at position 0) has a fitness of 13.5%. Data points are from Monte Carlo simulation. Parameters: $K = 256$, $b = 5$, $\omega = 3$, and the number of loci and the mutation parameters are as in Box 10.3. (b) The observed genetic variance in the simulations of Figure 10.4a.

Recently, more attention has been paid to the role of genetic variability in population survival in randomly varying environments. Charlesworth (1993b) and Lande and Shannon (1996) investigated fluctuating stabilizing selection on a quantitative trait by assuming that the optimum $\hat{P}_t$ in Equation (10.1) follows a linear stationary Markov process with mean zero, variance V_θ, and autocorrelation c between –1 and 1. They assumed the Gaussian phenotypic model of Box 10.2 and a constant genetic variance. Charlesworth (1993b) showed that the expected log mean fitness increases with increased genetic variance if

$$\frac{V_\theta}{V_S} > \frac{2(1-c)}{1+c} . \tag{10.12}$$

Thus, genetic variation is only beneficial if the variance of the fluctuations is high or if the process is highly autocorrelated, in which case adaptation, that is tracing the optimum, increases the mean fitness. For a continuous-time model (which precludes large instantaneous fluctuations), Lande and Shannon (1996) also showed that in autocorrelated environments more genetic variation is usually beneficial, but for uncorrelated environments they found that more genetic variation always decreases mean fitness.

These authors did not consider population extinction, but assumed an unlimited reproductive potential and a constant variance. Bürger (1999) employed the simulation model described in Box 10.3 to study population extinction for the case of no autocorrelation ($c = 0$). He found that, even for $V_\theta / V_S \approx 1$, higher levels of genetic variation can slightly enhance the mean persistence times. Nevertheless, with such an environmental change a high reproductive rate is much more effective at improving population longevity than a high genetic variance. Bürger and Lynch (1995) considered a mixture of the moving optimum model, Equation (10.2a), and the above model of environmental stochasticity with $c = 0$ by assuming that the optimum $\hat{P}_t$ evolves according to $\hat{P}_t = \kappa t + \varepsilon_t$, where ε_t is a normally distributed random number with a mean of zero and variance of V_θ. They showed that, relative to a smoothly moving optimum, superimposition of mild levels of stochasticity can reduce the mean time to extinction by one or more orders of magnitude. This reduction is most pronounced if, without stochastic fluctuations, the expected mean extinction time of the population is above several thousands of generations, so that from environmental change alone the extinction risk is small. This indicates that the synergistic interactions of environmental changes that separately cause only a minor risk can cause rapid extinction. The amount by which the critical rate of environmental change is reduced by small random fluctuations of the optimum has been calculated by Lynch and Lande (1993).

Single abrupt change

Gomulkiewicz and Holt (1995) investigated environmental change caused by a single shift of the optimum phenotype of a quantitative trait to a new constant value. Such an abrupt change reduces the mean fitness of the population, possibly to the extent that the multiplicative growth rate falls below 1. If the maladaptation caused by the shift is too severe, rapid extinction of the population will be the consequence. Survival of a population after such a shift may be possible if adaptive variation admits sufficiently rapid evolution toward the new optimum, thus leading to an increase in growth rate. Even if this is the case, the population size decreases for some time because of the low initial fitness. If it decays below a certain critical value, it may be highly endangered by demographic stochasticity. Therefore, "the problem of population persistence in a novel environment can be viewed as a race between two processes, one demographic, another evolutionary" (Maynard Smith 1998; see also Gomulkiewicz and Holt 1995).

10.4 Concluding Comments

What is the practical value of these models for the conservation biologist? Given the limitations conservation biology faces in practice, it may well be that such models only offer possible explanations for the extinction of populations rather than provide measures to ensure their survival. For small populations of up to a few hundred individuals, extinction is an almost certain event, even under very moderate, but sustained, changes of the environment that shift the optimal value of a trait by just a few percent of a phenotypic standard deviation per generation. If environmental change proceeds too fast (on the order of 10% of a standard deviation or more), even a large population size cannot guarantee survival for a long period, in particular if additional stochastic fluctuations occur in the environment, or if pleiotropically connected traits are under stabilizing selection (see Figures 10.2

and 10.3; and Bürger and Lynch 1995). This is a crucial difference to extinction risks caused by genetic factors, such as the accumulation of deleterious mutants (Chapter 9), or by demographic and environmental stochasticity (Chapter 2), in which the risk decreases rapidly with increasing population size. Population size is the parameter that can be influenced most directly by conservation biology efforts; however, as shown above, a large population size is not always sufficient for population survival during prolonged episodes of environmental change. The case might be more promising with populations that face one abrupt shift to a new constant environment. Here, direct measures may be taken to support the population during the period when it is below the critical density.

The consequences of genetic variability for population survival depend on the kind of environmental change. A high level of genetic variance improves population performance under continuous directional change, under a single abrupt shift of the optimum, under a cyclically varying environment (if the amplitude is not too small and the period is long), and under stochastic fluctuations with high variance or a high autocorrelation. In such cases, the only means to survive is adaptation to the environment, which requires genetic variance. In a constant environment, in a periodically changing one with a short period, and in a randomly fluctuating environment with small variance and no autocorrelation, more genetic variance may even be slightly detrimental, because it increases the load caused by stabilizing selection (see Slatkin and Lande 1976; Charlesworth 1993a, 1993b; Lande and Shannon 1996). In such environments, the production of many offspring may substantially increase population persistence (Bürger 1999).

In our models, several factors with possibly strong influences on population persistence were not considered. Populations were assumed to be panmictic, the environment was unstructured, the life cycle was very simple, rates of mutation were assumed to be constant, and unconditionally deleterious mutations were neglected. For a discussion of complex life cycles see Chapter 7, for the effects of a metapopulation structure and a spatially structured environment, see Part D. Chapter 8 discusses mechanisms for variable mutation rates.

Shifts in the population mean of more than several standard deviations have been observed in artificial selection experiments, which shows that at least for some characters there is enough genetic variation to respond to large changes [see Barton and Keightley (2002) for references]. These shifts, however, at least partially resulted from the spread of recessive deleterious alleles. Therefore, when selection was relaxed, the mean often returned to a value between its initial value and its maximum value. In addition, lines were lost in such experiments because of reproductive failure. Whether a particular trait has the potential for substantial evolution depends on a multitude of genetic and demographic details and, in general, is hardly predictable.

Acknowledgments This work was financially supported by the Austrian Science Foundation (FWF), Projects P10689-MAT and P12865-MAT.

11

Adaptive Dynamics and Evolving Biodiversity

Ulf Dieckmann and Régis Ferrière

11.1 Introduction

Population viability is determined by the interplay of environmental influences and individual phenotypic traits that shape life histories and behavior. Only a few years ago the common wisdom in evolutionary ecology was that adaptive evolution would optimize a population's phenotypic state in the sense of maximizing some suitably chosen measure of fitness, such as its intrinsic growth rate r or its basic reproduction ratio R_0 (Roff 1992; Stearns 1992). On this basis it was largely expected that life-history evolution would always enhance population viability. In fact, such confidence in the prowess of adaptive evolution goes back as far as Darwin, who suggested "we may feel sure that any variation in the least degree injurious would be rigidly destroyed" (Darwin 1859, p. 130) and, in the same vein, "Natural selection will never produce in a being anything injurious to itself, for natural selection acts solely by and for the good of each" (Darwin 1859, p. 228).

The past decade of research in life-history theory has done away with this conveniently simple relation between population viability and evolution, and provided us with a picture today that is considerably more subtle:

- First, it was realized the optimization principles that drive the evolution of life histories could (and should) be derived from the population dynamics that underlie the process of adaptation (Metz *et al.* 1992, 1996a; Dieckmann 1994; Ferrière and Gatto 1995; Dieckmann and Law 1996). In the wake of this insight, the old debate as to whether r or R_0 was the more appropriate fitness measure (e.g., Stearns 1992; Roff 1992) became largely obliterated (Pásztor *et al.* 1996).
- Second, we now understand that the particular way in which population densities and traits overlap in their impact on population dynamics determines whether an optimization principle can be found in the first place, and, if so, what specific fitness measure it ought to be based on (Mylius and Diekmann 1995; Metz *et al.* 1996b). It thus turns out that for many evolving systems no optimization principle exists and that the conditions that actually allow the prediction of life-history evolution by maximizing r or R_0 are fairly restrictive (e.g., Meszéna *et al.* 2001; Dieckmann 2002).
- Third, it became clear that, even when adaptive evolution did optimize, the process would not necessarily maximize population viability (Matsuda and Abrams 1994b; Ferrière 2000; Gyllenberg *et al.* 2002; Chapter 14). In addition, it has been shown recently that, even when adaptive evolution gradually

improves population viability, such a process could eventually lead to a population's sudden collapse (Renault and Ferrière, unpublished; Parvinen and Dieckmann, unpublished).

This chapter expounds in detail the intricate link between adaptive evolution and population viability. Section 11.2 reviews conceptual limitations inherent in the traditional approaches to life-history evolution based on optimization criteria, and Section 11.3 introduces adaptive dynamics theory to overcome these limitations. Adaptive evolution without optimization has intriguing consequences for the origin and loss of biodiversity, and these implications are reviewed in Sections 11.4 and 11.5, respectively. While the processes described there can unfold in a constant environmental setting, Section 11.6 provides an overview of how the viability of adapting populations can be affected by environmental change.

11.2 Adaptation versus Optimization

Life-history optimization in the form of maximizing r or R_0 has been applied widely to a variety of questions in evolutionary ecology, including the evolution of clutch size, age and size at maturation, sex ratio, reproductive systems, and senescence. Unfortunately, however, this approach faces several fundamental limitations. Since these restrictions are conceptually important and have wide-ranging significance for evolutionary conservation biology, we discuss them in some detail, before, in the next section, summarizing a framework with which to surmount the difficulties.

Optimization in earlier evolutionary theory

Despite repeated discussions about the limitations of optimizing selection (e.g., Lewontin 1979, 1987; Emlen 1987), it is surprising how long it has taken to account thoroughly for these limitations in the practice of evolutionary ecology research – to the extent that this process is still ongoing today. We thus start out with a brief sketch of some key earlier approaches that favored the idea of evolution as an optimizing process:

- Following a notion introduced by Wright (1932) early on in the modern synthesis, adaptive evolution is often envisaged as a hill-climbing process on a fixed-fitness landscape. Whereas Wright originally considered adaptive landscapes based on the dependence of mean population fitness on genotype frequencies, subsequent work extended Wright's concept by utilizing adaptive landscapes to describe the dependence of individual fitness on phenotypes. Yet, Wright himself recognized that the adequacy of his convenient metaphor was lost when selection was frequency dependent (Wright 1969, p. 121).
- The same conclusion applies to Fisher's so-called "fundamental theorem of natural selection" (Fisher 1930). This predicts mean population fitness to increase monotonically over the course of adaptive evolution – provided, however, that certain restrictive assumptions are fulfilled. It is not surprising that one of these assumptions is the constancy of fitness values, and thus the absence of

frequency-dependent selection (Roughgarden 1979, p. 168; Frank and Slatkin 1992). To reconcile this assumption with the fact that, in the long-term, the mean absolute fitness of a population must hover around zero, Fisher stipulated a balance between the "progress" of natural selection and a "deterioration" of the environment: "Against the rate of progress in fitness must be set off, if the organism is, properly speaking, highly adapted to its place in nature, deterioration due to undirected changes either in the organism [mutations], or in its environment [geological, climatological, or organic]" (Fisher 1930). The quote illustrates that when explaining the environment's "deterioration" Fisher did not appear to have thought of density- or frequency-dependent selection. Today, evolutionary ecologists realize that a phenotype possessing a relative fitness advantage when rare loses this advantage once it has become common. As we show below, the infamous environmental deterioration simply results from a changing composition of the evolving population itself. Therefore, density- and frequency-dependent selection are at the heart of reconciling the conflict between Fisher's theorem and long-term population dynamics.

- Also, the fitness-set approach developed by Levins (1962a, 1962b, 1968) still enjoys widespread recognition in life-history evolution (Yodzis 1989, pp. 324–351; Calow 1999, p. 758; Case 1999, pp. 175–177). It is based on the assumption that, within a set of feasible phenotypes defined by a trade-off (the "fitness set"), evolution maximizes fitness (referred to as the "adaptive function" by Levins). Since the adaptive function is assumed to remain constant in the course of evolution, selection is optimizing and frequency-dependent selection is excluded.
- Results presented by Roughgarden (1979) overcame the strict confines of selection on fixed-fitness landscapes. Yet Roughgarden's approach to adaptive evolution by maximizing a population's density is applicable only when selection is density dependent, and not when it is frequency dependent.
- The concept of frequency-dependent selection also continues to receive short shrift in contemporary textbooks on life-history evolution. For example, out of the 465 pages of Roff (2002), not more than five deal with the description and implications of frequency-dependent selection, while the corresponding percentage in the seminal textbook by Stearns (1992) is even smaller.

We now proceed with a detailed review of the reasons that preclude the application of optimality principles to realistic problems in evolutionary ecology. Complementary to the considerations below are long-standing debates about the roles of developmental constraints (e.g., Maynard Smith *et al.* 1985) and of accidental historical by-products of evolution (e.g., Gould and Lewontin 1979) in obscuring the match between observed evolutionary outcomes and underlying "fitness maxima".

The quest for suitable optimization criteria

Even evolutionary biologists who favor optimality approaches concede that it is not always obvious which specific optimization criteria ought to be applied. In particular, the results of maximizing r or R_0 usually are not equivalent. For instance, predictions about the evolution of reaction norms for age and size at maturation critically depend on whether R_0 (Stearns and Koella 1986) or r (Kozlowski and Wiegert 1986) is used as the optimization criterion. Consequently, the question as to which function should be viewed as the Holy Grail of fitness measures has led to heated debate, reviewed, for example, in Roff (1992), Stearns (1992), Charnov (1993), and Kozlowski (1993).

The key issue here, recognition of which resolves the earlier debate for good, is that the bi-directional interaction between an evolving population and its environment was missing from the discussion (Metz *et al.* 1992). Whereas few biologists would contest that fitness always depends both on an individual's phenotype and on the environment the individual experiences, classic fitness measures used as optimization criteria, like r or R_0, only capture the former dependence. From today's perspective it is self-evident that the drastic reduction in complexity implied by dropping from consideration the dependence of fitness on the environment can only be justified under rather restrictive conditions. In particular, this convenient simplification is warranted only if the environment of an evolving population stays fixed, instead of varying along with the evolutionary change. Most of the time, however, conspecifics form an integral part of the environment that individuals experience. Therefore, when the distribution of conspecific phenotypes changes, so does a focal individual's environment. This explains why to maximize classic fitness measures like r or R_0 cannot do justice to the richness of phenomena in life-history evolution.

Optimization arguments in evolutionary game theory

The crucial importance of envisaging fitness as a function of two factors, an individual's trait(s) and its environment, was highlighted early on by work in evolutionary game theory (Hamilton 1967; Maynard Smith and Price 1973; Maynard Smith 1982). The payoff functions employed in that approach, which depend on two (usually discrete) strategies, and the broader notion of feedback between an evolutionary process and its environmental embedment are linked because, at ecological equilibrium, a population's resident strategies determine crucial aspects of its environment. When characterizing fitness we can therefore often simply replace a set of environmental variables by a description of the trait values currently resident in the population, and thus arrive at the notion of strategy-specific payoffs in which the explicit consideration of environmental variables is suppressed.

With regard to optimization arguments in evolutionary game theory, some confusion has arisen over two important distinctions: one between local and global optimization, and another between particular and universal optimization. An evolutionarily stable strategy (ESS) is essentially defined as one that maximizes payoffs in the environment the ESS sets for itself, and thus it adopts a global, but particular, notion of optimization. First, alternatively an ESS can be construed locally as a strategy that cannot be invaded by any neighboring strategy, a notion that is especially relevant when quantitative characters or metric traits are considered – a ubiquitous situation in life-history evolution. Second, it is crucial to understand that an ESS obeys a particular, and not a universal, optimization principle: the ESS usually maximizes payoffs only in its own environment, and not in the many other environments set by alternative resident strategies. This is a significant restriction, since, unless the ESS is already known, it thus cannot be recovered from this particular optimization principle (Metz *et al.* 1996b). Again, it is therefore only under restrictive conditions that an ESS maximizes payoffs in some "standard" environment that is independent of which phenotype is currently prevalent in the population and can be applied universally throughout the evolutionary process. And it is only in still more restrictive cases that such an optimization criterion happens to coincide with maximizing r or R_0 (Box 11.1).

Limitations to the existence of optimization criteria

The preceding discussion shows that it is by no means clear that for a given system an optimization principle exists. Whether or not such a principle can be found critically depends on how an evolving population interacts with its environment. This interaction is characterized by what we refer to as the eco-evolutionary feedback loop. To describe this feedback loop involves specifying the genetically variable and heritable traits, their impact on the focal organism's life history, and the ecological embedding that determines how life-history traits affect and are affected by environmental conditions.

It turns out that when one departs from the simplest ecological embeddings (e.g., the case in which the effect of density dependence is equally felt by all individuals in a population, irrespective of their phenotypes) optimization criteria cease to exist. It can even be shown that this is always the case if the "dimension" of the eco-evolutionary feedback loop is larger than one, a situation that readily arises in many realistic models and implies that populations are experiencing frequency-dependent selection (Heino *et al.* 1997b, 1998; Box 11.1). From a mathematical point of view, the conditions under which an optimization criterion exists are clearly degenerate (Metz *et al.* 1996b; Heino *et al.* 1997b), with the technical term "degenerate" meaning "infinitely rare". This finding contrasts rather sharply with the widespread use of optimization arguments in current evolutionary ecology. It may well be that a limited perception of the range of feedback scenarios actually existing in nature biases our evolutionary models toward the simple subset that conveniently obey optimization principles (J.A.J. Metz, personal communication). In particular, while frequency-dependent selection is still treated as a special case by virtually every contemporary textbook on evolution, this mode of selection is increasingly being recognized as one that ubiquitously acts on many life-history traits involved with, for example, foraging or reproduction (e.g., Kirkpatrick 1996). Since optimization approaches are invalidated by all (non-trivial) forms of frequency-dependent selection (Heino *et al.* 1997b), the absence of optimization criteria from realistic models of life-history evolution must be accepted as the rule, rather than the exception.

A celebrated example of an evolutionary game in which no single quantity can be construed as being maximized by evolution is the rock–paper–scissors game (rock beats scissors by crushing, paper beats rock by wrapping, scissors beat paper by cutting). The intransitive dominance relation in this game has been used to explain the coexistence of three mating strategies – "territorial", "mate-guarding", and "sneaking" – in the side-blotched lizard *Uta stansburiana* (Sinervo and Lively 1996; Sinervo *et al.* 2000). In that system the population growth rate of each strategy was shown to depend on the composition of the established, or resident, population, in such a way that the territorial strategy beats the mate-guarding strategy in an environment where mate-guarding is prevalent, while the mate-guarding strategy wins against sneakers in the environment set by sneakers, and sneakers beat territorials in the environment set by territorials. In cases like this, characterized by the absence of an optimization principle, the study of life-history evolution must rely on evaluating which sequences of invasion are possible, and to which evolutionary outcome they lead.

Side-blotched lizard
Uta stansburiana

Evolutionary stability and attainability

Classic evolutionary game theory, as well as approaches of r or R_0 maximization, are based on the assumption that phenotypes predicted to be unbeatable or evolutionarily stable against all other possible phenotypes are those that we expect to find in nature as outcomes of past evolutionary processes. Two objections have been raised against this premise, and both are based on the observation that adaptive evolution can usually proceed only gradually by means of mutations of small phenotypic effect.

The critical question is whether a strategy identified as evolutionarily stable is actually attainable by small mutational steps from at least some ancestral states. A first issue, recognized early on in the modern synthesis and leading to Wright's shifting-balance theory (Wright 1931, 1932, 1967, 1988), is that global fitness maxima may often not be attainable, since the evolutionary process becomes stuck on a local fitness maximum. This lends weight to the notion of a "local ESS", already highlighted above. A second, and completely separate, issue arises from the presence of frequency dependence, under which evolutionary stability and attainability turn out to part company (Eshel and Motro 1981; Eshel 1983). This means that gradual evolution may lead away from ESSs, and that, even more disturbingly, outcomes actually attained by gradual evolution may not be ESSs. Only within the restricted realm of optimization approaches is this second problem absent (Meszéna *et al.* 2001; Box 11.2).

Box 11.1 Limitations of optimization in life-history evolution

Here we illustrate the critical consequences of environmental feedback, using the evolution of age at maturation as an example. By referring to models developed by Mylius and Diekmann (1995) and by Heino *et al.* (1997b) we make two important points: (1) when environmental feedback is one-dimensional and monotonic, evolution is optimizing – but even so only rarely can it be reduced to the maximization of r or R_0; and (2) optimization approaches lose their validity whenever the environmental feedback is more than one-dimensional.

Environmental feedback refers to the full description of the environment as it occurs in the feedback loop in the considered population dynamics. In general, for populations that attain stable equilibria, the dimension of the feedback environment is the minimal number of variables that, independently of the mutant trait value, are sufficient to characterize the environment established by a resident population for the dynamics of a rare mutant population (Metz *et al.* 1996b).

One-dimensional environmental feedback. We consider an organism's life history as follows (Mylius and Diekmann 1995). Juveniles mature into adults at age x, after which they produce offspring at a constant rate b. Juveniles and adults die at rates d_J and d_A, respectively. All of these parameters can be affected by the environment E, as a consequence of the feedback loop. We denote their values in the virgin environment E_V (the environment unaffected by the population) by the subscript V. The adaptive trait considered here is x_V. Postponed maturation leads to an increased adult reproductive rate, $b(x_V) = \max(0, x_V - 1)$. This means that b is 0 for $x_V < 1$ and that it equals $x_V - 1$ otherwise. Three alternative feedback loops are investigated: (1) E only affects juvenile and adult mortality rates by an equal additional term for both; (2) E only affects juvenile mortality rate additively; and (3) E only affects the age at maturation multiplicatively. For each feedback scenario, parameters not affected by the environment take on their value in the virgin environment. For fixed values of x_V and E, the basic reproductive ratio $R_0(x_V, E)$ is given by

$$R_0(x_V, E) = \frac{b(x_V)}{d_A(E)} e^{-d_J(E) x(x_V, E)} \,. \tag{a}$$

Also, the population's intrinsic rate of increase $r(x_V, E)$ can be obtained as the unique real root of the corresponding Euler–Lotka equation (e.g., Roughgarden 1979; Yodzis 1989),

$$\frac{b(x_V) e^{-[r(x_V, E) + d_J(E)] x(x_V, E)}}{r(x_V, E) + d_A(E)} = 1 \,. \tag{b}$$

It turns out that only for feedback scenario (1) does adaptive evolution maximize r. Consequently, one can determine the evolutionary optimum x_V^* by maximizing $r(x_V, E)$ with respect to x_V, either for $E = E_V$ or for any other fixed E. For feedback scenario (2), the quantity maximized by evolution turns out to be $[\ln R_0(x_V, E_V)]/x_V$. This is not equivalent to maximizing $R_0(x_V, E_V)$. Instead, the optimized quantity can be rewritten as $[\ln b(x_V)]/x_V$, which is also the quantity that is evolutionarily maximized for feedback scenario (3).

This first example thus highlights that the appropriate fitness measure maximized by evolution under a one-dimensional environmental feedback loop clearly depends on the mode of density dependence, and only under special conditions reduces to r or R_0.

continued

Box 11.1 *continued*

Two-dimensional environmental feedback. A multidimensional feedback environment can only occur when there is some structure in the considered population. This structure can be genetic, social, temporal, spatial, or physiological (i.e., age-, stage-, or size-structured) and enables different individuals to have a different influence on, as well as a different perception of, the environment. Thus, whether or not a particular population structure creates a multidimensional feedback environment depends on how these aspects of influence and perception are specified in the considered population dynamics model.

As a typical example, the following model – simplified from Heino *et al.* (1997b) – investigates a population structured in two age classes. The species is semelparous, and individual transitions between classes take one time unit (e.g., 1 year). Maturity can be reached within the first year of life, or delayed until the second year. The adaptive trait is the probability of maturing at age 1, denoted by x. The other life-history parameters – intrinsic age-specific survival s_i (i refers to ages 0 and 1) and intrinsic fecundity b_i (with $i = 1, 2$) – are potentially affected during any year t by a two-dimensional environment $\{E_1(t), E_2(t)\}$. Transitions between age classes are as follows. Recruitment into age 1 from age 1 and 2: the per capita number of recruited individuals at time $t + 1$ is given by $s_0 b_1 x/[1 + c_1 E_1(t)]$ and $s_0 b_2/[1 + c_1 E_1(t)]$, respectively, where c_1 is a scaling parameter. Survival from age 1 to age 2: the survival probability is given by $s_1(1 - x)/[1 + c_2 E_2(t)]$, where c_2 is a scaling parameter. If the population dynamics reach equilibrium, we denote the equilibrium sizes of age class 1 and age class 2 by N_1^* and N_2^*, respectively. Recruitment is assumed to decrease with the density of newborns, and survival at age 1 decreases with the density of non-reproducing adults. The considered environmental feedback $\{E_1, E_2\} = \left\{b_1 x N_1^* + b_2 N_2^*, (1 - x)N_1^*\right\}$ is thus two-dimensional.

The evolutionarily stable fraction x^* of individuals that mature at age 1 depends on the order of three quantities: $s_1 b_2 - b_1$, $(s_0 b_1 - 1)c_2/c_1$, and 0. All individuals are predicted to mature at age 2 (age 1) if $s_1 b_2 - b_1 \geq (s_0 b_1 - 1)c_2/c_1$ ($s_1 b_2 - b_1 \leq 0$). However, when both of these conditions are not satisfied, $0 < s_1 b_2 - b_1 < (s_0 b_1 - 1)c_2/c_1$, a stable polymorphism arises with $x^* = c_1(s_1 b_2 - b_1)/[c_2(s_0 b_1 - 1)]$: a fraction $0 < x^* < 1$ of individuals mature at age 1 and the remaining fraction $1 - x^*$ at age 2. Thus, when the dimension of the environmental feedback is greater than one, a stable phenotypic polymorphism in the age at maturity can evolve. Intuitively, this is possible because under density dependence fitness ought to vary with population density, and thus require one environmental variable; the addition of a second environmental variable makes it possible for fitness to depend also on the relative frequencies of trait values in the population. A two-dimensional feedback environment is, indeed, a necessary condition (although not a sufficient one) for the evolution of stable polymorphisms. Importantly, no optimization principle can be devised to predict the evolutionarily stable fraction x^* (Metz *et al.* 1996b).

The dimension of feedback environments is only sharply defined in the world of models. In reality, this dimensionality is often relatively large or even infinite, with the environmental variables involved decreasing in their importance and impact. This implies, in particular, that one-dimensional feedback environments are not actually expected to occur in nature – which means, in turn, that evolutionary optimization will almost never apply to natural systems.

Box 11.2 Pairwise invasibility plots

Pairwise invasibility plots provide a handy way to analyze which mutant can invade which resident populations (Matsuda 1985; Van Tienderen and de Jong 1986; Metz *et al.* 1992, 1996a; Kisdi and Meszéna 1993; Geritz *et al.* 1997; see also Taylor 1989). Pairwise invasibility plots portray the sign structure of the invasion fitness f across all possible combinations of one-dimensional mutant trait values x' and resident trait values x. Zero contour lines at which $f(x', x) = 0$ separate regions of potential invasion success ($f > 0$) from those of invasion failure ($f < 0$). An example is shown below (left panel).

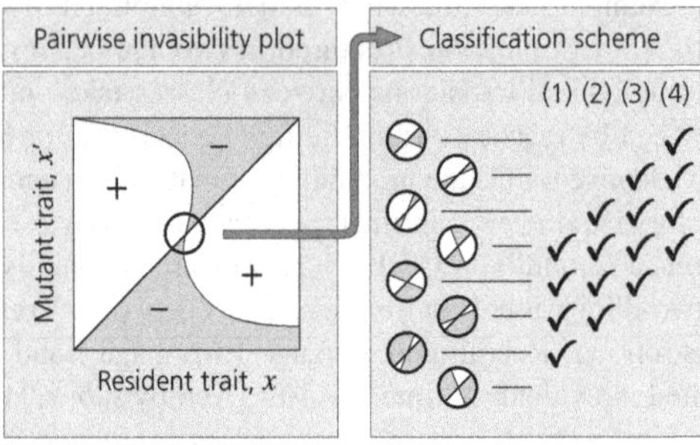

The resident trait value is neutral in its own environment, so one necessarily has $f(x, x) = 0$, and the set of zero contour lines therefore always includes the main diagonal. The shape of the other zero contour lines carries important information about the evolutionary process. In particular, intersections of zero contour lines with the main diagonal define the evolutionary singularities that are possible evolutionary end-points. Evolutionary singularities can be characterized according to four properties (Geritz *et al.* 1997):

1. evolutionary stability;
2. convergence stability;
3. invasion potential; and
4. mutual invasibility.

Whether each of these properties applies to a given evolutionary singularity can be decided simply by looking at the pairwise invasibility plot and reading the slope of the zero contour line at the singularity, as illustrated in the right panel above.

Four interesting types of evolutionary singularities are highlighted below. In each case, the staircase-shaped line indicates a possible adaptive sequence by which evolutionarily advantageous mutants repeatedly invade and replace residents.

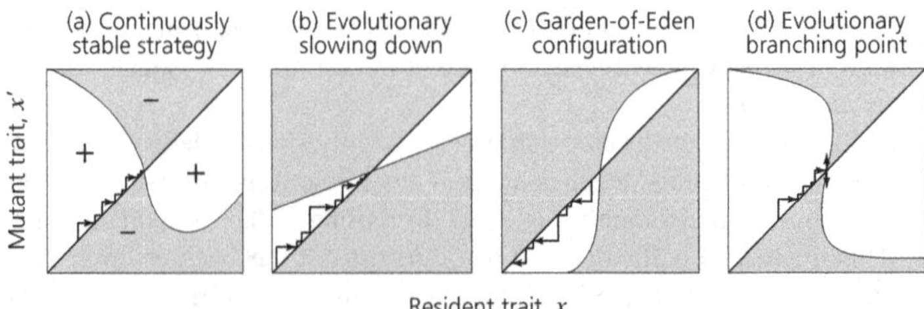

continued

Box 11.2 *continued*

Panel (a) above shows a situation in which the singularity is a so-called continuously stable strategy (CSS; Eshel and Motro 1981; Eshel 1983). A CSS is both evolutionarily stable and convergence stable, and thus serves as a likely endpoint of gradual evolutionary change. Panel (b) depicts a CSS that lacks invasion potential, which causes the evolutionary process to slow down algebraically as the population moves closer to the CSS (Dieckmann and Law 1996). Panel (c) illustrates a Garden-of-Eden configuration (Nowak and Sigmund 1989), an ESS that is not convergence stable and hence cannot be attained by small mutational steps. Panel (d) shows an evolutionary branching point (Metz *et al.* 1992, 1996a), in which the singularity is convergence stable, but not evolutionarily stable, and nearby mutants are mutually invasible. Such configurations cause disruptive selection and thus permit the phenotypic divergence of two subpopulations that straddle the branching point.

Optimization and population viability

Even when restricting attention to those models that allow evolutionary outcomes to be predicted through r or R_0 maximization, the assumption that population viability would be maximized as well is incorrect. This can be shown easily with a simple example.

For this purpose we consider a population of organisms with non-overlapping generations regulated by Ricker-type density dependence (Chapter 2). A life-history trait x influences the population's intrinsic growth rate r such that its dynamics are governed by the recursion equation $N_{t+1}(x) = r(x)\exp(-\alpha N_t(x))N_t(x)$, where N_t denotes the population size at time t and α measures the strength of density dependence. A mutant trait value x' can invade a resident population of x individuals if the mutant population's geometric growth rate in the environment set by the resident exceeds 1, that is, if $\left[\prod_{t=0}^{T-1} r(x')\exp(-\alpha N_t(x))\right]^{1/T} > 1$ for large durations T. The resident population is at ecological equilibrium if $\left[\prod_{t=0}^{T-1} r(x)\exp(-\alpha N_t(x))\right]^{1/T} = 1$ for large durations T, which, together with the previous inequality, yields the simple invasion criterion $r(x') > r(x)$. Thus, evolution in this model is expected to maximize r as a function of the trait x. The existence of such an optimization principle is the consequence of a one-dimensional eco-evolutionary feedback: all individuals perceive the same environment, characterized by the size of the whole population. It is readily shown that the average asymptotic population size of an x-population is $(1/\alpha)\ln r(x)$, which implies that this population size is evolutionarily maximized together with r. The same conclusion, however, does not extend to population viability: as r increases in the course of evolution, the population equilibrium becomes unstable and is replaced with oscillations (cycles or chaos) of increasing amplitude, with the lowest population size approaching zero (May and Oster 1976; Gatto 1993), thus increasing the risk of extinction through demographic stochasticity (Allen *et al.* 1993; Renault and Ferrière, unpublished). We must therefore conclude that, although evolution in this example follows an optimization principle, it nevertheless drives up the risk of population extinction.

This section shows that the conventional approach of maximizing r or R_0 to study life-history evolution is fraught with fundamental limitations. In the next section we introduce the theory of adaptive dynamics as an extended framework that overcomes these limitations, while it encompasses the classic theory as a special case.

11.3 Adaptive Dynamics Theory

Whenever an ecological system adapts, it affects its environment, which in turn can modify the selective pressures that act on the system: as the preceding section shows, the resultant eco-evolutionary feedback is critical for describing adaptive evolution.

Invasion fitness

The fitness of organisms can only be evaluated relative to the environment in which they live. Eco-evolutionary feedback means that this environment depends on the current adaptive state of the population under consideration. To assess the fitness of a variant phenotype, one must therefore specify the resident phenotype against which the variant is competing. In adaptive dynamics theory this is accomplished by the concept of invasion fitness (Metz *et al.* 1992). This quantity measures the long-term per capita growth rate f of a phenotype x in a given environment E, $f = f(x, E)$. The environment E is determined by externally fixed parameters and by the population density and phenotype of the resident population(s). Thus, the invasion fitness of a variant readily accounts for the consequences of frequency-dependent ecological interactions. If the variant has an advantage compared with the resident – that is, if it has positive invasion fitness – it can spread through the population; by contrast, if the variant has negative invasion fitness, it will quickly become extinct.

Remarkably, the analysis of invasion fitness provides important insights into the dynamics and outcome of adaptive evolution, as long as it is justified to assume that the environment E has settled to a stationary state determined by the resident set of phenotypes. Under that assumption, we can replace the dependence of invasion fitness on the current environment E with a dependence on the resident phenotypes $x_1, x_2, ..., f = f(x, x_1, x_2, ...)$. In general, these phenotypes can belong to the same species as the variant phenotype x does, or they can involve other, coevolving species (see Chapters 16 and 17 for applications of the adaptive dynamics framework in the context of coevolution). If the community of resident phenotypes possesses coexisting attractors, invasion fitness is usually multi-valued, as the environmental conditions engendered by the resident phenotypes then depends on which attractor is attained. For the sake of simplicity, it is often sufficient to characterize a population by its prevalent or average phenotype (Abrams *et al.* 1993). Although strictly monomorphic populations are seldom found in nature, it turns out that the dynamics of polymorphic populations (harboring, at the same time, many similar phenotypes per species) can often be well described and understood in terms of the simpler monomorphic cases.

Evolutionary singularities and their properties

For a single species we can thus consider the invasion fitness $f = f(x', x)$ of a variant phenotype x' in a resident population of phenotype x. The sign structure of these functions can be depicted graphically to produce so-called pairwise invasibility plots, which carry important information about the evolutionary process (Box 11.2).

In particular, pairwise invasibility plots clearly identify potential evolutionary endpoints at which selection pressures vanish. These potential endpoints are called evolutionary singularities and are characterized by the following four properties:

- *Evolutionary stability.* Is a singularity immune to invasion by neighboring phenotypes? This property defines a local version of the classic ESS that lies at the heart of evolutionary game theory (Hamilton 1967; Maynard Smith and Price 1973; Maynard Smith 1982).
- *Convergence stability.* When starting from neighboring phenotypes, do successful invaders lie closer to the singularity? Here the attainability of the singularity is under consideration, an issue separate from its invasibility (Eshel and Motro 1981; Eshel 1983).
- *Invasion potential.* Is the singularity able to invade populations of neighboring phenotypes (Kisdi and Meszéna 1993)?
- *Mutual invasibility.* If a pair of neighboring phenotypes lie on either side of a singularity, can they invade into each other? Assessment of this possibility is essential to predict coexisting phenotypes and the emergence of polymorphisms (Van Tienderen and de Jong 1986; Metz *et al.* 1992, 1996a).

Among the eight feasible combinations of these properties (Metz *et al.* 1996a; Geritz *et al.* 1997), some have striking implications for the adaptive process:

- *Convergence and evolutionary stability.* The first two properties in the list above characterize a so-called continuously stable strategy (CSS; Eshel 1983). Processes of gradual adaptation experience a considerable slowing down when they converge toward a CSS (Dieckmann and Law 1996); this deceleration is most pronounced in the absence of invasion potential.
- *Evolutionary stability without convergence stability.* Although the singularity is resistant against invasion from all nearby phenotypes, it cannot be attained by small mutational steps – a situation aptly referred to as a Garden-of-Eden configuration by Nowak and Sigmund (1989). The existence of this type of evolutionary singularity echoes one of the limitations of optimization approaches highlighted in the previous section.
- *Convergence stability without evolutionary stability.* Convergence stability does not entail that the singularity be evolutionarily stable. In the absence of evolutionary stability, selection becomes disruptive near a convergence-stable singularity. Two phenotypically distinct subpopulations can then diverge from around the singularity in a process called evolutionary branching (Metz *et al.* 1992, 1996a; Geritz *et al.* 1997).

Box 11.3 Models of adaptive dynamics

The theory of adaptive dynamics derives from consideration of ecological interactions and phenotypic variation at the level of individuals. Extending classic birth and death processes, adaptive dynamics models keep track, across time, of the phenotypic composition of a population in which offspring phenotypes are allowed to differ from those of their parents.

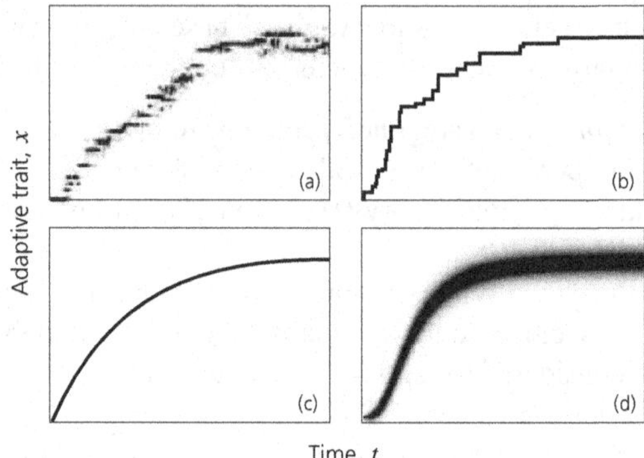

Four types of models are used to investigate adaptive dynamics at different levels of resolution and generality:

- At any time the population can be represented in trait space as a cloud of points, each point corresponding to an individual's combination of trait values. This polymorphic cloud of points stochastically drifts and diffuses as a result of selection and mutation (Dieckmann 1994; Dieckmann *et al.* 1995), see panel (a).
- In large populations characterized by a low mutation rate, evolutionary change in clonal species proceeds through sequences of trait substitutions (Metz *et al.* 1992). During each such step, a mutant with positive invasion fitness quickly invades a resident population, ousting the former resident. These steps can be analyzed through the pairwise invasibility plots introduced in Box 11.2. Concatenation of such substitutions produces a directed random walk of the type depicted in panel (b) above. Formally, such random-walk models are obtained from the process in panel (a) by considering the case of rare mutations (Dieckmann 1994; Dieckmann *et al.* 1995; Dieckmann and Law 1996).
- If, in addition, mutation steps are sufficiently small, the staircase-like dynamics of trait substitutions are well approximated by smooth trajectories, see panel (c) above. These trajectories follow the canonical equation of adaptive dynamics (Dieckmann 1994; Dieckmann *et al.* 1995; Dieckmann and Law 1996), which in its simplest form is

$$\frac{dx}{dt} = \tfrac{1}{2}\mu\sigma^2 N^*(x) \left.\frac{\partial f(x',x)}{\partial x'}\right|_{x'=x} ,$$

where x is the adaptive trait, μ is the probability for mutant offspring, σ^2 is the variance of mutational steps, $N^*(x)$ is the equilibrium size of a population with resident trait value x, and f is the invasion fitness. The partial derivative in the equation above is the selection gradient $g(x)$. Evolutionary singularities are trait values x^* for which the selection gradient vanishes, $g(x^*) = 0$.

continued

Box 11.3 *continued*

- In large populations characterized by high mutation rates, stochastic elements in the dynamics of the phenotypic distributions become negligible; this enables mathematical descriptions of reaction–diffusion type (Kimura 1965; Bürger and Bomze 1996; Bürger 1998), see panel (d) above. However, the infinitely extended tails that phenotypic distributions instantaneously acquire in this framework often give rise to artifactual dynamics that have no correspondence to processes that could occur in any finite population (Mollison 1991; Cruickshank *et al.* 1999).

At the expense of ignoring genetic complexity, models of adaptive dynamics are geared to analyze the evolutionary implications of ecological settings. This allows all types of density- and frequency-dependent selection mechanisms to be studied within a single framework, into which coevolutionary dynamics driven by interspecific interactions are also readily incorporated (Dieckmann and Law 1996; Chapters 16 and 17). Extensions are also available to describe the evolution of multivariate traits (Dieckmann and Law 1996) and of function-valued traits (Dieckmann and Heino, unpublished).

As long as the adaptive process stays away from evolutionary branching points, the evolutionary dynamics follow selection gradients determined by the first derivative of invasion fitness in the direction of the variant trait, and can be described by a simple differential equation known as the canonical equation of adaptive dynamics (Box 11.3).

In the next two sections we utilize adaptive dynamics theory to investigate two remarkable consequences of closing the eco-evolutionary feedback loop:

- Natural selection can play a major role in driving the diversification of communities.
- Natural selection can cause population extinction, even in the absence of environmental change.

11.4 Adaptive Evolution and the Origin of Diversity

The response of biodiversity to environmental changes is likely to span a continuum, from the immediate ecological consequences to longer-term evolutionary effects. Both ends of this continuum raise conservation concerns.

Conservation and speciation

On the ecological time scale, global biodiversity can only be lost. Locally, of course, biodiversity may be enhanced by the invasion of exotic species, but even that often leads to the subsequent loss of native species (Drake *et al.* 1989; Williamson 1996; Mooney and Hobbs 2000; Mooney and Cleland 2001; Perrings *et al.* 2002). By contrast, on the evolutionary time scale, not only can biodiversity be lost (Section 11.5), but also it can be generated, which thus has conservation implications: "Death is one thing, an end to birth is something else", in the words of Soulé (1980). The "birth" process in ecological communities is speciation, for which human activities are suggested to have at least three major repercussions (Myers and Knoll 2001):

- *Outbursts of speciation.* As large numbers of niches are vacated, there could be explosive adaptive radiations within certain taxa – notably small mammals, insects, and terrestrial plants – able to thrive in human-dominated landscapes.
- *Reduced speciation rates.* Biogeography theory suggests that speciation rates correlate with area (e.g., Rosenzweig 1995, 2001; Losos 1996; Losos and Schluter 2000). Therefore even the largest protected areas and nature reserves may prove far too small to support the speciation of large vertebrates. Even for smaller species, habitat fragmentation may severely curb speciation rates.
- *Depletion of evolutionary powerhouses.* The unrelenting depletion and destruction of tropical biomes that have served in the past as pre-eminent powerhouses of evolution and speciation (Jablonski 1993) could entail grave consequences for the long-term recovery of the biosphere.

The long-term, macro-evolutionary character of hypotheses like those above means they are notoriously difficult to evaluate empirically. Models that do justice to the underlying mechanisms have to be reasonably complex, which appears to deter theorists from tackling these questions. However, at least the first two notions in the list above have received some attention from modeling and theory. Below we summarize recent studies that bear on these issues.

Determinants of evolving biodiversity

Law (1979) introduced the "Darwinian Demon" as a hypothetical organism that has solved all challenges of life-history evolution – it starts to reproduce immediately after birth, produces very large numbers of offspring at frequent intervals, supplies each offspring with massive food reserves that ensure survival, possesses a high longevity, disperses well, finds mates at will, and it can achieve all these successes in any habitat. Evidently, such a super-organism would quickly take over the earth's biosphere and would thus eradicate all diversity. This illustrates that understanding biodiversity always entails understanding the life-history trade-offs that prevent such demons from arising: ecological coexistence is possible because of such trade-offs. In this vein, many biodiversity models (e.g., Hastings 1980; Tilman 1994; Tilman *et al.* 1994; May and Nowak 1994; Nowak and May 1994) focused on species assemblages that are ecologically stable. Yet most ecologically stable communities are not evolutionarily stable. To describe processes that go beyond short-term responses to environmental change, we must learn to understand the mechanisms and environmental determinants that generate and maintain diversity in evolving communities. The two models described next address this question by analyzing, respectively, evolution under trade-offs between competition and dispersal, and between growth and fecundity.

Modeling the exposure of a formerly nitrogen-poor community of terrestrial plants to a large increase in the rate of nitrogen deposition, Tilman and Lehman (2001) considered the community's response both at the ecological and the evolutionary time scale. Unsurprisingly, their model predicts that the short-term effect of the environmental change is the take-over of a few formerly rare but now fast-growing and rapidly dispersing species. The differential success of these plants is enhanced by asymmetric competition for light. After the initial ecological response, evolutionary processes come into play and reshape the entire community. Based on a trade-off between competitive ability and dispersal potential, the model predicts that the winners of the short-term round gradually reduce their capacity to disperse by evolving into progressively better local competitors. To justify their reaction–diffusion modeling of adaptive dynamics (see Box 11.3), Tilman and Lehman (2001) assumed that mutations are so frequent that, at any time, the community always features a wide range of phenotypes at low density. Under such conditions, evolution first establishes two distinct morphs: a good disperser that is a poor competitor and a good competitor that is a poor disperser. Afterwards, the former morph again evolves toward better competitive ability and thus allows a well dispersing third morph to invade with traits similar to those the first and second morph had both possessed initially. Thus, the range between the two extreme strategies successively fills with a collection of intermediate species. Tilman and Lehman (2001) describe this pattern as the result of a speciation process that eventually yields a local flora that is as species rich as that before the environmental change. The far-reaching conclusion from this theoretical study is that rapid speciation processes can confer high long-term resilience to the diversity of natural communities against the immediate negative impacts of habitat degradation.

A different model of biodiversity evolution was analyzed by Jansen and Mulder (1999; see also Mouquet *et al.* 2001) to describe a seasonal community of self-pollinating plants that inhabited a large collection of patches. Throughout the season, competing plant species grow within patches of equal carrying capacity. At the end of the season, the plant biomass thus accrued is converted back into seeds, which are then distributed randomly across all patches. Plant species differ in a single quantitative trait that describes their growth rate; fecundity is negatively correlated with growth and vanishes at a given maximal growth rate, while competitive ability and dispersal potential are independent of the trait. Evolution is enabled by a small probability that a seed is a mutant, in which case its growth rate slightly differs from its parent. Figure 11.1 shows how biodiversity in the evolved species assemblages depends on season length and environmental quality:

- Predicted biodiversity increases with season length. This is because longer seasons select for fast-growing but less fecund phenotypes, which results in a larger fraction of patches being unoccupied by fast-growing phenotypes and thus open to more slowly growing phenotypes. The finding is compatible with observed biodiversity, which increases toward the equator.

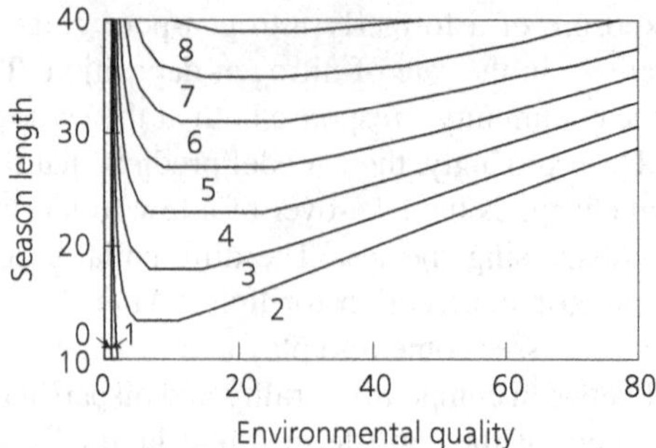

Figure 11.1 Patterns of biodiversity that emerge from the adaptive dynamics of a competitive plant community. Predicted biodiversity, measured as the number of species in the evolutionarily stable community, changes at the contour lines, increasing with season length and exhibiting a maximum for local environments of intermediate quality. *Source*: Jansen and Mulder (1999).

- Predicted biodiversity is maximal for environments of intermediate quality. Rich environments, here defined as featuring patches of high carrying capacity, lead to high total fecundity and thus to a saturated situation in which most patches are occupied by the types that grow fastest, which drives any other types to extinction. By contrast, poor environments lead to low total fecundity and thus to a situation in which diversity is "starved" by the rare colonization of patches. These antagonistic effects cause the model's biodiversity to peak at a medium environmental quality. Also this prediction is in accordance with observed productivity–diversity relations (Rosenzweig 1995).

We may thus expect diversity patterns to follow environmental conditions predictably, as these change over space or time. Once corroborated and complemented by more detailed ecological models, such insights may help to diagnose community-level disturbances caused by environmental change, and, where necessary, to devise recovery measures that restore the evolutionary potential and/or stability of affected species assemblages.

Adaptive speciation

The two models discussed above are based on a phenotypic representation of quantitative traits. Their utility lies in highlighting the ecological and environmental conditions conducive to adaptive radiation and necessary to maintain diverse communities. A critical element in both models is frequency-dependent selection, which allows, as shown in Section 11.3, evolving populations to converge through directional selection to fitness minima, at which selection turns disruptive. The key point to appreciate here is that under such circumstances, which cannot arise in models of life-history optimization, the splitting of a lineage trapped at a fitness minimum becomes adaptive. The resultant processes of adaptive speciation (Dieckmann *et al.* 2004) are very different from those stipulated by the standard model of allopatric speciation through geographic isolation, which dominated speciation research for decades (Mayr 1963, 1982). Closely related to adaptive speciation are models of sympatric speciation (e.g., Maynard Smith 1966; Johnson *et al.* 1996), of competitive speciation (Rosenzweig 1978), and of ecological speciation (Schluter 2000), which all indicate the same conclusion: patterns of species diversity are not only shaped by processes of geographic isolation and immigration, which can be more or less random, but also by processes of selection and evolution, which are bound to infuse such patterns with a stronger deterministic component.

When considering adaptive speciation in sexual populations, selection for reproductive isolation comes into play. Since at evolutionary branching points lineage splits are adaptive, in the sense that populations are freed from being stuck at fitness minima, premating isolation is expected to evolve more readily under such circumstances than previously believed. Any evolutionarily attainable or already existing mechanism of assortative mating can be recruited by selection to overcome the forces of recombination that otherwise prevent sexual populations from splitting up (e.g., Felsenstein 1981). Since a plethora of such mechanisms exist for assortativeness (based on size, color, pattern, acoustic signals, mating behavior, mating grounds, mating season, the morphology of genital organs, etc.), and since only one of these many mechanisms needs to take effect, it would be surprising if many natural populations remained stuck at fitness minima for very long (Geritz *et al.* 2004). Models for the evolutionary branching of sexual populations corroborate this expectation (Dieckmann and Doebeli 1999, 2004; Doebeli and Dieckmann 2000; Geritz and Kisdi 2000; Box 11.4).

In conjunction with mounting empirical evidence that rates of race formation and sympatric speciation are potentially quite high, at least under certain conditions (e.g., Bush 1969; Meyer 1993; Schliewen *et al.* 1994), the above considerations suggest that longer-term conservation efforts will benefit if attention is paid to how environmental change interferes with evolutionarily stable community patterns.

Area effects on adaptive speciation

Doebeli and Dieckmann (2003, 2004) incorporated spatial structure into models of evolutionary branching. They found that, even in the absence of any significant isolation by distance, spatial environmental gradients could greatly facilitate adaptive parapatric speciation. Such facilitation turned out to be most pronounced along spatial gradients of intermediate slope, and to result in stepped biogeographic patterns of species abutment, even along smoothly varying gradients. These findings are explained by observing that the combination of local adaptation and local competition along a gradient acts as a potent source of frequency-dependent selection. The investigated models allow substantial gene flow along the environmental gradient, so isolation by distance does not offer an alternative explanation for the observed phenomena.

Box 11.4 Sympatric speciation in sexual populations

Sympatric speciation in sexual populations necessarily involves a sufficiently high degree of reproductive isolation – otherwise hybrids occupy any potentially developing gap between the incipient species. Apart from chromosomal speciation, which involves incompatible levels of ploidy, reproductive isolation in sympatry is most likely to occur through a prezygotic mechanism that results in assortative mating. Unless assortativeness is already present for some reason, it thus has to evolve in the course of sympatric speciation.

Dieckmann and Doebeli (1999) considered a simple model with two adaptive traits: first, an ecological character exposed to selection pressures that would lead to evolutionary branching in an asexual population, and second, a variable degree of assortativeness on the ecological character. Both traits were modeled with diploid genetics, assuming sets of equivalent diallelic loci with additive effects and free recombination. Under these conditions, sympatric speciation happens easily and rapidly. This is illustrated by the sequence of panels below, in which both quantitative traits are coded for by five loci, thus giving rise to a quasi-continuum of 11 different phenotypes. In each panel, gray scales indicate the current frequencies of the resultant $11^2 = 121$ phenotypic combinations in the evolving population (the highest frequency in a panel is shown in black, with a linear transition of gray scales to frequency zero, shown in white).

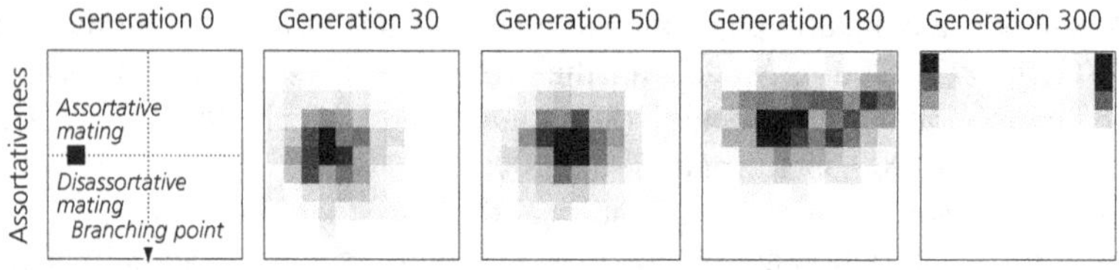

The above sequence of events starts out with random mating, away from the evolutionary branching point. After the population has converged to the branching point, it still cannot undergo speciation, since recombination under random mating prevents the ecological character from becoming bimodal. However, if the disruptive selection at the branching point is not too weak (Matessi *et al.* 2001), it selects for increased assortative mating. Once assortativeness has become strong enough, speciation can occur. Eventually, the ecological characters of the incipient species diverge so far, and assortive mating becomes so strong, that hardly any hybrids are produced and the gene flow between the two species essentially ceases.

In a second, related model, Dieckmann and Doebeli (1999) considered an additional quantitative character that is ecologically neutral and only serves as a signal upon which assortative mating can act. Numerical analysis shows that in this case sympatric speciation also occurs. Conditions for speciation are only slightly more restrictive than in the first model, even though a linkage disequilibrium between the ecological character and the signal now has to evolve as part of the speciation process.

These results support the idea that when frequency-dependent ecological interactions cause a population to converge onto a fitness minimum, solutions can often evolve that allow the population to escape from such a detrimental state. This makes the speciation process itself adaptive, and underscores the importance of ecology in understanding speciation.

These findings, which were obtained for models of both asexual and sexual populations, could have repercussions in terms of understanding species–area relationships, widely observed in nature. Species diversity tends to increase with the size of the area over which diversity is sampled, a characteristic relationship that is often described by power laws (Rosenzweig 1995). It is therefore noteworthy that the speciation mechanism highlighted by Doebeli and Dieckmann (2003, 2004) also lets the emerging number of species increase with the total area covered by the environmental gradient. Of course, a shorter gradient in a smaller area often covers a reduced range of environmental heterogeneity compared with an extended gradient in a larger area. So one component of species–area relationships is expected to originate from the enhanced diversity of environmental conditions that in turn supports a greater diversity of species. Interestingly, however, Doebeli and Dieckmann (2004) found that their model predicted larger areas to harbor more species than smaller areas, even when both areas featured the same diversity of environmental conditions. This suggests that a second component of species–area relationships originates because the evolutionary mechanism of adaptive speciation operates more effectively in larger than in smaller areas.

Other mechanisms are also likely to contribute to species–area relationships. MacArthur and Wilson (1967), for example, based a classic explanation on their "equilibrium model of island biogeography". This model relies on the assumption that equilibrium population sizes increase linearly with island size, so that species extinctions occur more rarely on larger islands. Adopting a purely ecological perspective, their argument makes no reference to the effect of island area on the rate at which species are being formed, rather than being destroyed. By contrast, Losos and Schluter (2000) argued that the greater species richness of Anolis lizards found on larger islands in the Antilles is because of the higher speciation rates on larger islands, rather than higher immigration rates from the mainland or lower extinction rates. Since the diversity of environmental conditions does not appear to be significantly lower on smaller islands in the Antilles (Roughgarden 1995), and since, nevertheless, the big islands of the Greater Antilles typically harbor many species of Anolis lizards compared to the smaller islands of the Lesser Antilles (which contain at most two species), the second component of species–area relationships as described above may have played an important role for anole radiations in the Antilles.

This brief discussion again underscores that traditionally envisaged ecological factors of diversity must be complemented by additional evolutionary factors (this also applies to understanding species–area relationships). The effect of habitat loss and habitat fragmentation on speciation rates might thus become an important focus of evolutionary conservation biology.

11.5 Adaptive Evolution and the Loss of Diversity

The notion that optimizing selection maximizes an evolving population's viability leaves no room for (single-species) evolution that causes population extinctions. An appreciation of evolution's role in culling biodiversity therefore requires that the narrow concept of optimizing selection be overcome, as discussed in Section 11.1.

Evolutionary deterioration, collapse, and suicide

Given the long tradition of describing evolutionary processes through concepts of progress and optimization, we must reiterate that no general principle actually prevents adaptive evolution from causing a population to deteriorate (Section 11.1). Even selection-driven population collapse and extinction are not ruled out and, in fact, these somewhat unexpected outcomes readily occur in a suite of plausible evolutionary models.

Evolutionary suicide (Ferrière 2000) is defined as a trait substitution sequence driven by mutation and selection that takes a population toward and across a boundary in the population's trait space beyond which the population cannot persist. Once the population's phenotypic traits have evolved close enough to this boundary, mutants that are viable as long as the current resident trait value abounds, but that are not viable on their own, can invade. When these mutants start to invade the resident population they initially grow in number. However, once they have become sufficiently abundant, concomitantly reducing the former resident's density, the mutants bring about their own extinction. This is not unlike the "Trojan gene effect" discussed by Muir and Howard (1999), although the latter does not involve gradual evolutionary change.

Two other adaptive processes are less drastic than evolutionary suicide. First, adaptation may cause population size to decline gradually in a process of perpetual selection-driven deterioration. Sooner or later, demographic and/or environmental stochasticity then cause population extinction. Second, the population collapse brought about by an invading mutant phenotype may not lead to population extinction, but only to a substantial reduction in population size. Such a collapse renders the population more susceptible to extinction by stochastic causes.

The three phenomena of population deterioration, collapse, and suicide have often been discussed in the context of evolving phenotypic traits related to competitive performance. A verbal and lucid example of evolutionary deterioration comes from overtopping growth in plants. Taller trees receive more sunlight while casting shade onto their neighbors. As selection causes the average tree height to increase, fecundity declines because more of the tree's energy budget is diverted from seed production to wood production. Under these circumstances it may also take longer and longer for the trees to reach maturity. Thus, arborescent growth as an evolutionary response to selection for competitive ability can cause population abundance and/or the intrinsic rate of population growth to decline. The logical conclusion of this process may be a population's extinction, as first explained by Haldane (1932).

Below we outline the analysis of several models that provide a mathematical underpinning to Haldane's considerations and that illustrate, in turn, processes of evolutionary deterioration, evolutionary collapse, and evolutionary suicide. All three models consider a single species with population dynamics influenced by a quantitative trait that measures competitive ability (e.g., adult body size). Variation in this phenotype is assumed to result in asymmetric competition: individuals that are at a competitive advantage by attaining larger body size at the same time suffer

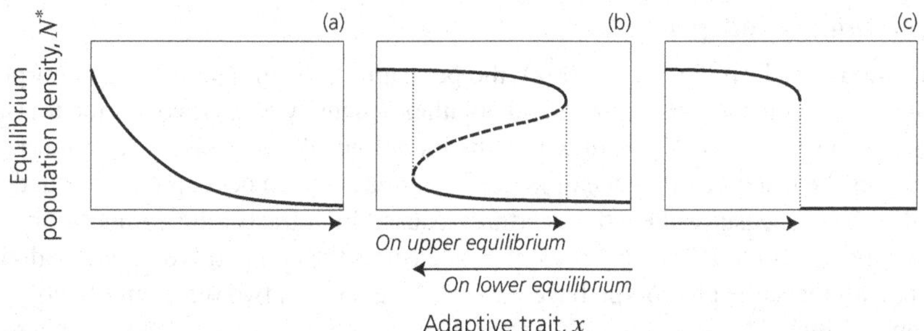

Figure 11.2 Evolutionary deterioration, collapse, and suicide. (a) Evolutionary deterioration as in the model by Matsuda and Abrams (1994a). (b) Evolutionary collapse as in the model by Dercole *et al.* (2002). (c) Evolutionary suicide as in the model by Gyllenberg and Parvinen (2001). In each case, continuous curves show how equilibrium population densities vary with the adaptive trait (body size), unstable equilibria are indicated by a dashed curve, and selection pressures on the adaptive traits are depicted by arrows.

from having to divert more energy to growth, which results in diminished reproduction or increased mortality. Asymmetric competition implies that in pairwise interactions the individual that is competitively superior to its opponent suffers less from the effects of competition than the inferior opponent.

Evolutionary deterioration

Matsuda and Abrams (1994a) analyzed a Lotka–Volterra model in which competing individuals experience asymmetric competition and a carrying capacity that depends on body size. In their model, the competitive impact experienced by an individual with body size x in a population with mean body size $\overline{x}$ is $\alpha(x, \overline{x}) = \exp(-c_\alpha(x - \overline{x}))$, and the carrying capacity for individuals with body size x is $K(x) = K_0 \exp(-c_K(x))$. Here c_α is a nonlinear function that preserves the sign of its argument, and c_K is a non-negative function that goes to infinity when its argument does.

Matsuda and Abrams (1994a) conclude that under these assumptions adaptive evolution continues to increase body size indefinitely – provided that the advantage of large body size (as described by c_α) is big enough and the cost of increased body size (as described by c_K) is small enough. Since large body sizes result in small carrying capacities, adaptive evolution thus perpetually diminishes population size (Figure 11.2a), a phenomenon Matsuda and Abrams call "runaway evolution to self-extinction". Notice, however, that in this model population size never vanishes, but just continues to deteriorate. This means that additional stochastic factors, not considered in the studied deterministic model, are required to explain eventual extinction.

For a related model that focuses on the evolution of anti-predatory ability in a predated prey, see Matsuda and Abrams (1994b). The actual extinction through demographic stochasticity, predicted by Matsuda and Abrams (1994a), is demonstrated in an individual-based model by Mathias and Kisdi (2002).

Evolutionary collapse

In a model by Dercole *et al.* (2002), the per capita growth rate in a monomorphic population with adult body size x and population density $N(x)$ involves the logistic component $r(x) - \alpha_0 N(x)$, in which the monotonically decreasing function $r(x)$ captures the negative influence of larger adult body size on per capita reproduction, and $\alpha_0 N(x)$ measures the extra mortality caused by intraspecific competition. As in the previous model, the coefficient α_0 measures the competitive impact individuals with the same phenotype have on each other. When two different phenotypes x and x' interact, the competitive impact of x on x' is $\alpha(x - x')N(x)$, where α increases with $x - x'$, $\alpha(0) = \alpha_0$, and $\alpha'(0) = -\beta$. Per capita growth is further diminished by a density-dependent term that accounts for an Allee effect. Such an effect may be caused either by reduced fecundity through a shortage of mating encounters in sparse populations, or by increased mortality because of the concentration of predation risk as density decreases (Dennis 1989; Chapter 2). Reducing the per capita growth rate by $\gamma N(x)/[1 + N(x)]$ captures both variants, with γ determining the Allee effect's strength. As described in Chapter 2, the resultant population dynamics can feature bistability: a stable positive equilibrium may coexist with a stable extinction equilibrium. Dercole *et al.* (2002) actually reduced the per capita growth rate by $\gamma N^2(x)/[1+N^2(x)]$ in an effort to add realism to the model by accounting for spatial heterogeneity in the chance of mating or predation risk. With this second choice, the population size can attain two stable equilibria $N^*(x)$, one at low and one at high density. When x is low, only the high-density equilibrium exists; when x is high, only the low-density equilibrium exists, while in-between the two stable equilibria coexist (Figure 11.2b).

The invasion fitness of a mutant x' in a resident population with phenotype x is then given by $f(x', x) = r(x') - \alpha(x - x')N^*(x) - \gamma N^{*2}(x)/[1 + N^{*2}(x)]$, which yields the selection gradient $g(x) = r'(x) + \beta N^*(x)$, with $N^*(x)$ determined by $f(x, x) = 0$. The selection gradient shows that two opposing selective pressures are at work: since fecundity decreases when adult body size increases, the term $r'(x)$ is negative and thus favors smaller adult body size, whereas the term $\beta N(x)$ is positive and selects for larger body size. Ecological bistability can make the balance between these two selective forces dependent upon which equilibrium the resident population currently attains: a specific resident phenotype that occupies the high-density equilibrium gives the positive selective pressure more weight and thus favors increased adult body size x, whereas the same resident phenotype at the low-density equilibrium promotes the reduction of x. If the reproductive cost of larger body size is not too large [i.e., if $r'(x)$ remains low], and/or if competitive asymmetry is strong [i.e., if β is large], an ancestral population characterized by small body size and high abundance will evolve toward larger and larger adult body size – up to a point where the population's high-density equilibrium ceases to exist (Figure 11.2b). There the population abruptly collapses to its low-density equilibrium and suddenly faces a much-elevated risk of extinction because of demographic and environmental stochasticity.

Evolutionary suicide

Also a model developed by Gyllenberg and Parvinen (2001) is based on asymmetric competition and incorporates an Allee effect. Their model is similar to the previous one, except for the following three features:

- Fecundity $b(x)$ peaks for an intermediate value of adult body size x;
- A trait- and density-independent mortality d is considered; and
- Rather than increasing mortality, the Allee effect reduces fecundity by the factor $N(x)/[1 + N(x)]$.

These features are reflected in the model's invasion fitness, which is obtained as $f(x', x) = b(x')N^*(x)/[1 + N^*(x)] - d - \alpha(x - x')N^*(x)$, with $N^*(x)$ again being determined by $f(x, x) = 0$.

From this invasion fitness we can infer that the extinction equilibrium $N^*(x) = 0$ is stable for all x. For intermediate values of x, two positive equilibria coexist, of which the high-density one is stable and the low-density one is unstable. The selection gradient $g(x) = b'(x)N^*(x)/[1+N^*(x)]+\beta N^*(x)$ is positive for any x, provided that $\beta = -\alpha'(0)$ is large enough (i.e., whenever competition is strongly asymmetric). It is thus clear that the adaptive dynamics of adult body size x must drive the population to the upper threshold of adult body size, beyond which the two positive equilibria vanish and only the stable extinction equilibrium remains. In this model, therefore, adaptive evolution not only abruptly reduces population density (as in the previous example), but also causes the population to become extinct altogether. The resultant process of evolutionary suicide is illustrated in Figure 11.2c.

Catastrophic bifurcations and evolutionary suicide

It is not accidental that the two previous examples both involved discontinuous transitions in population density at the trait values where, respectively, evolutionary collapse and evolutionary suicide occurred. In fact, Gyllenberg *et al.* (2002) have shown (in the context of a particular model of metapopulation evolution) that such a so-called "catastrophic bifurcation" or "discontinuous transition to extinction" is a prerequisite for evolutionary suicide. A simple geometric explanation of this necessary condition is given in Box 11.5.

This result allows us to distinguish strictly between mere evolutionary deterioration and actual evolutionary suicide:

- Evolutionary deterioration implies that evolution by natural selection gradually drives a population to lower and lower densities until it is eventually removed by demographic or environmental stochasticity. The first example above, by Matsuda and Abrams (1994a), is of this kind.
- By contrast, evolutionary suicide implies that evolution by natural selection drives a population toward a catastrophic bifurcation through which its density abruptly decreases to zero. Notice that it is the evolutionary time scale on which this extinction is abrupt; on the ecological time scale, of course, a decrease in

Box 11.5 Transcritical bifurcations exclude evolutionary suicide

Wherever a population goes through a continuous transition to extinction it cannot undergo evolutionary suicide. For simplicity, we show this for cases in which the population's density N and its adaptive trait x are both one-dimensional. The generic continuous transition to extinction is then the so-called transcritical bifurcation, in which a positive equilibrium and the extinction equilibrium exchange their stability at a critical trait value x_c.

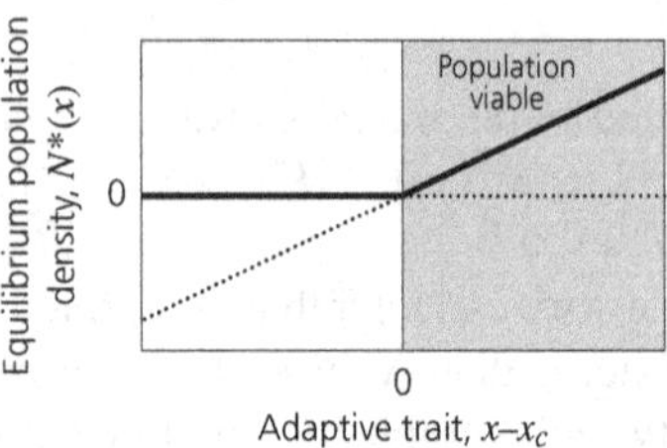

In the vicinity of the critical trait value x_c, population dynamics that exhibit a transcritical bifurcation can always be written as $\frac{d}{dt}N = r[(x - x_c) - N/K]N$, where $K > 0$ scales N and $r > 0$ scales $\frac{d}{dt}N$ (up to re-orientation of the direction of x; Guckenheimer and Holmes 1997, p. 145). With the per capita growth rate of a mutant trait value x' in an environment with population density N thus being given by $r[(x' - x_c) - N/K]$, and with the equilibrium population density of a resident population with trait value $x = x_c$ vanishing ($N = 0$), we obtain the invasion fitness $f(x', x_c) = r(x' - x_c)$ for the rare mutant that competes with the critical resident. In addition, the consistency condition $f(x, x) = 0$ for ecological equilibrium has to be fulfilled for all x (see Box 11.2). If we now make the generic assumption that $f(x', x)$ has a leading linear order around $x' = x_c$ and $x = x_c$, that is, $f(x', x) = c'x' + cx$, we can determine the coefficients c' and c from the two constraints (1) $f(x', x_c) = r(x' - x_c)$ for all x' and (2) $f(x, x) = 0$ for all x, which yield $f(x', x) = r(x' - x)$. For the selection gradient (Box 11.3) we thus obtain $g(x) = r$, which is always positive. This means that adaptive evolution takes x away from x_c by making it larger, and thus increases the equilibrium population size from $N^*(x_c) = 0$ to $N^*(x) = (x - x_c)K$. Therefore, adaptive evolution in this system cannot cause evolutionary suicide by driving x toward the critical trait value x_c.

The above reasoning can be collapsed to a glance at an illustration of the local geometry, as sketched below.

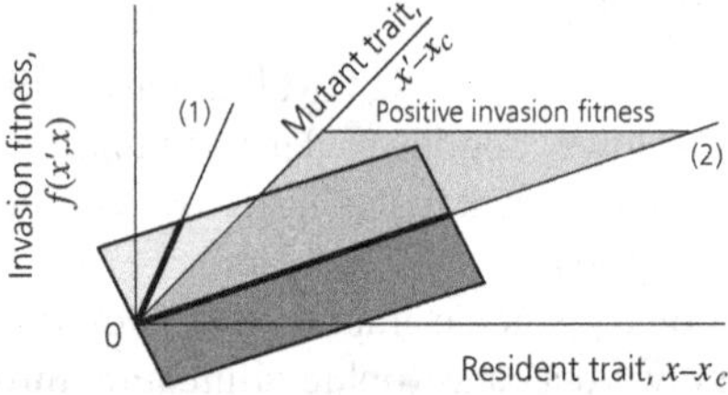

Since the plane that represents the invasion fitness $f(x', x)$ is constrained to pass through the two straight lines that represent constraints (1) and (2), the region $x' > x$ has a positive invasion fitness. Thus evolution increases x, moving it away from x_c.

population density always takes a while to result in extinction. The third example above, by Gyllenberg *et al.* (2002), is of this kind [as is the second example, by Dercole *et al.* (2002), although it involves a catastrophic bifurcation that does not lead to immediate extinction].

The important role played by demographic and environmental stochasticity in evolutionary deterioration means that such processes may also be referred to loosely as stochastic evolutionary suicide. The same applies to an evolutionary collapse that exposes a population to a high risk of accidental extinction. Another form of stochastic evolutionary suicide, driven by mutational stochasticity rather than by demographic or environmental stochasticity, and not discussed here, can occur in higher-dimensional trait spaces (Parvinen and Dieckmann, in press).

Further examples of evolutionary suicide

Another example of evolutionary suicide is driven by the evolution of dispersal rates (Gyllenberg *et al.* 2002). The ecological model of structured metapopulations that underlies this example was introduced by Gyllenberg and Metz (2001) and Metz and Gyllenberg (2001). It considers a large number of identical patches of habitable environment. Each patch can support a local population. Patches are connected by dispersal, and individuals leave their patch at a rate that can evolve through mutation and selection. Dispersal risk is defined as the probability that a dispersing individual will not survive until it settles down again in a patch. Local populations may go extinct as a result of catastrophes. At least two scenarios, which involve two different kinds of Allee effects, can then cause evolutionary suicide in this model. First, evolutionary suicide occurs when small local populations have a negative intrinsic growth rate and thus can only persist through immigration from other patches: when a high dispersal risk then selects for a low dispersal rate, adaptation drives the metapopulation to extinction. Second, evolutionary suicide can also occur when the rate at which local populations are wiped out by catastrophes increases as the population size decreases: again, a high dispersal risk makes dispersal unattractive for individuals, even though the population as a whole depends on this dispersal. This selection pressure results in an abrupt extinction of the metapopulation when the dispersal rates falls below a critical level. A more detailed discussion of this family of models is provided in Chapter 14.

For adaptive evolution that involves kin selection, Le Galliard *et al.* (2003) observed evolutionary suicide caused by the adaptive dynamics of altruism. In this model, three selective forces act on an adaptive trait that measures the level of altruistic investment:

1. Direct, physiological cost of investing more in the altruistic behavior;
2. Indirect benefit of locally interacting with more altruistic individuals; and
3. Indirect cost of locally saturating the habitat.

Since locally interacting individuals often share a common ancestry, the second selection pressure involves kin selection. The third selection pressure turns out to be negligible because demographic stochasticity and individual mobility tend to reduce local crowding. Mobility has a cost, however, and the population cannot sustain itself at a high mobility without a substantial degree of altruism between individuals. The combination of high mobility and a high ambient level of altruism creates the ideal conditions for "cheaters" to invade – phenotypes that invest slightly less in altruism and yet reap the full benefits provided by the resident, more altruistic individuals. This causes the population to evolve toward a state in which the population's level of altruism is no longer sufficient to ensure its persistence, resulting in evolutionary suicide.

Evolutionary suicide can also be expected when adaptive evolution involves sexual selection (Kirkpatrick 1996). Mating preferences can establish a trait even if it has negative side effects on an individual's survival. A gene for a preferred trait that is expressed in both sexes will spread if its fitness gain through male mating success more than offsets its survival cost evaluated over males and females (Kirkpatrick 1982). Thus, adaptive evolution can establish traits that have negative effects on female reproductive success, and hence on the population's reproductive output. We expect, and often see, the evolution of modifiers suppressing the expression of those genes in females that give a fitness advantage only to males, even though sexual selection can cause the evolution of traits that decrease population viability. This feature of sexual selection had already been realized by Darwin (and presumably was one of the reasons why he attributed so much emphasis to the distinction between natural and sexual selection).

A recent study by Ernande *et al.* (2002) shows that selection-induced extinction can, in principle, also happen in the context of exploited living resources, where these are modeled realistically. The model considers a physiologically structured population in which individuals continually age and grow in body size. On reaching a size threshold, they turn from larvae dispersed only passively into juveniles able actively to select their local environment. These local environments differ in the growth and mortality rates they induce. When the growth trajectories of individuals reach the maturation reaction norm, represented as a function that describes maturation size as dependent on maturation age, juveniles turn into adults and start to reproduce at a rate that increases with their body size. In this model the shape of the maturation reaction norm is the evolving trait. Ernande *et al.* (2002) show that when the evolving population is exposed to a harvesting regime that extracts biomass at a constant rate, the maturation reaction norm evolves so as to allow individuals to mature at younger ages and smaller sizes. At a certain point, this adaptation may cause the entire population to become extinct – a phenomenon of evolutionary suicide that is especially worrisome in the context of commercially exploited fish stocks.

Evolutionary suicide in sexual populations

A factor that could prevent evolutionary suicide in sexual populations is the premature depletion of additive genetic variance (Matsuda and Abrams 1994a, 1994b). If the additive genetic variance approaches zero as the trait value approaches the suicidal threshold, the evolutionary process will be much decelerated. However, unless mutations cease to induce genetic variance, eventual evolutionary suicide remains inevitable.

At the opposite end of the spectrum, a surplus of phenotypic variance may prevent evolutionary suicide. This can be understood as follows. When a broad phenotypic distribution approaches a suicidal threshold, it extends its head tail beyond the population's viability domain. The loss of individuals in this tail then affects the selective pressures that act on the rest of the population. In particular, the release of density regulation through very low reproductive success in the head tail may boost the reproductive success of individuals in the rear tail. It turns out that this source–sink dynamics across a population's viability boundary can allow the population to hover temporarily at the brink of extinction. The smaller the population's phenotypic variance, the closer it approaches extinction. This places an extra premium on maintaining the genetic variance of populations threatened by evolutionary suicide: once their variance is sufficiently depleted, their extinction is imminent.

Box 11.6 shows how evolutionary suicide is expected to occur in sexual populations and, in particular, how the underlying genetics could interfere with the ecology of evolutionary suicide as outlined above.

Extinction driven by coevolutionary dynamics

Also, coevolutionary dynamics can cause extinctions. Some early treatment, which still excludes the effects of intraspecific frequency-dependent selection, is given in Roughgarden (1979).

Dieckmann *et al.* (1995) considered an example of predator–prey coevolution in which the predator's extinction is caused by the prey's adaptation. In this model, the phenotype of a predator has to remain sufficiently close to that of its prey for the predator's harvesting efficiency to remain high enough to ensure predator survival. This may reflect the need for a match between, for example, prey size and the dimensions of the predator's feeding apparatus. Thus, whenever evolution in the prey takes its phenotype too far away from the predator's matching phenotype, harvesting efficiency drops below a critical level, and so causes the predator to become extinct.

Notice that in all cases in which such a transition to extinction is gradual (rather than discontinuous), evolutionary suicide cannot contribute to the extinction (Box 11.5). In addition, gradual extinction causes mutation-limited phenotypic evolution in the dwindling species to grind to a halt, since fewer and fewer individuals are around to give birth to the mutant phenotypes that fuel the adaptive process (Box 11.3). This stagnation renders the threatened species increasingly defenseless by depriving it of the ability to counteract the injurious evolution of its partner by a suitable adaptation of its own. For these two reasons, gradual coevolutionary extinction is driven solely by adaptation in the coevolving partner(s). The situation is different, of course, when a transition to extinction is discontinuous: in such cases, processes of evolutionary suicide and "coevolutionary murder" may conspire to oust a species from the coevolving community.

Box 11.6 Evolutionary suicide in sexual populations

The dynamics of sexual populations differ in several respects from those considered in asexual models of evolutionary suicide. In particular, sexual populations are typically polymorphic, which has two important implications:

- First, compared with a monomorphic asexual population that features the same mean value of the adaptive trait, variance corrections to the model's invasion fitness are bound to arise. These corrections can affect the population size predicted for a given trait value, the critical trait values at which evolutionary suicide is expected, and the selection gradient. Occasionally, a changed selection gradient may even enable evolution away from the extinction boundary.
- Second, a polymorphic population may hover at the brink of extinction, because the death of individuals in the population's tail that extends beyond the extinction boundary may enhance reproductive success in the remaining population.

The two illustrations below show results for a sexual model of evolutionary suicide; to our knowledge this is the first time such an analysis has been carried out. Based on an adaptive trait x, per capita birth rates are given by $b(x) = b_0/(1 + K_{1/2}/N)$, and per capita death rates by $d(x) = d_0 + N_{\text{eff}}(x)/K(x)$. Here, b_0 and d_0, respectively, denote the intrinsic birth and death rates, N the population's total size, and $K_{1/2}$ the population size at which b drops to $\frac{1}{2}b_0$ through an Allee effect. The death rate is increased by asymmetric competition, with the sum in $N_{\text{eff}}(x) = \sum_i \alpha(x_i - x)$ extending over all individuals, and the competitive effect of an individual with trait value x_i on an individual with trait value x given by $\alpha(x_i - x) = 2/[1 + e^{-(x_i - x)/w}]$, where w determines the degree of competitive asymmetry. Asymmetric competition thus favors individuals with larger values of the adaptive trait x. The population's carrying capacity is trait dependent and given by a normal function, $K(x) = K_0 \exp(-\frac{1}{2}x^2/\sigma_K^2)$, which thus favors individuals with intermediate values of the adaptive trait x. For the illustrations below, parameters are set to $b_0 = 1$, $d_0 = 0.2$, $K_0 = 2000$, $K_{1/2} = 200$, $\sigma_K = 1$, and $w = 0.2$. The adaptive trait x is polygenic, determined by $n = 10$ equivalent diploid loci with additive effects and free recombination. Loci can either be diallelic, with allelic values $+1$ and -1, or they can feature a continuum of alleles. The set of trait values in the diallelic model is scaled to $-2 < x < +2$, with an analogous scaling applied to the infinite-allele model. Mutations occur at a probability of $u = 10^{-3}$ per locus and, in the case of continuous allelic values, are distributed normally with standard deviation $\sigma = 0.2$.

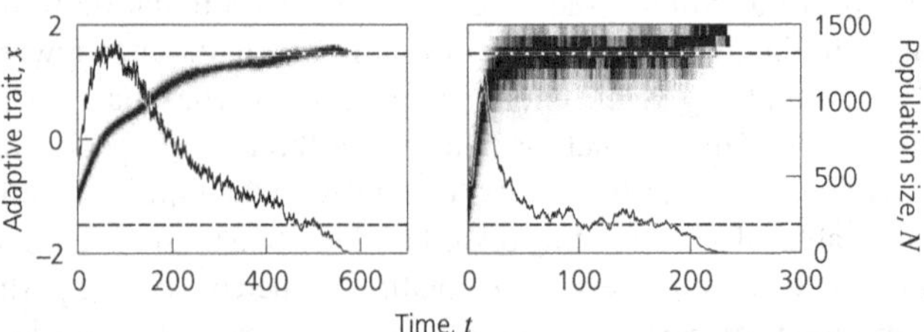

continued

Box 11.6 *continued*

The left panel on the preceding page is based on the infinite-allele model and shows how the polymorphic distribution of adaptive trait values, depicted by gray scales, starts out on one side of the carrying capacity's peak at $x = 0$ and then, driven by asymmetric competition, evolves toward and beyond this peak, until it reaches the extinction boundary at about $x = 1.5$. The model's two extinction boundaries are depicted by dashed lines (notice that, since selection drives the population away from the lower boundary, no evolutionary suicide can occur there). The black continuous curve shows the changes in actual population size that result from the trait's evolution. The dynamics of evolutionary suicide in this model is thus very similar to that predicted by the corresponding asexual model. The right panel shows exactly the same situation, except that the diallelic model is now being considered. The different genetic architecture that underlies the adaptive trait x imposes a much larger phenotypic variance on the population throughout all phases of its evolution. With just a few diallelic loci, this phenotypic variance is now so large that the population lingers for a while at the brink of extinction, before perishing eventually.

We can thus conclude that – except for some quantitative corrections and for the extra potential of populations to hover temporarily at the brink of extinction – the phenomenon of selection-driven extinction appears to apply just as well to sexual populations as it does to the asexual populations investigated in earlier studies.

Further examples of coevolution-driven extinction are provided in Chapter 16 for coevolving communities that exhibit both mutualistic and competitive interactions.

Summary

Evolutionary suicide occurs for a rich variety of ecological settings and appears to be robust to variations in the underlying system of inheritance. Even if evolutionary suicide does not occur, the related phenomena of persistent evolutionary deterioration or of an abrupt collapse toward perilously low densities are possible. Also, coevolution can bring about a species' demise. Thus, phenomena in which the adaptive process itself harms an evolving species or community are by no means peculiar outcomes of particularly rigged ecological models.

A question of acute interest in the context of population management is to identify the circumstances through which environmental change can expose a population to the threat of evolutionary deterioration, collapse, or suicide. We address this issue in the following section.

11.6 Adaptive Responses to Environmental Change

Populations exposed to environmental change usually experience altered selection pressures acting on their traits. If the population had enough time to adapt to the prevailing environmental conditions before the evolutionary change, with the result that selection had become stabilizing, it typically experiences a qualitative change to directional selection during and after the environmental change. Classic models of such situations are based on the notion of a fitness maximum that gradually shifts its position in trait space. The primary question is then whether or not the evolving population can respond quickly enough to the new directional selection pressures for it to track the shifting maximum and thus to persist despite the threatening change in its environment. Questions of this kind are best analyzed using techniques of quantitative genetics, and are discussed in detail in Chapter 10. Here we take a broader perspective and consider more general (and intricate) patterns of interplay between environmental change, adaptive evolution, and ecological viability.

Ecology–evolution–environment diagrams

A geometric approach to the interplay of ecology, evolution, and environment is facilitated by focusing attention on conditions that imply population extinction. For this purpose we consider those phenotypic values x that allow a population to be viable under environmental conditions e. Combinations (x, e) that do not allow for this imply population extinction. Such a focus on extinction conditions conveniently removes the population size N from the graphic considerations below, which renders the resultant diagrams much easier to read.

To describe the evolutionary dynamics for viable combinations (x, e) of phenotypes and environmental conditions, we can utilize the pairwise invasibility plots introduced in Box 11.2, which allow us to consider all kinds of density- and frequency-dependent selection. Figure 11.3a shows a sample sequence of pairwise invasibility plots that illustrates how they may change when environmental conditions are altered:

- Initially, an evolutionary attractor (technically speaking, a convergence-stable evolutionary singularity) coexists with an evolutionary repellor, both situated away from an extinction region of trait values that render the population unviable.
- As environmental conditions change, the two evolutionary singularities approach each other.
- Eventually, they collide.
- Directional selection then drives the evolving phenotype into the extinction region, causing evolutionary suicide.

Figure 11.3b shows how this very same sequence of events can be depicted in a single diagram, employing three characteristic features:

- Arrows show the direction of selection;
- Line styles indicate the different types of evolutionary singularity; and
- Shading shows the extinction regions.

We refer to such plots as ecology–evolution–environment diagrams, or E^3-diagrams for short: the environmental and evolutionary components of change

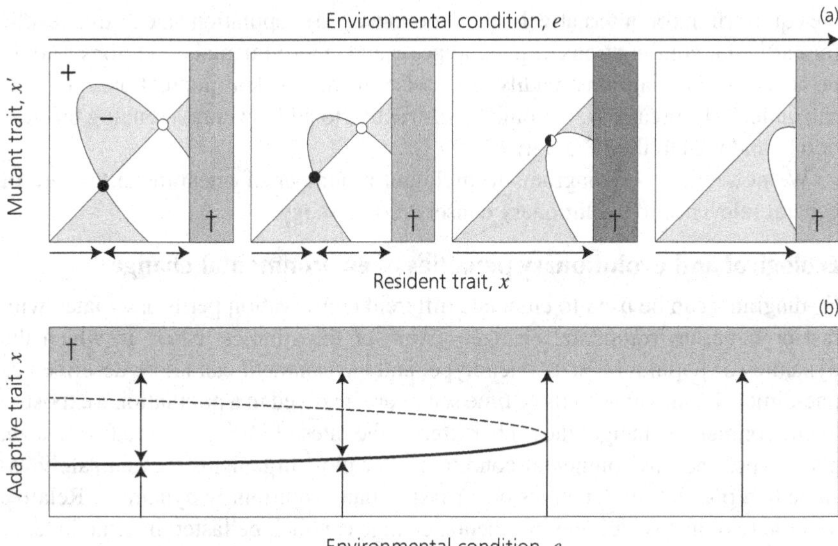

Figure 11.3 Interplay between ecological, evolutionary, and environmental change. (a) Pairwise invasibility plots that show the collision and resultant disappearance of an evolutionary attractor and repellor, which lead to induced evolutionary suicide as environmental conditions are varied toward the right. Light gray regions correspond to negative invasion fitness; the extinction region is shown in dark gray. Evolutionary attractors and repellors are depicted by filled circles and open circles, respectively. Short arrows indicate the direction of selection. (b) Ecology–evolution–environment diagram (E^3-diagram) that depicts the same situation as in (a). Arrows show the direction of selection, line styles indicate the type of evolutionary singularity, and shading shows the extinction region. Continuous (dashed) curves indicate evolutionary attractors (repellors), while thick (thin) curves indicate evolutionarily stable (unstable) singularities.

are represented along the horizontal and vertical axes, respectively, while the ecology furnishes the shown selection pressures and determines the extinction regions throughout which the population is not viable. Comparing Figures 11.3a and 11.3b suggests that a single E^3-diagram is more immediately comprehensible than sequences of pairwise invasibility plots, while, as long as we are content to consider gradual evolution, they contain the same salient information.

E^3-diagrams have much in common with those regularly used in the classic bifurcation theory of dynamic systems (e.g., Kuznetsov 1995; Guckenheimer and Holmes 1997) – yet they acquire essential extra complexity because of two additional features: first, the incorporation of the extinction region, and second, the distinction between evolutionarily stable and unstable singularities. In the classic theory, only convergence stability would be considered, and consequently only evolutionary attractors and repellors would be discriminated.

We notice in passing that, if, for a particular study, a need were to arise to retain more ecological information in E^3-diagrams, then contour lines of, for example, the equilibrium (or, alternatively, the time-averaged) population size N that results for particular combinations of phenotypes x and environmental conditions e could be added to the diagrams readily. Likewise, if the resident population dynamics can undergo bifurcations, it would be instructive to add the corresponding bifurcation boundaries to the E^3-diagram.

We now utilize E^3-diagrams to highlight a number of phenomena that are of general relevance to evolutionary conservation biology.

Ecological and evolutionary penalties of environmental change

E^3-diagrams can be used to elucidate different conservation perils associated with fast or large environmental change. Even for the simplest case – in which the dynamics of population size, phenotype, and environment can all be described as one-dimensional – at least three time scales are involved in a population's exposure to environmental change; these characterize the rates of change in population size, phenotype, and environmental condition. For most organisms we can safely assume that population dynamics occur faster than evolutionary dynamics. Relative to these two time scales, environmental change can then be faster, intermediate, or slower.

Figures 11.4a to 11.4c illustrate three different ways in which fast environmental change can cause population extinction:

- In Figure 11.4a, environmental change occurs so rapidly that it outpaces both the population dynamics and the evolutionary dynamics of the affected population. The figure shows how environmental change takes the population right into an extinction region (Point A), where the population gradually diminishes in size and eventually perishes.
- Figure 11.4a also shows what happens when environmental change occurs at an intermediate time scale, rendering it slower than the population dynamics, but faster than the evolutionary dynamics. Under these circumstances the population becomes extinct as soon as environmental change forces it to trespass into the extinction region: population extinction thus occurs right at the region's boundary (Point B).
- In Figure 11.4b, environmental change occurs at a time scale commensurable with that of evolutionary dynamics. The figure shows that even such a situation can still lead to population extinction: as soon as the environmental change forces the population across the separatrix curve that corresponds to the evolutionary repellor, the course of directional evolution is reversed and the population steers toward evolutionary suicide.
- Finally, in Figure 11.4c environmental change is sufficiently slow for evolutionary rescue (Gomulkiewicz and Holt 1995) to become feasible. Such a situation allows the population to track its evolutionary attractor, which in the illustrated case saves the population from extinction.

When environmental change is abrupt, the amount of change becomes key to predicting the fate of the exposed population. Such situations can also be analyzed

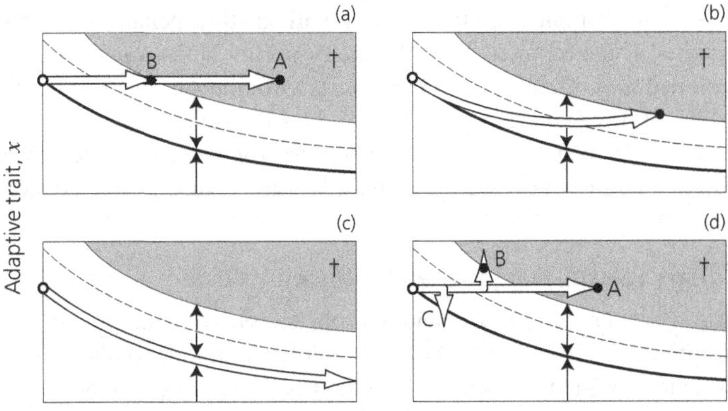

Figure 11.4 Ecological and evolutionary penalties of fast or large environmental change. Elements of the E^3-diagrams are as in Figure 11.3b. Open circles show the population's state before environmental change occurs, white arrows depict the resultant trajectories, and filled circles indicate where the population becomes extinct. (a) If environmental change is faster than the population dynamics, the population perishes in the interior of the extinction region (Point A). If environmental change is slower than the population dynamics, but faster than the evolutionary dynamics, the population vanishes at the border of the extinction region (Point B). (b) If environmental change occurs at the same time scale as that of the evolutionary dynamics, the population may still undergo induced evolutionary suicide once it is trapped beyond an evolutionary repellor (dashed curve). (c) If environmental change is sufficiently slow, evolutionary rescue may occur. (d) Consequences of abrupt environmental changes of different magnitudes. An ecological penalty occurs if the environmental change takes the population into the extinction region (Trajectory A), whereas an evolutionary penalty occurs if the environmental change takes it beyond the evolutionary repellor (Trajectory B). If the environmental change is small enough, the population may be rescued by adaptation (Trajectory C).

conveniently using E^3-diagrams. Retaining the same setting as for Figures 11.4a to 11.4c, Figure 11.4d illustrates two fundamentally different ways in which large and abrupt environmental changes can cause population extinction:

- A large environmental change settles the population right in the extinction region, which implies its demise through the ensuing population dynamics (Trajectory A).
- An intermediate environmental change moves the population beyond the separatrix given by the position of the evolutionary repellor, which causes its extinction through evolutionary suicide (Trajectory B).
- By contrast, a small environmental change allows the population to stay on the safe side of the evolutionary separatrix, and thus enables it to undergo evolutionary rescue (Trajectory C).

These simple examples highlight conceptually distinct penalties associated with environmental change: an ecological penalty occurs when a population's viability is forfeited as a direct consequence of environmental change (Points A and B in Figure 11.4a; Trajectory A in Figure 11.4d), whereas an – often less obvious – evolutionary penalty is incurred when environmental change compromises a population's ability to evolve out of harm's way (Figure 11.4b; Trajectory B in Figure 11.4d).

Evolutionary rescue, trapping, and induced suicide

E^3-diagrams are also useful to depict the phenomena of evolutionary rescue and trapping introduced in Box 1.4. In fact, the left and middle plots in Box 1.4 can be interpreted as E^3-diagrams if we take their horizontal axis to measure environmental condition, rather than time. Evolutionary rescue can occur when an evolutionary attractor escapes an encroaching extinction region as the environmental conditions are changed. Similarly, evolutionary trapping – in its simplest form (see below) – requires that an evolutionary attractor collide with an extinction region as environmental conditions are changed.

In contrast to evolutionary rescue and trapping, evolutionary suicide can occur in the absence of any extrinsic environmental change, as it is intrinsically driven by the feedback between an evolving population and its environment. The fingerprint of evolutionary suicide in E^3-diagrams is directional selection pointing toward an extinction region as, for example, in the right part of Figure 11.3b.

Evolutionary suicide, however, is involved critically in another phenomenon we need to understand to assess a population's response to environmental change. Figure 11.3b illustrates this scenario, which we call induced evolutionary suicide: an evolutionary attractor collides with an evolutionary repellor, such that a population that is tracking the attractor as environmental conditions are changing suddenly becomes exposed to directional selection toward the extinction region, and hence undergoes evolutionary suicide. Here it is the environmental change that abruptly creates the conditions that lead to evolutionary suicide.

More complex forms of evolutionary trapping

Figure 11.5a illustrates how induced evolutionary suicide can result in a more complex form of evolutionary trapping. Here an evolutionary attractor again vanishes in collision with a repellor. Now, however, there is a range of environmental conditions in which two attractor–repellor pairs coexist. This means that, if a large jump occurs in its phenotype, the population could survive environmental conditions that change toward the right by shifting to the lower attractor. Gradual phenotypic change, however, keeps the population trapped at the upper attractor, and results in its inevitable demise.

A much more benign (and simple) situation is depicted in Figure 11.5b. Here small environmental change results in a large shift of an evolutionary attractor, which obviously makes it difficult for gradual evolution to catch up with the required pace of phenotypic change. Such a situation could thus be described as an

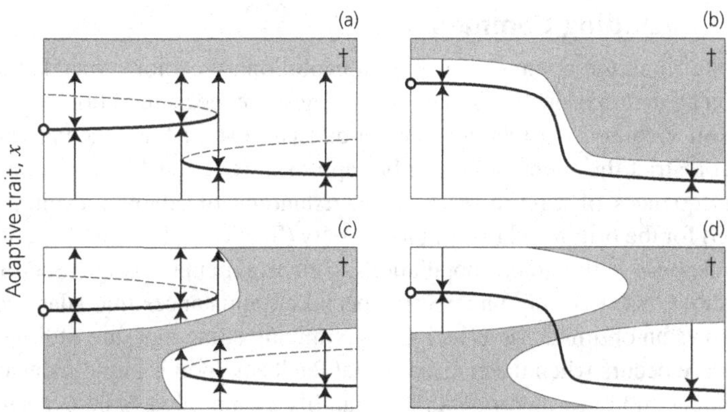

Figure 11.5 Examples of more complex forms of evolutionary trapping. Elements of the E³-diagrams are as in Figure 11.3b. Open circles show the population's state before environmental change shifts conditions to the right. (a) When one evolutionary attractor vanishes in a collision with a repellor, the population can only survive by a large phenotypic jump across a fitness valley to the alternative evolutionary attractor. (b) When the positions of the evolutionary attractor and the extinction region undergo substantial changes in response to a small environmental change, the population can survive only through a particularly swift evolution. In such situations, the population easily becomes trapped by limitations on its pace of adaptation. The settings in (c) and (d) are the same as those in (a) and (b), respectively, but with the two evolutionary attractors now separated by the extinction region.

evolutionary bottleneck, where, at some stage, only swift adaptation could rescue the threatened population.

The scenarios in Figures 11.5a and 11.5b can be exacerbated considerably if the extinction region takes a more complex and expansive shape. Such cases are illustrated in Figures 11.5c and 11.5d. In both cases, gradual evolution cannot rescue the population. The shift of the population's phenotype to the safe evolutionary attractor now not only requires it to trespass through a fitness valley (as in Figure 11.5a) or particularly rapid evolution (as in Figure 11.5b), but also gradual evolutionary change toward the safe attractor takes the population into the extinction region, and thus completely forestalls gradual evolution as a sufficient evolutionary response to the imposed environmental change.

The latter two scenarios may look complex, but are not as improbable as one perhaps would wish to think: Section 14.4 describes dispersal evolution in response to landscapes changes and presents, in Figure 14.10b, a result of the type depicted in Figures 11.5c and 11.5d. In Chapter 14, also the results shown in Figures 14.4, 14.11, 14.12, and 14.13 showcase the use of E³-diagrams in understanding the conservation implications of dispersal evolution. A particularly intriguing finding in this context is that induced evolutionary suicide can result from environmental conditions that become less severe, as illustrated in Figure 14.13b.

11.7 Concluding Comments

For a long time, the common wisdom in evolutionary ecology was that adaptive evolution by natural selection should maximize some measure of fitness, and hence population viability. In this chapter we discuss several fundamental shortcomings that restrict the scope of this earlier approach (Section 11.2), and investigate the consequences of a more realistic understanding of adaptive evolution (Section 11.3) for the origin and loss of biodiversity (Sections 11.4 and 11.5), as well as for the response of threatened populations to environmental change (Section 11.6).

We show that it is only under very special circumstances that adaptive evolution follows an optimization principle, maximizing some measure of fitness. This special case occurs when the environmental feedback loop is one-dimensional and monotonic – and even then population viability cannot always be expected to be maximized: as described in Section 11.2, adaptive evolution that follows an optimization principle can drive a population to extinction. Such evolutionary suicide turns out to be a common phenomenon in more realistic models that incorporate frequency- and density-dependent selection, and must therefore be expected to play a major role in the loss of biodiversity. The other facet of non-optimizing adaptive evolution is its role in promoting biodiversity by means of evolutionary rescue and evolutionary branching, which result in the maintenance or even enhancement of biodiversity. A suite of new theoretical tools is thus now in place to translate ecological knowledge of the interaction of populations with their environment into quantitative predictions about the evolving diversity of ecological communities.

The analysis of adaptive responses to environmental change raises new challenges for conservation biology and evolutionary theory. We have introduced E^3-analysis as a tool to investigate and predict the conservation perils associated with environmental changes that unfold on different time scales. E^3-diagrams summarize the salient features of series of pairwise invasibility plots obtained for gradually changing environmental conditions and enable easy graphic interpretation. The use of such E^3-diagrams and, more generally, of adaptive dynamics models in changing environments provide a synthetic approach to the dramatic consequences of adaptive evolution on biodiversity in a changing world. Indeed, it would seem advisable to extend medium-term conservation efforts based on traditional models of population extinction by taking advantage of the now-available new tools to link ecological and evolutionary insights.

Acknowledgments We are grateful to Fabio Dercole, Jean-François Le Galliard, Hans Metz, Kalle Parvinen, and Sergio Rinaldi for interesting discussions. Financial support for this study was received from the Austrian Federal Ministry of Education, Science, and Cultural Affairs, and the Austrian Science Fund to Ulf Dieckmann; from the French Ministère de l'Education Nationale de la Recherche et des Technologies ("ACI Jeunes Chercheurs 2000" and "ACI Bioinformatique 2001"), and from the French Ministère de l'Aménagement du Territoire et de l'Environnement (Programme "Invasions Biologiques") to Régis Ferrière; and from the European Research Training Network *ModLife* (Modern Life-History Theory and its Application to the Management of Natural Resources), funded through the Human Potential Programme of the European Commission, to Ulf Dieckmann and Régis Ferrière.

Part D

Spatial Structure

Introduction to Part D

Human population growth and economic activity convert vast natural areas to serve for settlement, agriculture, and forestry, which leads to habitat destruction, habitat degradation, and habitat fragmentation. These forces are among the most potent causes of species decline and biodiversity loss. Habitat destruction contributes to the extinction risk of three-quarters of the threatened mammals of Australasia and the Americas, and of more than half of the world's endangered birds. Populations confronted with the degradation of their local environment (in excess of the tolerance conferred by phenotypic plasticity) can exhibit two basic types of evolutionary response: either they stay put and adapt to the new local environmental conditions, or they adapt in ways that allow individuals to shift their spatial range efficiently in search of better habitats. In Parts B and C, attention is focused on the former type of adaptation. It is evident that to account for spatial heterogeneity in populations and habitats raises formidable empirical and theoretical challenges. Part D reviews the current achievements and challenges in understanding the role of spatial processes in the persistence of natural populations.

Increased fragmentation typically reduces the size of local populations and/or the flow of migrants between them. This enhances extinction risks, because of either a higher sensitivity of the isolated local populations to demographic stochasticity, or a diminished probability of rescue through immigration. Also, increased fragmentation may affect evolutionary processes in many ways, through a variety of conflicting genetic and demographic effects. First, isolation of population fragments increases the rate of inbreeding on the regional scale, which may in turn exacerbate inbreeding depression, while simultaneously relaxing any existing outbreeding depression. Second, recurrent migration into an open population can either foster or hamper evolutionary rescue: incoming dispersers provide an infusion of genetic variation upon which local selection can act, while, at the same time, potentially swamping existing local adaptation. To predict the overall direction and speed of evolutionary change in spatially structured populations thus requires the selection pressures that operate at all the relevant spatial scales to be accounted for. Chapters in this part follow a path toward increasing levels of complexity in the description of spatially structured populations and culminate in a review of empirical studies on the effects of spatial heterogeneity on population viability.

In Chapter 12, Gaggiotti and Couvet examine how the genetic structure of spatially extended populations is shaped by so-called isolation by distance. The chapter describes the effects of several factors that alter spatial differentiation, including the overall rate of local extinction, the intensity of selection (when traits for which differentiation is measured are not selectively neutral), and the local demographic balance. The authors also discuss the extent to which sink populations can participate in maintaining genetic variation.

The role of sink habitats for evolutionary rescue is probed further in Chapter 13 by Holt and Gomulkiewicz. The authors provide a review of the effect of local adaptation on population persistence in ecological settings for which processes of immigration cannot be neglected. The chapter presents two different modeling frameworks with which to address this issue. Source–sink models are particularly suited to fragmented populations, in which individual patches depend on immigration for persistence. By contrast, focus–periphery models assume continuously distributed populations, in which low-density fringes surround high-density strongholds. It is shown that a fundamental conflict between the supply of variant genetic material and a danger of swamping local adaptation in marginal habitats is central to understanding evolutionary change in both of these frameworks.

In Chapter 14, Parvinen presents metapopulation models that incorporate both individual parameters and environmental characteristics. These models are applied to predict evolutionary change in metapopulations, with a particular focus on the evolution of dispersal traits. A classification of evolutionary outcomes for different environmental conditions advances our understanding of how adaptive processes that operate on dispersal traits are affected by different types of environmental change, and thus influence metapopulation dynamics and viability. This helps to identify combinations of demographic profiles and types of environmental changes that are more likely to result in evolutionary rescue, evolutionary trapping, or evolutionary suicide.

Chapter 15 reviews empirical evidence for fast evolutionary change in response to landscape fragmentation. Colas, Thomas, and Hanski examine examples of metapopulation adaptation – involving local adaptations, adaptation of migration rates, or interactions of such effects – and assess the consequences of such adaptations for metapopulation viability. Examples involve studies of butterfly and plant populations, for which it is shown that the adaptive response of migration rates depends on the fine details of how the metapopulation fragmentation processes unfolds. Such processes may generate different patterns of correlation between patch size, local demographic dynamics, and rates of local extinction. The authors show that local adaptations can occur quite rapidly in response to habitat change. Empirical material presented in this chapter underlines that the evolution of dispersal traits can result in evolutionary rescue as well as evolutionary trapping or suicide.

The study of spatial effects in the context of evolutionary conservation biology is still in its infancy. Nevertheless, several implications for conservation practitioners have already emerged. First, the theory suggests that the absence of changes in the geographic range of a species must not be misconstrued as evidence for resilience to large-scale environmental deterioration. Second, to prioritize protection actions on the basis of local demographic wealth can easily go awry – "sink" habitats characterized by reproductive deficits may prove critical to maintain genetic variability on a regional scale. Third, when planning population reinforcement, the addition of individuals with new dispersal phenotypes (resulting, for instance, from artificial selection under conditions of captive breeding, or from importation from other populations evolved under different selection pressures) may have undesired effects on population viability.

12

Genetic Structure in Heterogeneous Environments

Oscar E. Gaggiotti and Denis Couvet

12.1 Introduction

Human activities dominate many of the world ecosystems and have resulted in the fragmentation of numerous habitat types and the animal and plant populations that inhabit them. As natural areas become smaller and more fragmented, it is increasingly important to understand the ecological and evolutionary dynamics of fragmented populations.

The amount of genetic variation maintained by a population influences the probability of its long-term survival, because genetic variation is a prerequisite for evolutionary adaptation to a changing environment (Lande and Barrowclough 1987). Thus, measurement of genetic variation should be a fundamental part of long-term population management programs. Box 12.1 discusses some of the types of genetic variation relevant to evolutionary processes and presents different measures used to monitor the level of genetic variability in natural populations.

The amount of genetic variability maintained by a population is determined primarily by the joint actions of mutation, selection, and genetic drift. The strength of genetic drift is not determined by the census size of the population, N, but by its effective size N_e (see Box 12.2). The effective population size N_e can be an order of magnitude smaller than the number of adults in the population for a host of different reasons. The best known are uneven sex ratios, temporal fluctuations in population size, variance in reproductive success among individuals, and nonrandom mating. However, as we show later, metapopulation dynamics can be considered one of the most important factors to determine N_e.

Fragmentation of the landscape leads to the original habitat being partitioned into isolated pieces of small size, which increases genetic drift. Genetic drift decreases within-population genetic variation and increases differentiation among populations. It also leads to inbreeding (mating between related individuals) and so influences fitness through inbreeding depression. Chapters 7, 9, and 15 describe how inbreeding depression may lead to genetic deterioration and an increased likelihood of the extinction of plant and animal populations that live in small habitat fragments. However, this information by itself is not enough to understand the effect of the fragmentation process on the genetic characteristic of a species as a whole. To achieve this, we need to know how the migration of individuals among different habitat fragments affects the population structure (i.e., local inbreeding and genetic differentiation among populations).

Box 12.1 Measures of genetic variability

It is generally accepted that most major phenotypic changes in evolution result from the accumulation of quantitative polygenic modifications of existing phenotypes. However, to measure genetic variation quantitatively involves heritability studies that require knowledge of parent–offspring relationships. This information may be readily available for captive populations, but it is extremely difficult to obtain for wild populations.

A variation that is much easier to measure is that present in single-locus selectively neutral polymorphisms (inferred from nuclear markers such as allozymes and microsatellites). Although this type of variability is presently nonadaptive, it may serve as the basis for future adaptation to environmental changes (Lande and Barrowclough 1987). The most widely used measure of population genetic variability at this level is heterozygosity H, defined as the mean percentage of heterozygous loci per individual (Avise 1994). Estimates of H can be obtained by a direct count of the number of heterozygous individuals for each locus, dividing this by the total number of individuals in the sample, and averaging over all loci. Heterozygosity can also be estimated from the observed frequencies of alleles, provided the population is in Hardy–Weinberg proportions. Thus,

$$h_j = 1 - \sum p_{k,j}^2 \,, \tag{a}$$

where h_j is the heterozygosity at locus j and $p_{k,j}$ is the frequency of allele k at locus j.

In subdivided populations, total genetic variability can be partitioned into variation that exists within and among subpopulations,

$$H_T = H_S + D_{ST} \,, \tag{b}$$

where H_S and D_{ST} are the within- and among-subpopulation components of the variance, respectively (Nei 1973). Partitioning of genetic diversity can be extended hierarchically to more complex situations. For example, for insular species,

$$H_T = H_S + D_{SI} + D_{IT} \,, \tag{c}$$

where D_{SI} and D_{IT} are the variances among subpopulations within an island and among populations on different islands, respectively.

Note that D_{ST} is an absolute measure of gene differentiation among subpopulations. A more widely used measure of population differentiation is

$$F_{ST} = D_{ST}/H_T \,, \tag{d}$$

the genetic differentiation relative to the total population. As shown in Box 12.3, most studies of population subdivision and of the effect of landscape fragmentation are cast in terms of F_{ST}.

Other measures of genetic variability for allozyme and microsatellite data include the mean number of alleles per locus and the percentage of polymorphic loci P. For data at the DNA level a useful statistic that summarizes heterozygosity is the nucleotide diversity (Nei and Li 1979),

$$P = \sum q_i q_j \delta_{ij} \,, \tag{e}$$

where q_i and q_j are the frequencies of haplotypes i and j, respectively, in the population, and δ_{ij} is the sequence divergence among haplotypes i and j.

Box 12.2 Effective population size

The effective population size N_e is a measure of how many individuals contribute genes to future generations. More formally, it is defined as the size of an "idealized Wright–Fisher population" that would give the same value for some specific property as a real population of size N (e.g., Crow and Kimura 1970). There are different forms of effective size depending on the specific property considered. The two most-used properties are:

- Amount of gene frequency change per generation as measured by its variance, which leads to the *variance effective size*;
- Probability that two randomly sampled genes have the same parent gene, which leads to the inbreeding effective size.

The idealized Wright–Fisher model consists of a finite, randomly mating population of constant size, generations do not overlap, and all individuals produce the same number of offspring.

Under these simple assumptions, the probability that two randomly sampled genes have the same parent gene is given by $f = \frac{1}{2N}$ (e.g., Ewens 1989), where N is the size of the ideal population. Thus, the inbreeding effective population size N_e^i is defined as $N_e^i = \frac{1}{2\tilde{f}}$, where, here and below, a tilde refers to the quantities measured in the considered system.

A central property of the Wright–Fisher model is that sampling of gametes is binomial and the variance of the change in gene frequency is

$$\sigma^2_{\Delta q} = \frac{q(1-q)}{2N}, \tag{a}$$

where q is the allele frequency of a gene in the ideal population. Thus, the variance effective population size is given by

$$N_e^v = \frac{\tilde{q}(1-\tilde{q})}{\tilde{\sigma}^2_{\Delta q}}. \tag{b}$$

Note that these two effective population sizes are not necessarily equal and in some cases can be very different from one another. It is therefore necessary to qualify the type of N_e being used.

12.2 Basic Models of Population Genetic Structure

The large body of theory developed by population geneticists to investigate the effects of population subdivision (or structure) on genetic variation (for reviews see Slatkin 1985, 1987) provided the initial tools for the study of the genetic effects of fragmentation. Most of the results were obtained using the so-called island model, which represents an infinite number of subpopulations of size N that exchange migrants among themselves at a constant rate m (Slatkin 1985, 1987). Box 12.3 presents a very general formulation of the island model by Crow and Aoki (1984).

Using the island model, it is possible to show that the subdivision of a large, widespread population into many small local fragments increases genetic drift, which leads to a reduction of within-population genetic variability and an increase

Box 12.3 The island model

Assume that a diploid monoecious species is subdivided into n subpopulations each of effective size N_e. In every generation each subpopulation exchanges migrants at a constant rate m. Immigrants into any colony are selected from all other subpopulations at random, and therefore the migration rate is independent of the geographic distance between colonies. Let u be the mutation rate from one allele to any of $k-1$ other alleles. Use f_0 to denote the probability that two randomly chosen homologous genes from a single colony are identical by descent (IBD), f_1 to denote the probability of IBD for two genes randomly sampled from different colonies, and f to denote the probability of IBD for two genes randomly sampled from the whole population. Under these assumptions and using primes for the next generation, the following set of recurrence equations hold,

$$\begin{aligned} f_0' &= v\{x[z+(1-z)f_0]+(1-x)f_1\}+w[x(1-z)(1-f_0) \\ &\quad +(1-x)(1-f_1)]\,, \\ f_1' &= v\{y[z+(1-z)f_0]+(1-y)f_1\}+w\{y[(1-z)(1-f_0) \\ &\quad +(1-y)(1-f_1)]\}\,, \\ f &= [f_0+(n-1)f_1]/n\,, \end{aligned}$$

where $v=(1-u)^2+\frac{u^2}{k-1}$, $w=\frac{1-v}{k-1}$, $x=(1-m)^2+\frac{m^2}{n-1}$, $y=\frac{1-x}{n-1}$, and $z=\frac{1}{2N}$.

Expressions for f_0 and f_1 at equilibrium are obtained by dropping the primes and solving the resultant equations. Note that this model leads to a partitioning of the total genetic variance into its within- and among-population components (f_0 and f_1, respectively). Thus, the degree of genetic differentiation within the population, F_{ST}, can be defined as the proportion of total genetic variation due to the among-population component,

$$F_{ST}=\frac{(1-f)-(1-f_0)}{1-f}=\frac{f_0-f}{1-f}\,. \tag{a}$$

F_{ST} can range in value from 0 to 1; a value of 1 indicates that different populations are fixed for different alleles, while a value of 0 indicates that the collection of subpopulations behaves as a single panmictic unit. The equilibrium value for F_{ST} can be derived by replacing the equilibrium values for f_0 and f in Equation (b). Assuming that $u \ll m$ and $1/N \ll 1$, this gives

$$F_{ST}=\frac{1}{4Nm\beta+1}\,, \tag{b}$$

where $\beta=(\frac{n}{n-1})^2$. Note that F_{ST} is independent of the mutation rate and therefore is a useful measure for population genetic analysis. For the infinite island model, $\beta \rightarrow 1$ and therefore, $F_{ST}=\frac{1}{4Nm+1}$.

The island model can be extended to consider more complex cases in which subpopulations are grouped into neighborhoods. Such models are referred to as hierarchical island models. Wright (1951) presented the first analysis of a hierarchical

continued

Box 12.3 *continued*

model of population subdivision, which was later further analyzed by Slatkin and Voelm (1991). To study this more complex model it is necessary to define the inbreeding coefficient of a subpopulation relative to its neighborhood, F_{SN}, and that of a neighborhood relative to the whole population, F_{NT}. Following the same principles as those used to obtain an equation for F_{ST} at equilibrium, it is possible to obtain equilibrium expressions for F_{SN} and F_{NT} [provided $Nm_1 \gg 1$; see Vigouroux and Couvet (2000) for the alternative case],

$$F_{SN} = \frac{1}{1 + 4N\varepsilon m_1}, \tag{c}$$

and

$$F_{NT} = \frac{1}{1 + 4Nc\zeta m_2}, \tag{d}$$

where N is the number of individuals within each subpopulation, $\varepsilon = (\frac{c}{c-1})^2$, c is the number of subpopulations within each neighborhood, $\zeta = (\frac{\eta}{\eta-1})^2$, η is the number of neighborhoods, m_1 is the migration rate between populations within the same neighborhood, and m_2 is the migration rate between populations from two different neighborhoods.

in genetic differentiation among populations. The effect of genetic drift, however, is counteracted by the gene flow that results from migration among local populations. Thus, the ultimate effect of fragmentation is determined by the relative strengths of genetic drift and gene flow, summarized by the combined parameter Nm. The degree of population subdivision decreases quite rapidly for increasing Nm (Figure 12.1), and therefore it seems only small amounts of gene flow are necessary to reestablish the connectivity of the landscape and avoid the detrimental effect of population fragmentation. Following this line of reasoning, different authors proposed a rule of thumb according to which one migrant individual per local population per generation is sufficient to counteract the disruptive effects of drift (e.g., Kimura and Ohta 1971). However, this rule is based on a very simplistic model that makes a host of unrealistic assumptions (Mills and Allendorf 1996).

Under the island model, the depletion of genetic variation within populations is not necessarily accompanied by a decrease in total genetic variance, since genetic drift increases the among-population component of genetic variation. Indeed, the inbreeding effective size of a subdivided population is (e.g., Whitlock and Barton 1997)

$$N_e = \frac{nN}{1 - F_{ST}}, \tag{12.1}$$

which clearly indicates that as F_{ST} increases, N_e also increases.

Although the island model provides some insights into the effects of fragmentation, its applicability is impaired largely by its many simplifying assumptions.

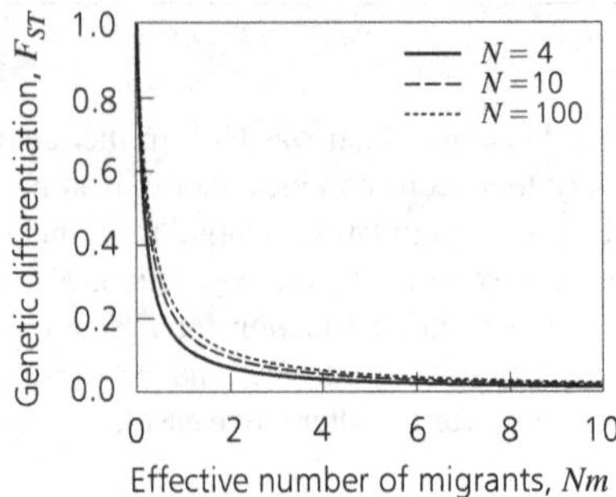

Figure 12.1 Genetic differentiation F_{ST} as a function of the effective number of migrants under the island model.

In real situations, the assumption of selective neutrality is violated and local populations differ in size. Moreover, local populations are liable to extinction through demographic and genetic factors, and therefore the system never reaches genetic equilibrium. These limitations led to the development of a more complex body of theory that describes more accurately the ecological and genetic properties of fragmented populations.

12.3 Adding Geography: The Stepping-stone Model

The island model of population subdivision ignores the fact that many species have limited dispersal capabilities. In cases of limited dispersability, it is more logical to assume that migration occurs only between adjacent demes. The stepping-stone models introduced by Malécot (1968) and Kimura (1953) address these situations. In the simplest version, the infinite one-dimensional stepping-stone model, an infinite number of colonies with constant effective size N_e is arranged along a linear habitat (Kimura and Weiss 1964). In this case, the correlation of allele frequencies between demes g steps apart is given by

$$\gamma(g) \propto \exp\left(-g\sqrt{2m_\infty/m_a}\right) . \tag{12.2a}$$

Here, m_∞ is the probability that a migrant comes from a very distant population and represents a long-distance migration parameter, while m_a is the probability of migration from an adjacent deme ($m_\infty \ll m_a$). Equation (12.2a) indicates that the correlation decreases approximately exponentially with distance, and therefore the genetic differentiation between populations increases quite rapidly. There are equivalent results for two- and three-dimensional stepping-stone models (Kimura and Weiss 1964). For two-dimensional stepping-stone models,

$$\gamma(g) \propto \frac{1}{\sqrt{g}} \exp\left(-g\sqrt{4m_\infty/m_a}\right) . \tag{12.2b}$$

For three-dimensional models,

$$\gamma(g) \propto \frac{1}{\pi g} \exp\left(-g\sqrt{6m_\infty/m_a}\right) . \tag{12.2c}$$

Thus, the correlation between alleles falls off much more rapidly as the number of dimensions increases. The results of stepping-stone models can also be presented in terms of F_{ST}, but the equations obtained are more convoluted (see Slatkin 1991). However, it is possible to obtain an expression for the relationship between F_{ST} and the correlation between adjacent demes from equation 3.5 in Kimura and Weiss (1964),

$$F_{ST} \approx \frac{1}{1 + 4Nm_a[1 - \gamma(1)]} . \tag{12.3}$$

As the number of dimensions increases, $\gamma(1)$ decreases [see Equations (12.2a–12.2c)] and so does F_{ST}.

Despite the large differences in the assumptions made by the island and stepping-stone models, there is an important similarity between the two-dimensional stepping-stone models and the island model. Populations will be strongly differentiated if $Nm < 1$ ($m = m_\infty + m_a$), whereas they behave essentially as a single panmictic unit if $Nm > 4$ (Kimura and Maruyama 1971).

Another issue of interest is the amount of genetic variability maintained by populations with a stepping-stone structure, one measure of which is the effective number of alleles n_e. Maruyama (1970) noted for the one-dimensional stepping-stone model that the effective number of alleles is $n_d\sqrt{\mu/2m_d} + 4Nn_d\mu$, where n_d is the number of demes, μ is the mutation rate, and m_d is the migration rate among adjacent demes. The equivalent quantity for the island model is $1 + 4Nn_d\mu + n_d\mu/m_d$. Both formulas are valid for large n_d. For a panmictic population of size n_dN, $n_e = 1 + 4Nn_d\mu$. Thus, genetic variability is increased by population subdivision and this increase is much greater in a linear stepping-stone structure.

Stepping-stone models lead naturally to the concept of "isolation by distance", first introduced by Wright (1943) to describe the accumulation of local genetic differences under restricted dispersal. A number of techniques are available to test for isolation by distance in natural populations. For example, Mantel's test of matrix correlation allows the comparison of genetic distance and geographic distance matrices [see Smouse and Long (1992) for a review]. Slatkin (1993) proposed another method to detect isolation by distance based on analytic theory that relates measures of isolation by distance to average coalescence times.

12.4 Metapopulation Processes and Population Differentiation

Extensive habitat fragmentation invariably results in the reduction of population sizes and, therefore, local populations become threatened by extinction through environmental and demographic factors (see Chapters 2 and 4). Nevertheless, new populations are continuously formed as colonists re-invade vacant habitat patches.

These dynamics of extinction and recolonization form the basis of metapopulation theory (for reviews, see Hastings and Harrison 1994; Harrison and Hastings 1996).

Effects of colonization–extinction processes

Wright (1940) first recognized that extinctions and recolonizations can play an important role in determining the genetic structure of natural populations. However, the first formal metapopulation genetic models were published almost four decades later by Slatkin (1977). In his analysis, Slatkin considers two possibilities regarding the formation of propagules:

- "Propagule pool" model (founders are all from the same existing population);
- "Migrant pool" model (founders are chosen at random from the entire metapopulation).

Slatkin's results are presented in terms of measures of genetic variability and seem to indicate that the extinction and recolonization process tends to decrease genetic differentiation among colonies.

Both models can be formulated using the principles presented in Box 12.3 for the island model. However, it is necessary to add terms for the effects of extinction and recolonization events. The resultant equations for f_0, f_1, and f are quite convoluted and were first presented by Slatkin (1977). Wade and McCauley (1988) used these equations and Equation (12.2a) to obtain an expression for F_{ST}. These authors noted that the effect of the extinction and recolonization process is to establish an age structure among the array of populations. This allowed them to develop expressions that describe the genetic variance among populations of each age class of the metapopulation. Averaging F_{ST} across all age classes, they obtained a measure of differentiation among all populations, which, for the "migrant pool" model, is given by

$$\overline{F_{ST}} = \frac{1 + e_0(N/K - 1)}{2N[1 - L(1 - e_0)]}, \tag{12.4}$$

where e_0 is the fraction of demes that go extinct, K is the number of founders, m is the migration rate, and L is an abbreviation of $(1 - \frac{1}{2N})(1 - m)^2$.

Taking the ratio of $\overline{F_{ST}}$ to the expected value under the standard island model, it is possible to obtain the conditions under which population turnover enhances or diminishes genetic differentiation among local populations. Genetic differentiation is enhanced when $K < 2Nm$, otherwise it is decreased.

Using a similar approach for the "propagule pool" model, Wade and McCauley (1988) obtained an equivalent expression for F_{ST},

$$\overline{F_{ST}} = \frac{a + \frac{1-a}{2N}}{1 - L(1 - a)}, \tag{12.5}$$

where $a = e_0/(2K)$. Here the ratio of $\overline{F_{ST}}$ to the expected value under the standard island model is always larger than 1, which indicates that in this case population turnover always increases population differentiation.

The "migrant-pool" and "propagule-pool" models of colonization represent extreme ends of what must be a continuum in nature (Wade and McCauley 1988). To consider intermediate cases, Whitlock and McCauley (1990) generalized Slatkin's models by including a parameter ϕ to represent the probability that two alleles in a newly formed population come from the same parental population. Thus, $\phi = 0$ describes the "migrant pool" and $\phi = 1$ the "propagule pool". This more general model shows that, at equilibrium, population turnover increases differentiation if

$$K < \frac{2Nm}{1-\phi} + \tfrac{1}{2} \, . \tag{12.6}$$

This condition is equivalent to the requirement that variance among newly founded populations be greater than variance among populations at equilibrium. Thus, any tendency for founders to move in groups, such as in the dispersal of seeds in fruits, increases the likelihood that population turnover will lead to greater genetic differentiation.

Effects of local population dynamics

The degree of genetic differentiation is also determined by the population dynamics of local populations. All the models discussed above assume that newly founded populations grow back to carrying capacity in a single generation and, therefore, ignore the effect of within-patch population dynamics. This unrealistic assumption was relaxed by Ingvarsson (1997), who used a simple exponential growth model to describe the population dynamics of newly recolonized subpopulations. In principle, a delayed period of population growth is expected to increase genetic drift because the size of recently recolonized subpopulations is small. However, whether genetic differentiation is increased or not depends on whether the migration rate is constant or not. If it is, then small populations will receive few migrants and genetic drift is important, leading to increased differentiation. But if the number of migrants is constant, then differentiation is decreased because small populations will be swamped by large number of migrants from large populations.

The relationship between migration rates and population size remains to be established, as widely divergent hypotheses have been proposed (see the review in Saether *et al.* 1999). For example, it is possible that the effective number of immigrants in large populations at or close to their carrying capacity is small, since competition for breeding sites and/or other resources will be strong. Thus, the successful settlement of new immigrants is impaired. Therefore, in real ecological situations, delayed population growth might lead to decreased differentiation. Clearly, migration patterns are important in determining the degree of genetic differentiation, a subject we address when discussing the particular case of source–sink metapopulations (see below).

Up to now, we have discussed the effect of population turnover on the degree of genetic differentiation, but we have not examined its effect on the total amount of genetic variability maintained by a species. Although, as explained before, extinction and recolonization events can increase among-population variability, they always decrease within-population variability because they act as local population bottlenecks. The above models help clarify the effect of population turnover on the partitioning of genetic variance, but they cannot clarify the effect of extinctions and recolonizations on the total genetic variability maintained by a species. To address this question, the effect of population turnover on the effective population size of a species must be considered.

12.5 Metapopulation Processes and Effective Population Size

The derivation of expressions for the effective size of a metapopulation that take into account all the relevant parameters is very difficult and leads to mathematically intractable models. For example, in the presence of variance of reproductive success between populations, V, the effective size of a metapopulation is given by (Whitlock and Barton 1997)

$$N_e = \frac{nN}{(1+V)(1-F_{ST}) + 2NVF_{ST}} , \tag{12.7}$$

which indicates, in particular, that N_e increases with F_{ST} only when V is low [compare Equations (12.1) and (12.7)]. By combining analytical and simulation approaches, Whitlock and Barton (1997) successfully provided a formulation of the inbreeding effective size that takes into account most of the relevant parameters. Their results indicate that metapopulation dynamics can greatly decrease N_e. Indeed, for simple models such as that used by Whitlock and McCauley (1990), when the number of founders is much smaller than the size of the local populations, the metapopulation N_e can be a very small fraction (smaller than 1%) of that without extinction. For more complex metapopulation models that incorporate variance in reproductive success between demes and explicit local dynamics, the results can be complex, but it is still possible to understand the effect of individual factors. In general, when migration rates are low, the resultant increase in genetic differentiation tends to increase the metapopulation N_e. However, when local population sizes are highly variable (e.g., under chaotic dynamics), the variation in deme reproductive success reduces N_e even when migration rates are low.

Hedrick and Gilpin (1997) used simulations of a much simpler model to study the effect of metapopulation dynamics on N_e. They used a generalization of Slatkin's models that allowed nonimmediate recolonization of extinct demes and decoupling of gene flow and number of colonists. An interesting result of this analysis is the effect of increasing local deme sizes on the effective population size. In principle, a proportional increase in N_e would be expected with increasing deme sizes. Their results show, however, that as deme sizes increase, N_e does not increase very much and asymptotes quickly. By setting the carrying capacity of local colonies to infinity, they show that population turnover can greatly reduce total genetic variability, even when genetic drift at the local population level is minimal.

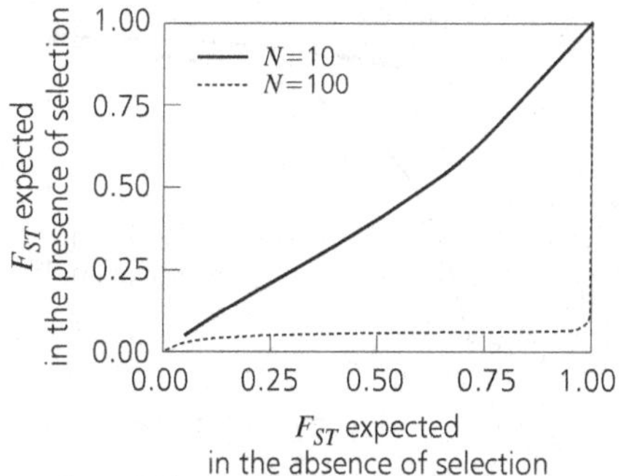

Figure 12.2 Expected F_{ST} value for a selected gene relative to that expected for a neutral gene for different population sizes, for the case of overdominance (fitnesses for the three genotypes are $1, 1+s, 1$). *Source*: Couvet (2002), with $s = 0.01$.

From the above discussion it is clear that for most ecologically realistic situations fragmentation of the landscape greatly reduces the effective size and level of genetic variability of most species.

12.6 The Effect of Selection on Differentiation: The Island Model

So far we have ignored the effect of selection on the spatial structure of populations. Selection, however, is an important factor that affects the pattern of genetic variability and differentiation in natural populations. There are three main reasons why we need to study the effect of selection:

- Neutrality of commonly used markers has been repeatedly questioned (e.g., Pogson *et al.* 1995).
- Although a particular marker might be neutral, it may be linked to a gene subject to selection.
- A significant fraction of the genome is composed of genes subject to selection, these genes being responsible for inbreeding depression, local adaptation, and interaction with other organisms. Inferring the spatial structure for these genes and comparing it to that of neutral genes leads to predictions on important issues such as the extent of heterosis and local adaptation.

Spatial structure of selected genes

The extent to which selection influences population structure is determined by its strength relative to that of other forces that modify gene frequencies (drift, migration, and mutation). Assuming that mutation is negligible when compared to the other forces, then for the selection coefficient $s \gg m, 1/N$, the effect of genetic drift can be disregarded and selection will be one of the main factors determining the degree of population structuring. If $s \ll m, 1/N$, then the effect of selection is negligible and the results of neutral models are still valid. In particular, the spatial structure of neutral genes is a good approximation to the structure of selected genes when the population size is small (Figure 12.2).

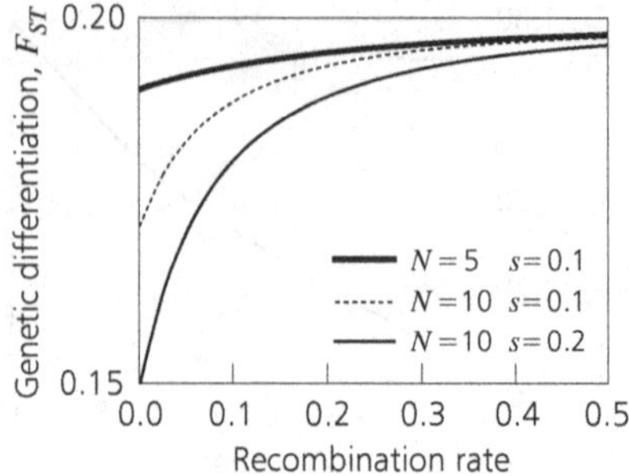

Figure 12.3 F_{ST} for a neutral bi-allelic locus as a function of its proximity to a selected gene (recombination rate) for the case of overdominance, depending on population size and the intensity of selection (s is as in Figure 12.2) In the absence of selection, the value of m used leads to $F_{ST} = 0.2$. The plot was obtained using a matrix approach, as for Figure 12.2.

The magnitude of the F_{ST} expected for selected genes depends on the pattern of selection being considered. In the case of overdominance, the expected F_{ST} value will be lower than that expected under neutrality. On the other hand, F_{ST} will be higher for genes responsible for local adaptations than for neutral genes. For a fitness differential comparable to that observed for most deleterious mutations ($s = 0.02$, see Chapter 9), the F_{ST} for overdominant genes is very close to that observed for neutral genes when $N = 10$, but is noticeably different when $N = 100$ (Figure 12.2).

The structure of selected alleles is a particularly important study for conservation biology, because it allows us to evaluate the effect of migration on population viability. This issue is empirically examined in Chapter 15.

Spatial structure of genes linked to selected genes

The study of the structure of neutral genes linked to selected ones concerns a large portion of the genome. Selected genes within the genome will affect, to various degrees, the spatial structure of one or more neutral genes, depending on the proximity of these to the selected genes.

As for the spatial structure of selected genes, the influence of selection is stronger when the local population size is large. For the case of genes linked to selected genes, however, the recombination rate also has to be considered; the higher the recombination rate, the less important is the effect of selection (Figure 12.3).

Some good evidence indicates that the indirect effect of selection on the structuring of neutral genes is not uncommon. The following examples provide details of the processes involved:

- F_{ST} values for markers located in low recombining regions of the genome are higher than those for markers located in regions where recombination is higher (see review in Charlesworth 1998). That F_{ST} in the low recombining regions is *higher* requires an explanation, one possibility being the overall presence of genes responsible for local adaptation.

- The decrease of genetic variability observed with selfing, relative to the case of random mating, is higher than that expected in the absence of selection (Liu *et al.* 1998). This could be because the influence of selected genes extends further for consanguineous reproductive systems, since the effective recombination rate is lower.
- Slightly deleterious mutations, which affect the survival of small populations (see Chapter 9), decrease genetic variability at nearby loci: such an effect has been called "background selection". The low selective differential of such mutations means that the effect on the regions that surround each selected gene will be low, but present on a large part of the genome as these mutations are supposedly numerous.

12.7 Structure and Selection in Source–Sink Metapopulations

Natural environments are heterogeneous and therefore species encounter variation in abiotic and biotic conditions. For many geographically extensive populations, a large fraction of the individuals may occur in depauperate habitats in which reproduction is insufficient to balance local mortality (e.g., Pulliam 1988). Populations in these depauperate habitats, generally termed "sink populations", may nevertheless persist, being maintained by continued migration from more productive "source" areas nearby.

Neutral genetic structure

Although source–sink metapopulations may occur naturally, human activities have led to an increase in their occurrence. Indeed, the most likely outcome of fragmentation of the landscape is a large tract of forest (the source) kept as a nature reserve or park and a series of small patches (the sinks) embedded in a landscape dominated by human development. That small tracts of forest are unable to maintain themselves may lead natural resource managers to assume that their protection is not warranted. It is, therefore, important to establish the extent to which sink populations can help maintain or increase the genetic variability of a species and, therefore, contribute to its long-term survival.

The population genetic models of source–sink metapopulations presented by Gaggiotti and Smouse (1996) and Gaggiotti (1996) help to clarify this problem. Their studies investigate the effect that different patterns of migration have on the genetic structure of source–sink metapopulations, specifically modeling the dynamics of local populations. They study a metapopulation that consists of a single source and a variable number of sink populations and consider the expected number of nucleotide differences between two genes drawn at random from the source–sink metapopulation. The migration patterns considered include constant migration and different types of stochastic migration.

In real ecological scenarios, migration is an inherently stochastic process and the results shows that stochasticity has an important impact on the genetic structure of the collection of sinks. Stochastic migration leads to long periods of zero migration events from the source, during which the effect of genetic drift is greatly enhanced, which decreases the amount of genetic variability maintained and increases the genetic differentiation between sinks. The importance of the stochastic effect on the genetic structure of the collection of sinks is determined by the rate of population decay and the variance in the number of migrants arriving at the sinks. The effect is maximized when the rate of population decay in the sinks is moderate, in which case the variance of the genetic parameters is large. The effect of stochasticity increases as the variance in the number of migrants arriving at any given sink increases. Thus, the genetic structure of the collection of sinks is most affected when stochasticity is introduced by sudden changes (catastrophic events) that preclude the arrival of new migrants to all sinks.

Gaggiotti and Smouse (1996) and Gaggiotti (1996) also show that genetic variability in the collection of sinks may be large even under stochastic migration, and that population subdivision among the subset of sinks may be significant. Thus, in a fragmented landscape sink populations may serve as temporary refugia that increase colonization and gene flow. These findings indicate that the protection of sink populations could help to alleviate (but not preclude) the detrimental effects of landscape fragmentation. However, a critical fact is the sort of genetic variability that accumulates in the sink population. If this variability is unconditionally deleterious (see Chapter 9), no overall advantages will be associated with such a higher genetic variability.

Fixation of beneficial alleles

Adaptive evolution of a population to the conditions of the habitat patch in which it lives is determined by the relative strength of opposing forces: natural selection favors locally fitter genes, and gene flow reduces the frequency of locally favored genes (Mayr 1963; Antonovics 1976). Peripheral (sink) populations may exist at densities considerably lower than more central (source) populations. Thus, immigration rates into sinks may be high relative to the local population abundance and may suffice to swamp out selection that favors locally fitter genes. This process has been invoked to explain the "niche conservatism" that seems to characterize many species (e.g., Holt and Gomulkiewicz 1997a; Kirkpatrick and Barton 1997; Chapter 13).

The above observations lead to the dismissal of the evolutionary potential of sink populations (e.g., Kirkpatrick and Barton 1997) and metapopulations in general (e.g., Harrison and Hastings 1996). These observations, however, are based on a highly unrealistic assumption, namely that there is a steady (constant) stream of migrants into the sink with each generation. In real ecological situations, however, migrants may or may not arrive at the sink in any particular generation, depending on environmental fluctuations or the vagaries of founder dispersal (Gaggiotti and Smouse 1996). A thorough investigation of the effect of stochastic migration on the genetic structure of source–sink metapopulations (Gaggiotti 1996; Gaggiotti and Smouse 1996) reveals that if the variance in the migration process is high, differentiation between source and sink is possible. This is particularly so when migration between source and sink is episodic. The model used by Gaggiotti

(1996) assumes selective neutrality. Preliminary results of a model that incorporates selection (Gaggiotti, unpublished) indicate that stochastic migration favors the spread of beneficial mutations in sink habitats. Much research (both theoretical and empirical) is needed before the evolutionary potential of sink populations can be thoroughly evaluated.

12.8 Concluding Comments

Under the most ecologically realistic scenarios, fragmentation of the landscape leads to a reduction of genetic variability at the species level. This is because population turnover increases genetic drift and therefore reduces the effective population size.

The outcome of habitat fragmentation is likely to differ depending on whether the original distribution of the species was continuous or fragmented. For continuously distributed species, fragmentation disrupts the existing patterns of gene flow and decreases the effective population size. For species with a spatially heterogeneous distribution, gene flow may not be strongly affected, but the increased likelihood of extinction of local populations still decreases the species' effective population size. In either case, the evolutionary potential of the species may be largely reduced.

It is important to realize that the use of simple approaches to study complex situations, such as those arising from the fragmentation of the landscape, can lead to serious misunderstandings. A good example is the notion that the best way to maintain genetic variability is by protecting a sufficient number of small populations instead of a single large population (e.g., Simberloff 1988; Hanski 1989). This idea is based on the results of the island model of population subdivision, and therefore rests on a large number of unrealistic assumptions. The results of more complex metapopulation models that have been developed over the past decade (see above) clearly indicate that this principle is flawed.

In the past ten years there has been a surge in the number of studies concerned with the genetic structure of metapopulations and the implications for conservation biology. Numerous aspects have been addressed and great progress has been made, but some outstanding issues still need to be considered. In particular, a paucity of studies address the effect of selection on the genetic structure of metapopulations. Some progress has been made in the study of local adaptation in source–sink metapopulations (Holt and Gomulkiewicz 1997a; Gomulkiewicz *et al.* 1999), but even in this area many questions still need to be addressed.

13

Conservation Implications of Niche Conservatism and Evolution in Heterogeneous Environments

Robert D. Holt and Richard Gomulkiewicz

13.1 Introduction

Species may, in principle, respond to environmental change in several different ways (Pease *et al.* 1989; Holt 1990; Chapters 10 and 11). Some species may track environmental states to which they are already well adapted and so shift in abundance and distribution. Other species may not evolve at all and so become extinct. Some species may evolve adaptively in ways that facilitate their persistence in changed environments. Yet other species may evolve in ways that hamper their long-term viability. A fundamental goal of the discipline of evolutionary conservation biology is to understand the factors that govern the relative likelihood of each of these outcomes.

Recognizing the importance of directional environmental change in driving extinctions in once-common species raises a profound puzzle. On the one hand, as ecologists we know that extinction risk emerges because directional environmental changes lead to lowered population abundances and/or restricted distributions; in effect, species are pushed outside their niches. On the other, as evolutionists we know that species often have abundant genetic variation, and so can adapt to novel circumstances. Conservation problems arise precisely because species do *not* adapt sufficiently to the new environments created by anthropogenic activity. In other words, conservation problems reflect a seeming *failure* of evolution by natural selection to adapt species to environmental change.

Such failures are examples of "niche conservatism". The history of life reveals examples of both niche conservatism (phylogenetic lineages that retain much the same ecological niche over substantial spans of evolutionary history) and niche evolution (Bradshaw 1991; Holt and Gaines 1992). Before proceeding any further, we should be clear about the meaning of "niche" (Schoener 1989). For a species with continuous, overlapping generations the intrinsic growth rate r is its expected per capita birth rate minus its expected per capita death rate, at low densities. Succinctly, if a habitat results in $r > 0$ for a given species it has conditions that are within that species' niche. By contrast, if $r < 0$, the habitat has conditions outside the niche. (For discrete generations, if the environment is such that the per capita growth ratio per generation $R_0 > 1$, the habitat lies within the niche, but if $R_0 < 1$, it is outside.) In effect, the niche of a species is an abstract mapping of the most fundamental attribute of that species' population dynamics – its persistence versus its extinction – onto environmental states. A population of a species should persist (in the absence of stochastic fluctuations) if it experiences conditions within its niche, but go extinct if forced to live outside its niche.

Many conservation problems arise because environmental change forces a species' population outside that species' ecological niche. Evolution that influences extinction risk often involves niche evolution, such that species expand or shift their niches to incorporate novel environments. We do not downplay the role of other evolutionary processes in conservation (e.g., mutational meltdown in small populations, Lynch *et al.* 1995a; Chapter 9), but we do contend that an understanding of niche conservatism and evolution is integral to an evolutionarily informed conservation biology. In this chapter, we review theoretical studies which show that an absence of evolutionary responses in changed environments, which at first seems puzzling, actually makes sense when the demographic context of evolution is considered. Recent theoretical studies provide elements of a conceptual framework that allow niche conservatism to be understood, and possibly predicted. The basic idea is that population dynamics can, at times, constrain evolutionary dynamics. At other times, population dynamics facilitate evolutionary responses. This chapter provides an overview of these studies and highlights their implications for conservation.

Patterns of environmental change are complex in space and time. For conceptual clarity, we focus on simple situations with a step transition between two environmental states, or on spatial flows of individuals between two discrete habitats. We also briefly discuss evolution along smooth gradients in time and space. These different scenarios illustrate how demographic asymmetries can channel and constrain the evolution of local adaptation. We provide partial answers to two essential questions:

- When does adaptive evolution mitigate extinction risk?
- Can we use our understanding of the dynamics of – and constraints on – adaptive evolution to guide practical conservation efforts?

13.2 Adaptations to Temporal Environmental Change

Consider a closed population, such as an insect species on an oceanic island or in an isolated habitat fragment. The population initially is found in a stable environment, at equilibrium within its niche. It then experiences an abrupt environmental change, and conditions shift to outside its niche (i.e., absolute fitness $R_0 < 1$). If the environment then stays constant, but the population does not evolve, extinction is certain (Figure 13.1a). Let us assume that the population has sufficient genetic variation to potentially persist in the novel environment. Selection increases average fitness, given the assumptions of Fisher's Fundamental Theorem (Fisher 1958; Burt 1995). However, as long as the average absolute fitness is less than 1, the population will continue to decline, though at a slowing rate. Eventually, mean fitness will exceed 1, and the population will start to increase. Such a species should display a characteristic U-shaped trajectory (Figure 13.1b).

Even so, this population may still experience a transient window of extinction risk. Given sufficient variation to adapt (i.e., evolve a positive growth rate) to

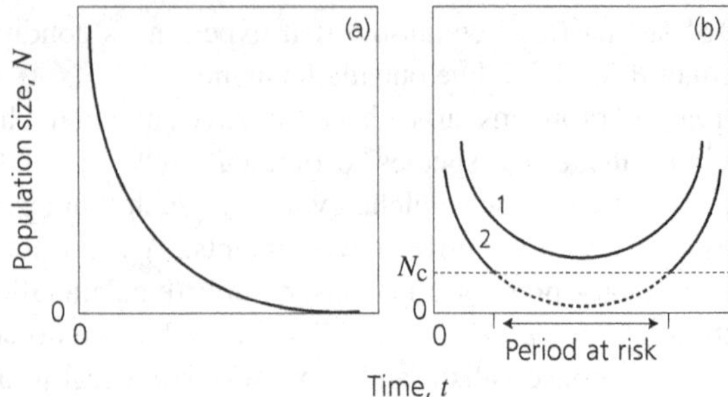

Figure 13.1 Population dynamics in a newly created sink habitat. (a) Without evolution: if evolutionary change is precluded (e.g., because relevant genetic variation is absent), population size declines toward extinction. (b) With evolution, if there is sufficient genetic variation, a population may adapt quickly (Trajectory 1), and so avoid exposure to the critically low sizes that are associated with high extinction risk (i.e., $N_t > N_c$ for all t). However, if the population adapts more slowly (Trajectory 2), it spends a period at low abundances, and thus incurs a high risk of extinction (i.e., $N_t < N_c$ for some span of time t). If by chance extinction is avoided and adaptation continues, the population may eventually rebound.

abrupt environmental change, in a deterministic world a population should eventually bounce back from low numbers. But when numbers become too low, even well-adapted populations may face extinction from demographic stochasticity. The dotted line in Figure 13.1b marks a critical population size N_c below which we assume demographic stochasticity to become a severe problem. For Trajectory 1 in Figure 13.1b the population starts at a high density and evolves a positive growth rate sufficiently fast to rebound before ever entering the "danger zone" of low numbers. This results in a process of "evolutionary rescue". By contrast, for Trajectory 2 the initial density is lower and the population evolves slowly, so it experiences a period of extinction risk. Figure 13.1b illustrates a race between an evolutionary process (improved adaptation to the novel environment, increasing mean fitness), and an ecological process (declining numbers, as long as mean fitness is less than 1). In Box 13.1 we present an analytic model that explicitly combines evolution in a quantitative character and population dynamics and results in U-shaped trajectories in numbers.

Given a critical population size N_c below which extinction is probable, we can use Equation (g) in Box 13.1 to determine whether or not a population trajectory includes periods of risk (i.e., $N_t < N_c$). Figure 13.2 summarizes whether or not a population experiences such extinction risk as a function of its initial density, the degree of initial maladaptation in the novel environment, and the heritability of the trait that undergoes selection. The basic messages are as follows:

- Populations that are initially rare are highly vulnerable to even moderate environmental change;
- Even large populations are vulnerable to strong environmental change;
- Evolutionary rescue is facilitated by increased genetic variability.

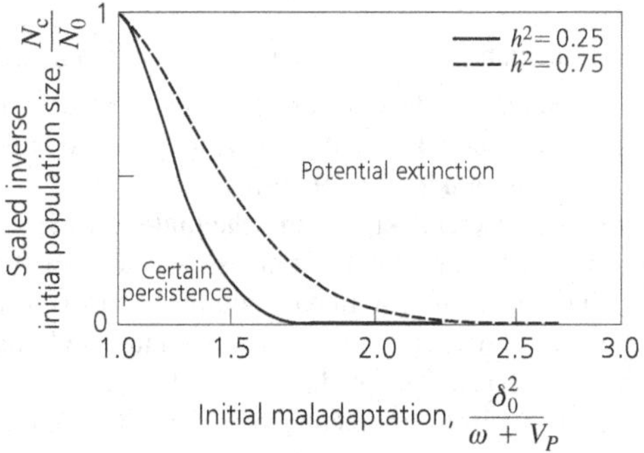

Figure 13.2 Combinations of initial population size and initial maladaptation that lead to high extinction risk. The vertical axis is an inverse measure of the initial abundance (scaled relative to a critical low size N_c). The measure of initial maladaptation used is $\delta_0^2/(\omega + V_P)$, where δ_0 is the distance between the optimal phenotype in the new sink environment and the initial mean phenotype of the population, ω is an inverse measure of the strength of selection, and V_P is the phenotypic variance for the character being selected. The quantitative genetic model presented in Box 13.1 is used to derive curves that separate those situations that lead to population sizes always above N_c (lower left region) from those in which population sizes fall below N_c for a period of time of high extinction risk (upper right region). Results are shown for two values of heritability, $h^2 = 0.25$ and 0.75. *Source*: Gomulkiewicz and Holt (1995).

The analytically tractable model described in Box 13.1 helps to clarify when evolution may rescue populations from extinctions. However, the model does not describe extinction directly, for it assumes continuous and deterministically variable densities, whereas individuals are discrete and numbers change stochastically. A direct assessment of extinction requires the use of models in which these features are respected, which implies counting individuals and alleles. Individual births, deaths, mating events, and mutations are all fundamentally stochastic processes (see Chapters 2 and 3). Analytical treatment of stochastic models that couple demographic and genetic dynamics for finite populations is a challenging task, even for simple one-locus situations (see Gomulkiewicz *et al.* 1999). An alternative approach that gives much insight is to use individual-based numerical simulations. We present results from such simulations, in which we track each individual and gene (locus by locus), and directly assess the probability of extinction by repeated simulations [extinction occurs when a population declines to zero abundance, and the probability of extinction in a given environment is the relative frequency of extinctions over a fixed time period for a large number of simulation runs; see Holt *et al.*, unpublished, for technical details of the simulation protocol, which follows that of Bürger and Lynch (1995)]. Box 13.2 describes the genetic, life-history, and ecological assumptions of these individual-based simulations. The two approaches – analytic treatment and individual-based numerical simulations – lead to mutually reinforcing insights about the potential for evolutionary rescue.

Box 13.1 Modeling adaptation and persistence after environmental change

Here an analytically tractable model to assess the propensity for evolutionary rescue is introduced. More details about the model can be found in Gomulkiewicz and Holt 1995 (see also Holt and Gomulkiewicz 1997b).

For a population with discrete generations, the finite rate of increase in population size N_t in generation t is just the mean fitness $\overline{W}_t$, $N_{t+1} = \overline{W}_t N_t$. We assume that a population is initially at an evolutionary equilibrium in a closed environment, and then experiences an abrupt change in environmental conditions. The average individual in the population is maladapted to the novel environment, so much so that in generation 0 after the environmental change $\overline{W}_0 < 1$. (We assume the density is low enough at this time to ignore density dependence.) If there is no evolutionary response, extinction results.

To couple population dynamics to evolution, we assume fitness depends upon a single trait z with polygenic autosomal inheritance, such that

$$W(z) = W_0 \exp\left(-\tfrac{1}{2}z^2/\omega\right) . \tag{a}$$

For convenience, we assume that the optimal phenotype in the new environment is at $z = 0$, and that the initial mean phenotype in the new environment is δ_0. The quantity ω is an inverse measure of the fitness cost of deviations from the optimum. Quantitative traits are often normally distributed, measured on an appropriate scale. We thus assume that the phenotypic distribution p_t in generation t can be described by

$$p_t(z) = (2\pi V_P)^{-1/2} \exp\left(-\tfrac{1}{2}(z - \delta_t)^2/V_P\right) . \tag{b}$$

Here V_P is the phenotypic variance, and δ_t is the distance of the mean phenotype in generation t from the new optimum at $z = 0$. The mean fitness in generation t is

$$\overline{W}_t = \int W(z)p_t(z)\,dz = \hat{W} \exp\left(-\tfrac{1}{2}\delta_t^2/(V_P + \omega)\right) , \tag{c}$$

where $\hat{W} = W_0\sqrt{\omega/(V_P + \omega)}$ is the growth rate attained when the mean phenotype is optimized.

In standard quantitative-genetic models of selection (Falconer 1989), the mean phenotype of a trait experiencing directional selection changes in accordance with

$$\Delta\delta_t = \delta_{t+1} - \delta_t = h^2 S , \tag{d}$$

where h^2 is the trait's heritability (a measure of faithfulness in genetic transmission of trait values across generations), and S is the selection differential (the difference in mean phenotype between individuals selected to be parents of the next generation and the mean phenotype of the current generation). For our model, we obtain

$$S = \int z[W(z)/\overline{W}_t]p_t(z)\,dz - \delta_t = -\frac{\delta_t V_P}{V_P + \omega} , \tag{e}$$

and hence

continued

Box 13.1 *continued*

$$\delta_{t+1} = \delta_t + \Delta\delta_t = \frac{\omega + (1-h^2)V_P}{V_P + \omega}\delta_t = k\delta_t\ . \tag{f}$$

The quantity k gauges evolutionary inertia; if h^2 is near zero, $k \approx 1$, and evolution is slow. As time passes, the mean phenotype approaches the local optimum, and the growth rate increases. Some algebra then shows the dynamics of population size to be described by

$$N_t = N_0 \hat{W}^t \exp\left(-\frac{\delta_0^2(1-k^{2t})}{2(V_P+\omega)(1-k^2)}\right)\ . \tag{g}$$

This expression gives U-shaped population trajectories, comparable to those in Figure 13.1b.

Figure 13.3 shows the probability of extinction over 1000 generations (averaged over 400 independent simulation runs, except for the $b = 3$ curve in Figure 13.3d, which is averaged over 1600 runs), as a function of the magnitude of the abrupt change in the phenotypic optimum caused by environmental change, and as influenced by genetic, life-history, and ecological parameters. Figure 13.3a depicts the influence of initial population size (carrying capacity) on extinction. Populations that are initially small (low K) or highly maladapted (large change in the optimum phenotype) have a high risk of extinction after an abrupt environmental change. Moreover, a large population does not, by itself, provide insurance against extinction if the degree of initial maladaptation is high. Changes in the number of loci that govern genetic variation in the trait have only a minor effect on the probability of extinction (although single-loci variation seems to hamper persistence with respect to polygenic variation, see Figure 13.3b). Populations with a higher mutational input of variation survive longer in the changed environment (Figure 13.3c). Species with high fecundities can tolerate more severe changes in the environment (Figure 13.3d). These conclusions qualitatively match those drawn from the extinction model described in Box 13.1 and summarized in Figure 13.2.

If the opportunity for niche evolution occurs primarily through sporadic colonization of novel environments outside a species' current niche, or because a species' entire population uniformly experiences severe, rapid environmental degradation, niche shifts would rarely save species from extinction. This is because populations that experience strong selection on niche characters are precisely those that face a severe risk of extinction. Species that are initially rare or have low fecundity may be particularly sensitive to environmental change. The degree to which such species persist will reflect their ability to disperse, tracking across space the shifting locations of environments to which they are already adapted.

Box 13.2 An individual-based model for analyzing niche evolution

The individual-based simulation model introduced here allows us to examine both closed populations after abrupt environmental change and spatially discrete scenarios in which stable sources are coupled with migration to sink habitats (Holt *et al.*, unpublished).

The model is based on assumptions made by Bürger and Lynch (1995), who studied adaptation to a continually changing environment for a single polygenic character:

- *Genetic assumptions*: (a) additive effects of loci, without dominance or epistasis (each allele contributes a fixed amount to the phenotypic value, and an individual's phenotype is the sum of this quantity over all loci, plus a random term); (b) mutational input maintains variation, following the "continuum-of-alleles" model, in which mutational effects are drawn from a normal distribution; (c) free recombination; and (d) in the spatial model, the source population is in mutation–selection–drift balance.
- *Life-history assumptions*: (a) discrete generations; (b) dioecious, hermaphroditic, monogamous, and random mating.
- *Ecological assumptions*: (a) in the spatial model, a constant number of immigrants per generation; (b) "ceiling" density dependence (i.e., population growth is density independent below the carrying capacity, at which growth stops abruptly); (c) constant fecundity per mated pair; (d) offspring survival probability is a Gaussian function of phenotype.

A census is made of the adults to determine the population size N_t in generation t. After the census, in the spatial model there is immigration at a per generation rate I, followed by random mating. The mating population is limited by a ceiling: if there are more than K adults, K individuals are sampled, without replacement from the pool, and are randomly assigned to mating pairs. Individuals produce gametes with free recombination among n loci. Mutation occurs on gametes, with a stochastic mutational input $n\mu$ per genome (distributed randomly over all loci). Each mated pair produces b offspring, which survive to adulthood with probability $p_i(z) = \exp\left(-\frac{1}{2}(z - \hat{z}_i)^2/\omega^2\right)$, where z is the realized phenotype of a given individual, $\hat{z}_i$ is the optimum phenotype in habitat i (1 = sink, 2 = source), and ω^2 is inversely proportional to the strength of stabilizing selection. Survival to adulthood is the life-history stage at which selection occurs. If the realized phenotype is too far from the optimum, the mean fitness is below 1, and the population tends to decline. Individuals that survive early mortality are adults at the next census, N_{t+1}. The population is assumed to be at carrying capacity K initially and at selection–mutation–drift equilibrium in the ancestral environment.

In this individual-based model, stochasticity enters at several stages:

- Mutation is stochastic;
- Gametic combinations and immigrants (in the spatial model) have multilocus allelic combinations that vary through random sampling;

continued

Box 13.2 *continued*

- Finally, survival is probabilistic, which leads to both genetic drift and chance fluctuations in population size.

To assess the summary statistics (e.g., the probability of extinction, or adaptation, over a given number of generations), many simulations are run, starting with identical initial conditions, and the relative frequencies of various outcomes are assessed.

One virtue of individual-based simulation models is that they enable comparison of the impact of various assumptions about the environment, and biology, for scenarios that are very difficult to tackle with analytical models. For example, does linkage among loci facilitate, or hamper, local adaptation? What is the effect of overlapping generations on niche evolution?

13.3 Adaptations in Population Sources and Sinks

A frequent scenario in real-world conservation crises is for a species to experience environmental degradation in only part of its geographic range. For instance, the localized dumping of toxins or invasion by exotic species could affect a species in certain areas, but not in others. Over time, many spatial patterns of habitat alteration could occur. For instance, degradation caused by the diffusion of a toxin could generate a smooth gradient in habitat quality, emanating from a point toxin source. By contrast, land clearance could lead to the abrupt juxtaposition of discrete habitat types in a complex mosaic.

A simple, but instructive, scenario is to assume that after habitat degradation there remains a discrete "source" habitat, in which the environment is unaltered and the species can persist at its ancestral evolutionary equilibrium. The degraded part of the species' range is represented by a discrete "sink" habitat, in which the conditions are so hostile that the species would go extinct but for recurrent immigration from the source. As used here, a "source" habitat is one in which births exceed deaths at low densities, so populations are expected to persist. By contrast, in a "sink" habitat there are fewer births than deaths, so populations decline to extinction in the absence of immigration (Pulliam 1996). Demographic sinks can also occur if immigration pushes the population size above the local carrying capacity, which gives rise to "pseudosinks" as discussed by Watkinson and Sutherland (1995).

Niche conservatism occurs if adaptive evolution to the sink habitat does not take place, even though the species is exposed to such evolution via immigration from the source. Should sink habitats concern conservationists? Sometimes the answer is surely "yes". Adaptation to poor environments may enhance the survival prospects for an entire species; indeed, it may be essential if a species' original habitat shrinks to pathetic fragments of a formerly extensive range. Models of adaptive evolution in the context of source–sink dynamics give insight into the potential for such adaptive responses.

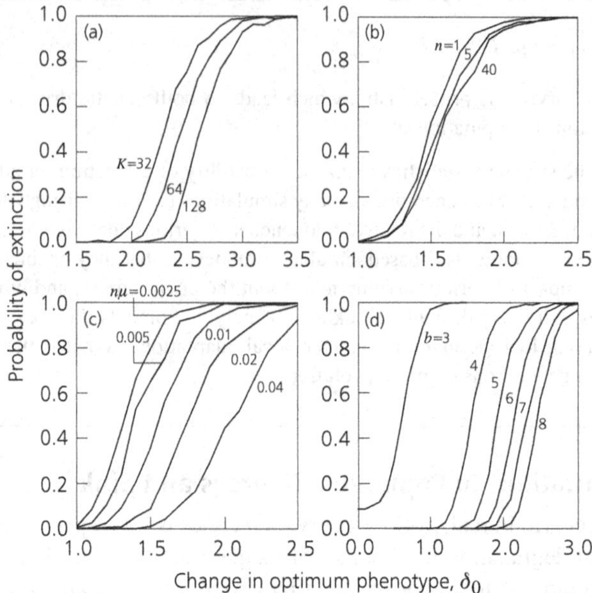

Figure 13.3 Probability of extinction over a span of 1000 generations for an individual-based model of a closed population that experiences an abrupt environmental transition. The initial genetic variability results from mutation–selection–drift balance for an ancestral population in a constant, favorable environment. When a mutation occurs in the simulations, a normally distributed random number with mean 0 and variance V is added to the current allelic value. (a) Effect of the initial density, which equals the carrying capacity K in all examples. Other parameters: $b = 8, n = 10, n\mu = 0.01, \omega^2 = 1, V = 0.05$. (b) Effect of numbers of loci n. Other parameters as in (a) with $K = 64$, except for $b = 4$. (c) Effect of mutational input of variation $n\mu$. Other parameters as in (b) with $n = 10$. (d) Effect of fecundity b. Other parameters as in (b) with $n = 10$.

To understand the interplay of migration and selection that determines local adaptation is a classic problem in population genetics (Hedrick 2000; see also Chapter 12), stemming back to J.B.S. Haldane (1930). However, this literature traditionally ignored the demographic context within which gene flow and selection occur. A very simple model that explicitly illustrates the importance of demography (Holt and Gomulkiewicz 1997a, 1997b) rests on the assumption that an asexual population with discrete generations is situated on a habitat patch. This population receives recurrent immigrants at a constant rate I (number of immigrants per generation), all fixed for an allele A_2. The absolute fitness of the immigrants is $W_2 < 1$. With these assumptions, the number of individuals on the patch follows the recursion $N' = NW_2 + I$, which implies the population equilibrates at $N^* = I/(1 - W_2)$. Now assume a novel mutant allele A_1 arises with higher fitness in the local environment, $W_1 > W_2$. Can this allele spread by natural selection? In each generation, the relative frequency of the allele increases because of local selection, but it is also reduced because of the dilution by immigration of individuals that carry the less-fit allele. The recursion that describes the net effect of these two processes on the frequency p of allele A_1 is $p' = (1 - m)(W_1/\overline{W})p$, where $\overline{W} = pW_1 + (1 - p)W_2$ is the mean fitness. The quantity m measures gene flow, which is the percentage of the island population that comprises immigrants; after immigration, $m = I/N'$. When the novel mutant is very rare, the population consists primarily of the less-fit immigrants, so the total population size is approximately N^*. After substitution, we find that the recursion for the frequency of the rare fitter allele becomes $p' = W_1 p$. The condition for the fitter allele to increase in frequency is that its absolute fitness exceeds 1, irrespective of the fitness of the less-fit allele or of the rate of immigration. The conclusions from this very simple model hold much more broadly (Gomulkiewicz *et al.* 1999). Box 13.3 describes a one-locus model for a "black-hole" sink: a habitat that receives immigrants, but does not export emigrants back to the source (which is assumed to have a reproductive surplus providing the flux of immigrants).

These genetic models lead to interesting conclusions in terms of conservation. When considering the fate of mutant alleles in a sink habitat, absolute – not relative – fitness is key to adaptive evolution. The "effect" of a mutation is measured relative to an ancestral condition (here, fitness of the immigrant). In a "harsh" sink environment, an immigrant has a fitness well below 1. In such environments, only mutants that have a large effect on fitness can be retained and swept to higher frequencies by selection. By contrast, in a mild sink the absolute fitness of an immigrant is less than, but close to, 1. In such an environment, mutants of small effect may be selected. If the rate of adaptation is limited primarily by the appearance of appropriate mutations, and mutants of small effect appear more frequently than do mutants of large effect (e.g., Orr 1998), then adaptation to a mild sink occurs more rapidly than does adaptation to a harsh sink. In effect, niche conservatism (the absence of adaptive niche evolution) is more likely given sharp contrasts in fitness between source and sink habitats.

What about immigration? If fitness is density *independent* in the sink environment, then because the rate of immigration I drops out of the recursion, the magnitude of immigration of maladapted genotypes does not directly influence the initial spread of the locally favored allele. If, instead, fitness is density *dependent*, declining with population size N, immigration can directly hamper adaptation, because increasing immigration increases local abundances, and thus depresses local fitness. Moreover, even if fitness is initially density independent, once a locally favored allele has arisen and spread, the population size will rise, and eventually the absolute fitness must become density dependent. At the new demographic equilibrium, recurrent immigration can lower the frequency of the fitter allele because of such density dependence. Moreover, in a diploid model, mating between immigrants and residents further lowers local fitness, because of the continued generation of less-fit heterozygotes. Thus immigration can hamper the degree of adaptation to the sink environment for both ecological and genetic reasons.

Box 13.3 A diploid, one-locus model for adaptation in a "black-hole" sink

To gain insight into how genetic structure affects adaptation in a "black-hole" sink (a habitat that receives immigrants, but does not export emigrants), we assume that fitness is governed by variation at a diploid locus with two alleles, A_1 and A_2. While allele A_1 is assumed to be favored in the sink, all immigrants are fixed for A_2. The population size in the sink is given by N, and p is the frequency of allele A_1. In each generation, I adults immigrate into the sink habitat (after selection, but before reproduction); subsequently, random mating occurs and a census is taken of the population. The viability of an individual with genotype A_iA_j is $v_{i,j}$, and all individuals have the same fecundity b. The fitness of genotype A_iA_j is thus $W_{i,j} = bv_{i,j}$. As the habitat is a sink for the immigrant genotype type A_2A_2, $W_{22} < 1$. We are interested in assessing the fate of the locally more favorable allele A_1; hence, we assume $W_{12} > W_{22}$.

With these assumptions, the number of breeding adults (after viability selection and immigration) is

$$N_b = v_{11}Np^2 + v_{12}N2p(1-p) + v_{22}N(1-p)^2 + I\ . \tag{a}$$

The first three terms describe the Hardy–Weinberg distribution of genotypes, as modified by differential mortality; the total population consists of survivors of selection plus immigrants (fourth term).

After reproduction, the density of newborns is

$$N' = bN_b = N\overline{W} + bI\ , \tag{b}$$

where $\overline{W} = W_{11}p^2 + W_{12}2p(1-p) + W_{22}(1-p)^2$ is the mean fitness. The frequency p' of the A_1 allele among newborns equals the frequency of A_1 in the breeding parents; thus $p' =$ (number of A_1 alleles in parents)/(N_b) and therefore

$$p' = \frac{1}{2N_b}[2v_{11}Np^2 + v_{12}N2p(1-p)]\ . \tag{c}$$

Multiplying both the numerator and denominator of this expression by b and using Equation (b) gives

$$p' = \frac{N}{N'}\overline{W}_1 p\ , \tag{d}$$

where $\overline{W}_1 = pW_{11} + W_{12}(1-p)$ is the mean fitness of individuals that carry allele A_1. Equations (b) and (d) describe coupled population and genetic dynamics.

Now consider the fate of the fitter allele when it is rare and the immigrant type is at demographic equilibrium. To a good approximation, $N \approx N'$, $\overline{W}_1 \approx W_{12}$, and Equation (d) reduces to $p' \approx pW_{12}$. Hence, allele A_1 increases in frequency if and only if $W_{12} > 1$. In other words, as in the simpler model described in the text, the initial spread of a locally favored allele depends upon its absolute – not relative – fitness. Moreover, the alleles that can deterministically increase when rare, can also permit the local population to persist without immigration.

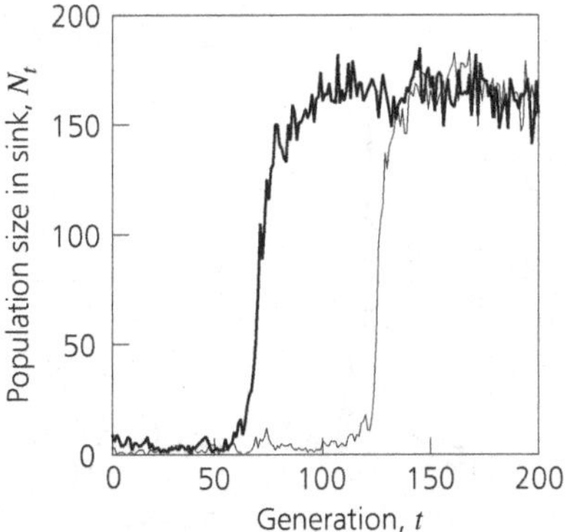

Figure 13.4 Examples of population dynamics in sink populations undergoing adaptive evolution. The abundances shown are measures after selection and before density regulation. Parameters of the individual-based model (see Box 13.2) are the same as in Figure 13.3a with $K = 64$, but there is now recurrent immigration. Two characteristic runs are shown, for identical initial populations, with constant immigration rates of $I = 4$ adults per generation. The optimum phenotype in the sink is $\hat{z}_1 = 2.8$; the mean phenotype of immigrants from the source is $\hat{z}_2 = 0$. Typically, the sink population stays at low abundance for a lengthy period of time, followed by a period of rapid increases to high abundance, which corresponds to a rapid shift in mean phenotype (see Figure 13.5 for two snapshots of this evolutionary process).

To counter the negative effect of immigration, given density dependence, is the potential role of immigration as a source of genetic variation for local selection, which may be quantitatively much more important than local mutation. Gomulkiewicz *et al.* (1999) examined this effect in detail for a stochastic model and concluded that the scope for local adaptation in a sink is often greatest at intermediate levels of immigration.

We complement these simple analytical one-locus models of adaptive evolution in sinks with individual-based simulation studies of multilocus evolution, using the model introduced in Box 13.2 to describe ongoing evolution in coupled sources and sinks (Holt *et al.*, unpublished). A large number of source–sink population pairs are tracked, in each of which a fixed number of immigrants per generation is drawn from a stable source population at its mutation–selection–drift equilibrium. Figure 13.4 shows two typical simulation runs. In these examples, the sink environment is harsh, so immigration maintains a population at low abundance only. A population stays in this state for a while (often a long while), but then increases in mean fitness and rapidly grows until limited at the local carrying capacity. Examination of the character states shows that evolution in this system is, in effect, "punctuational": the sink population is either maladapted or near the local optimum, and spends very little time between these two (Figure 13.5), unless immigration is very large relative to the local carrying capacity. Indeed, in these simulations, if immigration is cut off once a population is adapted, the population

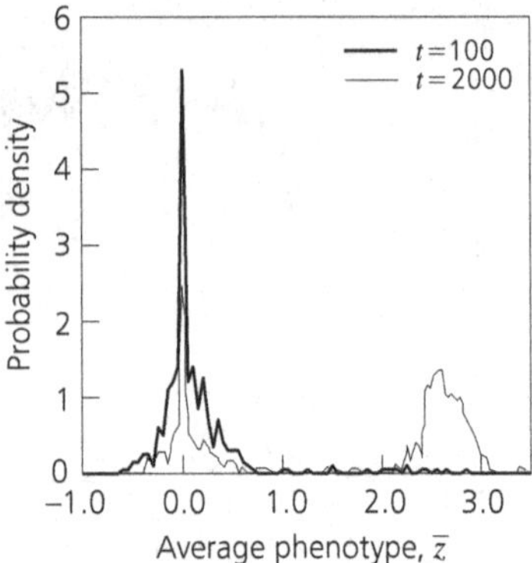

Figure 13.5 Punctuational evolution in a sink habitat. The distribution of average phenotypes $\bar{z}$ in 400 populations is shown at two different times. After 100 generations, most populations are still near the source phenotype $\hat{z}_2 = 0$. After 2000 generations, a substantial number of populations are near the sink optimum $\hat{z}_1 = 3$ (notice that mean phenotypes tend to be somewhat displaced from 3 because of recurrent gene flow from the source). Very few populations are in an intermediate state of adaptation. Unless migration is very high (relative to the local carrying capacity), populations in the sink can thus be dichotomized into being in a "maladapted" state (near the source optimum), or in an "adapted" state (near the sink optimum). Consequently, in a given generation sets of equivalent populations can be characterized by a "probability of adaptation". Other parameters are as in Figure 13.4.

continues to persist for a very long time. This near-dichotomy in degree of adaptation provides a convenient diagnostic with which to summarize large numbers of simulation runs by determining the probability that a population is "adapted" (with a mean genotype near the local optimum, and the population close to its carrying capacity), or "maladapted" (with a mean genotype near that of the source habitat, and a much lower population size), with few populations at intermediate levels of adaptation.

Figure 13.6 shows a typical example of the probability of adaptation that occurs over 1000 generations as a function of the magnitude of initial maladaptation (a measure of the difference between the source and sink environments), and at two different immigration rates. A species that does not adapt over this time scale exhibits niche conservatism. There are several things to note in Figure 13.6:

- We compare Figure 13.6 with Figure 13.3a (for $K = 32$). In Figure 13.3a, most closed populations exposed to a degree of maladaptation of 2.5 go extinct. By contrast, if these populations were open, drawing immigrants each generation from a source habitat, in each case they would eventually adapt. This illustrates a simple, but fundamental, role of immigration in heterogeneous environments – immigration sustains populations and thus provides an *opportunity for evolution*. Immigration, in essence, facilitates adaptation to the local environment by the repeated exposure of individuals to it.

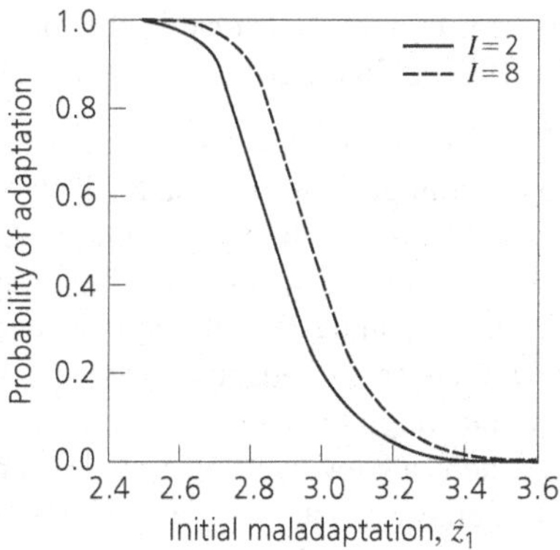

Figure 13.6 The probability of adaptation to the sink environment (within 1000 generations) as a function of sink maladaptation $\hat{z}_1$ relative to the source for two different immigration rates, $I = 2$ and 8. Whereas adaptation becomes more difficult the more different the sink environment is from the source environment, adaptation is facilitated by increased immigration. Results shown are averages over 400 populations. Other parameters are as in Figure 13.3a with $K = 32$.

- Note that the harsher the sink environment is (as measured by the maladaptation of immigrants), the less likely is adaptive evolution. This corresponds with the other theoretical results sketched above.
- Figure 13.6 shows that the probability of adaptation actually *increases* with an increasing number of immigrants per generation. Rather than gene flow swamping selection, in the initial phases of adaptation to a novel environment immigration facilitates adaptation. The reason for this is quite simple. Evolution requires variation. A sink population tends to be low in numbers and thus is not likely to retain variation, or generate much variation by mutation. The main source of variation in a low-density sink population is immigration from more abundant persistent sources. Increasing the immigration rate in effect increases the sample of variation drawn each generation from the source. Thus, a higher rate of immigration fuels adaptation by providing more raw material for selection to act upon in the local environment. [See Gomulkiewicz *et al.* (1999) for a detailed exploration of this effect in the one-locus model of Box 13.3.]
- Once a population has adapted and increased to the carrying capacity, the mean phenotype is displaced from the local optimum (see Figure 13.5, in which the mean genotype of adapted populations is lower than the local optimum of 3). Immigration has two distinct negative effects that hamper local adaptation. First, gene flow from the source (in which there is a different phenotypic optimum) introduces individuals with locally suboptimal phenotypes, who mate with better-adapted residents. This gene flow hampers the perfection of local adaptation. Moreover, at the carrying capacity immigrants compete with residents. This tends to lower absolute fitness, and thereby makes it

harder for locally superior mutants to spread in the local population (Holt and Gomulkiewicz 1997a, 1997b; Gomulkiewicz *et al.* 1999).

These recent theoretical results suggest a bias in the diverse roles immigration plays in local adaptation. Immigration facilitates adaptive evolution by exposing species to novel conditions, and also by providing a potent source of genetic variation. Immigration also constrains adaptive evolution, because gene flow can swamp locally favored variants, and because immigrants can compete with better-adapted residents. In the initial stages of adaptation to harsh sink habitats, however, the former effects seem to outweigh the latter.

The models discussed here all involve "black-hole" sink habitats, with unidirectional flows out of source habitats. Studies of comparable one-locus and individual-based models with reciprocal back-flows lead to results broadly consistent with the findings summarized here (Kawecki 1995, 2000; Holt 1996; Ronce and Kirkpatrick 2001; Kawecki and Holt 2002). In particular, the insight that the worse the sink environment, the less likely that adaptive evolution will occur there (over some defined time period) appears quite robust.

13.4 Adaptations along Environmental Gradients

So far we have focused on evolution in spatially discrete environments. Such settings offer a useful starting point, but it is important to consider a broader range of spatial scenarios. Another useful limiting case is to imagine that a species is distributed along a smooth environmental gradient that influences both population dynamics and adaptive evolution. Here, we provide a brief overview of work by other authors who have taken this approach, and relate their findings to the discrete–environment models presented in the preceding section.

In a seminal study, Pease *et al.* (1989) developed a model for a population that grows, adapts, and disperses along a unidirectionally shifting environmental gradient. The model splices a submodel for local population dynamics (exponential growth or decline) with a submodel for local adaptive evolution (for a single quantitative trait that affects fitness). Dispersal influences both the dynamics of abundances (in which individuals on average move from high-density to low-density sites) and the character evolution (in which such movements displace local populations away from their local phenotypic optima). The model is described in Box 13.4. It leads to several predictions, which broadly match the results described above for evolutionary rescue in abruptly changed environments (Figure 13.2):

- A species is more likely to persist when the environment changes slowly than when the environment changes rapidly;
- The scope for persistence is enhanced with greater genetic variation, and with a greater maximal growth rate.

The influence of movement upon persistence is more complex. Without dispersal, the model by Pease *et al.* (1989) predicts that local populations are doomed in a constantly changing environment, even if genetic variation is abundant. Persistence thus requires movement, so a species can track suitable habitats. However, movement also tends to move individuals from productive zones along the gradient into unproductive habitats. If sufficiently large, this reproductive drain can cause extinction. This implies an "optimal" rate of movement, measured by the maximal environmental change a species can tolerate. This optimal rate increases with the amount of genetic variation available for selection. The reason is that this increases the relative importance of local selection versus gene flow in determining local phenotypes, which increases the range of environments over which a species will be reasonably well adapted.

Important studies by Kirkpatrick and Barton (1997) and Case and Taper (2000) explore the interplay of gene flow and selection along fixed environmental gradients. Box 13.4 also contains a sketch of the Kirkpatrick–Barton model, which adds density dependence to the Pease *et al.* (1989) formulation. Also, in the Kirkpatrick–Barton model a species occupies a gradient, along which the optimal value for a phenotypic character changes. The mean phenotype of a population influences its growth rate, and hence its realized density; maladaptation depresses local population size. Gene flow from central populations can inhibit adaptation at the periphery, which depresses fitness and thus local population size; peripheral populations therefore tend to be demographic sinks, maladapted to their environment. This model predicts that if the gradient is sufficiently steep, a species' range can be sharply limited by gene flow. Indeed, a species may not be able to persist at all. By contrast, if the gradient is shallow, gene flow does not prevent local adaptation, and a species' range can expand to fill all available space.

There is an interesting implication of the Kirkpatrick–Barton model for conservation. Assume a species is limited in its range along a gentle gradient by gene flow and swamps local selection at the range margins. If human activity now further sharpens the abundance gradient, the swamping effect of gene flow is magnified relative to local selection. This leads to a degradation in local adaptation, and an evolutionary reduction in the range, as marginal populations become yet more maladapted to their local environments. This is an example of evolution hampering conservation goals.

There are similarities, and differences, between the predictions of the Kirkpatrick–Barton continuum model and those of the discrete-space source–sink models discussed above. In the discrete-sink model, the worse a sink environment is, the harder it is for local adaptation to occur. Likewise, in the continuum model, a sharper gradient implies larger differences in locally optimal phenotypes, so dispersers from sources experience lower fitness in the peripheral sinks. The sharper the gradient, the larger is this drop in fitness for immigrants, and the more likely it is that gene flow prevents local adaptation and thus constrains the range.

However, the models do differ in the predicted role of immigration. In the Kirkpatrick–Barton model, increased immigration can depress population size and even cause extinction (because too many individuals are drained from sources into peripheral sink habitats to which they are maladapted). In the discrete source–sink models, immigration could inhibit local adaptation for ecological reasons (given strong density dependence in the sink), but it can also facilitate local adaptation

Box 13.4 Modeling adaptation along smooth environmental gradients

Here we describe models used to analyze adaptation and range shifts of populations that live in habitats with smooth environmental gradients. We explain a model by Pease *et al.* (1989) before introducing its extension by Kirkpatrick and Barton (1997).

The Pease *et al.* (1989) model has three components. First, changes in population size N at position x along a gradient are modeled by the reaction–diffusion equation

$$\frac{\partial N}{\partial t} = \tfrac{1}{2}\sigma^2 \frac{\partial^2 N}{\partial x^2} + Nr \,. \tag{a}$$

The first term assumes individuals move at random over short distances (σ is the root-mean-square distance moved per time unit), whereas the second term describes local population growth at a per capita rate r that can depend on population size N and on the mean phenotype $\bar{z}$ of individuals at the considered location x. Evolution occurs at a single, quantitative character, of mean phenotype $\bar{z}$, influenced by many loci, each of small effect. The local evolutionary dynamics of $\bar{z}$ incorporate gene flow and selection

$$\frac{\partial \bar{z}}{\partial t} = \tfrac{1}{2}\sigma^2 \frac{\partial^2 \bar{z}}{\partial x^2} + \sigma^2 \frac{\partial \ln N}{\partial x}\frac{\partial \bar{z}}{\partial x} + V_G \frac{\partial r}{\partial \bar{z}} \,. \tag{b}$$

The first two terms describe how movement modifies the mean character value at a location along the gradient. The third term describes the response of the population to selection, which depends both upon genetic variability V_G and on the strength of the relationship between the character and fitness.

The final model component is the expression that describes fitness, which links the two above equations. Pease *et al.* (1989) were concerned with global extinction versus persistence, so they assumed density-independent growth described by the bivariate function of spatial position and mean phenotype

$$r = r_0 - \frac{(x - vt)^2}{2W_{11}(1-\rho^2)} + \frac{\rho\bar{z}(x - vt)}{(1-\rho^2)\sqrt{W_{11}W_{22}}} - \frac{\bar{z}^2}{2W_{22}(1-\rho^2)} \,. \tag{c}$$

This function describes how fitness depends jointly upon spatial position (in a time-dependent way), and deviations in mean phenotypes from local optima. The parameters describe how wide (or fat) the fitness function is along two dimensions, one being the phenotypic dimension (for W_{22}), and the other the spatial dimension $x - vt$ (for W_{11}). The maximal per capita growth rate is achieved only in a population at spatial position vt, given that the mean phenotype there is 0; v is the velocity of movement of the gradient. The magnitude of spatial variation in the phenotype optimum is given by ρ, the correlation between location and the value of the optimal character. The second term measures how fitness decays in space away from the (current) spatial optimum. The final term measures how character variation away from a local optimum translates into reduced fitness.

With these expressions at hand, and assuming that selection is weak, Pease *et al.* (1989) showed that the maximal rate of environmental change a species can withstand is

continued

Box 13.4 *continued*

$$v_{\max} \approx \sigma \sqrt{2r_0 + \frac{V_G \rho^2}{W_{22}(1-\rho^2)} - \sqrt{\frac{\sigma^2}{W_{11}(1-\rho^2)}}}\,. \qquad \text{(d)}$$

Inspection of this equation leads to the conclusions stated in the text.

Kirkpatrick and Barton (1997) use the same dynamic equations, but assume that local growth is density dependent and given by

$$r = r_0(1 - N/K) - \frac{1}{2\omega}\left[\hat{z}(x) - \bar{z}\right]^2 - \frac{I_S}{2}\,. \qquad \text{(e)}$$

The first term describes logistic population growth, the second term defines how population growth is depressed if the local mean phenotype $\overline{P}$ deviates from the local optimum $\hat{z}(x)$, with ω being an inverse measure of the strength of stabilizing selection, and the third term measures the intensity of selection in units of phenotypic variance V_P, $I_S = V_P/\omega$. The optimum is assumed to change linearly with space, $\hat{z}(x) = gx$, in which the quantity g determines the steepness of the environmental gradient.

Kirkpatrick and Barton (1997) considered a number of limiting cases of their model. When dispersal is high, and the population is well below carrying capacity, the maximal gradient slope that permits persistence is approximately $g = \sqrt{V_P}[4r_0 - (2-h^2)I_S]/(2\sigma\sqrt{I_S})$. This expression implies that persistence is facilitated if:

- The environmental gradient is shallow;
- Dispersal is low;
- Maximal growth rate is high;
- Selection is weak;
- Heritability is high.

by providing novel genetic variation on which selection can act. The latter effect is not dealt with in the continuum model, which assumes that heritability is fixed. Other differences between the models should be considered in future studies. The continuum model assumes homogeneous bidirectional dispersal, so source populations can be depressed by a net loss of emigrants into sinks. The rates of emigration examined in the discrete source–sink models discussed above were low enough for this effect to be ignored. An open challenge for future work is to develop models for evolution (along gradients comparable to those of the Kirkpatrick–Barton model) that also include the positive effect of migration as a source of novel genetic variation in peripheral populations.

Case and Taper (2000) recently combined the Kirkpatrick–Barton model with a Lotka–Volterra model of interspecific competition to examine the interplay of character displacement and range limitation along a gradient. Space limitations here preclude a full discussion of their results, but it is worth noting that, in effect, an interspecific competitor at one end of a gradient sharpens the gradient, which makes it more likely that gene flow can limit range size. In the source–sink models, the reason a given habitat becomes a sink may well be the presence of effective competitors or predators. Changes in community structure can thus lead to additional changes via evolutionary responses (see also Chapters 16 and 17). This is an important and largely unexplored dimension in evolutionary conservation biology.

13.5 Conservation Implications

There are several general conservation implications that emerge from this wide range of models. A robust result in all the models, involving changes in both time and space, is that populations exposed to "mild" environmental degradation may often be rescued from extinction by evolution. By contrast, severe changes are either likely to lead to global extinction or to the persistent restriction of a species to remnant habitats in which it is already well adapted, without adaptation to novel environments.

The specific models have additional implications. For the model of evolutionary rescue in an abruptly changing environment, the basic message for our rapidly changing world, alas, is sobering. If the environment changes sufficiently fast so that a species in its initial state reaches low densities over short time scales (e.g., tens of generations), natural selection will be rather ineffective at preventing extinction. However, the theory does suggest two distinct avenues through which we might conceivably influence evolution so as to foster conservation:

- Since population size is the product of density and area, populations in large areas take longer to decline to a given absolute abundance than do populations in small areas. This justifies the conservation of large fragments (beyond the usual reasons). Habitat fragments maintained as reserves are likely to continue to experience a broad range of secular changes in the environment, and species in large fragments, in effect, enjoy a "demographic buffer" against unanticipated future environmental changes that may require evolutionary responses by a species for it to persist.
- If the rate of decline can be slowed, populations have an enhanced "window of opportunity" in which to evolve adaptations to environmental stresses. So, if we cannot prevent environmental change, we may be able to reduce the magnitude of its impact upon a focal species. This lengthens the time scale available for evolutionary change and provides more opportunity for evolution by natural selection to alter the species' niche sufficiently to ensure persistence in the novel environment.

The models considered in this chapter highlight that we should not automatically assume that if a habitat is a demographic sink for a species, that habitat has no conservation value. Such habitats may provide sites within which adaptive evolution by the species could occur, and so facilitate its ultimate persistence over a wider landscape than that provided by the current source habitats alone. A conservationist faced with a choice of habitats should obviously attempt to save source habitats first. Without these, the species as a whole is doomed over short ecological time scales. When possible, it is also clear that the conversion of current source habitats into future sink habitats should be prevented. However, if such conversion has already occurred, and the landscape is static, it may still be useful to attempt to save or improve sink habitats, particularly if the sinks are "mild" sinks in which substantial populations can be sustained by a trickle of immigration from source habitats. According to the source–sink models discussed here, these are the sites within which significant adaptive evolution to novel environments may occur, buffering the species against further changes in the landscape that destroy or degrade its original required habitat. In the dynamic landscape of the Pease *et al.* (1989) model, the crucial leading edge of a species' range comprises sink populations that provide a toe-hold for a species to shift its range and track changing environmental conditions.

The source–sink models suggest a management strategy that may sometimes be feasible. What matters in adaptive evolution is the overall demographic contribution of a habitat, and the pattern of coupling by movement among habitats. We may indirectly be able to facilitate local adaptation to, say, a novel toxin encountered in a given habitat by improving resources or other habitat conditions, or by removing a predator. These environmental modifications increase the overall fitness and provide a demographic "boost", which in turn facilitates the efficacy of selection on traits that reduce the fitness impact of the toxin.

The range of models reviewed here reveals that dispersal has disparate effects upon species' survival. In continually changing environments, dispersal may be essential for persistence, but too much dispersal can lead to a substantial mortality load because individuals move into habitats to which they are maladapted. The latter effect can constrain species' ranges in constant environments, and may even threaten persistence when environmental gradients are steep. To counter the latter negative effect of dispersal, genetic variation can be increased and so the capacity of a species to respond to change is enhanced (see Figure 13.3c). The ultimate source of all variation is, of course, mutation, but in many local populations mutational input may be minor relative to another source of variation – gene flow from spatially separated populations. This effect has just begun to be explored, and the results provided above suggest that in some circumstances it could be an important avenue through which immigration facilitates adaptive evolution. This version of the effects of dispersal upon evolution and persistence complicates analyses of the ultimate conservation importance of different patterns of landscape connectivity and warrants further investigation.

13.6 Concluding Comments

Conservation problems exist because humans change the environment in ways that harm species (as measured in distribution and abundance), because species do *not* adapt by natural selection to these novel environments, and because actual adaptations eventually harm populations. The theoretical studies sketched above suggest that, in some situations, the evolutionary dynamics of populations may be harnessed to facilitate species' preservation. However, these same theoretical results suggest that evolution will not be particularly useful in promoting the persistence of species in radically changed environments, or in environments that comprise a spatial admixture of unchanged and greatly altered habitats. The basic message is that the demographic context of evolution matters greatly in determining the likelihood of conservatism versus rapid evolution in altered environments, and that this insight should be useful in applied conservation biology. In particular, we have seen that in spatially heterogeneous environments evolution by natural selection improves adaptation less effectively in local environments in which fitness and population size are both low (as, for example, in a sink habitat, or near the margin of a species' range). The sharper the difference in fitness between source and sink, or the steeper the gradient, the more likely is an absence of evolutionary response to spatial heterogeneity. Conversely, adaptive evolution is likely along gentle environmental gradients.

Acknowledgments We thank the National Science Foundation for its support (DEB-9528602). We particularly thank Michael Barfield, who has been a collaborator on much of the work summarized here, and assisted in figure preparation.

14

Adaptive Responses to Landscape Disturbances: Theory

Kalle Parvinen

14.1 Introduction

Habitat loss is probably the most important factor to cause species decline worldwide (Sih *et al.* 2000), but habitat fragmentation and deterioration are widespread also. The habitats of most species are no longer large homogeneous areas, but instead consist of small patches of habitable environment, often connected by migration paths. Some of these patches are inhabited, while others are empty. Inhabited patches may become empty because of demographic or environmental stochasticity, and empty patches may be colonized by dispersers. The collection of local populations forms a metapopulation, a concept originally introduced by Richard Levins (1969, 1970; see also Hanski and Gilpin 1997; Hanski 1999; and Chapter 4).

Dispersal is a key life-history trait in metapopulations, and the evolution of dispersal rates has attracted particular attention, both in the past (Roff 1974; Hamilton and May 1977; Comins *et al.* 1980; Motro 1982a; Motro 1982b; Hastings 1983; Levin *et al.* 1984; Frank 1986; Pease *et al.* 1989; Cohen and Levin 1991) and more recently (Olivieri *et al.* 1995; Cadet 1998; Gandon 1999; Gandon and Michalakis 1999; Parvinen 1999; Travis and Dytham 1999; Travis *et al.* 1999; Ronce *et al.* 2000a; Gyllenberg and Metz 2001; Heino and Hanski 2001; Metz and Gyllenberg 2001; Parvinen 2001b; Kisdi 2002; Nagy, in press). Changes in dispersal strategies provide an option for threatened populations to respond to the fragmentation of their habitats. Consequently, success or failure of such adaptation often determines whether a challenged population can persist.

Although the metapopulation model introduced by Levins gives much insight into the behavior of populations in heterogeneous landscapes, as explained in Section 4.3, it is based on several simplifying assumptions. Most importantly, individual dispersal behavior is not described. Therefore it is not possible to study the evolution of dispersal without using more detailed models (see also Section 4.7). Chapter 15 analyzes the evolutionary responses of metapopulation dispersal to habitat fragmentation based on empirical data. Here, the focus is on the theoretical analysis of dispersal evolution in metapopulations with local population dynamics.

Viability and persistence are important concepts from the point of view of conservation biology. As far as the ecological time scale is concerned, these topics are examined in Chapter 4. A metapopulation is persistent if the metapopulation extinction equilibrium is unstable. Therefore, if the total population size is close to zero, the growth rate is positive and the metapopulation does not go extinct. A metapopulation is viable if there exists a nontrivial attractor (i.e., an attractor other than the extinction equilibrium). If the metapopulation is not viable, it is doomed to extinction, and so evolution of dispersal cannot be studied. However, persistence or viability of the metapopulation at the current moment does not guarantee its persistence on the evolutionary time scale. It is possible that a metapopulation is viable on the ecological time scale, but that natural selection will force the metapopulation to change its dispersal strategy to a nonviable one, and thus cause evolutionary suicide (Ferrière 2000; Gyllenberg and Parvinen 2001; Gyllenberg *et al.* 2002; Chapter 11).

Evolution acts at the level of individuals. To study the evolution of dispersal, it is necessary to consider individual behavior, such as birth, death, immigration, and emigration, in local populations. By combining these events, local population growth or decline can be described. No local population can grow unlimitedly. Therefore, models in which local population growth is fully density regulated, either deterministically or stochastically, are studied. Real metapopulations comprise only a finite number of habitat patches, and each local population that lives in a habitat patch consists of a finite number of individuals. Any finite metapopulation will, however, go extinct in finite time, even though its persistence time can be long enough to enable evolutionary studies. Yet, for the purpose of theoretical investigation, it is convenient to make at least one of the following simplifying assumptions: either the number of patches is large (infinite) or the local population sizes are large enough to approximate local population dynamics with a deterministic model.

14.2 Selection for Low Dispersal

In nature, most species disperse. There are, however, examples of species whose evolution seems to have resulted in very low dispersal rates. As described in Section 15.4, the plant species *Centaurea corymbosa* lives in a highly fragmented habitat. Many suitable sites are available not too far from the extant populations, but dispersal is selected against because of the low colonization ability and unsuitable habitat that surrounds the existing populations. It is important to identify which sort of ecological scenarios can lead to selection for low dispersal, for in rapidly deteriorating environments species that have evolved reduced dispersal abilities may be at much higher risk of extinction.

In a metapopulation with all the local populations at their ecological equilibrium, and in the absence of environmental disturbances, there is no advantage to dispersal. A dispersing individual will never find a better patch than the one it left. Instead, a disperser encounters the possibility of death during dispersal, which is a direct cost to dispersal. If there are no benefits to disperse, only costs, selection enforces zero dispersal rate in the metapopulation.

This reasoning can be made mathematically precise. Assume that there are n patches of suitable habitat and that local populations are large. Population growth in patches can therefore be described either by a differential equation in continuous time, or by a difference equation in discrete time. The dispersal strategy is the emigration rate, the rate at which individuals leave the patches. Dispersers, which are exposed to a risk of mortality, choose the patch into which they immigrate at random, independently of patch quality and local population size. Details of this model are described in Box 14.1, which also shows that under such circumstances the selection gradient of the dispersal rate is always negative. Selection is therefore expected to take the dispersal rate to zero (Hastings 1983; Holt and McPeek 1996; Doebeli and Ruxton 1997; Parvinen 1999). The same phenomenon is observed in both continuous-time and discrete-time models, and is illustrated in Figure 14.1.

The ecological situation described so far is, however, unrealistic for at least three reasons, which are discussed in turn in the following three subsections. All three ecological conditions create a positive selection pressure on dispersal. In the perspective of population conservation, it is important to identify conditions under which these positive selective pressures on dispersal fail to compensate for negative pressures, and thus result in selection for low dispersal.

Deterministically fluctuating populations

Fluctuations, either deterministic or stochastic, in local population growth rates can make dispersal advantageous. Let B_i be the expected number of offspring that an individual in patch i will produce, and assume that these random variables are independent and distributed identically. A nondispersing strategy then experiences a long-term growth rate equal to the geometric mean of the random variables B. By contrast, a population that consists of always dispersing individuals will spread the risk, and grow according to the arithmetic mean. Since the arithmetic mean is larger than the geometric mean, dispersing individuals have a higher fitness than nondispersing individuals. Dispersal is thus selected for (Levin *et al.* 1984); see also Kisdi (2002).

Nonequilibrium local population dynamics also cause deterministic fluctuations in local population growth rates. Therefore, if a metapopulation can reach a nonequilibrium attractor, this can result in selection for positive dispersal rates. Even dispersal polymorphisms (i.e., the coexistence of several dispersal strategies) can be promoted by temporal variation caused by cyclic or chaotic local dynamics (Holt and McPeek 1996; Doebeli and Ruxton 1997; Johst *et al.* 1999; Parvinen 1999; Kisdi 2002), or can result from temporally and spatially varying carrying capacities (McPeek and Holt 1992; Mathias *et al.* 2001).

The type of the resident attractor has a substantial effect on dispersal evolution. Deterministically fluctuating population dynamics are easier to describe with discrete-time models. Thus, the discrete-time version of the metapopulation model mentioned in the previous section is of interest (Parvinen 1999); details are expounded in Box 14.2. If the resident attractor is a two-cyclic orbit, in which local population sizes are large in one year and small in the next year, the resident attractor is an in-phase cycle. An alternative is an out-of-phase cycle, in which some local populations are large and others are small, and in the next year the roles are

Box 14.1 Ecological conditions that select for vanishing dispersal

In a metapopulation with a finite number of patches, the local population sizes are assumed to be large and the population densities are denoted by N_i in patch i. Population growth rate $r_i(N_i)$ is the difference between the birth rate and death rate in patch i. Individuals emigrate out of patches with rate $m \geqslant 0$. Emigrants survive migration with probability $\phi = 1 - \rho$ and choose the patch into which they immigrate at random, independently of the local population size. The quantity ρ can be interpreted as the dispersal risk. The dispersal strategy under selection is the emigration rate m. The following differential equations for $i = 1, ..., n$ describe the system,

$$\frac{dN_i}{dt} = r_i(N_i)N_i - mN_i + \frac{\phi}{n}\sum_{j=1}^{n} mN_j \,. \tag{a}$$

We assume that the resident population has reached a stable equilibrium $\hat{N}_i$ in each patch. Then a mutant population $N' = (N'_1, y_2, ..., N'_n)$ with strategy m' initially grows linearly according to $\frac{d}{dt}N' = MN'$. The invasion fitness λ of the mutant is the dominant eigenvalue of the matrix M, the elements of which are

$$\begin{aligned} M_{ii} &= r_i(\hat{N}_i) - m' + \frac{\phi m'}{n}\,, \\ M_{ij} &= \frac{\phi m'}{n} \text{ for } i \neq j\,. \end{aligned} \tag{b}$$

According to matrix theory (Caswell 2001), we have Equation (c) for the selection gradient,

$$\frac{d}{dm'}\lambda(m')\bigg|_{m'=m} = \frac{v\frac{\partial M}{\partial m'}w}{v^T w}\,, \tag{c}$$

where v and w are the left and right eigenvectors of matrix M that correspond to the dominant eigenvalue. Since matrix M is symmetric, the eigenvectors ψ and ϕ are equal. When $m' = m$, the resident equilibrium satisfies $MN = 0$, and therefore the equilibrium $(\hat{N}_1, \hat{N}_2, ..., \hat{N}_n)$ is an eigenvector that corresponds to the eigenvalue 0, which is dominant because the metapopulation equilibrium is assumed to be stable. Using Equation (c) we obtain

$$\frac{d}{dm'}\lambda(m')\bigg|_{m'=m} = \frac{\frac{\phi}{n}\left(\sum_{i=1}^{n}\hat{N}_i\right)^2 - \sum_{i=1}^{n}\hat{N}_i^2}{\sum_{i=1}^{n}\hat{N}_i^2} \leqslant 0\,, \tag{d}$$

where the inequality follows from the Cauchy–Schwarz inequality. Equality holds if $\rho = 0$ and all $\hat{N}_i$ are equal. Otherwise, the selection gradient is always negative, so the dispersal rate m will evolve to zero. A similar result is also valid in the discrete-time case (Parvinen 1999).

Section 14.2 explains how deterministic population fluctuations, environmental disturbances, and demographic stochasticity, all ignored here, can change this result.

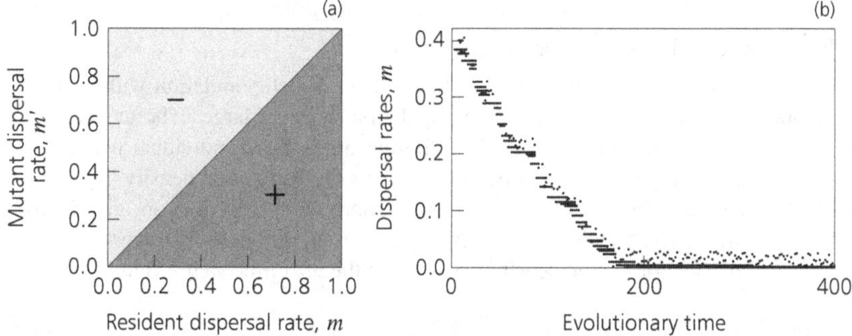

Figure 14.1 Dispersal evolution under equilibrium dynamics: (a) Pairwise invasibility plot (see Box 11.2) and (b) evolutionary dynamics with the currently resident dispersal strategies shown as points. For any resident strategy, a mutant strategy with a lower dispersal rate can invade; this results in dispersal strategies that converge to zero. *Source*: Parvinen (1999).

reversed. There are parameter values for which both types of attractors coexist. To calculate the fitness of a mutant, the attractor of the resident must be known.

In the model described in Box 14.1, the effects of habitat fragmentation can be examined by increasing the probability ρ to not survive dispersal. The quantity ρ can then be interpreted as the direct risk of dispersal. For extremely low dispersal risk ($\rho \approx 0$), evolutionary branching that results in dispersal polymorphism can be observed for in-phase cycles (Figure 14.2). However, when dispersal risk is increased only slightly, selection for low migration is observed. Therefore, a small deterioration in the environment can result in a large change in the evolutionary behavior of the metapopulation. As this change takes the dispersal rates of the metapopulation to zero, it exposes the metapopulation to chance extinction.

For out-of-phase cycles (Figure 14.3), positive dispersal rates can be observed for much larger dispersal risks than for in-phase cases, as explained in the following. Consider an individual in a patch in which, in the present year, the number of offspring will be high. If those offspring stay in the same patch, they will experience a situation with low fecundity in the next season. By contrast, a dispersing offspring has a good chance of entering a patch in which the fecundity is poor in the current season, but will be good in the next season. Therefore, there is a strong benefit to dispersal, which outweighs the direct cost; thus, dispersal is selected for.

Depending on the dispersal risk, selection is for low dispersal, evolutionary branching, or high dispersal in the out-of-phase cycle (Figure 14.3). If all individuals were to disperse, all local population sizes would be equal in the next time step. Therefore, if dispersal increases enough, the out-of-phase attractor disappears. When that happens, the metapopulation changes to the in-phase attractor (Figures 14.3b, 14.3c, and 14.3d), on which there is selection for low dispersal, unless the dispersal risk is very low (Figure 14.3d).

To conclude, increasing the direct cost of dispersal can cause selection for low dispersal. Patch synchrony is already known as an ecological factor of the extinction risk of a metapopulation (Allen *et al.* 1993; Heino *et al.* 1997a; Lundberg

Box 14.2 Coexisting metapopulation attractors

As in the model described in Box 14.1, we study a metapopulation with a finite number of patches in which the local population sizes are large. The model considered here is, however, described in discrete time. Each individual in patch i produces an average of $\lambda_i(N_i)$ offspring, and thus the population density in patch i in the next generation is $\lambda_i(N_i)N_i$ before migration. After reproduction, an individual in a patch migrates with probability m, and survives migration with probability $\phi = 1 - \rho$. The population density in patch i in the next time step is thus

$$N_{i,t+1} = (1-m)\lambda_i(N_{i,t})N_{i,t} + \frac{\phi}{n}\sum_{j=1}^{n} m\lambda_j(N_{j,t})N_{j,t}\ . \tag{a}$$

The choice $\lambda_i(N_i) = \lambda_{0i}e^{-\kappa_i N_i}$ corresponds to the Ricker model (Ricker 1954). Here λ_{0i} is the per capita number of offspring when there is no competition ($N_i \approx 0$), and κ_i is a measure of the strength of competition.

Such a metapopulation does not necessarily have only one feasible attractor. For some parameter values, there exists both an in-phase and an out-of-phase cycle. In the panel below this is illustrated for the case of two patches. The growth functions λ_i are the same as in Figures 14.2 and 14.3. The parameters are $\lambda_1 = 10$, $\lambda_2 = 9$, $\kappa_1 = 1$, $\kappa_2 = 1.1$, $\rho = 0.1$, and $m = 0.04$. The in-phase attractor (0.945, 0.994), (3.65, 3.00) is plotted with open circles and the out-of-phase attractor (0.985, 2.92), (3.63, 1.09) is plotted with filled circles. Combinations of the initial population densities from which the metapopulation state enters the in-phase cycle are plotted in gray.

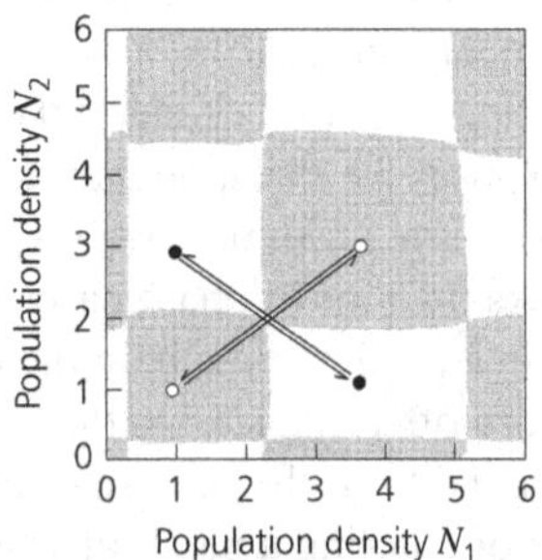

Since the simple deterministic metapopulation model considered here can feature coexisting attractors already, it is likely that real metapopulations also have several possible ecological attractors. The evolution of dispersal is strongly affected by the type of resident attractor, as discussed in Section 14.2.

et al. 2000). Notice that a metapopulation in an in-phase cycle is less likely to experience selection for increased dispersal than a metapopulation in an out-of-phase cycle. As dispersal almost always has some cost, it seems that the most probable scenario is that with selection for low dispersal in the in-phase case, and selection for high dispersal (or branching) in the out-of-phase case. At first sight, it appears

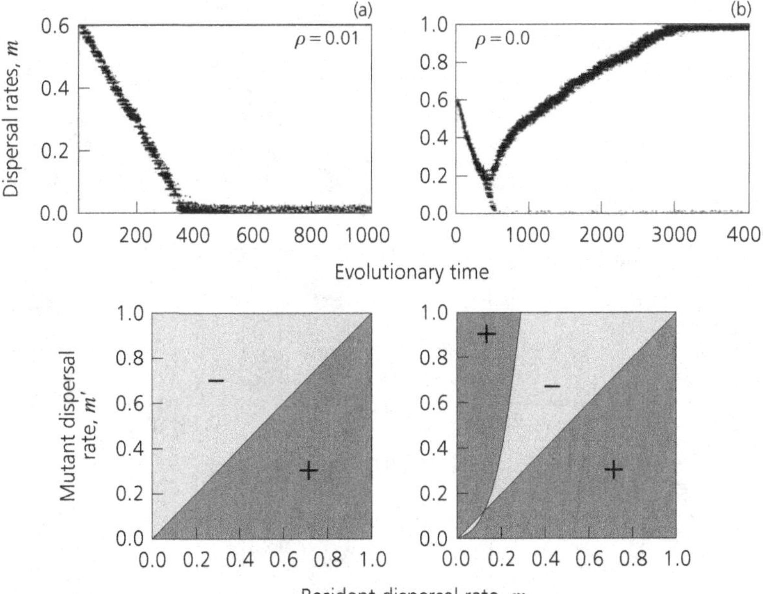

Figure 14.2 Dispersal evolution in the two-cyclic in-phase attractor case (upper panels, evolutionary dynamics; lower panels, pairwise invasibility plots). (a) Selection for low dispersal ($\rho = 0.01$). (b) Evolutionary branching ($\rho = 0.0$) in which the population is divided into two subpopulations, low dispersal and continual dispersal. *Source*: Parvinen (1999).

this would result in dispersal always being zero. However, a small demographic or environmental disturbance can cause a nondispersing metapopulation to change from an in-phase cycle to an out-of-phase cycle, which results in selection for dispersal. The end result could be a long-term evolutionary cycle that oscillates between decreasing dispersal in the in-phase case and increasing dispersal in the out-of-phase case. However, periods would still occur during which dispersal is dangerously low, and the metapopulation would be exposed to extinction because of environmental or demographic disturbances. In the next two subsections the direct effects of environmental and demographic disturbances on dispersal evolution are discussed.

Environmental disturbances

Environmental disturbances, or catastrophes that result in habitat patches becoming empty (and thus open to recolonization), are an essential feature of the classic Levins metapopulation model. The effect of such environmental disturbances on the evolution of dispersal are considered here. It is clear that, in the presence of catastrophes, a metapopulation cannot persist if individuals do not disperse (Van Valen 1971). Therefore, it was believed that dispersal always increases when environmental disturbances become more frequent.

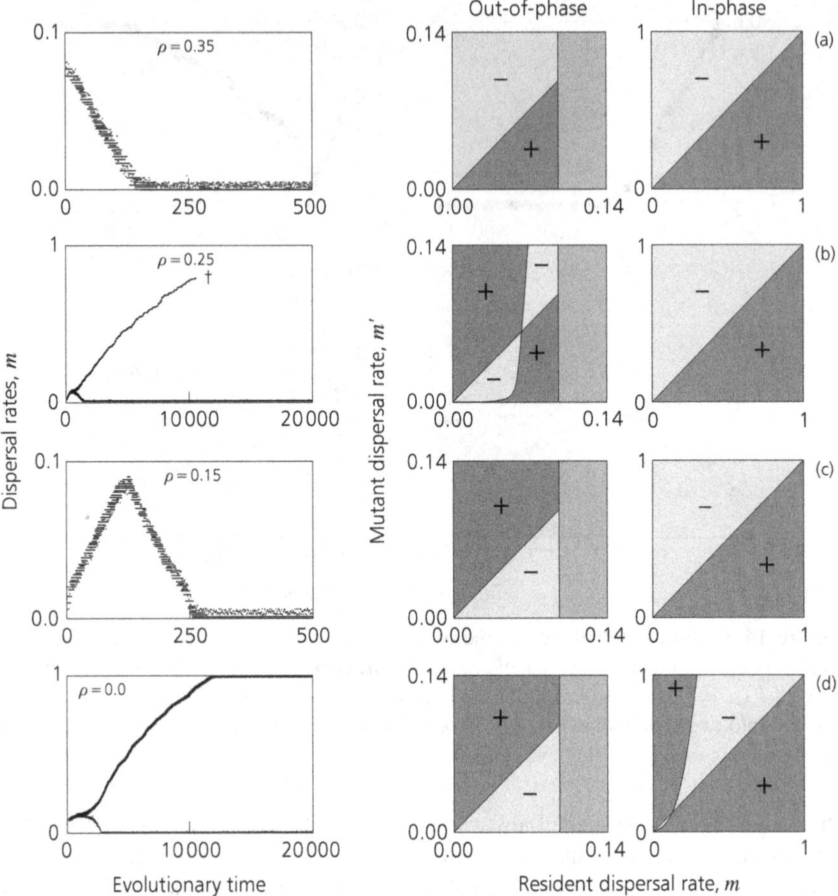

Figure 14.3 Dispersal evolution in the two-cycle out-of-phase attractor case (left column, evolutionary dynamics; right column, pairwise invasibility plots). (a) Selection for low dispersal ($\rho = 0.35$). (b) Evolutionary branching ($\rho = 0.25$) in which the lower branch approaches zero, and the upper one increases until it becomes extinct at $t \approx 10\,000$. The metapopulation then switches to the in-phase cycle, and dispersal strategies stay low. (c) Selection for high dispersal ($\rho = 0.15$) until the out-of-phase cycle disappears ($t \approx 120$). The metapopulation then changes to the in-phase cycle, for which selection for low dispersal is observed. (d) With $\rho = 0.0$, the situation is initially similar to that in (c), but when the metapopulation changes to the in-phase cycle ($t \approx 1000$), evolutionary branching occurs. *Source*: Parvinen (1999).

To incorporate catastrophes into an evolutionary metapopulation model that also takes local population dynamics into account, it is convenient to assume that the number of patches is infinite. Here, a metapopulation model is examined in continuous time under the simplifying assumption that all patches are ecologically equal and equally coupled by dispersal (Gyllenberg and Metz 2001; Metz

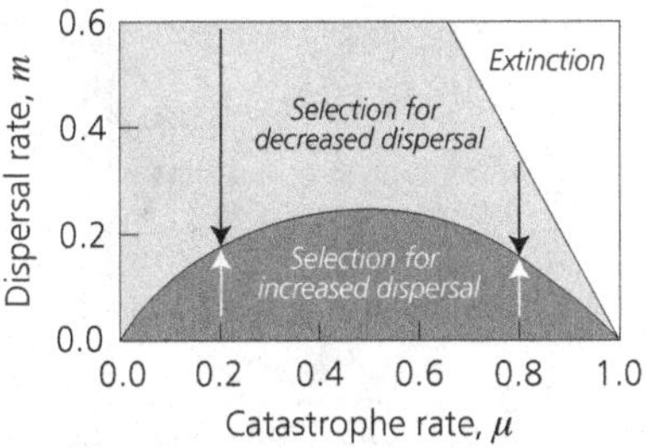

Figure 14.4 Effect of environmental disturbances on dispersal evolution. The region of parameter values in which lower dispersal is selected for is plotted in light gray, and selection for higher dispersal occurs in the region plotted in dark gray. The metapopulation is not viable in the white area. Parameters: $r(N) = 1 - N$, $\rho = 0.55$.

and Gyllenberg 2001). Local population growth through birth and death events is described by a per capita growth function $g(x)$, where x is the local population size, measured as a density. Individuals emigrate to a disperser pool at a per capita rate m. Dispersal is costly in the sense that a migrating individual may die before reaching a new patch: individuals in the disperser pool die at the per capita rate ν. Dispersers immigrate to habitat patches at a per capita rate δ. The probability of death during dispersal is thus $\rho = \nu/(\delta + \nu)$, which is a direct cost to dispersal. Immigrants choose their patch at random, independently of local population size. In patches of population size N, catastrophes occur at the rate $\mu(N)$. A catastrophe wipes out the local population in the patch, but the patch remains habitable. Empty patches can be recolonized by immigrants from the disperser pool. Details of this metapopulation model are presented in Section 4.5, in which persistence and viability are studied on the ecological time scale. Here, the effect of dispersal evolution is described (Gyllenberg and Metz 2001; Metz and Gyllenberg 2001; Gyllenberg *et al.* 2002; Parvinen 2002; Parvinen *et al.* 2003).

Figure 14.4 illustrates the effect of an increased level of environmental disturbances, as described by increasing the (density-dependent) catastrophe rate μ. In the absence of catastrophes, the strategy not to disperse is both evolutionarily stable and convergence stable (Section 11.3). For low catastrophe rates, dispersal rates do increase, as expected, with higher rates of local extinction. For high rates of local extinction, however, selected dispersal rates start to decrease again when local extinctions become more frequent. In other words, evolutionarily stable rates of dispersal are maximal for intermediate catastrophe rates; around this maximum, selected dispersal rates fall off when the catastrophe rate is either increased or decreased. Such an intermediate maximum of dispersal rates has also been observed in a discrete-time metapopulation model with natal dispersal and a density regulation that imposes a ceiling on local population sizes (Ronce *et al.* 2000a). If, in such a situation, catastrophes become more frequent, the metapopulation is subject to a high extinction risk, unless it is able to evolve quickly to lower dispersal rates.

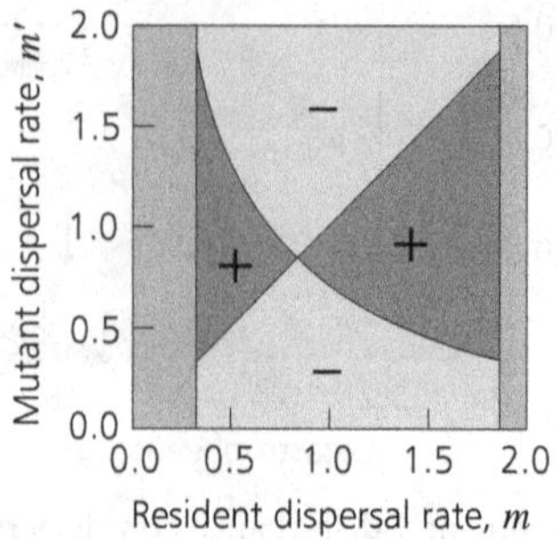

Figure 14.5 Pairwise invasibility plot for dispersal evolution under demographic stochasticity. Parameters: $K = 5$, $\mu = 0.3$, $\rho = 0.2$.

Demographic stochasticity

If local populations are large, deterministic models can be used to describe local population dynamics. However, if local populations are small (which may often be the case for species of conservation concern), demographic stochasticity cannot be ignored (Cadet *et al.* 2003; Parvinen *et al.* 2003). In such a situation, dispersal is selected for even in the absence of catastrophes. Local population sizes then fluctuate around the carrying capacity, and individuals always have a chance to find a better patch when dispersing. In particular, dispersal is beneficial for individuals in overpopulated patches. As there are more individuals in the overcrowded patches than in the less inhabited patches, dispersal is selected for.

In the model explained above with small local population sizes (Metz and Gyllenberg 2001; Parvinen *et al.* 2003), population growth is described by a Markov chain in continuous time with per capita rates for birth b_n, death d_n, and emigration m. At carrying capacity K, birth and death rates are equal. Figure 14.5 shows a pairwise invasibility plot in which, if the dispersal rate is small, larger dispersal rates are selected for. Therefore, the evolutionarily stable dispersal rate is positive.

Study of the effect of increased disturbances through catastrophes under demographic stochasticity reveals interesting details. Evolutionarily stable dispersal rates never become zero. However, it turns out that demographic stochasticity allows a wider range of possible responses to increased catastrophe rates. Evolutionarily stable dispersal rates can increase, or have an intermediate maximum, as explained before (Figures 14.6a and 14.6b). Also, patterns of decrease or that feature an intermediate minimum are possible responses [Figures 14.6c and 14.6d; see Parvinen *et al.* (2003) for details].

The results presented in Figure 14.6 demonstrate that demographic stochasticity must not be ignored when dispersal strategies are predicted. In particular, for small rates of local extinction, systematic qualitative departures from the results presented in the previous subsection arise: under such conditions, selection favors much higher dispersal rates than expected without demographic stochasticity.

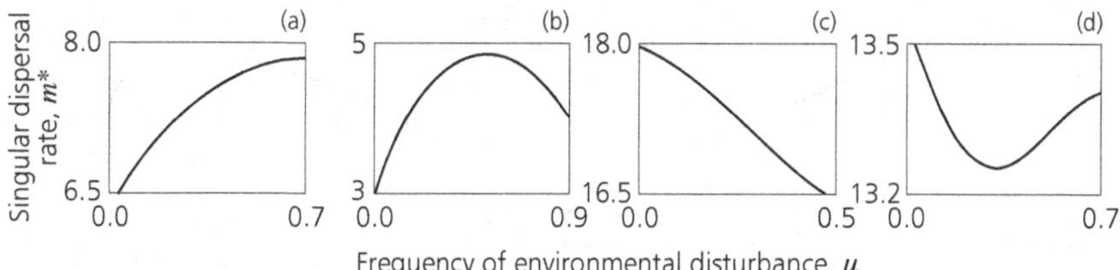

Figure 14.6 Different responses in a continuously stable strategy (CSS) of dispersal rate to increased catastrophe rate μ: (a) monotonic increase, (b) intermediate maximum, (c) monotonic decrease, and (d) intermediate minimum. In all four panels, the variation of the catastrophe rate μ extends over the range of viable metapopulations. *Source*: Parvinen *et al.* (2003).

14.3 Dispersal Evolution and Metapopulation Viability

Olivieri and Gouyon (1997) defined the optimal dispersal strategy as the migration rate that maximizes a metapopulation's carrying capacity. In the model studied by Olivieri and Gouyon (1997), the number of patches is infinite and there is a finite number of states in which a patch can be. If the local population in a patch is disturbed by a catastrophe, the state in the next season will be 0, which corresponds to an empty patch. If the population is not disturbed, the state will increase by one for the next season, unless the maximum state has been reached. Each patch can support a local population of K individuals at maximum. The disturbance probabilities depend on the current state. At the beginning of a season, adults reproduce and some of them die. Juveniles either remain in the patch or disperse. Dispersers form a migrant pool and immigrate uniformly to the patches. Dispersed and nondispersed juveniles compete for the space not occupied by adults. In this model, the optimal strategy is always larger than the evolutionarily stable strategy (ESS).

An ESS is therefore, in general, different from the optimal strategy. This has been noted already by Comins *et al.* (1980) and Motro (1982a). Below, the difference between the optimal strategy and the ESS is analyzed in several models. Furthermore, an extreme example of this phenomenon is discussed, evolutionary suicide, in which evolution itself causes metapopulation extinction.

Effect on population size

The effect of the evolution of dispersal on the average metapopulation size in some of the models studied in this chapter is examined here. In Figure 14.7 the average metapopulation size in the discrete-time metapopulation model with two patches is plotted for a situation in which there are two coexisting attractors. If the metapopulation is in an in-phase cycle, selection for low dispersal occurs, which maximizes the average metapopulation size (Figure 14.7a). However, if the metapopulation is in an out-of-phase cycle, selection for high dispersal occurs. The optimal strategy (at about $m = 0.062$) is therefore not evolutionarily stable (Figure 14.7b).

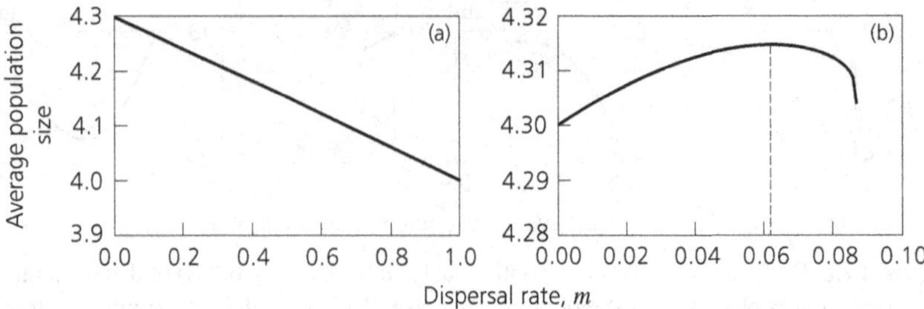

Figure 14.7 Average population size in the discrete-time metapopulation model with two patches, (a) in-phase and (b) out-of-phase. The optimal strategy is marked with a dashed line in (b). Parameters are as in Figure 14.3c.

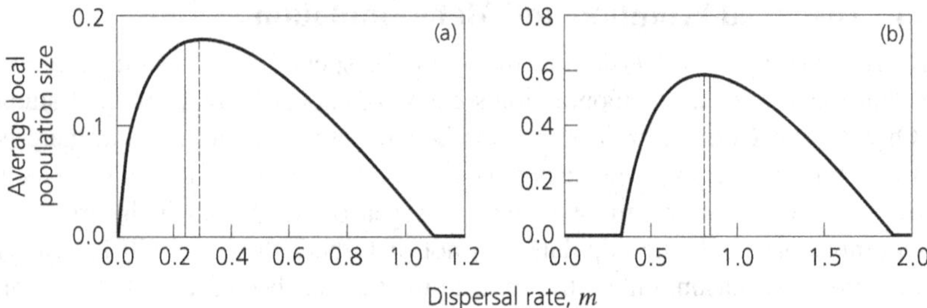

Figure 14.8 Average local population size in metapopulation models with (a) large or (b) small local populations. The optimal strategies are marked with dashed lines, and the evolutionarily stable strategies with continuous lines. Parameters: (a) as in Figure 14.4 with $\mu = 0.4$; (b) as in Figure 14.5.

In the metapopulation model with large local populations, ESS dispersal rate is not the optimal strategy. This is illustrated in Figure 14.8a, which shows that the evolutionarily stable dispersal rate is smaller than the optimal rate, just as in the study by Olivieri and Gouyon (1997). The situation is the opposite in the metapopulation model with finite local populations. In Figure 14.8b, the ESS dispersal rate of $m \approx 0.843$ is greater than the dispersal rate of $m \approx 0.809$ that maximizes the average population size.

Evolutionary suicide in dispersal evolution

A metapopulation is not necessarily viable for all possible dispersal rates. Evolutionary suicide results from an initially viable metapopulation that adapts in a way that it can no longer persist (Ferrière *et al.* 2000; Gyllenberg and Parvinen 2001; Gyllenberg *et al.* 2002; Chapter 11). Evolutionary suicide happens if selection takes the dispersal rate to the extinction boundary, beyond which the metapopulation is no longer viable and so is driven to extinction.

At first sight this seems an impossible scenario. How could such disadvantageous dispersal strategies ever be favored by natural selection? As we know,

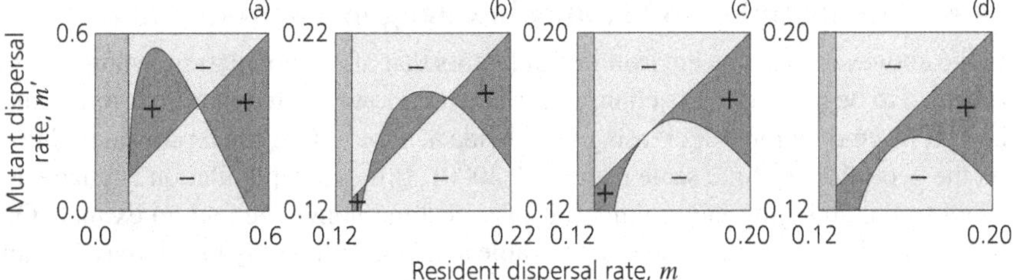

Figure 14.9 Pairwise invasibility plots that illustrate an evolutionary bifurcation to evolutionary suicide: (a) $\rho = 0.05$, (b) $\rho = 0.11$, (c) $\rho = 0.1175$, and (d) $\rho = 0.12$. *Source:* Gyllenberg *et al.* (2002).

evolution operates at the level of individuals – but what is good for an individual is not necessarily good for its population (Chapter 11). Natural selection is said to be frequency dependent if a strategy's advantage varies with its overall frequency within a population. Typically, in such a setting selection does not maximize the size of the evolving population. Selection pressures at the individual level may cause a metapopulation to decrease its dispersal rate below a critical threshold, beyond which the metapopulation becomes extinct, as is shown next.

For this, we return to the model with an infinite number of patches and large local populations (Gyllenberg and Metz 2001; Metz and Gyllenberg 2001). Previously, the catastrophe rate was assumed to be independent of local population size. It may, however, be possible that a large population is less vulnerable to catastrophes caused, for example, by predation. Therefore, settings in which the catastrophe rate $\mu(N)$ decreases with local population size N can be of biological relevance. It turns out that evolutionary suicide can be observed in such settings also (Gyllenberg *et al.* 2002).

In the following example, an evolutionarily stable and convergence-stable dispersal rate exists for a low dispersal risk ρ (Figure 14.9a). When the dispersal risk increases, this ESS moves toward the lower boundary of viability. Once the dispersal risk is high enough, another ESS appears, which is not convergence stable (Figure 14.9b). When dispersal risk increases further, these two strategies collide (Figure 14.9c) and finally disappear (Figure 14.9d). In the situation depicted in Figure 14.9d, a mutant with a lower dispersal rate than that of the resident can always invade the resident. This leads to a decreasing dispersal rate until the extinction boundary is reached. Even then, a mutant with a lower dispersal rate can invade. This mutant, however, cannot persist. Instead, the mutant, which could be called a "kamikaze mutant", takes the resident away from its attractor and to the metapopulation extinction equilibrium, which results in evolutionary suicide. Evolutionary suicide has been observed only in the model associated with selection for dispersal rates that are too low. For more examples of evolutionary suicide, see Gyllenberg and Parvinen (2001), Gyllenberg *et al.* (2002), and Chapter 11.

14.4 Metapopulation Viability in Changing Environments

In the above sections the environmental factors that affect population dynamics are assumed to be constant or to change very slowly. However, habitats deteriorate and habitat fragmentation is increasing worldwide at a speed that can be considered fast on the evolutionary time scale (Sih *et al.* 2000). If the metapopulation is unable to adapt to the changing environment, it may become nonviable and go extinct. On the other hand, if the metapopulation is able to adapt sufficiently fast, it may remain viable and persist. Such a process is called evolutionary rescue (Gomulkiewicz and Holt 1995; Ferrière 2000; Heino and Hanski 2001); see also Box 1.4, as well as Chapters 11 and 13.

To study these effects, we focus on a metapopulation model with an infinite number of patches and large populations. Increasing habitat fragmentation can be modeled by increasing the dispersal risk ρ, and habitat deterioration by increasing the catastrophe rate μ or by combining elementary landscapes into a heterogeneous one. The effects of changing these environmental factors are discussed in the following subsections.

Landscape heterogeneity

Most metapopulation models described above assume that habitat patches are identical and differ only in population size. In reality, patches are never exactly equal. One way to model patch heterogeneity is to assume that the actual habitat landscape is made of a combination of n_L elementary landscapes of type i, $i = 1, 2, ..., N$ with proportions p_i, such that $\sum_{i=1}^{n_L} p_i = 1$. Brachet *et al.* (1999) studied this generalization of a metapopulation model by Olivieri and Gouyon (1997), which featured two elementary landscapes. Brachet *et al.* (1999) found that, depending on the viable ranges of dispersal in the elementary landscapes, several scenarios are possible. If these ranges overlap, the metapopulation persists in any mixture of the elementary landscapes. Otherwise, there exists a combination of the landscapes for which the metapopulation becomes extinct for any dispersal rate, or the metapopulation is viable for two distinct ranges of dispersal rates. In Figure 14.10 the domain of parameters for which the metapopulation is viable is illustrated for two scenarios. In both cases, with low dispersal, the metapopulation is viable in landscape 2, but not in landscape 1. In Figure 14.10a, if we start with a viable metapopulation with a low dispersal rate, and then increase the proportion p_1 of elementary landscape 1, the metapopulation becomes extinct, unless it is able to adapt and increase dispersal, and thereby experience evolutionary rescue. If the same procedure is repeated in Figure 14.10b, evolutionary trapping occurs. The dispersal rate of the metapopulation is trapped into one gray region that disappears as p_1 increases, which results in metapopulation extinction, unless large mutational steps are possible that allow a jump to the other gray region.

Let us now look at an analogous generalization of the model with an infinite number of patches and large local populations. This generalized model (Parvinen 2002) incorporates several different types of patches with different growth conditions and catastrophe rates. In Figure 14.11 an example of dispersal evolution

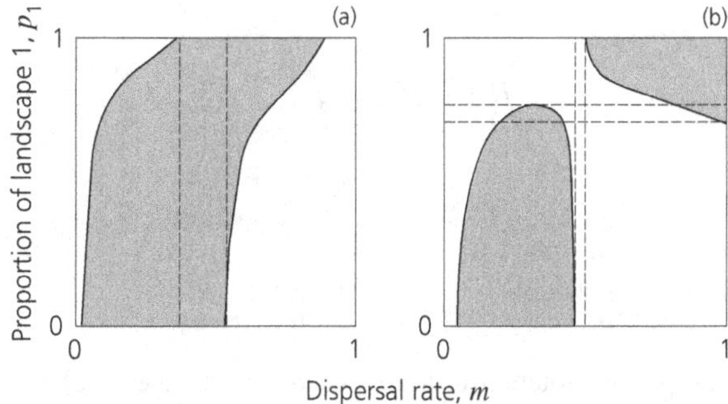

Figure 14.10 Examples in which dispersal evolution can cause (a) evolutionary rescue and (b) evolutionary trapping. Shaded areas indicate parameter combinations for which the metapopulation is viable. *Source*: Brachet *et al.* (1999).

with two types of patches is illustrated. The viable ranges of dispersal in the two elementary landscapes largely overlap, and therefore the metapopulation is viable for any proportion p_1 of landscape type 1. The evolutionarily singular strategy is not necessarily found between the two singular strategies for metapopulations that consist of only one type of patches. Furthermore, evolutionary branching can occur for some values of the proportion p_1.

In Figure 14.11b, a pairwise invasibility plot of such a case is plotted. Strategies approach the singular strategy $m^* \approx 0.17$, which is a branching point. The metapopulation then becomes dimorphic and moves into the domain of protected dimorphisms (Figure 14.11c). The strategies finally reach the evolutionarily attracting dimorphism at approximately $(0.11, 0.235)$ or $(0.235, 0.11)$. This unique convergence-stable dimorphism is evolutionarily stable, and is therefore the final outcome of the evolutionary process. This means that spatial heterogeneity, involving different patch types with sufficient proportions, together with temporal variation caused by catastrophes, can result in a dispersal polymorphism through evolutionary branching.

Increased fragmentation

As mentioned above, increasing habitat fragmentation can be modeled by increasing the dispersal risk ρ. In the previous section, evolutionary suicide is discussed in a setting in which the catastrophe rate $\mu(N)$ decreases with local population size N. The parameter values for which such suicide happens are now studied in more detail. Figure 14.12 illustrates the direction of selection on dispersal rates for different values of dispersal risk. When dispersal risk increases, we observe an evolutionary bifurcation from an ESS to evolutionary suicide, as illustrated in Figure 14.9. In such a scenario, the metapopulation becomes extinct long before the dispersal risk increases beyond the boundary of ecological viability (Figure 14.12). Such a collapse as a result of slow environmental change is called evolutionary trapping (Chapters 1 and 11).

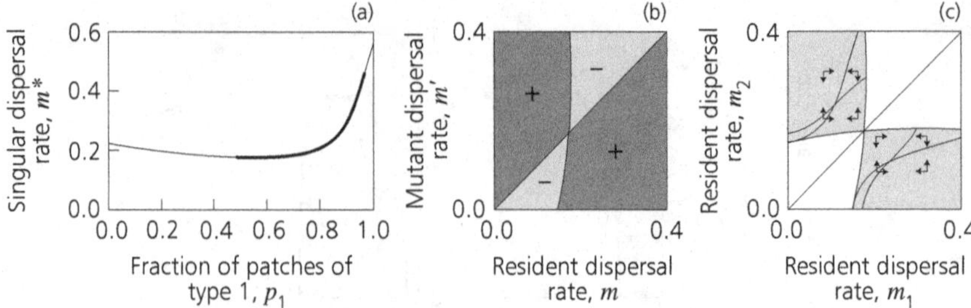

Figure 14.11 Dispersal evolution in heterogeneous landscapes. (a) Singular dispersal rates m^* in dependence on p_1. The thin curve corresponds to ESSs and the thick curve to evolutionary branching points. (b) Pairwise invasibility plot for $p_1 = 0.6$. (c) Domain of protected dimorphisms (plotted in gray) and direction of evolution for $p_1 = 0.6$. Parameters: $n_L = 2$, $\rho = 0.1667$. Patch properties: $r_1(N) = 1 - N$, $\mu_1(N) = 0.5$, $r_2(N) = 0.7(1 - N/1.5)$, $\mu_2(N) = 0.1$. *Source*: Parvinen (2002).

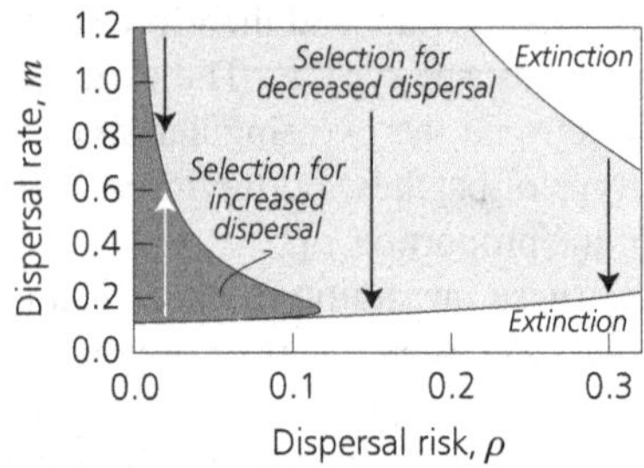

Figure 14.12 Evolutionary suicide resulting from dispersal evolution. The region of parameter values that selects for lower dispersal rates is plotted in light gray; selection for higher dispersal rates occurs in the regions shown in dark gray. In the white regions the metapopulation is not viable. The catastrophe rate $\mu(N)$ decreases with the local population size x: evolutionary suicide occurs when the dispersal risk ρ increases beyond a threshold value. *Source*: Gyllenberg *et al.* (2002).

Catastrophe rate and temporal uniformization

The effect of changing the catastrophe rate, while assuming an infinite number of patches and large local populations, is now described. Figure 14.4 shows that if the initial catastrophe rate is approximately $\mu = 0.76$, and the metapopulation has adapted to this situation, the dispersal rate is approximately $m = 0.18$. If the environment experienced by the metapopulation deteriorates through an increase in the catastrophe rate, and the metapopulation does not adapt to the new condition, it goes extinct when the catastrophe rate exceeds about $\mu = 0.9$. If, however, the metapopulation adapts, evolution decreases the dispersal rate such that the metapopulation stays viable and persistent (at least until the catastrophe rate exceeds $\mu = 1.0$, after which the metapopulation cannot persist with any dispersal strategy). This is an example of evolutionary rescue that results from dispersal evolution.

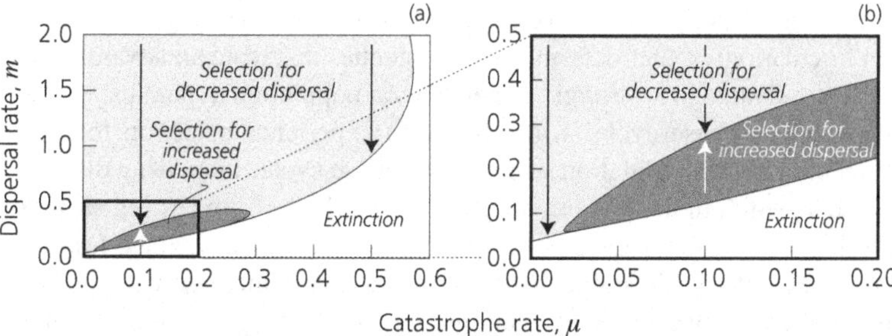

Figure 14.13 Evolutionary suicide resulting from dispersal evolution in the presence of an Allee effect. The region of parameter values that selects for lower dispersal rates is plotted in light gray; selection for higher dispersal rates occurs in the regions shown in dark gray. In the white regions the metapopulation is not viable. Notice that evolutionary suicide occurs when the catastrophe rate μ is either increased or decreased. *Source*: (a) Gyllenberg *et al.* (2002).

In a second example, we consider an Allee effect in the local growth rate, which means that small local populations have a negative intrinsic growth rate and can only persist by immigration. In this example, we again assume that the catastrophe rate is independent of local population size. Under these assumptions, increasing the catastrophe rate results in evolutionary trapping, or induced evolutionary suicide (Figure 14.13a).

Under some circumstances, decreasing the catastrophe rate can also cause population extinction. When catastrophes are rare, practically all local populations reach their carrying capacity. Under such temporal uniformization, selection for low dispersal rate occurs because all the patches are ecologically equal; consequently, the metapopulation is at a high risk of extinction. Decreasing the catastrophe rate can then cause evolutionary trapping (Figure 14.13b). When catastrophes are rare, and the dispersal rate is high, there are enough dispersers to colonize empty patches and overcome the negative growth rates that result from the Allee effect. Individuals that disperse less avoid the cost of dispersal and therefore have a higher reproductive output, as long as such less-frequent dispersers are rare. Therefore, dispersal is not selected for. At some point, however, dispersal becomes so limited that most colonizers in empty patches are exterminated by the Allee effect. The metapopulation is then doomed to extinction, and evolutionary suicide occurs (Gyllenberg *et al.* 2002).

14.5 Concluding Comments

In this chapter the adaptive responses of dispersal rates to landscape disturbances are discussed. In the metapopulation models considered here, a deteriorating environment can be modeled either by increasing dispersal risk or by altering the temporal pattern of environmental disturbances.

It is shown that if all local populations are at their ecological equilibrium, and there are no environmental disturbances, there is no advantage to dispersal, and a nondispersing strategy is both evolutionarily stable and convergence stable. Population fluctuations, either deterministic or stochastic, create an advantage for dispersal. This advantage is strongly dependent on population dynamics. A metapopulation in an in-phase cycle is less likely to experience selection for increased dispersal than a metapopulation in an out-of-phase cycle. Increasing dispersal risk can cause a switch in the evolutionary attractor, which results in selection for low dispersal. Also, the environmental disturbances create an advantage for dispersal. Therefore, evolutionarily stable dispersal rates initially increase with environmental disturbances. However, this monotonic relation no longer applies when the rate of environmental disturbance is increased further (Ronce *et al.* 2000a; Parvinen *et al.* 2003). Demographic stochasticity is a third mechanism that makes dispersal advantageous. In the presence of demographic stochasticity (i.e., when local populations are not very large), positive dispersal rates evolve even if there are no environmental disturbances.

Traditionally, metapopulation viability and persistence were studied on the ecological time scale, and this remains an important task for conservation biology. However, evolution natural selection can result in metapopulation extinction, even though the metapopulation would be viable with such evolution. Selection may favor strategies that are initially beneficial for the individuals, but turn out to be harmful for the metapopulation by making it nonviable. Such phenomena of evolutionary suicide (Ferrière 2000; Gyllenberg and Parvinen 2001; Gyllenberg *et al.* 2002; Chapter 11) have been observed in dispersal evolution under two ecological scenarios. In one scenario, the rate of environmental disturbances decreases with local population size. In the other, an Allee effect occurs in the local population growth rate. Such scenarios may apply to many metapopulations, and therefore evolutionary aspects of metapopulation conservation should no longer be ignored.

This conclusion is strongly supported by the results given in Section 14.4. We are used to situations in which environmental deterioration causes extinction, as shown in Figures 14.4 and 14.12, in which either the rate of environmental disturbances or dispersal risk is increased. In Figure 14.4, if the environmental change is slow enough to allow the metapopulation to adapt, such that evolutionary rescue can keep the metapopulation viable. In Figure 14.12, evolutionary suicide drives the metapopulation to extinction prematurely. However, in the presence of an Allee effect (Figure 14.13), improving the environment by decreasing the rate of environmental disturbance can result in metapopulation extinction through evolutionary suicide. Before measures thought to benefit endangered species are put into effect, it is therefore necessary to clarify their evolutionary consequences.

Metapopulation dynamics are influenced by many life-history parameters. Two directions for future research thus deserve to be highlighted:

- Evolving dispersal rates may imply changes in other life-history traits, and thus induce additional costs and benefits to dispersal. For example, the production of wings consumes resources that could otherwise be allocated to maintenance or reproduction, yet the availability of wings is also likely to have beneficial side effects on resource acquisition and predator avoidance.

- Dispersal strategies may evolve jointly with other life-history traits. Studies of such more complex evolutionary dynamics, which should also involve investigations of the underlying life-history trade-offs, are clearly needed; for interesting work in this direction see Ronce *et al.* (2000b) and Kisdi (2002).

Further research on these extensions will help evolutionary conservation biology to develop predictive tools with which to study metapopulation evolution.

Acknowledgments The author thanks Ulf Dieckmann, Mats Gyllenberg, and Hans Metz, who have been collaborators in much of the work summarized here, and also Odo Diekmann, Stefan Geritz, and Éva Kisdi for discussions, and the Academy of Finland for financial support.

15

Adaptive Responses to Landscape Disturbances: Empirical Evidence

Bruno Colas, Chris D. Thomas, and Ilkka Hanski

15.1 Introduction

Previous chapters of this book examine the risks of extinction for small and/or isolated populations. Here we try to analyze, from empirical data, how species and metapopulations can respond to habitat fragmentation. During the past 30 years, metapopulation biology has become recognized as an invaluable perspective in ecology, genetics, and evolution (Olivieri *et al.* 1990), and a considerable body of theoretical and empirical research has been carried out (for reviews see Harrison and Hastings 1996; Hanski and Gilpin 1997; Hanski 1999).

The abundance of a species depends not only on the availability of habitats that match the ecological requirements of the species, but also on the life-history characteristics that influence the persistence and colonization ability of the local population, that is its metapopulation dynamics (see Chapter 4). Colonization ability, in particular, is critical, and depends on traits related to migration and on the ability to establish a new population from a small propagule size (Box 15.1). Rapid colonization may compensate for a low probability of local persistence (e.g., Van der Meijden *et al.* 1985). Besides determining colonization abilities, migration also affects local population viability, especially when local population dynamics are greatly influenced by spatially uncorrelated stochasticity (e.g., Stacey and Taper 1992).

The evolved characters of populations, metapopulations, and species also depend on rates of gene flow among populations and on extinction–colonization dynamics (Olivieri *et al.* 1995; Storfer 1999). Migration creates gene flow, which may result in outbreeding depression (e.g., Waser and Price 1989; Fenster and Galloway 2000; Quilichini *et al.* 2001) and prevents the evolution of local adaptations (Storfer 1999), but also decreases inbreeding along with its associated deleterious effects (Table 15.1). The migration rate itself will evolve. In a landscape characterized by long-term and large-scale stability, the metapopulation is expected to reach a large-scale steady state, within which selective processes operate at the local and metapopulation levels and depend on the different demographic parameters, rates of migration, local population size, and extinction and colonization rates.

Interactions between ecological and evolutionary processes in fragmented landscapes are likely to have been impacted by the recent growth of human populations and the associated development that has altered land-use patterns. Natural habitats

Box 15.1 Factors involved in species' distribution and abundance

The geographic range of any species is limited by (Gaston 1994):

- The availability of sites in which the required ecological conditions are met;
- The species' ability to colonize these suitable sites.

The colonization ability of any species is determined by:

- Its dispersal;
- Its ability to establish a population from one or a few individuals.

This establishment ability depends on several life-history traits, given that one or a few colonizing individuals reach an unoccupied and suitable site (i.e., one within which a population may occur given the ecological niche of the species):

- The reproductive system: for example, an inbreeding species or a plant species able to reproduce vegetatively will colonize a new site more easily from a base of one or a few individuals than will a strictly outbreeding species;
- The life cycle: for example, an iteroparous colonizer is more likely to produce offspring than a semelparous one.

have become more fragmented by intensive agriculture, forestry, urban development, etc. The properties of habitat fragments, such as size, shape, quality, and isolation, have become radically altered over large parts of the globe.

Habitat fragmentation has three major components (Wilcox and Murphy 1985; Andrén 1994):

- Reduction of the pooled area of habitat (habitat loss);
- Reduction in the average fragment size;
- Increase in the average distance between fragments.

The immediate consequences of habitat fragmentation are that the average size of the local populations decreases and the average number of immigrants that arrive at particular sites also decreases (Hanski 1999). Small isolated populations are more prone to extinction than large connected populations for various reasons, which include a greater impact of environmental and demographic stochasticities, Allee effects (Chapter 2), and edge effects. At the level of the metapopulation, after habitat fragmentation, individual migrants have reduced chances of encountering a suitable site, which increases the rate of mortality during migration. In a metapopulation context, fragmentation may thus increase local extinctions and decrease colonization rates, and, consequently, may doom a species to extinction (Hanski 1989).

Later consequences of fragmentation concern the genetic effects associated with smaller population size and lower migration, which can lead to a higher rate of inbreeding (see Chapter 12). The deleterious effects related to inbreeding can significantly impair population survival (Box 15.2). However, lower migration does increase the possibilities of adaptation to marginal habitats (see Chapter 13).

Table 15.1 Examples that suggest deleterious genetic effects associated with higher levels of fragmentation in natural populations.

Genetic effects	Species	Reference
Lower survival Lower developmental homeostasis	*Bufo bufo*	Hitchings and Beebee (1998)
Lower larval survival Lower larval growth rate	*Bufo calamita*	Rowe *et al.* (1999)
Lower amount of phenotypic variation	*Salvia pratensis* *Scabiosa columbaria*	Ouborg *et al.* (1991)
Lower seed size Lower germination success Higher susceptibility to environmental stress	*Ipomopsis aggregata*	Heschel and Paige (1995)
Lower survival	*Gentianella germanica*	Fischer and Matthies (1998)

Thus, the extent to which inbreeding will impair population survival depends on the relative importance of inbreeding depression and local adaptations [see the discussion in Mills and Allendorf (1996)].

The effects of fragmentation have been much discussed in conservation biology [for a review see Hanski and Simberloff (1997)], but how metapopulations and species cope with fragmentation and evolve thereafter is much less certain. Fragmentation modifies the type of selection that acts on migration, depending on how fragmentation has already affected the demographic and genetic characters of the metapopulation. That leads to the possibility of many interactions.

The goal of this chapter is to show empirical evidence for the evolutionary consequences of habitat fragmentation, and their effects on metapopulation viability. First, the effects of fragmentation on migration are examined (Section 15.2), followed by discussion of a further complexity in which local adaptations are also involved (Section 15.3). Section 15.4 deals with a threatened species in terms of its long-term evolution in a naturally highly fragmented landscape. In the final section, a summary is attempted of the knowledge concerning the evolutionary consequences of habitat fragmentation in natural populations.

15.2 Responses of Migration to Landscape Fragmentation

Predictions concerning the evolution of migration in response to fragmentation are presented in Box 15.3 (for explorations of the mechanisms involved, see Chapter 14). Different outcomes can occur, depending on the response to fragmentation of local population extinction, survival of migrants, and the proportion of empty patches. Keeping in mind these results, necessary to interpret the observations, empirical studies are now examined.

Box 15.2 Inbreeding depression in the metapopulation of *Melitaea cinxia*

The case study of a butterfly metapopulation living in a highly fragmented landscape has recently demonstrated genetic causes of population extinction. The large metapopulation of *M. cinxia* in Finland consists of some hundreds of mostly very small local populations (Hanski 1999). Local populations have a high risk of extinction for many reasons, including demographic and environmental stochasticity and parasitism. The caterpillars live in large sibling groups, and often a population comprises one group of full siblings. Mating among close relatives must therefore occur commonly in local populations. Inbreeding reduces heterozygosity and, if inbreeding depression increases the risk of extinction, the expectation is that, other things being equal, reduced heterozygosity is associated with an elevated risk of population extinction.

Saccheri *et al.* (1998) tested this prediction by genotyping a sample of butterflies from 42 local populations, of which seven went extinct in one year. Previous studies had shown that those populations most likely to become extinct are characterized by small size, are isolated (small numbers of butterflies in the neighboring populations, and hence no rescue effect), and the habitat patch has a low density of nectar flowers (which increases emigration and decreases immigration). In their study, Saccheri *et al.* (1998) found that, in addition to these ecological factors, the level of heterozygosity in the population also made a highly significant contribution to the model that explained the observed extinctions. A recent experimental study by Nieminen *et al.* (2001) confirmed that inbreeding does, indeed, increase the risk of local extinction in this species. Laboratory studies have shown that inbreeding affects several fitness components in *M. cinxia*, including the egg hatching rate, weight of post-diapause larvae, pupal period (inverse relationship), and adult longevity (Saccheri *et al.* 1998; Haikola *et al.* 2001). A single round only of brother–sister mating, which must occur commonly in the small populations of the *M. cinxia* metapopulation, was sufficient to reduce the egg hatching rate by about 30%.

These results suggest that this metapopulation maintains a large genetic load. Apparently, selection against deleterious recessives exposed by local inbreeding is relatively inefficient, most likely because of a high probability that slightly deleterious alleles are fixed in small populations and because of gene flow among neighboring small local populations. The study by Saccheri *et al.* (1998) is the first to demonstrate an effect of inbreeding on the extinction of natural populations. It remains an open question as to how widespread extinction through inbreeding is in metapopulations in general.

Effects of patch isolation

The effects of habitat patch isolation on migration rate have been described for several butterfly populations. Dempster *et al.* (1976) and Dempster (1991) found that two species of butterflies undertook phenotypic changes in body morphology following habitat fragmentation and population isolation. For the swallowtail butterfly, *Papilio machaon*, and the large blue, *Maculinea arion*, Dempster examined the morphology of museum specimens that had been collected over many years.

Box 15.3 Predicting responses of migration to landscape fragmentation

Generally, migration is selected against locally because the probability of finding favorable breeding conditions is often higher within the natal patch than moving outside it. However, at the level of the metapopulation, migration can be selected for if the turnover rate of populations is sufficiently high: the increased availability of empty habitat patches, generated by local extinctions, may lead to a metapopulation-wide increase in migration rate (Olivieri *et al.* 1995; Leimar and Nordberg 1997). This is known as the metapopulation effect (Olivieri and Gouyon 1997, see Chapter 11), the scope of which was reviewed in Ronce *et al.* (2000a).

Heino and Hanski (2001) have constructed an individual-based simulation model that can be used to study the evolution of migration rate in real landscapes with particular spatial configurations of the suitable habitat. Individual-based models have their own problems, in particular a large number of model assumptions and parameters, and it is difficult to arrive at conclusive predictions via simulations. The authors were able to use two accurately parametrized submodels [on adult movements (Hanski *et al.* 2000) and on extinction–colonization dynamics (Hanski 1994)] as building blocks of the evolutionary model.

Heino and Hanski (2001) further parametrized their individual-based model of the evolution of migration to compare two closely related species of checkerspot butterflies, *M. cinxia* and *M. diamina*. Using these parameter values, but allowing the emigration rate parameter to evolve in the model, they ran the model in the real landscape for *M. diamina* and predicted a value of 0.106 (standard deviation 0.006) for this parameter (the parameter gives the daily probability of leaving a habitat patch of 1 hectare). This value does not differ significantly from the value estimated in an empirical mark–release–recapture study (0.130, 95% confidence limits of 0.104 and 0.171; Hanski *et al.* 2000). The great advantage of this model is that it allows quantitative predictions to be made both for particular individual habitat patches and for networks of patches. It remains a great empirical challenge to collect sufficient data to test such model predictions.

Heino and Hanski (2001) found that if the primary consequence of habitat fragmentation was an increase in the local extinction risk through decreased local population sizes, the evolutionary response was an increased migration rate. If the quality of the matrix habitat deteriorated, leading to increased mortality during migration, a more complex evolutionary response was detected. In this case, as long as habitat-patch occupancy did not decrease greatly with increased migration mortality, a reduced migration rate evolved. However, once mortality became so high that empty patches remained uncolonized for a long time, evolution tended to increase the migration rate, essentially because of the "metapopulation effect" referred to above (see also Leimar and Nordberg 1997). Heino and Hanski (2001) present scenarios in which the increased migration rate in response to habitat fragmentation leads to an "evolutionary rescue" (i.e., a persisting metapopulation) if evolution of the migration rate occurs, but metapopulation extinction if no evolution occurs. (See Chapters 11 and 14 for a more theoretical treatment of the notion of evolutionary rescue.)

Table 15.2 Inferring possible selective pressures from associations between habitat characteristics and observed traits in butterfly species.

Habitat characteristics	Observed butterfly traits	Possible selective pressure	Species (reference)
Isolated patches	Small thoraxes	Selection against dispersal	*Papilio machaon*, *Maculinea arion* (Dempster *et al.* 1976; Dempster 1991)
Small patches	Large thoraxes	Selection for the ability to defend mating territories	*Plebejus argus* (Thomas *et al.* 1998)
New habitats	Large thoraxes compared to abdomens	Selection for dispersal/ colonizing ability	*Hesperia comma*, *Pararge aegeria* (Hill *et al.* 1999a, 1999b)

Through time, the thorax shapes of both species became thinner in isolated populations. Furthermore, Dempster *et al.* (1976) showed that swallowtails with thin thoraxes flew more slowly than swallowtails with wide thoraxes. Since the thorax of a butterfly is almost entirely filled with flight muscles, Dempster suggested that these two butterfly species might have evolved reduced flight capacity following population isolation. Strong-flying individuals with fat, muscle-filled thoraxes would tend to emigrate, whereas relatively puny insects would stay in the habitat in which they emerge (Table 15.2).

Body resources saved by not developing the flight muscles could be allocated to reproduction, which might be as important as, or more important than, selection on migration itself. Most of the resources allocated to reproduction in adult butterflies are located in the abdomen, and so it is important to assess whether isolated populations possess relatively large abdomens as well as small thoraxes. Museum specimens are not ideal for this purpose, for various reasons. These include that the age and reproductive history of adults is unknown, as are the weather conditions (which influence thorax and abdomen shapes and weights) under which they developed. Therefore, it is better to measure the relative allocation of resources to different body parts in freshly emerged specimens, before they have flown, mated, fed, and laid eggs. Thus, Hill *et al.* (unpublished) followed up Dempster's (1991) suggestion that British swallowtail butterflies had evolved smaller thoraxes by rearing British and Spanish swallowtails in a common environment in the laboratory. Spanish swallowtails are widely distributed, fly fast, and occur in many habitats. These authors confirmed that isolated British swallowtails have relatively smaller thorax sizes and larger abdomen sizes than their more dispersed Spanish counterparts. It is difficult to prove that habitat isolation causes this difference, rather than some other difference between the two study populations, but habitat isolation is a plausible explanation.

Effects of patch size in relation to other life-history traits

In similar vein, Thomas *et al.* (1998) studied a third species of butterfly, *Plebejus argus*, in isolated metapopulations. Some of the metapopulations inhabited small and isolated fragments of heathland vegetation, whereas others inhabited much more extensive areas of heathland. Within each heathland the butterflies themselves had quite localized distributions, being restricted to areas with high densities of mutualist ants (the caterpillars are tended and protected by ants, which obtain sugars and amino acid secretions from them). In a large area of heathland, adult *P. argus* butterflies that leave their ant-enriched habitat would almost certainly find another area of suitable breeding habitat somewhere else within the heathland. In contrast, a wandering butterfly might be more likely to leave a small fragment of heathland vegetation completely, eventually perishing in the surrounding agricultural land. Surprisingly, butterflies reared in the common environment that originated from small heathland fragments were relatively large – indicative of relatively strong flight capacity. This was contrary to the original expectations, which were that weak fliers would be found in the small heathlands. Of the various reasonable explanations for this result, one possibility is that large males might be better at defending mating territories, so small males displaced from territories might be more likely to emigrate completely, the impact being greatest in small areas (Thomas *et al.* 1998, Table 2). The important lesson is that the evolution of flight ability and migration is often a complex compromise between different activities (adult feeding, mating, finding egg-laying sites, avoiding predators, and true migration), and any or all of these activities may also be affected by habitat area and isolation.

Thomas *et al.* (1998) studied one metapopulation of *P. argus* in more detail. This metapopulation inhabited a different type of vegetation, limestone grassland. In this case, all of the patches of habitat were within the migration range of at least one other patch, with a measured exchange rate of 1.4% per generation (Lewis *et al.* 1997). In this system, adult butterflies from small patches had relatively large thoraxes and small abdomens. Two equally plausible explanations are:

- Individuals with large-thoraxes are able to defend territories, so they can stay within small patches and/or territories, whereas weaker adults are forced out of small patches;
- Small patches receive higher fractions of immigrants or have been recolonized more recently than large patches, such that small patches are populated mainly by strong-flying immigrants and by the offspring of immigrants.

The first of these explanations predicts that migrants between patches will have small thoraxes, whereas the second predicts that migrants between patches will have large thoraxes. These predictions have yet to be tested.

Effects of colonization opportunities

In plant species of the Asteraceae family, Cody and Overton (1996) showed that dispersal can evolve within a few generations only. They studied the evolution of pappus length of achenes on the mainland and in newly established island populations in British Columbia. They showed that island populations were established by achenes with the longest pappus (increasing the ability to be dispersed by wind) in the mainland populations. However, once on an island dispersal can be selected against, because islands are surrounded by an unfavorable habitat (the sea). After less than five generations of the establishment, pappus length had significantly reduced in at least one species (Cody and Overton 1996).

A similar sequence in the results of selection pressures might be expected in terrestrial habitat islands (Olivieri *et al.* 1990, 1995). In butterflies, indeed, there are some indications that large thoraxes and small abdomens may characterize colonizing populations, the opposite of the syndrome predicted for truly isolated populations. Dempster (1991) noted an increase in thorax width in *Ma. arion* during a period of population expansion, a reversal of the trend toward small thoraxes in isolated populations. Further evidence comes from common-environment rearings of colonizing and noncolonizing metapopulations of two other butterfly species. Human activities have generated empty habitat networks for the silver-spotted skipper butterfly, *Hesperia comma* (Thomas and Jones 1993). This butterfly is restricted to fragmented grasslands on calcareous soils in Britain, and it requires very short and sparse vegetation. *H. comma* declined steeply when humans introduced the disease myxomatosis, which killed virtually all of the rabbits that had been responsible for maintaining the short turf favored by this butterfly. As this habitat rapidly became overgrown, the skipper became rare and localized. More recently, both the recovery of rabbits from myxomatosis and climate warming (Thomas *et al.* 2001a) has restored fragmented habitat networks for *H. comma*, which is gradually recolonizing them. A metapopulation in East Sussex, characterized by large thoraxes and small abdomens, is rapidly expanding its distribution into an empty habitat network, whereas insects from a stable metapopulation elsewhere in southern England have relatively small thoraxes and large abdomens (Hill *et al.* 1999a).

The same pattern is seen in expanding populations of the woodland butterfly *Pararge aegeria*. This species is expanding its northern margin in Britain, almost certainly in response to anthropogenic climate warming (Hill *et al.* 1999b). The expansion is taking place in agricultural landscapes that contain scattered woodlands, which provide ample opportunity for repeated founder events and selection for increased dispersal rate. As predicted, the expanding northern population was found to possess larger thoraxes and smaller abdomens than a more stable population further south (Hill *et al.* 1999c, Table 2). Interestingly, the difference was much greater in females (the colonizing sex) than in males (Hill *et al.* 1999c) – male morphology is probably largely determined by mate-location strategy and thermoregulation (van Dyck *et al.* 1997).

The above examples illustrate the complexity of potential responses to landscape pattern, which result in both increased and decreased migration rates depending on the exact changes that take place in the landscape (Heino and Hanski 2001) and on the temporal and spatial scales observed. The impacts of other traits and selection pressures on migration rate (mating strategy, predator evasion, etc.) further complicate the consequences of habitat fragmentation. Nonetheless, these examples do indicate that evolutionary changes in migration rates and in life-history traits associated with dispersal are likely to be extremely common in modern landscapes. How often this is of practical concern is another matter. In any case, the average dispersal traits of organisms that survive in modern, fragmented landscapes are likely to be different from those of the original fauna and flora (Thomas 2000).

15.3 Fragmentation, Migration, and Local Adaptation

Habitat fragmentation has the potential to disrupt any former balance between gene flow and local selection. Local adaptations might initially be expected to increase following habitat fragmentation and isolation, because locally adapted traits will no longer be diluted by gene flow from populations adapted to other habitats (Dhondt *et al.* 1990; Storfer 1999). However, in the long term successful colonists may be favored at the metapopulation level, a situation analogous to that already described for the evolution of migration rates. Selection within individual populations may favor local specialists, but generalists may be more successful colonists (Thomas *et al.* 2001a). Evolutionary changes may also take place as a result of the changing qualities of habitat remnants, caused by disturbance, or because of changes in the identities of other species that inhabit the same fragments. Some species may develop adaptations to anthropogenic habitats beyond the natural vegetation fragments, but continue to exchange genes with populations that still reside in nearby natural habitats. In the extreme, this could disrupt local adaptations to natural habitats.

Adaptation to local hosts

An example of the reciprocal influence of extinction–colonization dynamics in a highly fragmented landscape and the evolution of host-plant preference in the butterfly *M. cinxia* is provided by the recent study of Hanski and Singer (2001). *M. cinxia* has two larval host plants in the Åland Islands, SW Finland, *Plantago lanceolata* and *Veronica spicata*. A distinct west–east gradient exists in the relative abundances of the two host plants, a gradient that runs parallel to a gradient in the genetically determined preference of female butterflies for the different host plants within a distance of 30 km (Kuussaari *et al.* 2000). Such small-scale adaptation to the regionally more abundant host plant raises an interesting question about metapopulation dynamics. Assume that butterflies in a particular region have evolved a preference for *V. spicata* (or vice versa). The proportion of host species varies from one habitat patch to another and, although most patches in a particular region are dominated by *V. spicata*, some patches have only *Po. lanceolata*. Given the evolved preferences for the different host plants in female butterflies, will the proportions of different host plants in the habitat patches influence the metapopulation dynamics? The answer turns out to be conclusively yes. Habitat patches that are dominated by the host plant to which the metapopulation in a given region has adapted have a substantially higher rate of colonization than patches dominated by the regionally rare alternative host plant (Hanski and Singer 2001). The effect of the female preference for a particular host on colonization probability is likely caused by the effects of female preference on emigration and immigration in relation to the local proportions of host plants, rather than by spatial variation in larval performance.

The reverse question may also be asked; does the extinction–colonization dynamics influence the evolution of a preference for a particular host plant? Model and empirical results indicate such an effect, as the model-predicted average host-plant preference in a particular habitat patch network explains a significant amount of within-network variation in preference (Hanski and Heino, unpublished). Thus, in this example colonization probability depends upon a genetically determined trait of the colonizers, which is selected both at the level of local populations and at the level of the metapopulation. Other studies elucidate how local adaptation and gene flow promote genetic variation in metapopulations in general (Karban 1989; Wade 1990; Antonovics *et al.* 1994) and how local adaptation may be influenced by gene flow (Holt 1996; Dias 1996; Pulliam 1996). The example of *M. cinxia* suggests that gene flow and the establishment of new populations may also be influenced by local adaptation.

Adaptation to anthropogenic habitat change

One form of habitat change is represented by the arrival of a new species. When the European plant *Po. lanceolata* arrived in North America, it spread rapidly and became established in irrigated meadows along the eastern slopes of the Sierra Nevada mountains. The native checkerspot butterfly *Euphydryas editha* was already present in the area. The caterpillars of this butterfly fed naturally on *Collinsia parviflora*, a plant that shares certain iridoid glycosides with *Po. lanceolata*. Some females started to lay batches of eggs on *Po. lanceolata* (Thomas *et al.* 1987). Larval survival was higher on the introduced plant than on the native *C. parviflora*, and female choice of host plant had a genetic basis (Singer *et al.* 1988). Over a 10-year period the proportion of females that lay on *Po. lanceolata* was observed to increase, and resulted in the exclusion of *C. parviflora* from the diet in at least one population of *E. editha* (Singer *et al.* 1993). This is either a great bonanza for this rare butterfly – *Po. lanceolata* is widespread – or a serious risk. Within the study region, *Po. lanceolata* is restricted mostly to irrigated land. The newly evolved *Plantago*-feeding population might have doomed itself to extinction if irrigation had ceased to be economic. In fact, it became extinct when the meadow was converted into a golf course (M.C. Singer, personal communication).

The logging of native forests that are then allowed to regenerate is another globally widespread form of habitat modification. Logging in a different part of the Sierra Nevada mountains, in California, provided another new opportunity for another metapopulation of *E. editha*. This metapopulation was quite different from the one described above. The butterflies naturally laid their eggs on *Pedicularis semibarbata* (mainly) and *Castilleja disticha*, two species of Scrophulariaceae:

the population was restricted to natural granitic outcrops in an otherwise forested region (Singer 1983). Logging created major new openings in the forest, and rendered a third species of Scrophulariaceae, *Ca. torreyi*, suitable as a host plant (following the disturbance, the plant grew to be sufficiently large for the caterpillars to complete their life cycle on it). Huge populations of the butterfly established in these new clearings (Singer 1983; Thomas *et al.* 1996). Vast numbers of butterflies emerged in the logged habitat and flew back to the natural habitat, with a resultant intense competition in the natural habitat. This set up an asymmetric gene flow from the logged habitat to the natural habitat. Adaptations to the new environment, involving the acceptance of *Ca. torreyi* plants by females, started to develop. However, these incurred the cost of a reduced level of adaptation (a reduced preference for *Pedicularis* and *Castilleja*) in nearby populations that still inhabited the natural rocky outcrops (Singer *et al.* 1993; Singer and Thomas 1996; Thomas *et al.* 1996).

If new populations in an anthropogenic habitat risk extinction, any reduced levels of adaptation to undisturbed environments are potentially worrying. In the case described above, the forest clearings eventually become overgrown and populations were bound to become locally extinct. But something much more dramatic happened. An aseasonal summer frost killed all of the *Ca. torreyi* plants in the forest clearings, without harming either of the natural host plants in the rocky outcrops. *E. editha* larvae were resistant to the frost, but larvae in the forest clearings starved because the *Collinsia* had died, and a year later all the clearing populations of the butterfly had become extinct. Fortunately, the butterflies survived in their natural habitat. These two examples from *E. editha* populations illustrate how the evolutionary adoption of habitats that are created and maintained by humans both enables species to cope with human activities and, simultaneously, puts them at risk from future changes in land management. Such changes in management practices and agricultural economics may take place too quickly for the species to cope, a serious problem once the original adaptations to the native habitats have been lost (see the discussion of niche conservationism in Chapter 13).

15.4 The Example of *Centaurea* Species

Studies of the metapopulation dynamics of species that live in naturally highly fragmented habitats, as for *Centaurea corymbosa* (Asteraceae), can improve our understanding of the long-term consequences of habitat fragmentation.

Evolutionary trapping in *Ce. corymbosa*

The combined results of population genetic structure, achene dispersal distances, reproductive system, seed sets, and survival following experimental introductions allow a general view of the species' population biology: the unique metapopulation of *Ce. corymbosa* has much demographic and evolutionary inertia (see Box 15.4). Many suitable sites (cliffs) are available near the extant populations, as shown by the experimental introductions, but they remain empty because of the very low colonization ability. Achene dispersal by gravity and wind is restricted to a few tens of centimeters from the mother plant, maybe because of the landscape structure. The unsuitable habitat surrounding the existing populations selects against dispersal (Olivieri *et al.* 1995; Cody and Overton 1996): any achenes that leave the cliff almost certainly settle in vegetation within which they are unable to grow and/or compete successfully. One species of ant (*Crematogaster scutellaris*) disperses some *Ce. corymbosa* seeds, but probably no farther than several meters; in any case, they cannot take a seed from one cliff to another. In addition to low dispersal, Colas *et al.* (1997) suggested that semelparity and self-incompatibility reinforce the low colonization ability. Indeed, reproductive success of an individual plant mainly depends on its neighbors, and the closer they are, the greater the probability that they will set seeds (Colas *et al.* 2001). If, by chance, a few seeds disperse longer distances to a suitable cliff, germinate, and develop in to flowering plants, they are likely to die without giving rise to any surviving progeny unless they flower in the same year and are compatible.

Centaurea
Centaurea corymbosa

Some convincing examples show that bottlenecks may induce the dissolution of systems (heterostyly, self-incompatibility) of selfing-avoidance in plants (Reinartz and Les 1994; Eckert *et al.* 1996; Barrett 1998). It also seems clear that founder events might promote the evolution of dispersal (Cody and Overton 1996). However, in the absence of colonization, it seems unlikely that these traits will evolve

Box 15.4 Natural history and genetic structure of *Centaurea corymbosa*

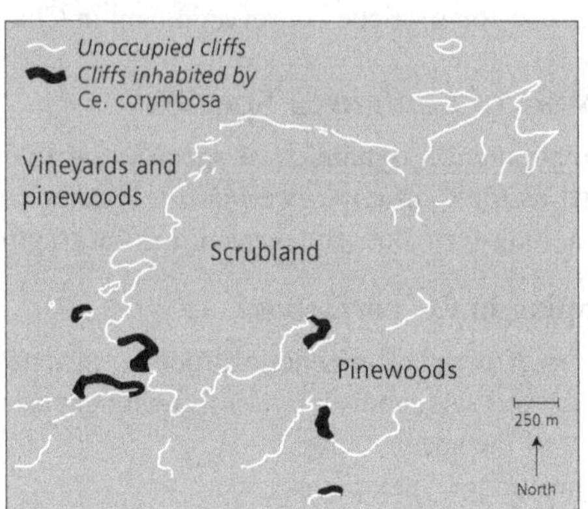

Ce. corymbosa is a self-incompatible, monocarpic perennial plant, endemic to a tiny 3 km^2 area situated within a 50 km^2 limestone plateau (Massif de la Clape) in southern France (Colas *et al.* 1996). *Ce. corymbosa* lives in clefts of rocks on the top of cliffs with very open vegetation and almost no soil. Apparently, it cannot stand competition: no plant can be found either on the plateau, in the middle of the scrubland, or in the pinewood, down in the depressions. Over all the six extant populations, 350 to 650 flowering plants were counted per year from 1994 to 2000 (from five to 250 per year per population). Each population is geographically clearly delimited.

Although populations are only 0.3 to 2.3 km apart, gene flow is very restricted among them, as evidenced by the very strong genetic differentiation observed on allozymes (Colas *et al.* 1997) and on microsatellites (Fréville *et al.* 2001). The correlation between geographic and genetic distances among the populations suggests that gene flow occurs mainly between adjacent populations. The population genetic structure for quantitative traits was similar to the genetic structure for molecular markers (Petit *et al.* 2001). This suggests the absence of heterogeneous selection among populations (see Bonnin *et al.* 1996 and Chapter 12 for a discussion of the relationship between neutral and selected characters in methods for measuring population genetic differentiation).

Although populations may stretch out along 670 m, each population can be considered as a panmictic unit (no departure from Hardy–Weinberg expectations). This is the consequence of pollen dispersal within populations in which about 20% of the mating pairs are separated by more than 50 m, as shown by paternity analyses (Hardy *et al.*, unpublished). However, the amount of pollen dispersed to long distances is not sufficient to prevent the lower fertilization rate of ovules of isolated flowering plants compared to plants situated in dense patches (Colas *et al.* 1997, 2001). Pollen flow seems to be limited mainly by gaps of unoccupied pieces of land between populations.

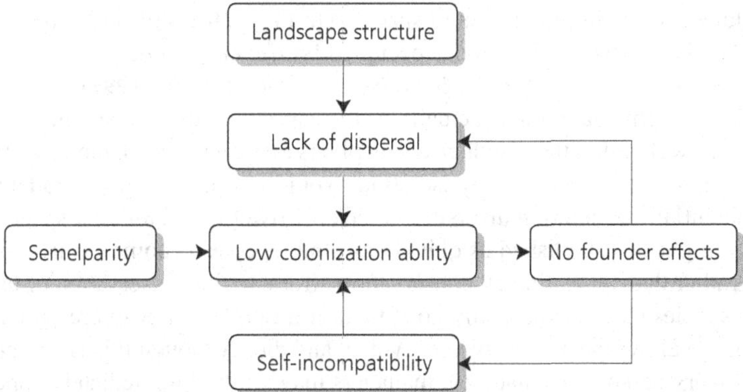

Figure 15.1 Factors that select against the colonization ability of *Ce. corymbosa*.

in *Ce. corymbosa*. Therefore, this species seems to be held within a vicious circle (Figure 15.1), an evolutionary spiral that prevent colonization and might lead to extinction. However, on the basis of historical data (Colas *et al.* 1997), five populations have been known from 100 to more than 200 years ago, and none have disappeared. Since the mean generation time is about 5.5 years (Colas *et al.*, unpublished), this represents 20 to 40 generations. The populations are very small, but the extinction rate may be no greater than the rate of colonization of new cliffs in this quite stable habitat. Thus, *Ce. corymbosa* could be well adapted to its fragmented habitat.

Fragmentation and *Centaurea* diversification

Ce. corymbosa belongs to the *Maculosa* group, which comprises 23 taxa (species and subspecies), of which most are narrowly endemic (Tutin *et al.* 1976). Many taxonomic groups within the huge *Centaurea* genus [500 species with many subspecies around the Mediterranean (Bremer 1994)] show this pattern: one widespread and several narrowly distributed endemics. Fréville *et al.* (1998) compared allozyme diversity in *Ce. corymbosa* and two related taxa that shared the same type of habitat in other places of southern France: the widespread *Ce. maculosa maculosa* and *Ce. maculosa albida*, which occurs in only one population of about 200 flowering individuals. Diversity of the two endemic taxa represented a sample of the diversity of the widespread one, from which they are probably derived. As in *Ce. corymbosa*, high allozymic differentiation was found among populations of *Ce. maculosa* spp., but the amount of differentiation among populations of different taxa was not different from that among populations within taxa. Fréville *et al.* (1998) suggest that ecological specialization following bottlenecks or founder effects associated with colonization events in *Ce. maculosa maculosa* resulted in the establishment of new taxa. Although probably very rare, long-distance achene dispersal of these species may occur by adhesion of the barbed and sticky bristles in the pappus to sheep or to the feathers of birds. In the Massif de la Clape, several cliffs rocks occupied by *Ce. corymbosa* are also those on which pellets from the eagle owl *Bubo bubo* are found (Colas 1997).

Habitat fragmentation may actually be the cause of wide diversity in *Centaurea* species, as well as in other Mediterranean plants (see review by Thompson 1999a). Some colonization, followed by isolation, would permit divergence (adaptive or not), potentially generating diversity. In the cliff-dwelling *Centaurea* species, new species might be established as quickly as other species become extinct. In such cases, rather than allocating all conservation efforts to the conservation of one particular species that may naturally go extinct, it might be better to preserve habitat diversity to allow species to diverge, evolve, and die. Although this is a long-term, evolutionary perspective that local managers may not find immediately appealing, the role that fragmentation may play in both the generation of diversity and the decline of genetic diversity should be considered seriously.

Interestingly, it appears from these studies on *Centaurea* species that the definition of a metapopulation depends on the scales of space and time considered. *Ce. corymbosa* may be considered, as stated above, a unique metapopulation of six panmictic populations among which restricted (medium distance) gene flow occurs with very slow extinction–colonization dynamics (probably observable over tens to hundreds of generations). However, it appears from a demographic study, using 40 permanent quadrats on rocks since 1994, that to understand the persistence of each population requires the processes of migration, extinction, and recolonization among quadrats (i.e., patches of plants) to be considered (Colas *et al.*, unpublished). Every patch may persist for one to several generations, and achene migration between adjacent suitable rocks (a few meters apart) occurs by gravity, wind, or ant dispersal. Thus, on a time scale of a few generations, each of the six panmictic units may be considered as an isolated metapopulation itself. Each has extinction–colonization processes that occur among patches because of the effects of environmental and demographic stochasticities and ordinary dispersal over short distances (Colas *et al.*, unpublished). Also, the *Maculosa* group of *Centaurea* species might be considered as a kind of meta-species with rare long-distance dispersal events that result in isolated populations, which may differentiate from each other through drift and/or heterogeneous selection. Extinction–colonization processes on this large scale in space and time are closely related to evolutionary processes of speciation and to species' extinction.

15.5 Concluding Comments

Almost all habitats have changed dramatically over recent decades and centuries, throughout the world. Changes in directional selection, new evolutionary optima, and changes in the balance between gene flow and selection must be widespread phenomena. The example of *Centaurea* species described in Section 15.4 suggests that natural fragmentation is probably an important cause of species diversity. However, this is a long-term evolutionary consequence of stable landscape heterogeneity. As suggested in Section 15.3 using the butterfly examples, recent changes in land use might be too fast for any evolutionary processes to occur. A key factor in future studies of the effects of fragmentation will probably be whether fragmentation occurs progressively or suddenly, and whether either allows adaptive evolution or not.

The examples described in this chapter illustrate some evolutionary consequences of these changes. The response of migration rates to fragmentation is idiosyncratic, and depends on how the local extinction or fitness of migrants varies with patch size. Local adaptations have mixed effects on the metapopulation viability, as they might reduce colonization abilities, as well as migration, in particular between different habitats. The response may be inefficient because of the many constraints that could prevent the metapopulation from responding optimally to landscape changes, as the interaction between local adaptations and migration in butterflies show. A common feature of most relevant studies is the large number of processes involved. Further understanding of the evolutionary responses to fragmentation in terms of inbreeding, local adaptation, or migration is necessary to predict the effect that fragmentation can have on the different species present in a landscape.

It is probably utopian to think that we can determine the critical thresholds for habitat fragmentation above which most of the biodiversity in a landscape can be conserved:

- First, the relevant scale at which habitat fragmentation becomes important depends on the size of the home range of the species, which varies from a few square centimeters to several square kilometers (Mönkkönen and Reunanen 1999).
- Second, the minimum patch areas for viable local populations and the maximum isolation distances that allow colonization are species specific, since they depend on many life-history traits.
- Third, a landscape does not always contain suitable and unsuitable patches, and intermediate habitats may serve as corridors for migration to occur (e.g., Kirchner *et al.* 2003). Levin (1995) illustrated the role of plant outliers for gene flow among populations. Depending on the habitat specificity of the species, the matrix surrounding a particular patch may be usable or completely hostile and impossible to cross.

A better understanding of the ecological, genetic, and evolutionary issues related to the thresholds of long-term persistence in fragmented landscapes remains one of the major challenges in conservation biology. This is not only because of the complexity of these thresholds, but also because of their relevance to a large portion of the biodiversity that remains.

Acknowledgments We are particularly thankful to Jane Hill. Both Ilkka Hanski and Chris D. Thomas were supported by the TMR FRAGLAND network. Thanks also to I. Olivieri and H. Fréville for their comments.

Part E

Community Structure

Introduction to Part E

"No man is an island", and no species of conservation interest exists in isolation from others. Ecosystems define the ecological theater, not only for any evolutionary play, but also for unfolding tales of population decline or rescue. This book therefore ends with a part on the community-level aspects of evolutionary conservation biology. In this way, we establish several additional perspectives to our general question on the ecological, demographic, and genetic conditions that enable or hinder populations to overcome extinction threats through adaptation.

On the ecological side, many density-dependent and frequency-dependent selection pressures emerge from interspecific interactions. It is the dependence of focal populations on the abundance of their resources, preys, predators, mutualistic partners, and competitors that creates complex webs of fitness effects and thus determines the strength and direction of natural selection. An interesting level on which to express such dependences is the flow of nutrients through ecological systems. Measures of nutrient cycling not only serve as indicators of ecosystem functioning, but also provide a convenient platform for resolving the interaction between organisms and their environment and for characterizing the implications of anthropogenic change.

On the demographic side, density regulation in communities is often nonlinear and can result in the coexistence of multiple demographic attractors. Anthropogenic change can, temporarily or permanently, tilt the established balance between regulating forces, and so lead to shifts of or even switches between demographic equilibria. Attractor switches are bound to bring about qualitatively new selection pressures and will often prove to be particularly resilient to conservation efforts directed at their reversal. Such complex dependences explain why ecosystems can react to altered conditions by displaying intricate cascades of ecological and evolutionary responses. Since such responses tend to act on different time scales, their study is critical to understand the expected ultimate impact of detrimental perturbations as well as conservation interventions.

By definition, genetic factors almost always act within species, not within communities. Exceptions can occur through interspecific gene flow. Hybridization is a prime example of such a process and can either exacerbate or ameliorate the perils of extinction experienced by small populations. However, because of the introgression of variant genetic material, hybridization may sometimes rescue only the demographic and ecological, but not the genetic, identity of challenged populations.

To consider issues of conservation and adaptation at the community level necessitates that the primary target of conservation efforts be clarified. Are we most interested in the conservation of life-history patterns, evolutionary lineages, networks of ecological interactions, current equilibrium states of ecosystems, their diversity, or their ecological function? Different targets mandate different types of interventions.

It is clear that conservation biology, and especially evolutionary conservation biology at the community level, is stretching current empirical and theoretical knowledge beyond its limits. The contributions to this part must therefore be appreciated as attempts to push these limits forward. The analyses presented are, of necessity, more tentative and hypothetical than those offered in the earlier parts of this volume.

In Chapter 16, Bronstein, Dieckmann, and Ferrière point out why a single-species focus may often be too narrow when investigating ecological and evolutionary responses to extinction threats. This is especially obvious for tightly coupled ecological associations, such as those involving mutualists. After a review of the range of anthropogenic threats to which mutualists are exposed today, the authors show how the reduction of an established species or the invasion of a non-native species can have dramatic repercussions for mutualistic partners. Expected responses range from benign resilience through the addition or replacement of mutualistic partners to the linked extinction of populations. The potential for such complex reaction patterns underlines that environmental threats can induce ecological and evolutionary effects that cascade through entire ecological communities.

Chapter 17 describes a framework for conducting analyses of adaptive responses from a whole-ecosystem perspective. Loreau, de Mazancourt, and Holt explain how to extend the "classic" view of environments being external to evolving populations and constant during their adaptation, by incorporating, first, organism–environment feedbacks for the operation of natural selection and, second, sufficient ecological resolution to describe such environments. One level on which to resolve evolutionarily relevant feedbacks is that of nutrient transport: the indirect ecological and evolutionary effects of nutrient cycling are sometimes so strong as to modify or even prevail over the selective pressures that arise from direct ecological interactions. Based on examples from plant–herbivore evolution and extending their findings to coevolution in other exploiter–victim systems, the authors argue that a comprehensive approach to evolutionary conservation biology will have to merge population-level and ecosystem-level perspectives to predict the responses of ecological systems to environmental degradation.

Turning from the ecological to the genetic implications of community-level interactions, Chapter 18 shows how the fate of endangered species can depend on hybridization with sexually compatible individuals of another species. Illustrated by a variety of empirical examples, Levin points out that the contact between an endangered species and its congener can lead to the rapid disappearance of the former. Hybrids may happen to be sterile or the congener itself may possess a higher fitness than the endangered species. From a conservation point of view, intermediate situations, in which hybrid derivatives are stabilized, have to be regarded as mixed successes, because the original genetic identity of the endangered species is sacrificed, an outcome that is most likely for insular species. At the opposite end of the spectrum, hybridization may also rescue an endangered species, by conferring the required local adaptations or by alleviating the deleterious effects of inbreeding.

Integration of insight at the community level will eventually enable conservation managers to interact, reliably and effectively, with our ecological environment. Years of inventive empirical and theoretical research still separate us from this ultimate goal.

16

Coevolutionary Dynamics and the Conservation of Mutualisms

Judith L. Bronstein, Ulf Dieckmann, and Régis Ferrière

16.1 Introduction

The vast majority of studies in conservation biology focus on a single species at a time. However, many of the anthropogenic threats that species face occur via disrupted or enhanced interactions with other organisms. According to one recent analysis, interactions with introduced species, such as predators, parasites, and pathogens, are the eighth leading cause of species endangerment worldwide; they are the primary cause of endangerment in Hawaii and Puerto Rico (Czech and Krausman 1997). Altering interactions not only has ecological effects, but also it can generate selective pressures and evolutionary responses, which may either favor or disfavor the evolutionary persistence of species and interactions. An increased focus on interspecific interactions will thus enlighten our efforts to conserve species and, more pointedly, our ability to understand when species will and will not respond evolutionarily to conservation threats. Such a focus is also critical for efforts to conserve communities as units, because interactions are the crucial and poorly understood link between threatened species and threatened species assemblages.

Different types of interspecific interactions are subject to, and generate, somewhat different ecological and evolutionary threats. Predator and pathogen introductions can lead to reduction, local exclusion, or extinction of native species (Savidge 1987; Schofield 1989; Kinzie 1992; Steadman 1995; Louda *et al.* 1997). Rapid evolution in the enemies and/or the victims may also result (Dwyer *et al.* 1990; Singer and Thomas 1996; Carroll *et al.* 1998). Conversely, the disappearance of enemies (or the introduction of a species into a habitat that lacks enemies) can have consequences that extend across the population, community, and ecosystem (Thompson 1996; Fritts and Rodda 1998). The effects of altering competitive interactions appear to be qualitatively similar, although smaller in magnitude (Simberloff 1981; Williamson 1996). Introducing competitors can reduce populations of native species, with the possible effects being local exclusion, extinction, or evolutionary change of one or both species (Schofield 1989; Moulton 1993; Cohen 1994; Dayan and Simberloff 1994).

Antagonistic interactions have been relatively well studied from the evolutionary, ecological, and conservation perspectives. In contrast, our understanding of mutualisms – interactions that are mutually beneficial to both species (Box 16.1) – is at a much earlier stage of development (Bronstein 1994, 2001a). The ecological

Box 16.1 Mutualistic interactions

Mutualisms are interspecific interactions in which each of two partner species receives a net benefit. Well-known examples include interactions between plants and mycorrhizal fungi, plants and pollinators, animals and gut bacteria, and corals and zooxanthellae (Herre *et al.* 1999; Bronstein 2001a). Mutualisms generally involve the exchange of commodities in a "biological market": each species trades a commodity to which it has ready access for a commodity that is difficult or impossible for it to acquire (Noë and Hammerstein 1995; see also Douglas 1994). For instance, plants provide carbon to their mycorrhizal fungi in return for phosphorus, and plants provide nectar to many animals in return for pollen transport. Although a great deal is known about the natural history of diverse mutualisms, relatively little effort has yet been invested in the study of ecological and evolutionary similarities among them (Bronstein 1994). This is particularly surprising in light of their perceived importance in nature. All organisms are currently believed to associate with mutualistic species at some point in their lives. Furthermore, mutualisms are thought to lie at the core of major transitions in the history of life, including the origin of the eukaryotic cell and the invasion of land.

To understand mutualism in an evolutionary conservation context, it is important to distinguish it from related phenomena with which it is often confused. Mutualism is an association between different species; it involves somewhat different evolutionary forces and poses different conservation challenges than does cooperation within species (Dugatkin 1997). Not all mutualisms are symbioses (intimate physical associations; Douglas 1994); many involve free-living organisms that associate for only part of their lives. Free-living organisms are likely to be vulnerable to somewhat different anthropogenic threats, which raises the interesting problem of how these mutualisms persist when one, but not both, of the partners is at risk. Conversely, not all symbioses are mutualistic. Hence, this chapter does not consider how anthropogenic change might affect the evolution of diseases (which are antagonistic symbioses). Finally, not all mutualisms have long evolutionary or coevolutionary histories. For instance, pairs of invasive species can sometimes form highly successful mutualisms (Simberloff and von Holle 1999). Evolution may well occur after the association has formed, however (Thompson 1994). Such evolution can change the specificity of the interaction (from more specialized to more generalized, or vice versa), as well as its outcome (from mutualistic to antagonistic, or vice versa).

The large majority of mutualisms are rather generalized: each species can obtain the commodities it requires from a wide range of partner species (Waser *et al.* 1996; Richardson *et al.* 2000). Furthermore, many mutualisms are facultative, in the sense that at least some of the commodity can be obtained from abiotic sources. However, many extremely specialized mutualisms do exist: they are species-specific (i.e., there is only a single mutualist species that can provide the necessary commodity), and may be obligate as well (i.e., individuals cannot survive or reproduce in the absence of mutualists). Box 16.3 provides details of one such specialized mutualism. Note that the degree of specificity is not necessarily symmetrical within a mutualism. For instance, many orchid species can be pollinated by a single species of orchid bee, whereas these bee species visit many different orchids, as well as other plants (Nilsson 1992). The evolutionary flexibilities that result from these asymmetries in specificity remain almost entirely unexplored.

effects of disrupting mutualism are known from only a handful of case studies, which have largely involved a single form of mutualism, plant–pollinator interactions (see the excellent reviews by Bond 1994, 1995; Allen-Wardell *et al.* 1998; Kearns *et al.* 1998). The evolutionary consequences of such disruptions remain virtually unexplored. This gap in knowledge is of particular concern because mutualisms are now believed to be a focus around which diversity accumulates, on both ecological and evolutionary time scales (e.g., Dodd *et al.* 1999; Wall and Moore 1999; Bernhard *et al.* 2000; Smith 2001).

We begin this chapter with a discussion of processes that foster the ecological and evolutionary persistence of mutualisms. We go on to discuss the sequence of events that can endanger species that depend on mutualists, in the context of some prominent forms of anthropogenic change. With this background, we outline three scenarios for the possible outcomes when the mutualists of a species of interest become rare – linked extinction, ecological resilience, and evolutionary response – and distinguish the likelihood of each outcome based on whether the mutualism is relatively specialized or generalized. As we show, simple evolutionary models can generate quite useful predictions relevant to the conservation of mutualisms and other species interactions. Furthermore, we show that modeling pairwise associations can form an excellent first step toward addressing the fascinating, but much less tractable, problem of coevolution at the community scale.

16.2 Factors that Influence the Persistence of Mutualisms

The persistence of mutualisms has long been a puzzle. From the ecological perspective, the positive feedback inherent to mutualisms led May (1976) to characterize mutualisms as an unstable "orgy of mutual benefaction". Yet, at the same time, dependence on mutualists also raises the likelihood of Allee effects (see Chapter 2), in which the low abundance of one species can doom its partner to extinction. From the evolutionary perspective, the major threat to mutualism is the apparent selective advantage that accrues to individuals who reap benefits from partner species without investment in costly commodities to exchange with them (Axelrod and Hamilton 1981; Soberon Mainero and Martinez del Rio 1985; Bull and Rice 1991; Bronstein 2001b). Slight cheats that arise by mutation could gradually erode the mutualistic interaction, and lead to dissolution or reciprocal extinction (Roberts and Sherratt 1998; Doebeli and Knowlton 1998). Although cheating has been assumed to be under strict control, recent empirical findings (reviewed by Bronstein 2001b) indicate that cheating is rampant in most mutualisms; in some cases, cheaters have been associated with mutualisms over long spans of evolutionary time (Després and Jaeger 1999; Pellmyr and Leebens-Mack 1999; Lopez-Vaamonde *et al.* 2001). Recent theoretical advances have increased our understanding of the ecological and evolutionary persistence of particular forms of mutualism (e.g., Holland and DeAngelis 2001; Law *et al.* 2001; Yu 2001; Holland *et al.* 2002; Morris *et al.* 2003; Wilson *et al.* 2003).

Below, we introduce and discuss a simple general model to describe the ecological and evolutionary dynamics of a two-species, obligate mutualism in a constant

Box 16.2 Ecology and evolution of specialized mutualisms: a simple model

We describe the obligate and specialized mutualistic interaction between species X (density N_X) and species Y (density N_Y) by a simple Lotka–Volterra model,

$$\frac{dN_X}{dt} = [-r_X(x) - c_X N_X + y N_Y (1 - \alpha N_X)] N_X \,, \tag{a}$$

$$\frac{dN_Y}{dt} = [-r_Y(y) - c_Y N_Y + x N_X (1 - \beta N_Y)] N_Y \,. \tag{b}$$

The mutualistic traits x and y are measured as per capita rates of commodities traded (a visitation rate by a pollinator, for example); thus, xN_X and yN_Y represent the probabilities per unit time that a partner individual receives benefit from a mutualistic interaction. Intraspecific competition for commodities provided by the partner species is expressed by the linear density-dependent factors $(1 - \alpha N_X)$ and $(1 - \beta N_Y)$, as in Wolin (1985). The terms $-c_X N_X$ and $-c_Y N_Y$ measure the detrimental effect of intraspecific competition on other resources. The mutualism being obligate, the intrinsic growth rates $-r_X(x)$ and $-r_Y(y)$ are negative, and $r_X(x)$ and $r_Y(y)$ increase with x and y, respectively, to reflect the costs of mutualism.

Ecological dynamics. A standard analysis of the thus defined ecological model shows that the situation in which both species are extinct, $N_X = 0$ and $N_Y = 0$, is always a locally stable equilibrium. Depending on the trait values x and y, two inner equilibria may also exist in the positive orthant, one being stable (a node) and the other being unstable (a saddle). The transition between the two cases (zero or two equilibria in the positive orthant) is caused by a saddle–node bifurcation. The corresponding bifurcation curve is the closed, ovoid curve depicted in Figures 16.1a to 16.1c, which separates a region of trait values that lead to extinction from the domain of traits that correspond to viable ecological equilibria.

A mathematical approximation of mutation–selection processes. By assuming that ecological and evolution processes operate on different time scales and that evolution proceeds through the fixation of rare mutational innovations, the rates of change of traits x and y on the evolutionary time scale are given by (Dieckmann and Law 1996)

$$\frac{dx}{dt} = \varepsilon_X N_X^* \left.\frac{\partial f_x}{\partial x'}\right|_{x'=x} \,, \tag{c}$$

$$\frac{dy}{dt} = \varepsilon_Y N_Y^* \left.\frac{\partial f_y}{\partial y'}\right|_{y'=y} \,. \tag{d}$$

Parameters ε_X and ε_Y denote evolutionary rates that depend on the mutation rate and mutation step variance (see Box 11.3 for further details); N_X^* and N_Y^* are the equilibrium population densities of resident phenotypes x and y (these factors occur because the likelihood of a mutation is proportional to the number of reproducing individuals); $f_X(x', x, y)$ and $f_Y(y', x, y)$ are the invasion fitnesses (defined as per capita rates of increase from initial rarity; Metz *et al.* 1992) of a mutant phenotype x' of species X and of a mutant phenotype y' of species Y in a resident association x, y.

continued

Box 16.2 *continued*

Evolutionary dynamics under symmetric versus asymmetric competition. Competition between two individuals is symmetric if the detrimental effect of their competitive interaction is the same on both individuals; otherwise, their competition is asymmetric. With symmetric competition, we have $\partial f_X = -r'_X(x)\partial x$ and $\partial f_Y = -r'_Y(y)\partial y$. Therefore, from any ancestral state, the process of mutation and selection causes the monotonic decrease of the traits x and y toward zero. Thus, all evolutionary trajectories eventually hit the boundary of ecological viability. Asymmetric competition between two phenotypes of species X that provide commodities at different rates is modeled by replacing the constant competition coefficient α with a sigmoid function of the difference in the rate of commodity provision (Matsuda and Abrams 1994c; Law *et al.* 1997; Kisdi 1999). With such a function, a large positive difference implies that α approaches its minimum value, whereas a large negative difference results in a value of α close to its maximum. The absolute value of the slope of this function at zero difference then provides a measure of the degree of competitive asymmetry. Likewise, we can define an asymmetric competition function β for species Y. The first-order effect on fitness induced by a small difference ∂x in the rate of commodity provision is then equal to $\partial f_X = [-r'_X(x) + \alpha' y N^*_X N^*_Y]\partial x$, where N^*_X and N^*_Y are the population equilibria that are solutions of $-r_X(x) - c_X N_X + y N_Y[1 - \alpha(0) N_X] = 0$ and $-r_Y(y) - c_Y N_Y + x N_X[1 - \beta(0) N_Y] = 0$, and $\alpha' = |\alpha'(0)|$ is the degree of competitive asymmetry. Likewise, we obtain $\partial f_Y = [-r'_Y(y) + \beta' x N^*_X N^*_Y]\partial y$, with $\beta' = |\beta'(0)|$. The intersection point of the isoclines $\partial f_X/\partial x = 0$ and $\partial f_Y/\partial y = 0$ defines a so-called evolutionary singularity (Geritz *et al.* 1997; Chapter 11). To investigate the existence and stability of this point, we performed an extensive numerical bifurcation analysis with respect to the degrees of asymmetry α' and β'; these parameters have the convenient property that they do not influence the ovoid domain of traits (x, y) that ensure ecological persistence. In general, there is a wing-shaped region of parameters α' and β' in which the evolutionary singularity exists as a stable node within this domain (see gray area in Figure 16.1d). Interestingly, the effect of changing the evolutionary rates ε_X and ε_Y is confined to the "tips" of this wing-shaped region – neither the front edge nor the back edge is affected by these parameters, whereas increasing (decreasing) the ratio $\varepsilon_X/\varepsilon_Y$ shifts the tips toward the upper left (lower right).

environment, first proposed and analyzed by Ferrière *et al.* (2002). Details of the model are presented in Box 16.2.

Ecological persistence

The ecological component of the model extends standard Lotka–Volterra equations for mutualisms. Each mutualistic species is characterized by:

- Its intrinsic growth rate;
- The rate at which it provides commodities to partners (e.g., services such as pollination and rewards such as nectar, see Box 16.1);

- Parameters that measure the strength of intraspecific competition for the commodities that partners provide in return, as well as for other resources.

The direct cost of producing commodities impacts the intrinsic growth rate of each species, an effect modeled by discounting a baseline intrinsic growth rate by a cost function for a specific commodity.

The model predicts that the ecological persistence of a mutualism is determined by three types of factors (Box 16.2):

- Individual life-history traits: the baseline intrinsic growth rates and the shape of the commodity cost functions.
- Interaction traits: the specific rates of commodity provision, and the strength of intraspecific competition for commodities provided by partners and for other resources.
- Species abundance: an Allee effect occurs that results in thresholds on each species' population size below which mutualism cannot persist.

Individual and interaction traits combine in a complex manner to determine the ecological viability of mutualisms and the minimum thresholds that each population size must exceed for the association to persist. Yet, in general, for fixed individual and competition parameters, ecological viability is achieved provided the rates of commodity provision are neither extremely low nor too high. At the boundary of the set of commodity provision rates that permit ecological persistence, the system undergoes a catastrophic bifurcation and collapses abruptly.

Evolutionary persistence

The model described here (Ferrière *et al.* 2002) provides a general explanation for the evolutionary origin of cheaters and the unexpected stability of mutualistic associations in which cheating occurs. To identify factors that promote the evolutionary persistence of mutualism, we incorporate an evolutionary dimension within the ecological model by assuming that the partners' rates of commodity provision can be subject to rare mutation. The resultant coevolutionary dynamics follow the selection gradients generated by the underlying ecological dynamics (Box 16.2; Hofbauer and Sigmund 1990; Abrams *et al.* 1993; Dieckmann and Law 1996; Chapter 11), and can have a dramatic impact on the long-term persistence of the association. If individuals compete with equal success for the commodity provided by the partner species, regardless of how much those competing individuals invest in mutualism (symmetric competition), long-term evolutionary dynamics will always drive the association toward the boundary of the ecologically viable region of the trait space, irrespective of the ancestral state; this results in evolutionary suicide (Chapter 11). The mutualism erodes because cheating mutants that invest less in mutualism are under no competitive disadvantage and thus are always able to invade, which ultimately drives the partner species to extinction. However, as a rule, competition in nature is asymmetric (Brooks and Dodson 1965; Lawton 1981; Karban 1986; Callaway and Walker 1997). Clearly, if any competitive asymmetry within either species gives an advantage to individuals that provide fewer

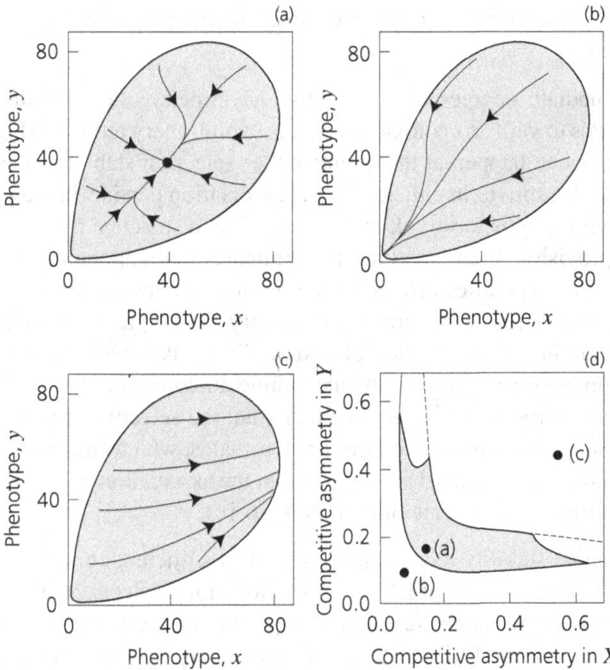

Figure 16.1 Competitive asymmetry and the evolutionary persistence of mutualism. The ovoid domain delineates the adaptive rates x and y of commodity provision by each species that make the mutualistic association ecologically viable. Each oriented curve depicts an evolutionary trajectory that starts from a different ancestral phenotypic state. (a) Convergence toward an evolutionary attractor that is ecologically viable (filled circle). Specific degrees of competitive asymmetry are $\alpha' = 0.035$ and $\beta' = 0.035$. (b) Evolutionary suicide through selection of ever-reduced mutualistic investments ($\alpha' = 0.01$ and $\beta' = 0.01$). (c) Evolutionary suicide by runaway selection for ever-increased mutualistic investments ($\alpha' = 0.20$ and $\beta' = 0.40$). (d) Dependence of the adaptive dynamics regime on the degrees of competitive asymmetry in species X and Y, as measured, respectively, by $\sqrt{\alpha'}$ (horizontal axis) and $\sqrt{\beta'}$ (vertical axis). The gray area shows the convergence to an evolutionary attractor that is ecologically viable; the blank area shows evolutionary suicide. The curves (continuous and dashed) that delineate the wing-shaped gray area are bifurcation curves obtained from the numerical analysis of Equations (c) and (d) in Box 16.2 (see Ferrière *et al.* 2002 for further details). Points (a), (b), and (c) correspond to the panels (a), (b), and (c). Parameters: $r_X(x) = 0.01(x + x^2)$, $r_Y(y) = 0.01(y + y^2)$, $c_X = 1$, $c_Y = 2$. *Source*: Ferrière *et al.* (2002).

commodities, the evolutionary suicide described above would be unavoidable. By contrast, individuals often discriminate among partners according to the quantity of rewards they provide, and associate differentially with higher-reward producers (e.g., Bull and Rice 1991; Christensen *et al.* 1991; Mitchell 1994; Anstett *et al.* 1998). Such a competitive premium, in effect, generates a selective force that can counter the pressure to reduce the provision of commodities.

Three outcomes are then possible (Figure 16.1d), depending on the strength of the asymmetry:

- At intermediate degrees of competitive asymmetry, the mutualistic association evolves toward an ecologically viable evolutionary attractor (Figure 16.1a). Two things can happen at this point: either selection stabilizes the mutualism or it turns disruptive. In either case, the association persists in the long term.
- If the asymmetry is too weak in either species, a selective pressure that favors a lower provision of commodities predominates in that population. As the total amount of commodities offered to the partner species decreases, the selective pressure induced by competitive asymmetry in the partner weakens, and selection to reduce the provision of commodities takes over on that side of the interaction also. Extinction is the inexorable outcome (Figure 16.1b).
- If the asymmetry is too strong on either side, the selective pressure that favors the provision of more commodities predominates, which causes runaway selection until the costs incurred are so large that the association becomes non-viable. Again, extinction is the outcome (Figure 16.1c).

Thus, ecological stability alone cannot provide a sufficient condition for the evolutionary persistence of a mutualism subject to natural selection. According to the analysis above, evolutionary suicide is expected to be a general property of mutualisms that involve too little or too much asymmetry in intraspecific competition for commodities provided by partners.

16.3 Anthropogenic Threats to Mutualisms

The ability of mutualisms to persist both on a short-term ecological time scale and on a longer-term evolutionary time scale, therefore, is closely related to the partners' life histories, behaviors, and abundances. Hence, any form of anthropogenic change that impacts these variables will threaten mutualisms. Below we address the known effects on mutualisms of two of the most serious anthropogenic threats, habitat fragmentation and biological invasions.

Habitat fragmentation

One of the more striking effects of human land use, and one that has increased dramatically in recent decades, is habitat fragmentation. Fragmentation, a phenomenon treated in depth in Part D, creates small populations from large ones by weakening or severing their linkage through dispersal. At the species level, problems caused by habitat fragmentation include increases in genetic drift, inbreeding depression, and demographic stochasticity (Chapter 4). As discussed in Chapter 14, diverse adaptive responses to fragmentation can be expected. Habitat fragmentation is of major concern beyond the species level also, since organisms can experience the effects of fragmentation indirectly, via its effects on the species with which they interact either positively or negatively.

Habitat fragmentation can impact all the factors that promote the ecological and evolutionary persistence of mutualisms. Reductions in the population size of one species caused by fragmentation can lead to failure of their mutualists as well, with a resultant local ecological instability. Aizen and Feinsinger (1994), for example, documented that the loss of native bee pollinators from forest fragments in Argentina reduced the seed production of about 75% of plant species within those fragments; reproduction of some species ceased almost entirely. Habitat loss and edge effects may reduce habitat quality for mutualists, and thus mutualist population sizes as well (Jules and Rathcke 1999). Intrinsic life-history traits and behaviors of mutualists may also be disrupted by fragmentation. For instance, habitat patches may become so isolated that mobile species become unable or unwilling to travel between them (Goverde *et al.* 2002); this affects the degree to which they provide mutualistic services and potentially alters the mode and intensity of intraspecific competition for these services. Ultimately, persistent isolation of local populations caused by fragmentation may lead to evolutionary changes in life-history traits linked to mutualism (e.g., Washitani 1996), although many other outcomes are also possible (see Sections 16.4 and 16.5).

Biological invasions

If the loss of partners can raise a major ecological threat to mutualisms, the reverse phenomenon – the addition of new species – can be at least equally problematic. A useful rule of thumb is that roughly 10% of the introduced species become established and 10% of these become troublesome pests (Williamson and Brown 1986), commonly in the context of interspecific interactions in their new habitat.

Biological invasions pose a number of threats to mutualisms. Predatory, parasitic, and pathogenic invaders can greatly reduce native populations or alter their life-histories and behaviors, with strong ecological impacts on the mutualists of those natives. For example, the Argentine ant, a particularly successful invader worldwide, can decimate populations of ground-dwelling insects (Holway 1998). In Hawaii, these ants substantially reduce insect-pollinator abundance, with potentially disastrous consequences for the persistence of native plants (Cole *et al.* 1992). Invaders can sometimes outcompete and displace native mutualists, generally to the detriment of their partner. Bond and Slingsby (1984) documented how the Argentine ant replaced native ant species as the seed disperser of South African Proteaceae, which has led to a reduced seedling establishment. Its preference for seeds that bear relatively small food bodies (elaiosomes) has resulted in a shift in local plant communities toward dominance by species with seeds that contain the preferred rewards (Christian 2001). Perhaps the most important case of mutualist replacement is the honeybee, intentionally transported by humans worldwide, but often a rather poor pollinator compared to the native insects they displace competitively (Buchmann and Nabhan 1996; Kearns *et al.* 1998). Invaders may ultimately induce evolutionary modifications in interactions within and between

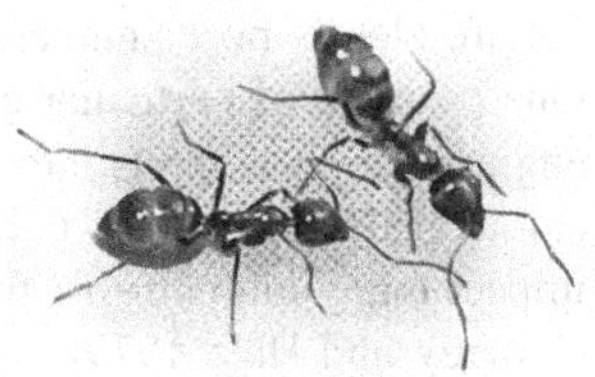

Argentine ant
Linepithema humile

species. Such effects have not yet been demonstrated for mutualisms, although they are well-documented for certain other kinds of interaction (e.g., Singer *et al.* 1993).

Not all introduced species have negative impacts, however. Certain invaders join native mutualist assemblages with no measurable negative effects on the residents, and probably some positive effects (Richardson *et al.* 2000). Furthermore, invaders can fill the gap created when a native mutualist has been driven to extinction, saving its partner from a similar fate. For example, an introduced opossum is now an effective pollinator of *Freycinetia baueriana*, a New Zealand liana that has lost its bat pollinator (Lord 1991). In the same vein, Janzen and Martin (1982) argued convincingly that numerous tree and shrub species in Central America still exhibit traits for seed dispersal by gomphotheres, large herbivorous mammals extirpated about 10 000 years ago, quite possibly through overhunting by humans (Martin and Klein 1984). Yet many of these plants thrive today, dispersed effectively by introduced livestock not too different ecologically from their extinct, coevolved dispersers.

Honeybee
Apis mellifera

Other anthropogenic threats

Other anthropogenic threats to mutualism are well known, but have been investigated less, so that their impact on factors that promote the ecological and evolutionary persistence of these interactions remains unclear. For example, agriculture clearly poses numerous problems for native plant–pollinator interactions. One problem of particular evolutionary interest is introgression from genetically engineered crop plants into related wild species (Snow and Palma 1997), which can alter the floral traits that attract pollinators (Lee and Stone 1998). Pollutants impact many mutualisms: the effects of automobile exhaust on lichen symbioses (Lawrey and Hale 1979), agrochemicals on pollinators (Buchmann and Nabhan 1996), and acid rain on endophytic fungi (Helander *et al.* 1996; Lappalainen *et al.* 1999) are particularly well documented. Finally, it has been recognized increasingly that global change impacts diverse species interactions (Kareiva *et al.* 1993). For example, elevated levels of CO_2 have both direct and indirect effects on mutualisms between plants and their root symbionts (Thomas *et al.* 1991; Diaz 1996; Staddon and Fitter 1998).

Which mutualisms are most at risk?

A major goal of conservation biology is to turn isolated case histories, like those summarized above, into testable predictions as to which species can be expected to be most vulnerable to anthropogenic change. One prediction has been cited repeatedly: organisms that are obligately dependent on a single species have the most to lose from the disruption of that mutualism. In contrast, organisms dependent on a broader array of species, or that succeed to some extent without mutualists at all, are believed to be somewhat buffered from the effects of such disruption. In the following two sections we consider first how specialized mutualisms, and then how more generalized mutualisms, are expected to respond to anthropogenic change.

16.4 Responses of Specialized Mutualisms to Threats

To summarize so far, we have seen that any force of anthropogenic change that drives down the numbers of one species can reduce indirectly the success of organisms dependent upon that species. What are the likely consequences for species with narrow or strict dependences on threatened species? (A classic example of a species-specific, obligate mutualism commonly thought to be at great risk from anthropogenic change is the interaction between fig trees and their fig-wasp pollinators, described in Box 16.3.) We consider three scenarios here:

- An ecological vortex in which both species dwindle to extinction;
- Ecological resiliency that buffers organisms from a fate similar to their mutualists;
- Evolutionary responses that rescue organisms from their mutualists' fate.

Rarity of coextinction?

We have already cited several examples in which anthropogenic impacts to one species have reduced indirectly the success of its mutualists. Interestingly, however, there is no documented case in which such joint failure has led to a linked extinction, at either the local or global scale. The case of the dodo and the tambalacoque tree is often cited in this context, but mistakenly. The dodo, a bird endemic to Mauritius, was hunted to extinction in the 1700s; this has supposedly driven to near extinction an endemic tree with seeds that could be dispersed by the dodo only (Temple 1977). However, more recent investigations show that over the past 300 years new individuals have been recruited into the tree population, which implicates another disperser or dispersers. Furthermore, morphological evidence suggests that the dodo was probably more of a seed predator than a mutualistic seed disperser (Witmer and Cheke 1991).

Mauritian dodo
Raphus cucullatus

What explains the apparent rarity of coextinction? We can offer three possibilities. First, it is perhaps only very recently that ecological conditions conducive to this phenomenon have appeared. This seems highly unlikely. Although the current biodiversity crisis is apparently generating a higher extinction rate than any previous mass extinction event (Wilson 1992), probably 99% of all species that have ever existed on Earth are now extinct, which

Box 16.3 Is the fig-pollinator interaction a threatened mutualism?

The mutualism between fig trees (about 750 *Ficus* species) and their pollinator wasps (family Agaonidae) has long served as a model for the intricate adaptations and extreme specialization that coevolution can produce. Most fig species are pollinated exclusively by a single species of fig wasp, which in turn is associated with a single fig species. The female wasps pollinate fig inflorescences, then deposit their eggs in certain flowers. Their offspring feed on the developing seeds. When the wasps are mature, they mate; the females collect pollen and then depart in search of an oviposition site. Trees within a fig population generally flower in tight within-tree synchrony, but out of synchrony with each other, which forces the wasps to depart their natal tree. Hence, figs sacrifice some proportion of their seeds to guarantee that their pollen will be dispersed effectively among individuals (Bronstein 1992; Anstett *et al.* 1997a; Weiblen 2002).

Figs are thought to be of exceptional conservation significance, yet subject to exceptional threats from anthropogenic change (McKey 1989; but see Compton and McCormack 1999). Their significance is linked to their unusual phenology. Since trees flower out of synchrony with each other, they also fruit out of synchrony; this provides one of the only year-round food sources for vertebrates in tropical habitats (Shanahan *et al.* 2001a). Their vulnerability, however, is an outcome of this same phenology. Fig wasps are tiny and fragile, and live a day or two at most as adults. During this brief period, they must transit large distances in search of their single mutualist species. Simulation models indicate that fairly high numbers of trees must be present within their flight range to give them any chance of locating a flowering individual (Bronstein *et al.* 1990; Anstett *et al.* 1995, 1997b). Habitat alterations that reduce their chances further – removal of fig trees, fragmentation of their habitats, pesticide spraying, etc. – are likely to lower the success of fig fruiting, with potentially disastrous consequences for vertebrate populations.

However, a number of very recent discoveries about this mutualism suggest that it exhibits more resilience than once thought:

- First, its level of specificity is lower than commonly believed: some figs have different pollinator species in different parts of their range, or even multiple pollinators at a single site (Rasplus 1996).
- Second, figs have remarkable adaptations to attract pollinators from long distances (Gibernau and Hossaert-McKey 1998), as well as adaptations that allow the inflorescences to persist for weeks while waiting for pollinators to arrive Khadari *et al.* 1995).
- Finally, fig wasps regularly travel far longer distances than they were once given credit for (Nason *et al.* 1998).

These traits help account for situations in which fig–pollinator mutualisms have been re-established rapidly after major disruptions (Bronstein and Hossaert-McKey 1995; but see Harrison 2000). They may also explain why, although species-specific pollination is certainly an important limit to range extension [since figs cannot occur where their pollinator is unable to persist or disperse (Kjellberg and Valdeyron 1990)], figs can also be surprisingly effective colonizing species (Shanahan *et al.* 2001b), as well as aggressive invaders in some habitats (McKey 1989).

indicates that the risk of coextinction is certainly not a new problem. It may well be a growing problem, however.

Second, and much more likely, an absence of evidence may not be evidence of absence: coextinction may actually occur, but be extremely difficult to detect. To determine the underlying cause of any extinction is problematic, in part because, as discussed in Part A of this book, many factors interact to doom populations once they are critically small. Also, the ability of paleontological data to reveal linked extinctions is limited: the fossil record rarely offers evidence as to why a given species has disappeared, and its temporal resolution is nearly always too crude to test an ecological hypothesis such as this. Our best hope to document coextinction may be to observe it in the field while it is happening, although if we see it, it is likely that we would attempt to *prevent* it.

The final possibility as to why coextinctions have not been documented is that mutualisms might be more resilient to change than we have given them credit for. The evolutionary past may have endowed mutualisms with some capacity to respond, ecologically or evolutionary, to current and future challenges – even in situations that, logically, we might expect would doom them. We wish to stress that we do not intend to trivialize the risk of coextinction. However, by investigating the kinds of mutualisms that seem to have evolved some ecological or adaptive resilience against coextinction, we can better focus our most intensive conservation efforts on those that do not.

Past adaptations that promote ecological resiliency

Generalization (Section 16.5) is often considered as a characteristic that buffers mutualisms from anthropogenic change. When specialized mutualisms are examined closely, however, it is discovered that they, too, exhibit adaptations that confer some resiliency. (Some of these are summarized in Box 16.3, for the fig pollination mutualism.) The explanation for the existence of these traits seems fairly straightforward: even in the absence of anthropogenic change, most natural environments are extremely variable. Surely, the only highly specialized and/or obligate mutualisms that have been able to persist to the present day are those able to persist in the face of variability. Below we consider three kinds of adaptations that help specialists survive in fluctuating environments: an ability to wait, an ability to move, and an ability to generalize. [See Bond (1995) for an expanded discussion of these and other such traits.] We then consider the degree to which these traits can rescue species from anthropogenic change.

We can find no examples of mutualism in which *each* of the two species has a *single* opportunity in its life to attract the *single* partner upon which it depends. In the face of anything other than an extremely constant environment and high population sizes of mutualists, such a relationship seems doomed to failure. Rather, at least one of the two species has the opportunity to acquire mutualists either continuously or at repeated intervals. In either case, that organism possesses some ability to persist for a while without mutualists (although it possibly experiences reduced success while it waits). For instance, certain flowers can persist in a receptive state for days or weeks until pollinators arrive (Primack 1985; Khadari *et al.* 1995), and orchid seeds do not germinate until they are invaded by their obligate beneficial mycorrhizae (Dressler 1981). Finally, many organisms can experience at least minimal success even when mutualists are entirely absent. That is, their mutualisms may be specialized, but they are not obligate. For example, plants may reproduce largely by self-pollination during intervals when pollinators are absent, though the offspring that result are likely to be genetically inferior to those produced in the presence of mutualist pollinators.

Organisms that can wait for mutualists are, as a rule, relatively immobile. Mobile species exhibit other suites of traits that increase the success of their mutualisms. Certain organisms, both terrestrial and aquatic, show remarkable abilities to track species-specific volatile substances released by physically distant, immobile mutualists (Ware *et al.* 1993; Brooks and Rittschof 1995; Elliott *et al.* 1995; Takabayashi and Dicke 1996). Larger and more cognitively advanced species learn where mutualists are likely to be found, and can shift to new areas when this distribution changes (Bronstein 1995).

Finally, specialists often have greater potential to associate successfully with the "wrong" mutualist than we usually realize, because the switch only occurs (or is only obvious) when the "right" mutualist is rare or absent. For example, many bees, termed oligolectic species, visit only one or a very few plant species for pollen. When flowering of the usual host fails, many of these bees can shift successfully to plant species with which they are almost never associated under normal conditions (Wcislo and Cane 1996). At a very low, but detectable, frequency native fig wasps visit fig species that have been introduced without their own pollinators (McKey 1989; Nadel *et al.* 1992); if the native and exotic figs are related closely enough, both partners within the mismatched relationship are able to reproduce, although generally at reduced rates (Hossaert-McKey, unpublished data).

Fig wasp (on fig)
Courtella wardi

What is the significance of these adaptations for life in environments that vary naturally, in a conservation context? They allow organisms with specialized mutualisms to cope with anthropogenic change at the mesoscale (i.e., change that is relatively local and relatively short in duration). They eliminate the risk of failing catastrophically in response to a brief absence of partners, and they permit populations to persist for some time when mutualists are in decline. On the other hand, this situation cannot necessarily continue for protracted periods. Fitness is likely to decline eventually and, with it, population sizes; as populations decline, inbreeding and other detrimental genetic effects follow. Ultimately, the degree of resiliency offered by these traits depends on:

- The nature and spatiotemporal scales of human disturbance, particularly with reference to the nature and scales of variation that the species of interest has experienced historically.
- The species' ability to evolve further in response to environmental change.

Evolutionary responses

There is abundant evidence that anthropogenic change initiates evolutionary responses within species involved in antagonistic interactions. For example, native animals can evolve to feed efficiently on novel food items (Singer *et al.* 1993) and to resist novel pathogens (Dwyer *et al.* 1990). Phenomena like these have barely been investigated in mutualistic interactions, although it seems probable that they exist. In the only such study that we know, Smith *et al.* (1995) demonstrated that a Hawaiian honeycreeper (whose coevolved nectar plant was driven to extinction) has evolved a bill shape within the past 100 years that allows it to feed from a more common native species.

Hawaiian honeycreeper (Iiwi) *Vestiaria coccinea*

The model introduced in Section 16.2 yields some insights into the ecological and evolutionary dynamics of specialized mutualisms in a slowly changing environment. Although a comprehensive analysis lies beyond the scope of this chapter, a graphic interpretation of Figure 16.1d suffices to illustrate the potentially dramatic consequences on the viability of a mutualism's evolutionary response to environmental change. Environmental change that affects the degree of competitive asymmetry in one or the other species is likely to lead to "evolutionary trapping" (Chapters 1 and 11): as the coefficient of asymmetry in one species slowly decreases or increases, the association tracks an evolutionary attractor that eventually becomes unviable. This can be seen in Figure 16.1d: given that the asymmetry coefficient is fixed for one species, there is a bounded range of asymmetry coefficients for its mutualist species that permits evolutionary stabilization at an ecologically viable equilibrium. When environmental change causes this parameter to hit the limits of its range, coextinction occurs through rapid evolutionary suicide of the kind depicted in Figure 16.1c (when the asymmetry coefficient hits the upper threshold), or in Figure 16.1b (when the asymmetry coefficient reaches the lower threshold). Interestingly, the range of asymmetry coefficients that one population may span without compromising the evolutionary persistence of the whole association is larger if the degree of competitive asymmetry and/or the level of genetic variability in the partner species is low.

At present, empirical data that would allow direct assessments of whether potentially disastrous evolutionary trajectories are occurring or will occur are lacking. However, we can offer one suggestion of a likely situation in which such a development may have already started. It has recently been shown that elevated CO_2 levels and global warming can alter flowering phenology and flower nectar volumes in certain plant species (Erhardt and Rusterholz 1997; Ahas *et al.* 2002; Fitter and Fitter 2002; Inouye *et al.* 2002; Dunne *et al.* 2003). Phenologies of different species appear to be shifting to different degrees, and in different directions: for example, Fitter and Fitter (2002) report that while 16% of British flowering plants are flowering significantly earlier than in previous decades (with an average advancement of 15 days in a decade), another 3% of species are flowering significantly later than they once did. This is likely to result in novel groups of plant species blooming simultaneously, between which individual pollinator species are becoming able to choose for the first time. Plants that are currently highly preferred and relatively specialized nectar resources may progressively become disfavored by their pollinators, as more rewarding plant species previously matched with other pollinators come into competition for the first time. Conversely, previously disfavored plants may slowly gain competitive advantage among newly coflowering species that are even less rewarding. It would thus seem wise to initiate studies of changing mutualisms within changing communities now, so as to be able to predict and possibly prevent incipiently suicidal evolutionary trajectories.

Emerald toucanet
Aulacorhynchus prasinus

16.5 Responses of Generalized Mutualisms to Threats

In generalized mutualisms, species gain benefits from multiple partner species rather than a single one. For example, in contrast to the obligate species-specific mutualism between figs and fig wasps (Box 16.3), figs are involved in facultative and highly generalized mutualisms with the birds and mammals that disperse their seeds. There are at least three ways in which generalization can buffer mutualisms from a changing environment:

- Rarity or extinction of one species is unlikely to drive the reproductive success of its mutualist to zero, because other beneficial partners are still present. Even in relatively undisturbed habitats, one commonly sees great year-to-year and site-to-site variation in the diversity of mutualist assemblages (e.g., Horvitz and Schemske 1990; Jordano 1994; Alonso 1998); quite commonly the success of individual species that benefit from these assemblages does not track that variation closely.
- Loss of one partner can spur increases in the abundance of alternative partners that might previously have been excluded or suppressed competitively; these alternative partners can be equally effective, or even more effective, mutualists (e.g., Young *et al.* 1997).
- Finally, the fairly generalized traits involved in the attraction and reward of diverse mutualists can function to attract and reward partners that may have no common evolutionary history with that species. One such adaptation is the elaiosome, a small lipid-rich body attached to certain seeds, which has evolved multiple times and which appeals to diverse seed-dispersing ants worldwide

(Beattie 1985). Invasive plants with elaiosomes are commonly dispersed by native ants (Pemberton and Irving 1990), while native plants with elaiosomes can be dispersed (although often comparatively poorly) by invasive ants (Bond and Slingsby 1984; Christian 2001).

Despite such buffering, there is no doubt that in recent years generalized mutualisms have suffered major impacts from anthropogenic change. Three examples of the disruption of generalized plant–pollinator mutualisms should suffice to make this point:

- Aerial spraying of herbicides in Eastern Canada during the 1970s catastrophically decreased populations of generalist bee pollinators. Subsequent reproductive failures in both native and crop plants have been well documented (Thomson *et al.* 1985).
- Shrinking and increasingly isolated plant populations may fail to attract pollinators, which leads to Allee effects that draw populations downward toward extinction (Groom 1998; Hackney and McGraw 2001).
- Invasive plants can outcompete native species for pollination services, which results in the local decline of native populations. For example, purple loosestrife, a weed introduced to North America, has been reducing both the pollinator visitation and subsequent seed set of a native congener (Brown *et al.* 2002).

What kinds of evolutionary dynamics in response to anthropogenic change can be expected in generalized mutualisms like these? To address this question, it becomes clear that one must adopt a perspective that goes beyond the purely pairwise approach that has characterized most theoretical work on mutualism (Stanton 2003). Here we introduce a simple adaptive dynamics model to illustrate the disturbing potential for evolutionary ripple effects to cascade through more complex ecological communities. More generally, we can look upon this model as a contribution toward elucidating the importance of community context when addressing questions in evolutionary conservation biology (Chapter 17).

Flexible mutualistic coadaptation

We focus on an ecological community that comprises two pairs of mutualistic species. This setup is chosen because switches between alternative mutualistic partners are important (as highlighted in Section 16.3), and also in an effort to keep matters tractable. In Figure 16.2 species 1 and 2, as well as species 3 and 4, are coupled through mutualistic interactions. In addition, species 2 and 3 can also engage in mutualism, as can species 1 and 4; thus, all four species potentially are generalists, within the bounds of this simple community structure. We can think, for example, of species 1 and 3 as two plants and of species 2 and 4 as two pollinators: species 2 can then pollinate both plants, and species 3 can be pollinated by both pollinators. The alternative couplings (i.e., 2 with 3, and 1 with 4) are, however, less efficient than the primary couplings (i.e., 1 with 2, and 3 with 4) in enabling the mutualistic exchange of commodities such as pollen and pollination. Intraspecific competition is present in all four species, and we also

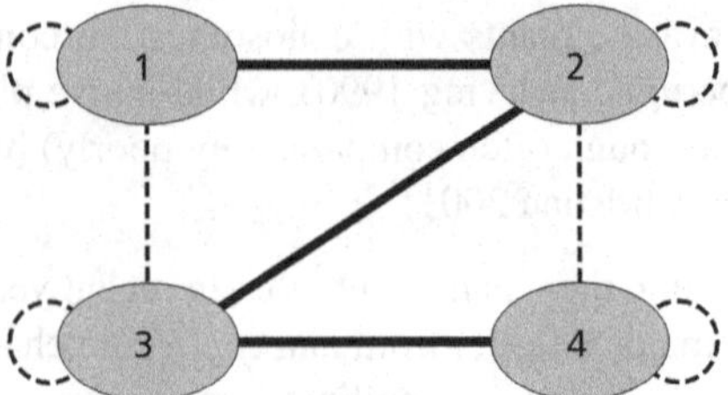

Figure 16.2 A pairwise mutualistic community. The strengths of mutualistic interactions (continuous lines) between four species depend on their level of coadaptation. Dashed lines depict competitive interactions. In this model, the mutualistic interactions can evolve, such that species 2 and 3 can gradually switch between their primary (species 1 and 4, respectively) and secondary mutualistic partners (species 3 and 2, respectively).

consider interspecific interactions between species 1 and 3 on the one hand, and between species 2 and 4 on the other. Representing this basic setup in terms of Lotka–Volterra systems leads to the model described in Box 16.4.

Coevolutionary responses to environmental disturbances

We can now utilize this four-species model to explore the evolutionary and co-evolutionary implications of changing environmental conditions. We start from a situation in which all the species are adapted so as to be maximally efficient in exchanging commodities with their primary partners, and thus much less efficient when associated with their alternative partners. We then change a single parameter of the model, equivalent to reducing the carrying capacity of species 1 by a factor of 10. This effectively models a situation in which anthropogenic change has altered species 1's environment in a way that makes it less suitable for these organisms. In response, we can observe one of the following three dynamical patterns of community reorganization (Figure 16.3):

- *Primary reorientation and primary extinction.* The reduction in species 1's carrying capacity makes it a much less attractive mutualistic partner for Species 2, so species 2 adapts to maximize its coupling with its alternative partner, species 3. We refer to this initial evolutionary response to the imposed environmental change as primary reorientation, and use analogous terms to refer to the subsequent events. Now that the benefit of mutualism received by species 1 from species 2 has been withdrawn, species 1 becomes extinct (Figure 16.3a). Notice that this extinction is not a direct consequence of the imposed environmental change, but, instead, is caused by the evolutionary dynamics that are triggered by the imposed environmental change.
- *Primary reorientation, primary extinction, and secondary reorientation.* After the imposed environmental change has reduced the abundance of species 1, species 2 specializes on species 3. Further evolutionary change may then ensue. In particular, because of its reorientation, species 2 becomes a more attractive partner for species 3, which may induce species 3 to switch from specializing on species 4 to specializing on species 2 (Figure 16.3b). This causes the newly

Box 16.4 Modeling eco-evolutionary responses of generalized mutualisms to threats

A simple adaptive dynamics model for an ecological community that comprises two pairs of mutualistic species (Figure 16.2) can be constructed as follows. Using a basic Lotka–Volterra model architecture (Box 16.2), the per capita growth rates in species $i = 1, ..., 4$ are given by $r_i + \sum_j a_{ij} N_j$ with intrinsic growth rates r_i and population densities N_i. The symmetric community matrix a contains elements $a_{11} = -c_{11}, ..., a_{44} = -c_{44}$, which describe intraspecific competition, and elements $a_{13} = a_{31} = -c_{13}$, $a_{24} = a_{42} = -c_{24}$, which describe interspecific competition. We assume that species 1 and 4 do not interact, $a_{24} = a_{42} = 0$. The remaining six elements of a describe mutualistic interactions and are determined as follows.

Each species possesses an adaptive trait x_i, bounded between 0 and 1, that describes its degree of adaptation to its primary partner (i.e., of species 1 to species 2, of species 2 to species 1, of species 3 to species 4, and of species 4 to species 3), while $1 - x_i$ describes the degree by which species i is adapted to its secondary partner (i.e., of species 2 to species 3 and vice versa; species 1 and 4 have no secondary partner). In the case of plant–pollinator interactions, the adaptive traits could represent morphological or phenological characters. The strength of mutualistic interactions is $a_{ij} = c_{ij} m_{ij}$, where j is either the primary or secondary partner of species i and $c_{ij} = c_{ji}$ scales the strength of their interaction. We assume that the level of coadaptation, which describes how well the relevant adaptations in species i and j match, is given by $m_{ij} = e_{ij} x_{ij} x_{ji} + (1 - e_{ij})[1 - (1 - x_{ij})(1 - x_{ji})]$. Here, x_{ij} is the degree of adaptation of species i to species j, which equals x_i if j is the primary and $1 - x_i$ if j is the secondary partner. The parameters $e_{ij} = e_{ji}$ measure how essential mutual adaptation is to the strength of the mutualistic interaction. When e_{ij} is high, the first term in m_{ij} dominates, such that both x_{ij} and x_{ji} have to be high for the interaction to be strong. By contrast, when e_{ij} is low, the second term in m_{ij} allows the interaction to be strong if *only one* species is adapted to the partner, regardless of how well the partner itself is adapted. Variations in the resultant levels of matching are illustrated in the figure below.

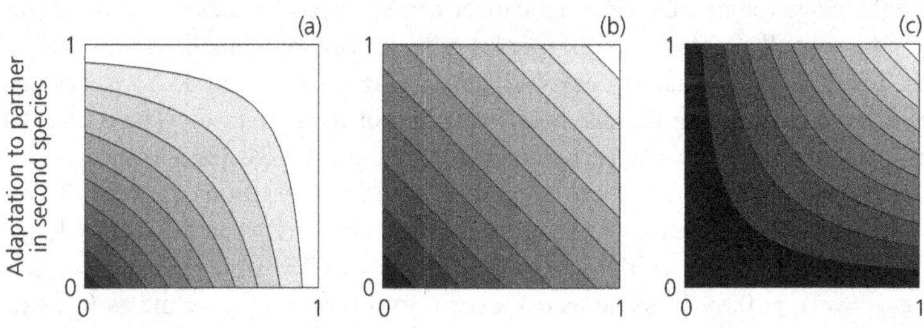

A high level of matching may already be present if just one partner is sufficiently adapted, panel (a), or it may more strictly require both partners to adapt to one another, panel (c). Such a continuum is described by the model parameters $0 \leq e_{ij} \leq 1$; the three cases shown correspond to $e_{ij} = 0$ (a), $e_{ij} = 0.5$ (b), and $e_{ij} = 1$ (c). High levels of coadaptation are indicated in white, and low levels in black.

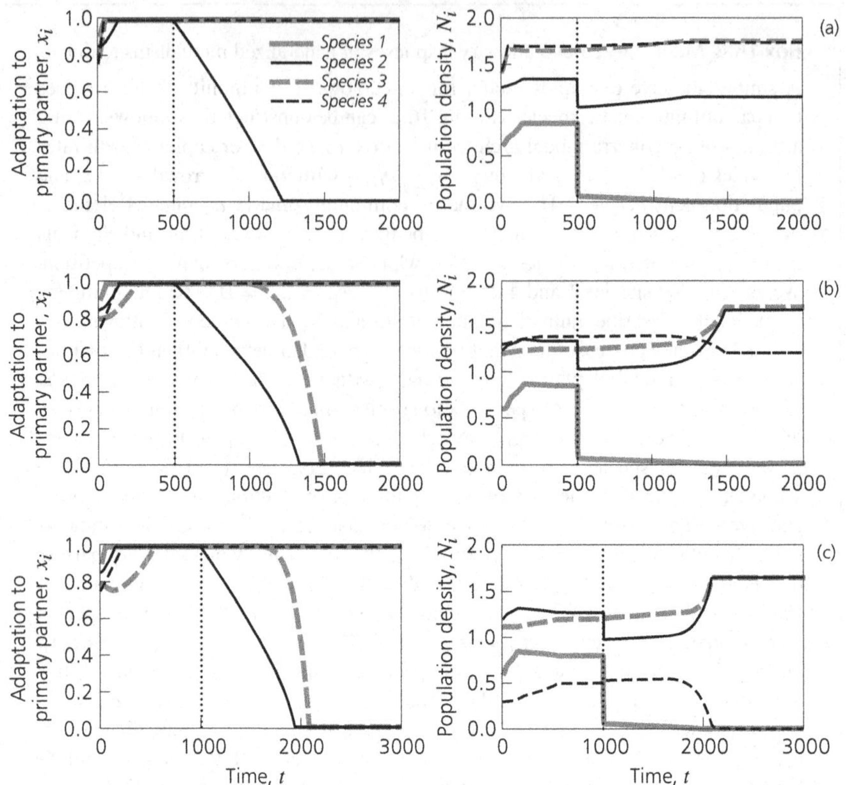

Figure 16.3 Coevolutionary ripple effects of environmental disturbances. The panels show how adaptive trait values (left column) and population densities (right column) change over evolutionary time. At the moments in time indicated by the dotted lines, an abrupt environmental change reduces the carrying capacity of species 1 by a factor of 10. Before that, selection favors full adaptation of all species to their primary mutualistic partners, whereas after the disturbance alternative coevolutionary responses can unfold. (a) Reorientation of species 2 to species 3 with the resultant extinction of Species 1. (b) The reorientation of species 2 and extinction of species 1 triggers reorientation of species 3 to species 2. (c) The reorientation of species 2, extinction of species 1, and reorientation of species 3 eventually lead to the extinction of species 4. Parameters: (a) $r_1 = -0.1$, $r_2 = r_3 = r_4 = 1$, $c_{11} = 0.5$ changing to $c_{11} = 5$, $c_{22} = c_{33} = c_{44} = 1$, $c_{13} = c_{24} = 0$, $c_{12} = c_{23} = c_{34} = 0.4$, $e_{12} = e_{23} = e_{34} = 0.8$; (b) same as (a), except for $c_{34} = 0.2$; (c) same as (a), except for $g_4 = 0.15$ and $c_{24} = 0.1$. All evolutionary trajectories are based on the canonical equation of adaptive dynamics (Dieckmann and Law 1996).

formed alliance between species 2 and 3 to thrive, and allows it to dominate the community.

- *Primary reorientation, primary extinction, secondary reorientation, and secondary extinction.* The ripple effects of the initial environmental change may propagate even further through the community. After species 2 and 3 have maximized their level of coadaptation, species 4, now that it has essentially lost its mutualistic partner species 3 to species 2, may perish (Figure 16.3c). This illustrates how environmental change that directly affects only one species can cascade relatively easily through a community and induce ecological and evolutionary change in species that are several interaction tiers away from the original perturbation.

Much remains to be studied before we will truly understand the likelihood and implications of coevolutionary cascades in threatened ecological communities. Yet the simple model considered here already cautions against ignoring the potentially wide-ranging consequences of such cascades to the structure and stability of ecosystems exposed to environmental change. Since many mutualistic interactions link pairs of species relatively tightly, they present a good starting point for these explorations. However, we can be quite certain that the likelihood and severity of coevolutionary cascades will not be fundamentally different when we extend our view to competitive or exploitative ecological interactions.

16.6 Concluding Comments

Only recently have mutualisms been subject to the same level of attention from evolutionary biologists as antagonistic interactions have received (Bronstein 2001a). As a result, it is not surprising that our understanding of how they might respond evolutionarily to anthropogenic change remains rudimentary. This is alarming, because mutualisms appear to be both an ecological and evolutionary nexus for the accumulation of diversity within ecosystems. Further studies of the fate of mutualisms in response to environmental change are therefore essential if the goal is to conserve higher units of biological organization.

Both empirical and theoretical studies are needed. On the theoretical side, there is a dire need for more detailed explorations of eco-evolutionary models of the type tentatively introduced and analyzed in this chapter. A better understanding of the adaptive dynamics that result from mutualistic interactions (Section 16.2) will provide crucial insights with which to probe deeper into the corresponding evolutionary and coevolutionary responses to environmental threats (Sections 16.4 and 16.5). Modeling such environmental threats more specifically, rather than merely through their effects on compound parameters (as done here), will be vital to understand the long-term conservation implications of habitat fragmentation, biological invasions, and genetic introgressions. Eventually we will need to consider models that describe complex webs of interactions realistically, to allow us to assess the dangers of both ecological and coevolutionary ripples cascading through entire communities. No doubt, many surprises are still lurking in the intricate interplay of mutualistic, competitive, and exploitative interactions [see, e.g., the so-called Red King effect, whereby slower evolution leads to a greater selective advantage (Bergstrom and Lachmann 2003)]. To the extent feasible, we should anticipate such surprises by means of careful modeling studies, rather than letting them jeopardize expensive and conservation-critical efforts in the field.

On the empirical side, we need information on where, when, and how mutualistic interactions are under natural selection in the context of anthropogenic change, and what the likely outcomes (increased generalization; partner shifts; extinction?) appear to be. In this regard, it is important to point out that, to date, the large majority of field studies, as well as nearly all the broad conceptual work on the conservation of mutualism, focus on a single type, plant–pollinator mutualisms. Pollination is undoubtedly of critical importance: perhaps 90% of angiosperms are animal-pollinated, and it has been estimated that half the food we consume is the product of biotic pollination (Buchmann and Nabhan 1996). However, the responses of a variety of other mutualisms critical to community functions are virtually unknown. For example, the health of some entire marine ecosystems, including coral reefs and hydrothermal vents, depends on mutualistic bacterial and algal symbionts, some of which are clearly sensitive to human activities (Smith and Buddemeier 1992; Knowlton 2003). Thus, in seeking a deeper understanding of the evolutionary conservation biology of mutualisms, it will be essential to take a broader natural history perspective than current knowledge allows.

17

Ecosystem Evolution and Conservation

Michel Loreau, Claire de Mazancourt, and Robert D. Holt

17.1 Introduction

A major problem in conservation biology is to decide the target of conservation: should conservation efforts aim to preserve species or ecosystems? The traditional approach has, by necessity, focused on particular species threatened by extinction. With the increasing attention on preserving biodiversity at large, for which the species-by-species approach falls short, a trend is now emerging that centers on ecosystems or habitats as the conservation targets (Schei *et al.* 1999). These two approaches, however, should not be opposed. Species and ecosystems are bound together by mutual ecological constraints and a shared evolutionary history, so that in the long term it may be impossible to conserve one without conserving the other (Loreau *et al.* 1995). Species' traits and their evolution are ultimately constrained by ecosystem processes, just as ecosystem properties are constrained by the ecological and evolutionary history of interacting species (Holt 1995). It is the web of interactions at the heart of an ecosystem that maintains both species and ecosystems as they are, or (more exactly) as they are evolving.

Another way to address this problem is to phrase it in terms of a basic issue in evolutionary biology: what are the constraints within which natural selection operates? Traditionally, evolutionists considered these constraints to arise internally, such as from allocations among competing physiological needs. However, feedbacks via ecosystem processes can also act as constraints, and can channel selection in directions that are different from those expected in the absence of such constraints. Box 17.1 contrasts three views of how natural selection operates: the "classic" view of a constant environment, the "modern" view of an organism–environment feedback, and the "ecosystem" view of a web of interactions among organisms and abiotic factors. Although inherent in the very definition of an ecosystem, rarely has this third view been applied consistently to evolutionary problems so far. Recognizing the ecosystem as the proper context within which natural selection, and hence evolution, operates is a major challenge for ecology today, with important implications in both basic science and applied areas, such as conservation biology and ecosystem management. This challenge emphasizes the need to overcome the barrier that has increasingly separated population ecology and evolutionary ecology, on the one hand, from ecosystem ecology, on the other hand.

In this chapter we show the potential importance of this perspective using plant–herbivore interactions to illustrate:

Box 17.1 Three views of the operation of natural selection

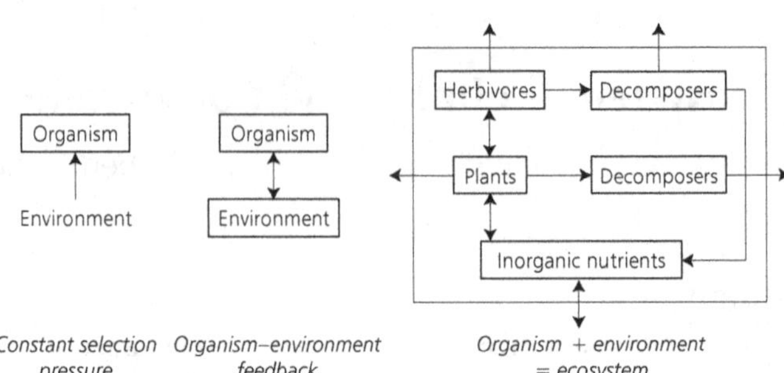

Natural selection is based on the selective multiplication of types in a population through environmental constraints on organisms:

- In the "classic" view, the environment is regarded as external to the organism and constant. Although most evolutionary biologists today would probably agree that this view is an oversimplification of reality, for simplicity's sake it has been, and still is, widely used in theoretical evolutionary biology as an implicit conceptual framework.
- The "modern" view recognizes that organisms modify and interact with their environment, which generates an organism–environment feedback in the operation of natural selection. This feedback is formalized, for instance, in the theory of adaptive dynamics (see Chapter 11).
- A further step is necessary to understand the full implications of this feedback: an organism's environment has to be resolved into its real physical, chemical, and biological constituents and their interactions. This is what we call the "ecosystem" view, because an ecosystem is defined as a locally interacting system of abiotic and biotic components.

Since the environment of each organism or component comprises other organisms or components, the ecosystem concept contains both the organisms and their environments. In this sense, it provides a higher-level perspective that transcends the duality between organism and environment (Loreau 2002).

- How incorporating organism–environment feedbacks (moving from the first to the second view of natural selection in Box 17.1) may change the direction of evolution compared with classic predictions for a constant environment;
- How explicit considerations of nutrient cycling as a key whole-ecosystem process (moving from the second to the third view of natural selection in Box 17.1) can further alter our view of the very nature of species' interactions, from both ecological and evolutionary perspectives;
- Some potential interactions between local evolution and biological invasions and their implications for conservation biology.

17.2 Evolution under Organism–Environment Feedback

If organisms collectively have a significant impact on their environment, to ignore the organism–environment feedback may lead to serious flaws in predictions of the qualitative direction of evolution and expected species' traits in ecological systems. We focus on the evolution of plant defense against herbivores as an example, assuming for the time being that herbivores only consume plants and do not provide them with any indirect benefits (see Section 17.3 for such indirect effects).

Understanding the evolution of plant antiherbivore defense

The classic "resource availability hypothesis" (Coley *et al.* 1985) proposes that low resource availability favors plants with inherently slow growth rates, which in turn favor large investments in antiherbivore defense. This hypothesis has been influential and attractive, because it seemed to explain patterns of plant defense and herbivory in a wide range of ecosystems. It hinged, however, on a very simple theoretical argument with a number of simplifying assumptions, in particular that the quantity of available resources is unaffected by plants. This implicit assumption of a constant environment led Coley *et al.* (1985) and subsequent authors (de Jong 1995; Yamamura and Tsuji 1995) to measure fitness by what they called the plant "realized growth rate", which in effect is a potential growth rate that ignores the feedback generated by plant resource consumption.

This fitness measure may make sense for pioneer species colonizing temporary environments, but is inappropriate for species competing for limited resources in more stable environments. Whenever plants have accumulated enough biomass to affect the amount of resources in their environment, they compete for these resources, and their growth hinges on their ability to tolerate low concentrations of the resource that is limiting. If the environment is homogeneous, fitness is determined by the ability to deplete the limiting resource (Tilman 1982). If the environment is structured spatially, fitness is determined by the basic reproduction ratio (Loreau 1998a). All these fitness measures can be derived as special cases of the more general concept of "invasion fitness" in the theory of adaptive dynamics (Metz *et al.* 1992; Dieckmann 1997; Geritz *et al.* 1998).

To explore the effects of this organism–environment feedback, we constructed a simple model of evolution of plant allocation to antiherbivore defense in a system that incorporates plant–resource dynamics (Loreau and de Mazancourt 1999). Assume that plants allocate a constant fraction x of a limiting resource to defense, and the remainder $1 - x$ to growth. Thus, x measures the level of defense investment. The dynamics of total plant biomass N_p can be described by

$$\frac{dN_p}{dt} = r(N_n, x)N_p \, , \tag{17.1a}$$

$$r(N_n, x) = r_{\max}\phi(N_n)(1 - x) - \psi(x) - m \, , \tag{17.1b}$$

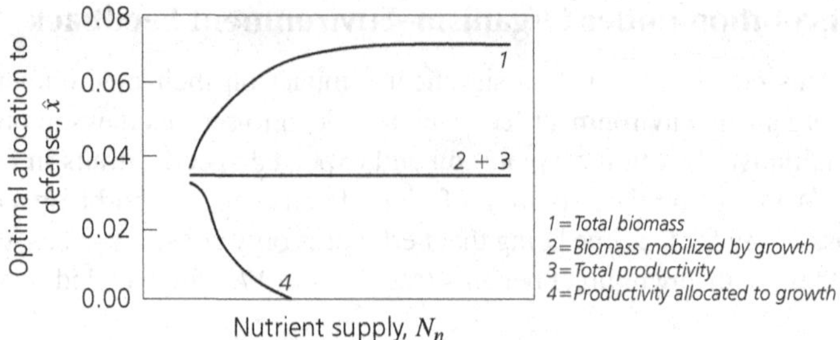

Figure 17.1 Evolutionary optimal plant allocation to defense $\hat{x}$ as a function of nutrient supply in a spatially structured environment in which fitness is determined by any of the following quantities, total biomass, the part of biomass that is mobilized for growth, total productivity, or the part of productivity that is allocated to growth. Plants cannot survive below a threshold nutrient supply; the curves start after this threshold. *Source*: Loreau and de Mazancourt (1999).

where $r(N_n, x)$ is the plant population growth rate per unit biomass, $r_{\max}$ is the maximum rate of resource uptake by plants, $\phi(N_n)$ is the plant functional response to resource concentration N_n [$\phi(N_n)$ increases monotonically with N_n and is scaled so that $0 \leq \phi(N_n) \leq 1$], $\psi(x)$ is the rate of herbivory (which is a monotonic decreasing function of x), and m is the loss rate of biomass. Resource concentration also changes with time in dependence on N_p, but its dynamic equation is irrelevant to the argument that follows, and so is ignored here.

In the long term, plant biomass reaches an ecological equilibrium such that the population growth rate $r(N_n, x)$ in Equation (17.1) is zero. Plants then control resource concentration at a level N_n^* set by Equations (17.1). But this ecological equilibrium itself changes gradually because of the natural selection that acts on x. The evolutionary equilibrium is attained when the population growth rate at the ecological equilibrium can no longer be increased, that is, when

$$\left.\frac{\partial r(N_n, x)}{\partial x}\right|_{N_n = N_n^*} = 0\,. \tag{17.2a}$$

(It can be shown that r is then indeed maximal.) The solution of Equation (17.2a) provides the evolutionary optimal allocation to defense $\hat{x}$. Using Equations (17.1), Equation (17.2a) reduces to

$$\psi'(\hat{x})(1 - \hat{x}) + \psi(\hat{x}) + m = 0\,, \tag{17.2b}$$

where $\psi'(\hat{x})$ is the derivative of ψ with respect to x evaluated at $\hat{x}$.

It is evident from Equation (17.2b) that the optimal defense investment is affected by features of the plant–herbivore interaction, encapsulated in the function ψ. In general, the higher the intrinsic herbivore voracity, the higher the plant defense investment. This is independent of the maximum rate of resource uptake $r_{\max}$, which is also a measure of the maximum growth rate and, indeed, of any conceivable measure of resource availability.

These results are based on the assumptions that the environment is homogeneous and plants have unrestricted global access to the limiting resource. However, usually plants have only local access to resources such as soil nutrients (Huston and DeAngelis 1994; Loreau 1996, 1998b). A homogeneous environment may be viewed as one extreme in the range of possibilities, the other extreme being a perfectly structured environment in which each plant occupies an isolated site (Loreau 1998a). A model for the latter case, in which competition obeys a "competitive lottery" for vacant sites, shows that the outcome is strongly dependent on the factor that determines a plant's ability to produce successful propagules that establish in vacant sites, which itself determines fitness. The optimal defense investment may then either increase, stay constant, or decrease with nutrient supply (Figure 17.1). The effects of maximum growth rate, as measured by r_{max}, are identical to those of nutrient supply on the optimal defense investment in the various scenarios.

Clearly, the resource availability hypothesis fails to describe evolution in a system in which plants and their limiting resources reach an ecological equilibrium. Resource supply and maximum growth rate may increase, decrease, or (in most cases) have no effect on the optimal investment in defense. A common argument used to justify this hypothesis is that herbivory is more costly in resource-poor environments because lost biomass is more costly to replace. However, this argument ignores that investment in defense is also costly. It is the balance between the two costs that determines the optimal investment, and in most cases this does not change in the way assumed in the resource-availability hypothesis. Indeed, the dynamics of coevolution between plants and herbivores, which we have not considered here, may contribute an increase to the intensity of their antagonistic interaction, and hence to plant investment in antiherbivore defense when resource availability increases (see Section 17.4).

Conservation implications

Current species' traits result from the evolution of a dynamic interaction between organisms and their environment. A neat empirical example that shows the

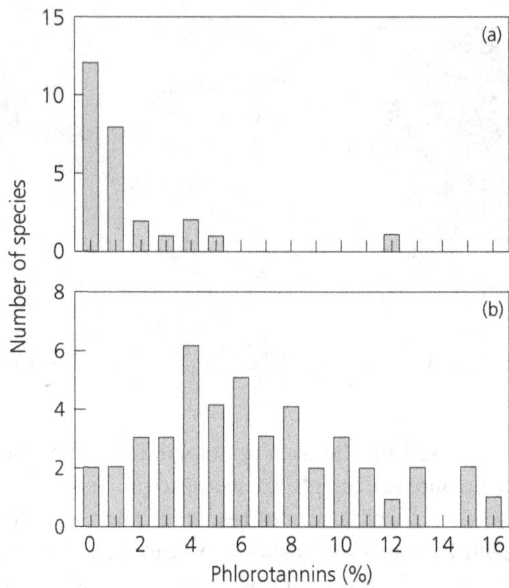

Figure 17.2 Frequency distribution of mean levels of phlorotannins, the principal secondary metabolites, in brown macroalgae. (a) From the North Pacific Ocean, where the predation of sea otters on invertebrate herbivores is important. (b) From Australasia, where sea otters are absent. The figure shows that Australasian seaweeds have been under strong selection to evolve chemical defenses. Data are average percentage dry weight phlorotannins. *Source*: Steinberg *et al.* (1995).

importance of evolutionary dynamics in plant defenses is provided by the work of Steinberg *et al.* (1995). In the North Pacific, sea otters keep invertebrate herbivores like sea urchins at low levels of abundance, which allows the establishment of luxurious algal beds. By contrast, in comparable environments in Australasia, sea otters are absent and herbivore pressure is high. Corresponding to this, macroalgae have much higher levels of secondary defensive compounds in Australasia (Figure 17.2).

One potential implication of this model for conservation is an initial asymmetry between the indirect effect of predator removal and exotic predator introductions. If a predator has had a strong impact on herbivore abundance over evolutionary time scales, plants in these systems should have a low investment in defense against herbivory. This makes them vulnerable to increased herbivore numbers following predator removal. The introduction of exotic predators can be devastating for herbivores, but plants may show a more muted initial response to this reduction in herbivory, for they have already experienced low herbivory because of a high investment in defense. Following predator removal or addition, over a longer time scale further changes in the plant communities are expected because of a shift in the optimal allocation to defense.

17.3 Evolution in an Ecosystem Context

The organism–environment feedback examined above is simple and direct. In reality, ecosystems are complex dynamic systems potentially capable of generating a multitude of indirect interactions among their components (Puccia and Levins 1985; Wootton 1994; Menge 1995) and hence of indirect feedbacks between an organism and the rest of the ecosystem. Some of these indirect effects are weak or unpredictable (Yodzis 1988), but some can be strong and predictable. In particular, material cycling is a key ecosystem process that drives a circular causal chain in ecosystems and transmits predictable indirect effects to their components (Loreau 1998a). Therefore, it is likely to affect the evolution of component species.

Indirect ecological effects of material cycling

Traditionally, in ecology plant–herbivore interactions have been considered as antagonistic because herbivores have a negative direct effect on plants through biomass consumption. This assumption has been challenged by the so-called grazing optimization hypothesis, which states that primary productivity increases with grazing and reaches a maximum at a moderate rate of herbivory (Owen and Wiegert 1976, 1981; McNaughton 1979, 1983; Hilbert *et al.* 1981; Dyer *et al.* 1986). This hypothesis is supported by some empirical data, notably from the Serengeti grassland ecosystem (Figure 17.3). One mechanism that could produce such a beneficial effect is nutrient cycling, which mediates positive indirect effects among ecosystem components. Should the traditional view of antagonistic plants and herbivores be changed, can these even be mutualistic, and under what conditions? These questions, which have important consequences for both ecosystem functioning and the evolution of plant–herbivore interactions, have been much debated over the past 20 years (Silvertown 1982; Belsky 1986; Paige and Whitham 1987; Bergelson and Crawley 1992; Paige 1992; Belsky *et al.* 1993; Mathews 1994; Bergelson *et al.* 1996; Gronemeyer *et al.* 1997; Lennartsson *et al.* 1997, 1998).

Given the ambiguity in interpretations of empirical data, we attempted to answer these questions theoretically using mathematical models. We first identified the ecological conditions under which herbivores increase primary production and achieve grazing optimization through recycling of a limiting nutrient (Loreau 1995; de Mazancourt *et al.* 1998). These conditions are:

- The proportion of nutrient lost while flowing along the herbivore recycling pathway must be sufficiently less than the proportion of nutrient lost while flowing in the rest of the ecosystem;
- Nutrient inputs into the system must exceed a threshold value, which depends on the sensitivity of plant uptake rate to an increase in soil mineral nutrient.

Contrary to traditional assumptions, nutrient turnover rates have no impacts on the long-term equilibrium primary production. These results are very general: they do not depend on the structure of the ecosystem or on the functional form of herbivore consumption (de Mazancourt *et al.* 1998). They are also potentially relevant to

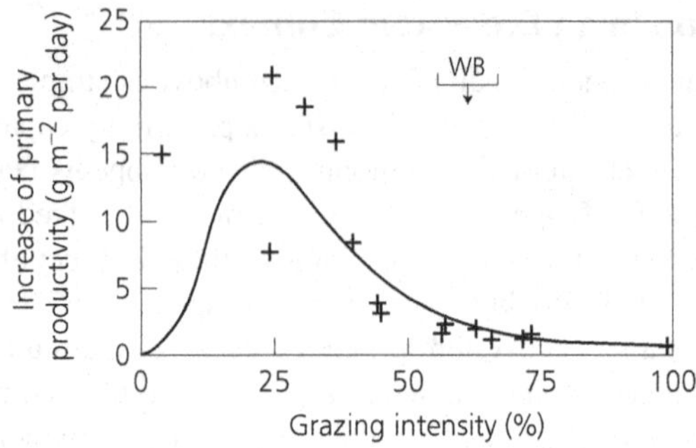

Figure 17.3 Relationship between the stimulation of above-ground grassland productivity and grazing intensity in the Serengeti National Park, Tanzania. Above-ground net productivity was calculated from positive biomass increments in temporary exclosures. Grazing intensity was calculated as $1 - N_{p,g}/N_{p,ng}$, where $N_{p,g}$ is the plant biomass in grazed areas unprotected by fencing and $N_{n,ng}$ is the plant biomass in the permanent exclosure. The effect of soil moisture was not incorporated into the curve (incorporating this extra factor reduced the unexplained variance by 9%). WB indicates the mean and 95% confidence interval of wildebeest grazing intensity in these grasslands during the wet season. *Source*: McNaughton (1979).

natural ecosystems: grazing optimization was found to be likely for an African humid savanna (de Mazancourt *et al.* 1999) and it can occur even if herbivory results in the replacement of a productive plant species by a less productive one (de Mazancourt and Loreau 2000b).

Evolutionary consequences of grazing optimization

The existence of a positive indirect effect of herbivory on primary production does not automatically lead to an indirect mutualism between plants and herbivores, for two reasons. First, increased plant productivity does not necessarily translate into increased plant fitness. It is still unclear which plant traits determine fitness. If a plant's fitness is mainly determined by its biomass, because a greater biomass means a greater nutrient stock available for seed production at the end of the season, then no mutualistic interaction with herbivores is possible, because plant consumption by herbivory always decreases plant biomass. Alternatively, if a plant's fitness is mainly determined by its productivity, because a higher productivity means a larger nutrient flow that is constantly allocated to seed production or vegetative propagation [as assumed in physiological models of plant resource allocation; see Mole (1994)], then herbivory can increase plant fitness through increased productivity. Reality probably lies between these two extremes, and thus we may expect herbivory to increase plant fitness in some cases. Second, it is not absolute fitness, but relative fitness that is important. If two plant types (species or genotypes) are mixed, one of them being tolerant ("mutualistic") and the other resistant ("antagonistic") to herbivory, the resistant type is expected to outcompete the tolerant type because it benefits from the positive indirect effect of increased nutrient cycling, but does not suffer the negative direct effect of herbivore consumption. As a result, tolerance should not evolve, even though it is indirectly beneficial. This might seem to put to an end the idea of any plant–herbivore indirect mutualism, indeed of any evolved indirect interaction, as some authors have suggested (Belsky *et al.* 1993).

This conclusion is premature, however, as two factors counteract this advantage of antiherbivore defense. First, the spatial structure of the plant–herbivore system can generate spatially heterogeneous nutrient cycling. If herbivores recycle nutrient in the vicinity of the grazed plants, or plants from the same type are aggregated, herbivores tend to recycle proportionally more nutrient onto the plants that are grazed more heavily, and thus augment the indirect benefit of grazing for the grazed plants. In such conditions, evolution is governed by the balance between two conflicting levels of selection, just as in the evolution of altruism (Wilson 1980); individual selection within patches, which favors the resistant type over the tolerant one because it has a higher relative fitness, and group selection between patches, which favors patches with a higher proportion of the tolerant type because they have a higher average absolute fitness. The outcome of evolution then depends on the strength of spatial aggregation and patch size; tolerance to grazing evolves provided that spatial aggregation is strong enough or patch size is small enough (de Mazancourt and Loreau 2000a).

Evolution toward plant–herbivore mutualism

Another factor that counteracts the advantage of antiherbivore defense is its cost. Although the empirical evidence for costs of defense is still equivocal (Simms and Rausher 1987; Mole 1994; Bergelson and Purrington 1996; Strauss and Agrawal 1999), some cost seems inevitable in many cases because the production of defense diverts resources from other functions, such as growth and reproduction. This factor is investigated in the previous section, and its consequences are explored further here in an ecosystem context. To this end, we constructed a simple model of a material cycle in a spatially structured ecosystem (Box 17.2). In this model, different plant strategies have different abilities to take up nutrient and to resist herbivory, and there is a trade-off between these two traits. Evolution of the plant traits is analyzed using the theory of adaptive dynamics.

Two major conclusions emerge from this analysis (de Mazancourt *et al.* 2001). First, for most ecologically plausible trade-offs between nutrient uptake and antiherbivore defense, evolution in plants leads to a single continuously stable strategy (CSS), that is to a strategy toward which evolution converges and that cannot be invaded by any other strategy (Eshel and Motro 1981; Eshel 1983). By ecologically plausible trade-off, we mean a trade-off such that plants cannot build defenses that are completely efficient, even when they allocate all their resources to defense, and such that they cannot increase their nutrient uptake rate beyond a maximum value, even when they allocate all their resources to nutrient uptake. The possibility of a single CSS has interesting implications for plant coexistence. Previous studies proposed that the presence of herbivores allows the coexistence of several

Box 17.2 Modeling the evolution of plant defense in an ecosystem context

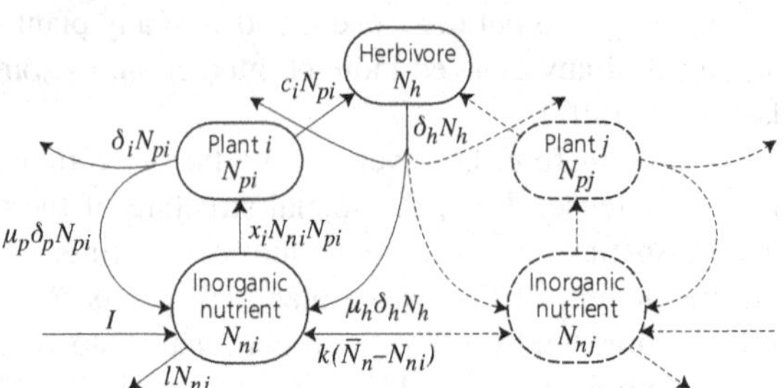

As an example of a model of evolution in an ecosystem context, consider the problem of the evolution of a costly plant antiherbivore defense when herbivory contributes to recycling a limiting nutrient in the ecosystem.

First, the ecological model setting the scene for evolutionary changes must be constructed. We assume a spatially structured ecosystem in which each plant occupies a site during its lifetime and absorbs mineral nutrient in a local resource-depletion zone around its roots at this site (Huston and DeAngelis 1994; Loreau 1996, 1998b). Mineral nutrient migrates laterally in the soil among the resource-depletion zones through diffusion, transport, or other processes. The flow of mineral nutrient into a local depletion zone is proportional to a migration coefficient k and the difference between the mean concentration in the soil and the local concentration. Each resource-depletion zone is replenished with a constant nutrient inflow I, and loses nutrient to the external world at a rate l. The total number of occupied sites is assumed to be constant, of which plants with strategy i occupy a fraction p_i. Herbivory is assumed to be donor controlled – it is determined by the plant's strategy, and does not depend on herbivore density. Plant strategies differ in the rates at which they absorb mineral nutrient and are consumed by herbivores. There is a trade-off between the ability of plant strategy i to take up nutrient and grow, as measured by its nutrient uptake rate r_i, and its ability to resist herbivory, as measured by its herbivore consumption rate c_i. Plants are assumed to be equivalent in all other respects. The nutrient stock N_{pi} in plant strategy i is recycled through two different recycling pathways, those of the plant and the herbivore. Part of the nutrient is not consumed by herbivores and follows the plant pathway; plant detritus is produced at a rate δ_p, of which a fraction μ_p is mineralized and recycled to the local nutrient pool (with nutrient stock N_{ni}) around the plant. The rest of the plant nutrient stock is consumed by

continued

plant species for some trade-offs (e.g., Holt *et al.* 1994). However, coexistence on an ecological time scale does not imply that coexistence can be maintained in the course of evolution. Although there are differences between the two types of models [in particular, coexistence in Holt *et al.*'s (1994) ecological model requires herbivore numerical response, which is not considered in de Mazancourt *et al.*'s

Box 17.2 *continued*

herbivores (with nutrient stock N_h per plant) and follows the herbivore pathway; herbivore detritus is produced at a rate δ_h, of which a fraction μ_h is recycled and distributed uniformly among sites.

The model is thus given by the equations

$$\frac{dN_{ni}}{dt} = I - lN_{ni} + k\left(\overline{N}_n - N_{ni}\right) - x_i N_{ni} N_{pi} + \mu_p \delta_p N_{pi} + \mu_h \delta_h N_h \,, \quad \text{(a)}$$

$$\frac{dN_{pi}}{dt} = x_i N_{ni} N_{pi} - \delta_p N_{pi} - c_i N_{pi} \,, \quad \text{(b)}$$

$$\frac{dN_h}{dt} = \sum_i p_i c_i N_{pi} - \delta_h N_h \,. \quad \text{(c)}$$

The productivity of plant strategy i is here measured by its nutrient inflow, $rN_{ni}N_{pi}$.

This model is intended to describe ecological interactions on a relatively short time scale – say, within a year – such that the spatial distribution of plants can be regarded as constant. On a longer time scale, however, this spatial distribution changes; the plant strategy with the highest reproductive ability increases its occupation of space at the expense of other strategies. The dynamics of the fraction of sites occupied by plant strategy i from year t to $t+1$ obeys a "competitive lottery" for vacant sites,

$$p_i(t+1) = (1-\alpha)p_i(t) + \alpha \frac{p_i(t)R_i(t)}{\sum_j p_j(t)R_j(t)} \,, \quad \text{(d)}$$

where α is the fraction of sites made vacant by mortality each year, and R_i is the reproductive ability of strategy i. We consider two plausible scenarios for the determination of R_i:

- It is proportional to biomass, hence to the plant nutrient stock;
- It is proportional to plant productivity, as measured by the plant nutrient inflow.

We assume that within each year nutrient concentrations attain equilibrium.

To investigate the evolution of plant traits in this model ecosystem, we use the theoretical framework of adaptive dynamics based on the invasion success of a rare mutant strategy invading a resident strategy (Dieckmann 1997; Geritz *et al.* 1998; Chapter 11). Here, the condition for the mutant strategy to invade the resident is $p_{\text{mut}}(t+1) > p_{\text{mut}}(t)$, which, according to Equation (d), is equivalent to $R_{\text{mut}} > R_{\text{res}}$, this condition being evaluated for $p_{\text{mut}} \to 0$ and $p_{\text{res}} \to 1$. This invasion condition requires simply that in an environment determined by the resident strategy, the mutant has a higher reproductive ability than the resident (de Mazancourt *et al.* 2001).

(2001) evolutionary model], this analysis suggests that herbivore-mediated plant coexistence may not be an evolutionarily robust phenomenon.

A second major conclusion concerns the nature of the plant–herbivore interaction. Our model can be used to explore different ecological and evolutionary scenarios of herbivore addition or removal, which leads immediately to a problem of definition: what is called "mutualism"? The classic approach to identify mutualistic interactions in ecology is based on removal experiments or press perturbations (Schoener 1983; Bender *et al.* 1984; Krebs 1985): if each of two populations is affected negatively (in density, biomass, or production) after the other has been depressed or removed, the interaction between them is considered to be mutualistic. Despite its interest, however, this approach has a number of limitations, in particular that the effects of a removal or a perturbation may be different on ecological and evolutionary time scales. It is therefore useful to distinguish two types of mutualism: an ecological mutualism, in which each species gains a benefit from the presence of its partner in the absence of any evolutionary change (as revealed, e.g., by an ecological press perturbation), and an evolutionary mutualism, in which the mutual benefit persists even after evolution has occurred. The conditions for evolutionary mutualism are generally more stringent than those for ecological mutualism, because interacting species may have evolved a mutual dependence, so that the removal of one species may have a negative impact on the other in the short term, but this negative impact may disappear after each species has had the opportunity to evolve and adapt to the new conditions created by the absence of its partner (Douglas and Smith 1989; Law and Dieckmann 1998).

This happens in our model. Not surprisingly, when a plant's reproductive ability is determined by its biomass, herbivory cannot have a positive effect on plant performance. In contrast, herbivore removal can have a negative effect on plant productivity, on both ecological and evolutionary time scales, provided that herbivore recycling efficiency (as measured by the fraction μ_h of nutrient flowing along the herbivore pathway that is recycled within the ecosystem) be sufficiently greater than plant recycling efficiency (as measured by the fraction μ_p of nutrient flowing along the plant pathway that is recycled within the ecosystem). Thus, when a plant's reproductive ability is determined by its productivity, herbivory can have a positive effect on plant performance and thus generate a mutualistic interaction. The requirements on herbivore recycling efficiency, however, are more stringent for an evolutionary mutualism than for an ecological mutualism. A surprising result in this case is that, as herbivore recycling efficiency is increased, the plant–herbivore interaction becomes increasingly mutualistic (first ecologically, then evolutionarily), but at the same time plants evolve to increase their level of antiherbivore defense because they gain a higher benefit from not being consumed relative to plants defended less well (Figure 17.4). Thus, mutualism can go hand-in-hand with increased conflict between partners. Although paradoxical at first sight, such evolutionary conflicts are also known to occur in other mutualistic interactions (Anstett *et al.* 1997b; Law and Dieckmann 1998).

Conservation implications

The preceding considerations show that an ecosystem process such as nutrient cycling can alter the very nature of species' interactions, both in an ecological and in an evolutionary sense. We are not aware of direct empirical evidence for the new

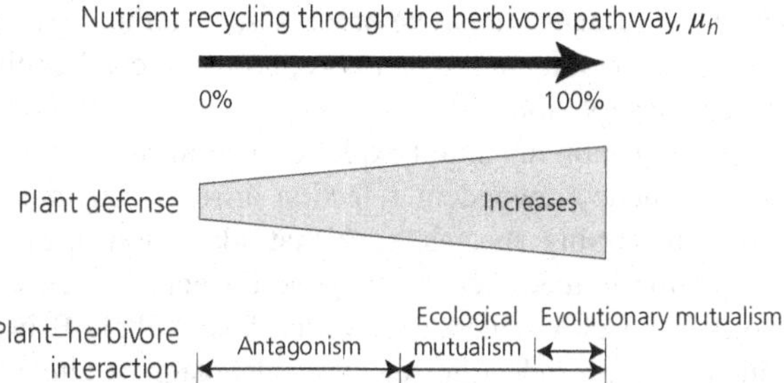

Figure 17.4 Changes in plant–herbivore interaction along a gradient of increasingly efficient herbivore nutrient recycling (μ_h), with the assumption that the plant reproductive ability is proportional to plant productivity. Plant defense increases as herbivores are more efficient at recycling the nutrient, but at the same time the interaction becomes more and more mutualistic. *Source*: de Mazancourt *et al.* (2001).

theoretical insights presented above, but their potential implications for conservation biology are profound. Extinction or introduction of herbivores, for instance, can have different effects on plants in different ecosystems and on different time scales. If herbivores recycle nutrients within the local ecosystem less efficiently than the plants do, their effect should be simple and consistent: their extinction should stimulate plant biomass and productivity. However, if they are more efficient at recycling nutrients within the local ecosystem, their extinction may lead to a cascade of different effects. Plant productivity may increase as a result of a physiological response shortly after herbivore extinction, then decrease because of an ecological response in the medium term, and finally either increase or decrease through species replacement or evolutionary adjustments in the long term. In the process, some plant species will become extinct and others will evolve different traits, so that ultimately the community may be very different in ways that cannot be anticipated from a simple consideration of the immediate, direct plant–herbivore interaction.

17.4 Coevolution in Other Exploiter–Victim Interactions

One important limitation of the models discussed so far is that they assume that the rate of herbivory is controlled by plant traits. More realistically, herbivores may be expected to show numerical and evolutionary responses to plants. Adding these extra dimensions can further change the predicted outcome of evolution. Recent years have seen a tremendous growth in theoretical studies of coevolution in exploiter–victim systems, considered more broadly to include host–parasitoid and predator–prey interactions (e.g., Abrams 1986; Brown and Vincent 1992; Seger 1992; Dieckmann *et al.* 1995; Hochberg and Holt 1995; Gandon *et al.* 1996; Abrams and Matsuda 1997). A full analysis of this problem is beyond the scope of this chapter, and as yet little attention has been devoted to the relationship between coevolution and ecosystem or landscape processes, or to the implications for conservation efforts. Some results in the literature, however, can be reinterpreted in a fashion relevant to conservation.

One general phenomenon in natural exploiter–victim coevolution is that a form of cross-species frequency-dependent selection arises when there are opposing tactics in the two interacting species, such that adaptation in one favors a precise counter-adaptation in the other (as in "gene-for-gene" systems, Frank 1993). Roughly, allele A in species 1 increases, which favors allele B in species 2, the increase of which in turn erodes the selective advantage of allele A, which now declines, and allele B in turn declines with a time lag. The time lag inherent in the cross-species interaction means that models of coevolution often lead to a sustained cycling in allele frequency whenever antagonistic species interact through complementary phenotypes or genotypes (Eshel and Akin 1983).

Often these evolutionary cycles (or more complex patterns of fluctuations) are large in amplitude, which means that in finite populations alleles can be lost when they are rare (Seger 1992). In spatially extended populations with limited dispersal, this is not a problem. Different local populations can be at different evolutionary phases and, with migration, recurrent gene flow will replenish the loss of genetic variation (Gandon *et al.* 1996). However, if habitat destruction and fragmentation are imposed on a system like this, the populations left behind in the habitat remnants tend to lose genetic variation. The exact impact of this loss depends upon a number of details, but we can readily imagine cases of conservation concern. Consider a plant species infected by a fungus. Local populations of the plant are likely to have lower effective population sizes, and the fungus may be more effective at long-distance dispersal. If so, the fungus can maintain its local genetic diversity, even as the plant loses its genetic pool. This means that remnant plant populations face a long-term risk of severe epidemics, and even extinction, because of a stream of novel pathogen genotypes that immigrate and little genetic reservoir from which the plant can mount an adaptive response. More broadly, coevolutionary systems often display geographic mosaics (Thompson 1997). Habitat fragmentation disrupts spatial coupling and so is likely to impact ongoing evolution in many coevolutionary interactions.

Hochberg and van Baalen (1998) examined coevolution of exploiter–victim systems along gradients in victim productivity. For a broad range of models, the interaction evolves toward greater intensity when productivity is high (e.g., as measured by the investment of each species in attack and defensive strategies). The reason is that productivity passes from lower to higher trophic levels, which results in greater impacts of the higher trophic levels when productivity is higher. This translates into a greater selection intensity on the trophic interaction relative to other selective factors. Further, because different genotypes are favored at different productivities, genetic diversity as a whole is maximized when there are viable populations present along all the productivity gradient. This is one rationale for conserving marginal habitats in addition to core productive habitats. It also suggests that anthropogenic impacts upon ecosystem processes could indirectly influence coevolution of natural exploiter–victim systems in a variety of ways. For instance, with carbon enrichment plants might have higher carbon:nitrogen ratios, and thus be lower in quality for herbivorous insects. All else being equal, this would reduce the productivity of these insects for their own specialist parasitoids and other natural enemies, which then become less important as limiting factors, and evolve toward lower effectiveness.

17.5 Local Evolution versus Biological Invasions

Biological invasions represent, after habitat loss and fragmentation, one of the most important environmental changes and threats to biodiversity. How does local evolution in ecosystems interact with biological invasions? The two processes bear some resemblance, since in both processes a rare local mutant or external immigrant progressively invades a community. However, the two processes do not necessarily obey the same constraints, and thus can have distinctly different consequences. To illustrate this, we discuss some potential implications of our evolutionary analysis of plant–herbivore interactions for situations in which there is a trade-off in plants between growth and resistance to herbivory.

Our analysis shows that plant evolution usually leads to a single CSS, and thus that herbivore-mediated plant coexistence may be ecologically, but not evolutionarily, stable. Since a CSS is a strategy that cannot be invaded by any close mutant strategy, once local evolution has produced this CSS, the community is likely to be resistant to invasion by another plant species that obeys the same trade-off (Figure 17.5). Species that originate from the same regional pool are likely to share a common history of environmental constraints, selective pressures, and phylogenetic relationships, and hence are more likely to obey the same trade-off than exotic species. Thus, local evolution should result in resistance to invasion by species from the same regional pool.

In contrast, if the immigrant is an exotic species that is not subject to the same trade-off as the resident, the community is much less likely to be resistant to invasion. Various scenarios are possible, with either invasion failure, ecological coexistence, or competitive displacement of the resident community (Figure 17.5). In particular, if the exotic species escapes herbivory because local herbivores are not adapted to consume it, it is generally better able to deplete the limiting nutrient, and hence to outcompete resident plants, which require higher nutrient availability to compensate for their additional losses to herbivory. Since extinction of the resident plant also entails extinction of the resident herbivore, a catastrophic outcome with displacement of the resident community by the invader may be likely. Local evolution of the invader after its establishment in the community makes this catastrophic outcome even more likely, because, being free from the selective pressure of herbivory, the invader can evolve toward a pure strategy of allocating all its resources to nutrient uptake, which further increases its competitive ability. Several examples of successful invasion by exotic species may conform to this theoretical scenario (Blossey and Nötzold 1995).

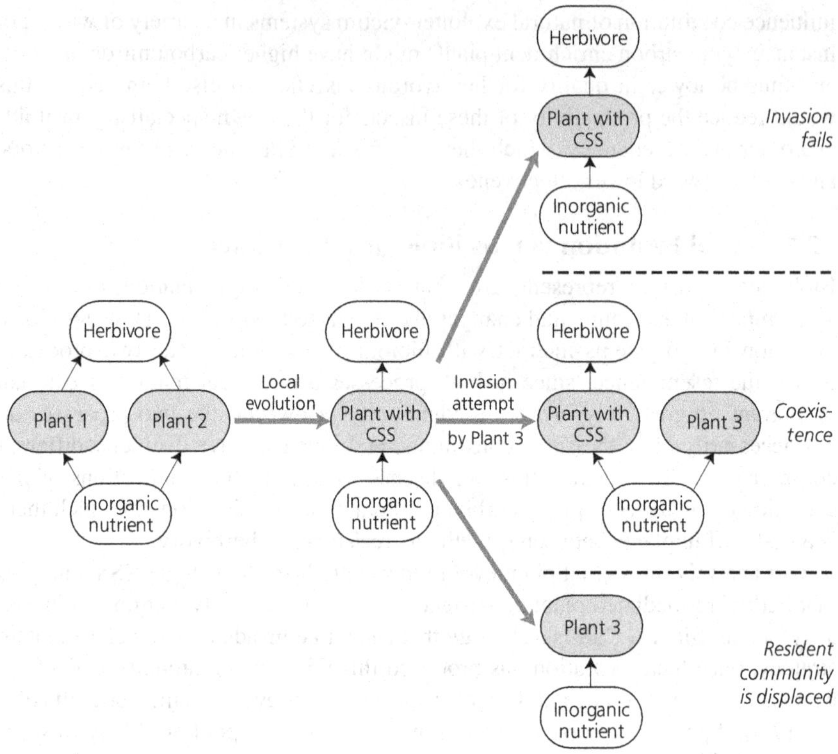

Figure 17.5 Theoretical scenarios for the ecological and evolutionary dynamics of simple nutrient–plant–herbivore communities in which local evolution and biological invasion of plants occur with a trade-off between growth and resistance to herbivory. If the potential invader obeys the same trade-off as the resident species and is consumed by the herbivore, invasion usually fails (top right). If, by contrast, the potential invader escapes herbivory, three cases are possible: invasion fails, the invader coexists with the resident (middle right), or the resident community is displaced by the invader (bottom right).

This suggests that local evolution may increase resistance to invasions by species from the same historical and biogeographic origin, and at the same time be impotent against invasions by exotic species that do not share the same evolutionary history, in particular the same history of herbivory, as the resident species. Local evolution is no guarantee against the disruption of local communities by some invasive exotic species, which are, indeed, a major threat to biodiversity.

17.6 Concluding Comments

Focusing on plant–herbivore interactions as major determinants of ecosystem patterns and processes, we show that the ecosystem is the proper context within which evolution shapes species' traits. In particular, nutrient cycling is a key ecosystem process that transmits predictable indirect effects in ecosystems. These indirect ecological effects can be so strong as to prevail over direct effects and exert effective selective pressures on the species involved, provided that there is sufficient spatial heterogeneity in the system or trade-offs exist between traits associated with the direct and indirect effects. Such effects are even able to change the nature of plant–herbivore interactions from antagonistic to mutualistic, both in an ecological and in an evolutionary sense, under some predictable conditions. We also argue that local evolution in ecosystems is likely to increase the resistance to invasions by species from the same historical and biogeographic origin, but at the same time be impotent against invasions by exotic species that do not share the same evolutionary history.

An evolutionary perspective on conservation is useful for two basic reasons. First, an understanding of evolutionary history can provide organizing principles that are useful to identify the sensitivity of species to different components of environmental degradation (Holt 1995). Second, evolutionary dynamics themselves can lead to a dramatic transformation in a species' ecological properties over short time scales (Thompson 1998). Feedback with the environment can also be important over short time scales. A comprehensive evolutionary conservation biology has to merge ecosystem- and population-level perspectives to predict the responses of ecological systems to environmental degradation.

This strongly suggests that conservation efforts should not only aim to preserve species, but also to preserve the rich web of interactions in which species are embedded in natural ecosystems, and which determine their current traits and persistence. Awareness of this need, however, provides no guarantee against species' extinction and ecosystem disruptions that are likely to result from environmental changes such as biological invasions.

18

The Congener as an Agent of Extermination and Rescue of Rare Species

Donald A. Levin

18.1 Introduction

Species with small population sizes and narrow distributions are represented in floras and faunas throughout the world. These species are vulnerable to extinction by natural processes and human disturbance. Many rare species contain ten populations or less, and some or all of these populations have only a few individuals (Box 18.1; see Schemske *et al.* 1994; Rhymer and Simberloff 1996).

Whereas there is considerable information on the roles of habitat alteration, interspecific competition, pest pressure, and inbreeding in species extinction (Chapters 5, 6, 9, and 15), only recently have we realized that hybridization also may jeopardize the existence of species. This possibility was recognized first by Harper *et al.* (1961). Its potential importance for both plants and animals was discussed by Rieseberg (1991) and Ellstrand (1992), and further considered by Levin *et al.* (1996) and Rhymer and Simberloff (1996).

The primary objective of this chapter is to discuss the role of abundant congeners in the extinction and salvation of rare species. First, I consider the factors that contribute to species contact (Section 18.2), and then interactions between rare species and congeners (Section 18.3). Examples of species threatened by hybridization (Section 18.4) follow. The outcome of hybridization is not always extinction, as we see in the stabilization of hybrid derivatives (Section 18.5) and the rescue of rare species through gene flow (Section 18.6).

18.2 Habitat Change and Species Contact

Contact between rare and common species may be facilitated by habitat alterations on different spatial scales. I consider some scales in order of increasing size.

Species contact may occur through local disturbance. Disturbance reduces competition between the parental species and hybrids, as well as between these entities and their ecological associates. Classic examples of disturbance-related hybridization include two Ozarkian species of spiderworts (*Tradescantia canaliculata* and *T. subaspera*) and two Louisiana species of *Iris* (*I. hexagona* and *I. bicolor*; Anderson 1949).

The creation of corridors between habitats often enables the expansion of aggressive congeners. For example, roadbuilding has allowed many plant species to

Box 18.1 Attributes of rare species

Rare species typically have low abundance and/or small ranges (Gaston 1994). As such, rare species might be expected to differ from common ones in that the former may have more difficulty in finding mates, are more likely to be fed upon by generalist parasites and predators, and are more likely to have heterospecific neighbors (Orians 1997). Rare species tend to differ from common species in that they (Gaston and Kunin 1997):

- Have lower reproductive investment;
- Have breeding systems that lean toward self-fertilization or asexual reproduction;
- Have poorer dispersibility;
- Utilize a narrower range of resources or less common resources.

All rare species are not the same, and neither are they necessarily similar. Rabinowitz (1981) proposed alternative forms of rarity based on habitat specificity (narrow or broad), geographic distribution (narrow or wide), and local population size (everywhere small or large). With the exception of broad specificity and broad distribution (and either population size), all combinations of alternative states of specificity, distribution, and population size are forms of rarity. Most rare species have wide distributions, and nearly all have large populations somewhere in their ranges.

encroach upon the habitats of rare relatives, as discussed below for *Argyranthemum* in the Canary Islands. The alteration of water courses also provides opportunities for range expansion and hybridization between fish taxa, such as an endemic and a widespread subspecies of *Cobitis taenia* in the Dongjin River, Korea (Kim and Yang 1993).

Regional landscape alterations allow the spread of one species into the range of another. For example, extensive cultivation followed by the abandonment of old fields allowed sympatry between the previously allopatric golden-winged (*Vermivora chrysoptera*) and blue-winged (*V. pinus*) warbler. The blue-winged warbler is expanding its range to the north, as a result of competitive superiority, hybridization, and/or habitat changes to which it is better adapted (Gill 1987).

Species contact may be made by jumps over geographic and ecological barriers. Some species have been transported hundreds or thousands of miles by humans, especially in post-Columbian times. For example, the movement of ballast water in ships throughout the world has increased the incidence of contact between closely related, but distantly distributed, near-shore marine mammals (Carlton 1996). Hundreds of plant species introduced from distant continents have escaped from gardens and crossed with indigenous species (Heywood 1979).

A notable example of long-distance vertebrate transport is the mallard duck (*Anas platyrhynchos*), which was introduced into New Zealand, Hawaii, and Australia from the Northern Hemisphere. Hybridization with *A. wyvilliana* in Hawaii may have contributed to the decline of this species (Rhymer *et al.* 1994). In New Zealand *A. platyrhynchos* hybridizes with the indigenous *A. supercilosa* ssp. *supercilosa*, and in Australia the former hybridizes with *A. supercilosa* ssp. *rogersi.*

In summary, contact between related species may result from the movements of species over terrain undisturbed by humans. This is a normal process that may result in the formation of hybrid zones. However, disturbance, our movements, and commerce greatly enhance the opportunities for contact.

18.3 Interactions between Rare Species and Congeners

Three interactions between a rare species and a congener threaten the former's existence: ecological interference; reproductive interference; and hybridization.

Ecological interference

Competition between related species has occurred in many plant and animal genera, but it seems not to be a prime factor in the extinction of rare species (Carlquist 1974; Simberloff and Boecklen 1991; Williamson 1996). However, there are some notable exceptions to the rule.

Competition increases proneness to extinction of a salamander in the genus *Plethodon* (Griffis and Jaeger 1998). Each population of the federally endangered *P. shenandoah* is surrounded by *P. cinereus*, which defends territories against conspecifics and heterospecifics. This behavior enhances the potential for *P. shenandoah* to become extinct by inhibiting its movements from source to distant sink populations.

Even products of hybridization may be a competitive threat, as found in the frog *Rana.* The hybrid derivative *R. esculenta* was introduced into Spain with its parental species *R. lessonae* and *R. ridibunda.* It is likely that *R. esculenta* will outcompete the native endemic *R. perizi* (Arano *et al.* 1995).

Ecological interference can happen between species at different trophic levels. Rare species may be free of the pathogens, predators, or parasites that plague their abundant relatives. Species contact may place the rare species in jeopardy, because related species are often susceptible to attack by the same predators, pathogens, or parasites (Harlan 1976; Fritz *et al.* 1994).

Domesticated cats may be a reservoir for diseases that can be transmitted to other species. For example, cheetahs display extremely high morbidity and mortality from outbreaks of a nearly benign domestic cat virus (feline infectious peritonitis virus; Heeney *et al.* 1990). Their susceptibility may be related in part to their depauperate gene pool, especially immunological loci (O'Brien *et al.* 1985).

Parasite shifts from a common to a rare species, or the potential for such, have been documented also in fish. For example, helminth parasites associated with the introduction of the sturgeon (*Acipenser stellaturs*) devastated populations of the native sturgeon (*Ac. nudiventris*) in the Aral Sea (Bauer and Hoffman 1976). Translocations of guppies (*Poecilia*) may expose native relatives to exotic trematodes (Harris 1986).

Reproductive interference

Reproductive interference is expressed as a reduction in conspecific progeny per capita as a result of interspecific matings (pollinations). Interference may arise from failed matings or by the wastage of gametes on nonviable or sterile hybrids.

Conspecific seed production may be reduced by interspecific cross-pollination. Alien pollen may block sites for conspecific pollen (*Polemonium*; Galen and Gregory 1989). Moreover, the presence of heterospecific pollen tubes in styles may inhibit conspecific pollen tube growth (*Petunia*; Gilissen and Linskens 1975) or may cause the abortion of "pure" seed (*Erythronium*; Harder *et al.* 1993).

In animals, interspecific mating *per se* may reduce the reproductive effectiveness of the species that serves as the female. In *Drosophila* the progeny of promiscuous females may be destroyed by virus-like P-elements (Engels 1983). In the tsetse fly *Glossina*, females are killed during coitus by males of another taxon as a result of mechanical mating incompatibility (Vanderplank 1948). A similar sort of incompatibility causes the mortality of female cimicid parasites of birds when *Hesperocimex coloradensis* females copulate with *H. sonorensis* males (Ryckman and Ueshima 1964).

The introduction of American mink (*Mustela vison*) into Europe has been followed by a decline in the European mink (*M. lutreola*; Rozhnov 1993). Females of the European mink mate with the larger American mink, and become averse to mating with males of their own species. However, promiscuous females have no offspring that year, because embryos resorb at an early stage. In the mean time, American mink females mate with conspecific males and have offspring.

The effect of reproductive interference on survival is seen in the plant genus *Clarkia*. In natural and artificial mixtures of *C. biloba* and its rare derivative *C. lingulata*, the numerically superior *C. biloba* consistently eliminates its derivative (Lewis 1961). The species cross readily and the proportion of hybrid seed produced by the minor species increases as it becomes less numerous. Hybrids are sterile.

Hybridization

Hybridization may foster the extinction of a rare species through genetic amalgamation or fitness decline. Amalgamation involves the formation of fertile F1 hybrids and backcross hybrids, and an increase in the size of the hybrid subpopulation relative to the rare entity. As the hybrid subpopulation grows, an increasing proportion of the progeny of the rare entity are hybrids. As a result, the minority entity has a reduced chance of replacing itself, and eventually is assimilated by the more abundant relative (Box 18.2).

The inflow of alien genes may cause a decline in population fitness. In plants there are many cases in which F1 hybrids are fertile and vigorous, but the F2 segregates or backcross hybrids are either weak or partially sterile (Stebbins 1958; Levin 1978). This advanced-generation breakdown results from the disruption of coadapted genes or chromosomes (Templeton 1986). In the plant genus *Zauschnaria*, F1 hybrids between *Z. cana* and *Z. septentrionalis* are vigorous and semifertile.

Box 18.2 Minority disadvantage

If species are present in unequal numbers, interference that involves gametic wastage or the production of nonviable or sterile offspring affects the minority species more than the majority one. When mating is at random, the minority species is always at a disadvantage, all else being equal, because the proportion of "pure" (= nonhybrid) offspring is a function of the proportional representation of a species:

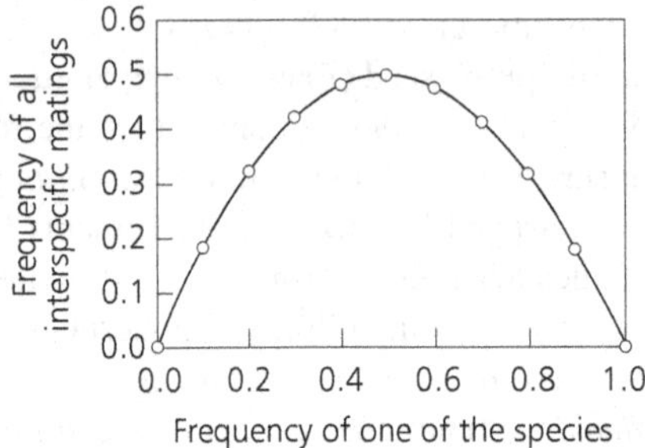

As shown in the figure above, the lower the proportion, the lower is the percentage of offspring that are "pure", and thus that take part in species replacement from generation to generation. If total progeny number is constant, then hybrids are produced at the expense of "pure" individuals.

However, the F2 generation consists principally of weak, small, disease-prone plants (Clausen *et al.* 1940).

The influx of alien genes may be accompanied by an increased susceptibility to pathogens or predators (herbivores). The literature for plants was reviewed recently by Fritz (1999).

18.4 Species Threatened by Hybridization

Many examples exist of rare species threatened by interspecific hybridization. Nearly all episodes of hybridization involve some form of human intervention, the most common being the introduction of congeners and habitat disturbance (Box 18.3).

One species in which the extinction process is nearly complete is the California mahogany, *Cerocarpus traskiae*. The species is restricted to the Santa Catalina Island off southern California, and is known only from a single population. This species is hybridizing extensively with *Ce. betuloides* (Rieseberg *et al.* 1989; Rieseberg and Gerber 1995). Few pure plants remain.

A rare invader also may endanger a large local population. A prime example involves the common native California cordgrass (*Spartina foliosa*) and the rare invader, smooth cordgrass (*S. alterniflora*). The site is San Francisco Bay. The invader produces about 21 times as much pollen per square meter as the native

Box 18.3 The threat to rare *Argyranthemum*

The composite *Ar. coronopifolium*, which inhabits the island of Tenerife (Canary Islands), is known from only seven populations. Four are isolated, two are in various stages of hybridization with the weedy and abundant *Ar. frutescens*, and one is approached by this species (Brochmann 1984; Bramwell 1990); see the figures below. The latter migrated into habitats of *Ar. coronopifolium* along corridors established by roadbuilding. We can see stages in the assimilation of this species at "Risco del Fraile", because the site has been monitored for over 30 years. Species contact occurred in 1965 (Humphries 1976). Hybridization ensued, and by 1981 only a few morphologically pure *Ar. coronopifolium* were embedded in a large hybrid swarm skewed toward *Ar. frutescens* (Brochmann 1984). By 1996, *Ar. coronopifolium* had disappeared. The population contained only hybrids and *Ar. frutescens* (Levin *et al.* 1996).

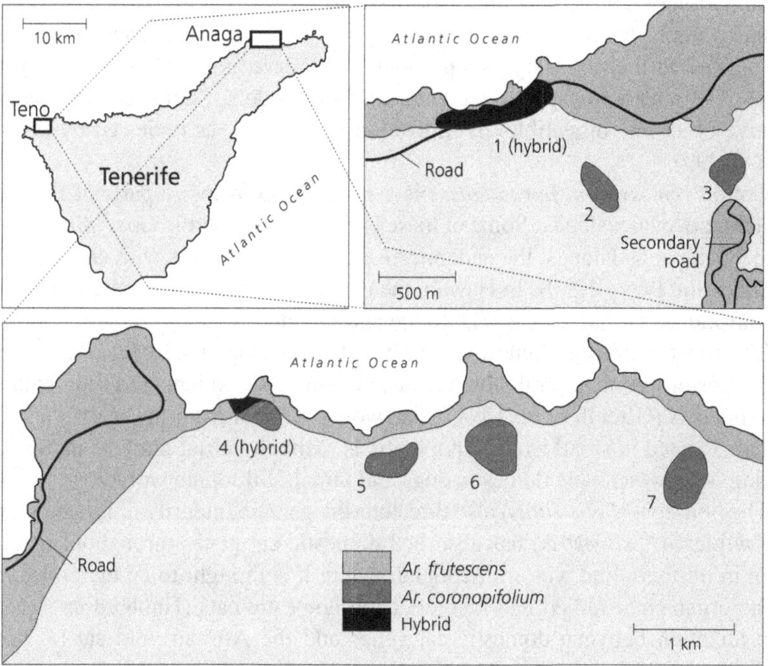

Distribution of *Ar. coronopifolium* in relation to *Ar. frutescens* in Tenerife. The former species occurs only in the Teno and Anaga peninsulas. *Source*: Levin (2000).

Other rare *Argyranthemum* species on Tenerife are also threatened by *Ar. frutescens* (J. Francisco-Ortega, personal communication). One of them, *Ar. sundingii*, is known only from one population that contains 12 flowering plants. The abundant relative was observed recently less than 2 km from it. The species hybridize readily. The other endangered rarity is *Ar. vincentii*, which is known from two populations, in one of which hybridization is occurring.

(Antilla *et al.* 1998). Moreover, the pollen germination rate of the invader is about twice that of the native. This is quite important, because the two species are cross-fertile.

The introduction of related species may endanger the existence of rare species, especially on islands. In the British Isles *Linaria repens* hybridizes with the rare *L. vulgaris*, and *Pinguicula grandiflora* hybridizes with the rare *Pi. vulgaris* (Stace 1975). In the Canary Islands, the introduced *Arbutus unedo* crosses with the rare *Ar. canariensis* (Salas-Pascual *et al.* 1993); and the endemic *Senecio teneriffae* interbreeds with the introduced *Se. vulgaris* (Gilmer and Kadereit 1989).

Hybrids have the upper hand in the crayfish *Orconectes* (Roush 1997). Hybrids of the Kentucky native *O. rusticus* (rusty crayfish) and the native *O. propinquus* (blue crayfish) are not only fertile, but also outcompete both species. About 30 of the 340 crayfish species found in North America may soon be eliminated by some combination of competition and hybridization.

Indigenous plants may be threatened by congeners when vast acreages of crops are planted in their vicinity. Crop populations serve as major sources of pollen, some of which are transported well beyond their borders. Should the crop and wild relative be cross-compatible, hybridization may ensue, as reviewed by Ellstrand *et al.* (1999).

Cotton (*Gossypium barbadense*) is a major crop in many parts of the world, including oceanic islands. Some of these islands have endemic *Gossypium* species. In the Galapagos Islands, the endemic *G. darwinii* hybridizes with *G. barbadense* (Wendel and Percy 1990). In Hawaii, the endemic *G. tomentosum* hybridizes with the introduced *G. barbadense* (DeJoode and Wendel 1992).

The most pervasive domesticated animal is the dog (*Canis familiaris*). Wild dogs hybridize with several other canids, including the endangered Ethiopian wolf *Ca. simensis* (Gottelli *et al.* 1994). The wolf is known from fewer than 500 individuals located in small isolated populations. Mitochondrial markers indicate that mating is between male domestic dogs and female Ethiopian wolves.

The housecat (*Felis catus*) also threatens the genetic integrity of related species. The wildcat (*F. silvestris*) has absorbed domestic cat genes throughout its range. Even in northern and western Scotland, where it is thought to be the purest, 80% of the organisms studied had markers from domestic cats (Hubbard *et al.* 1992). Hybridization between domestic cat genes and the African wild cat (*F. libyca*) similarly threatens the existence of the latter (Stuart and Stuart 1991).

Many animal species (especially fish) have been introduced as part of restocking programs. For example, the introduction of spotted bass (*Micropterus punctulatus*) into the drainage of the native smallmouth bass (*Mi. dolomieui*) resulted in the introgressive swamping of the native (Avise *et al.* 1997). In the Lake Chatuge (Georgia) population, 95% of the remaining smallmouth bass mitochondrial DNA (mtDNA) haplotypes and a similar percentage of nuclear genes reside in animals of hybrid origin.

The time to the extinction of local populations (through either reproductive interference and/or swamping) is a function both of the relative abundances of the rare and abundant species and of the breeding systems of the species. The smaller the proportional representation of the minority species, the sooner the minority species becomes extinct. Also, the closer the population is to random mating, the sooner the minority species becomes extinct. Partial self-fertilization and asexual reproduction mitigate the minority disadvantage, because some percentage of offspring are pure regardless of the species' proportions.

Marguerite daisy
Argyranthemum frutescens

Ellstrand *et al.* (1999) show how rapidly extinction may occur. If 900 individuals of a common species cross randomly with 100 individuals of the rare entity, the latter may become extinct within just two generations. If the rare species crosses with a conspecific five times more frequently than expected with random mating (assortative mating), the time to extinction will be twice as long as that with random mating.

Instead of two species meeting *en masse*, an abundant congener may gradually invade populations of the rare species. Let us assume that invaders enter a small population at some rate, invaders have a fitness advantage, and either no hybrids are formed or they are sterile. The time to replacement is the shortest when the immigration rate and fitness differential are high – less than 30 generations in some scenarios (Huxel 1999). The production of hybrids with fitness that exceeds that of the native accelerates the pace of assimilation.

Thus far local populations have been considered here. However, rare species may have several populations. The extinction of a rare species proceeds most rapidly if all the populations hybridize with the congener. The greater the percentage of populations, the higher the extinction rate as the proportion of nonhybridizing populations declines (Burgman *et al.* 1993).

18.5 Stabilization of Hybrid Derivatives

Instead of hybridization being followed by the extinction or assimilation of one of the participants, it may be followed by the stabilization of hybrid derivatives. Stabilization may occur with or without a change in ploidal level.

One rigorous demonstration of hybrid speciation involves *Helianthus annuus* and *He. petiolaris*. The former occurs on heavy clay soils, and the latter on dry sandy soils. Rieseberg *et al.* (1996) demonstrated that these species contributed to the origin of *He. anomalus* and *He. deserticola*, which are more xerophytic than *He. petiolaris*. They also parented *He. paradoxus*, which is confined to brackish or saline marshes. The disparity in habitat tolerances of these derivatives relative to each other and the parental species is quite extraordinary, and indicates that hybridization can foster unexpected and diverse habitat shifts from the same basic genetic milieu.

The stabilization of the *Helianthus* hybrid derivatives occurred even though F1 hybrids between *He. annuus* and *He. petiolaris* have pollen viabilities less than 10% and seed viabilities less than 1%. These species differ by at least three inversions and seven reciprocal translocations, which are the prime sources of sterility (Rieseberg *et al.* 1996).

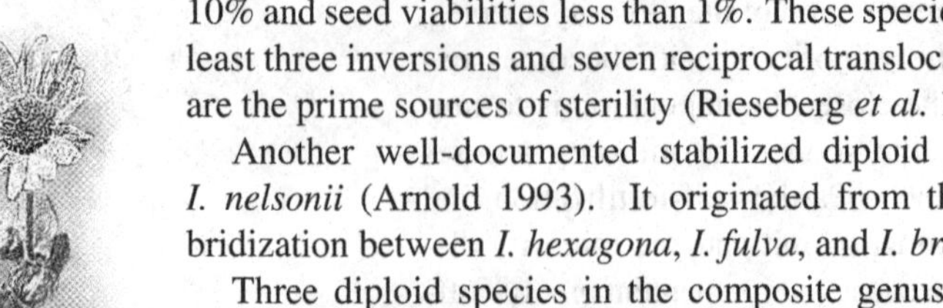

Sunflower
Helianthus annuus

Another well-documented stabilized diploid derivative is *I. nelsonii* (Arnold 1993). It originated from three-way hybridization between *I. hexagona*, *I. fulva*, and *I. brevicaulis*.

Three diploid species in the composite genus *Tragopogon* (*Tr. pratensis*, *Tr. porrifolius*, and *Tr. dubius*) were introduced from Europe into North America in the early 1900s (Ownbey 1950). Hybridization between *Tr. dubius* and *Tr. porrifolius* followed by chromosome doubling gave raise to *Tr. mirus*; and hybridization between *Tr. dubius* and *Tr. pratensis* followed by chromosome doubling gave rise to *Tr. miscellus* (Soltis and Soltis 1993). It is especially notable that each polyploid species had multiple independent origins. *Tr. miscellus* may have originated as many as 21 times (Soltis *et al.* 1995).

Another polyploid derivative of recent vintage is the British allohexaploid *Se. cambrensis*, which is a product of hybridization between *Se. vulgaris* and *Se. squalidus* (Ashton and Abbott 1992). Contact between the parental species did not occur until approximately 1910; and the allohexaploid was first discovered in North Wales in 1948.

Turning to animals, a prime example of a sexual hybrid species is the cyprinid fish *Gila seminuda.* Its derived from *G. elegans* and *G. robusta* (DeMarais *et al.* 1992). Some populations of *G. seminuda* are quite distinct from one another, which suggests diversification within the species or multiple origins.

Another animal thought to be of hybrid origin is the red wolf (*Ca. rufus*). The species is thought to have originated through hybridization between the coyote (*Ca. latrans*) and the grey wolf (*Ca. lupus*; Wayne and Jenks 1991; Roy *et al.* 1994).

Many stabilized derivatives among vertebrates are unisexual, that is, female (Dawley and Bogart 1989). Most are polyploid. Some may require fertilization, others not. The sperm-dependent species must coexist with a sexual relative, and thus are likely to have some overlap in ecological tolerances. The parthenogenetic species are not so restricted, and many have rather divergent tolerances relative to their progenitors (Vrijenhoek 1994).

Some of the better known animal hybrid taxa are in the fish *Poeciliopsis.* Hybridization between *Po. monacha* females and *Po. lucida* males produced a parthenogenetic species (Vrijenhoek 1994). The presence of genetically distinct lines indicates that this entity has had multiple, independent origins (Quattro *et al.* 1991).

About one-third of the 50 *Cnemidophorus* (lizard) species are unisexual, hybrid derivatives (Dessauer and Cole 1989). Densmore *et al.* (1989) investigated the

origin of nine morphologically distinct species from the *Cn. sexlineatus* species group. All but one were triploid. MtDNA cleavage maps showed that these species were most alike *Cn. inornatus* ssp. *arizonae*, which thus was judged to be the female parent. Either *Cn. costatus* or *Cn. burti* were the male parents.

18.6 Rescue of Rare Species through Gene Flow

The gene pools of endangered species often are depauperate. By augmenting these pools with novel genes from conspecific populations or related species, the persistence of endangered species may be prolonged. The introduction of genetic novelty may modify the adaptive gestalt of populations and/or increase genetic variation and heterozygosity.

Several claims have been made that the distribution, ecological tolerances, or strategies of species are influenced by hybridization. However, even if introgression can be demonstrated, we typically lack *prime facie* evidence that a transfer of adaptations has indeed occurred, because the character of the recipient populations prior to hybridization is unknown.

One way to resolve this problem is to study species in which introgression is recent and pure populations still exist. Domesticated plants and their wild relatives offer the best system for doing this. In many species, gene flow from wild populations into domesticates has fostered the genesis of weedy races within a short time span (Harlan 1992).

A classic example of weed origin through introgression involves the cultivated radish *Raphanus sativus* and the wild *Ra. raphanistrum* (Panetsos and Baker 1968). Introgression of *Ra. raphanistrum* was a prime factor in converting the erstwhile crop into a weed.

A shift in ecological tolerance in association with hybridization has occurred in the finch genus *Geospiza* on the Galapagos Island of Daphne Major (Grant and Grant 1992, 1996). A drought from 1977 to 1982, followed by unusually high

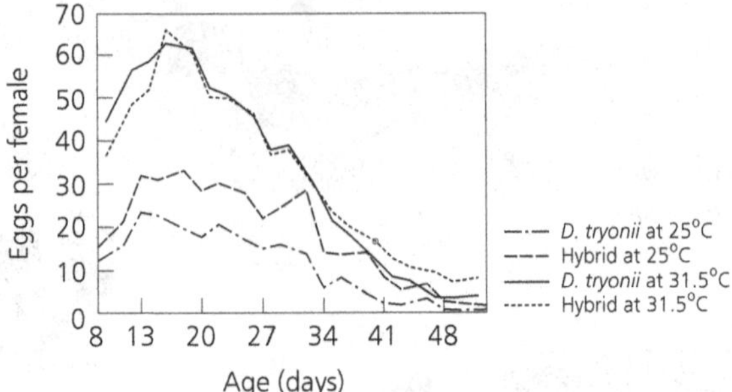

Figure 18.1 The fecundity schedules of *D. tryonii* and populations derived from hybridization between that species and *D. neohumeralis* grown at two temperatures. *Source*: Lewontin and Birch (1966).

precipitation in 1983, greatly reduced the number of large and hard seeds available to birds and other seed eaters. Hybrids between *Ge. fortis* and *Ge. scandens* were at an advantage relative to the parental species, because their beak morphology allowed more efficient foraging for small seeds. The frequency of hybrids increased significantly.

The enhancement of genetic variation within small populations allows more substantive responses to selection and an escape from inbreeding depression. Hybridization may even be an evolutionary stimulus that enables populations to undergo rapid change (Anderson and Stebbins 1954). The Queensland fruit fly species *Dacus tryonii* has extended its range sharply in the past 100 years and invaded habitats with higher temperature regimens. Lewontin and Birch (1966) argued that introgression from *D. neohumeralis* promoted this shift. They measured population growth in the two species and in hybrid populations under different temperature regimes for several generations. Eventually, populations derived from F1 hybrids shifted in the direction of *D. tryonii*, but were more tolerant of high temperatures than nonintrogressed populations of this species (Figure 18.1). This tolerance evolved from the enriched gene pool afforded by hybridization rather than from introgression.

Whereas the infusion of novel genes in general may be advantageous for a small population, novelty at some loci may be crucial for its survival. An example of such is the self-incompatibility (S) locus in plants. Small populations may have a paucity of variation at this locus, a condition that restricts the number of successful mating combinations. This restriction is apparent in the rare composites *Hymenoxys acaulis* (De Mauro 1990) and *Aster furcatus* (Reinartz and Les 1994). Byers and Meagher (1992) demonstrated that populations with fewer than 25 plants are unable to retain large numbers of S-alleles. The introduction of new S-alleles would increase the reproductive success of populations as a whole. They may come from conspecific or heterospecific sources.

18.7 Concluding Comments

The fate of rare taxa is dependent on many biotic and abiotic variables. Only recently have we begun to appreciate that the presence of related taxa may also influence this fate. The genetic swamping of rare taxa by their abundant congeners is known in an increasing number of plant and animal genera. This process is facilitated by reproductive and ecological interference. Such interference may have an immediate impact on population size, whereas the negative effects of hybridization are likely to be evident only after several generations of hybridization.

The species most threatened by hybridization are likely to have very restricted ranges. In plants, insular species are particularly vulnerable to hybridization because of weak crossing and postzygotic barriers (Levin *et al.* 1996). Moreover, they often have poorly differentiated floral architecture and unspecialized pollinators, which enhance the incidence of interspecific pollination. Rare species are also vulnerable because hybridization is not likely to be counterbalanced by immigration from conspecific populations.

Gene flow from a related species may bolster the adaptive posture of a rare species or provide it with the genetic wherewithal to respond to selection. However, little evidence indicates that rare species have, indeed, been rescued by this mechanism. Their small overall population size makes swamping a more likely outcome than an altered course of evolution. However, evolutionary novelty associated with genomic mixing is evident in the many stabilized hybrid derivatives.

19

Epilogue

Régis Ferrière, Ulf Dieckmann, and Denis Couvet

"Ecologists traditionally have sought to study pristine ecosystems to try to get at the workings of nature without the confounding influences of human activity. But that approach is collapsing in the wake of scientist's realization that there are no places left on Earth that don't fall under humanity's shadow" (Gallagher and Carpenter 1997).

19.1 Introduction

Indeed, the preoccupation of evolutionary ecologists with the pristine reflects a long tradition in western culture and a philosophy that separated humanity and nature (Latour 1999; Gould 2000; Western 2001).

As highlighted by the quote above, currently a large fraction of the world's ecosystem structure and dynamics is dominated by human effects (Vitousek *et al.* 1997; Palumbi 2001). By the 20th century, domestic production and settlement had visibly transformed nearly half of the world's land surface, and as we enter the 21st century, human activity is altering biogeochemical cycles and climate on a global scale (Hammond 1998; Western 2001). As a consequence, we must face the prospect of large-scale extinctions in the near future. While this could become comparable in magnitude to some of the catastrophic mass-extinction events of the past, the current biodiversity crisis has a unique feature: humankind as the primary cause. The threat is intrinsic, and because the originator of the trauma has a presumed capacity to mitigate its own deleterious impact, conservation action may be warranted (Novacek and Cleland 2001).

In this closing chapter we argue that evolution in the wake of human-induced environmental change should be the default prediction and should therefore be part of every thorough conservation analysis. By appreciating the potential speed and pervasiveness of anthropogenic evolutionary change, by predicting evolutionary trajectories where possible, and by managing evolutionary threats and responses with foresight, evolutionary conservation biologists can help to reduce or steer our evolutionary impact on the biosphere and thus ameliorate the economic and social costs of altered eco-evolutionary processes.

19.2 Humans as the World's Greatest Evolutionary Force

The ecological role humans now play in the world and the industrialization of our agriculture, medicine, and landscape mean that humankind has an overwhelming impact on the evolutionary processes that produce, maintain, and sometimes doom biodiversity (Palumbi 2001). One striking feature of contemporary human activities is that they raise highly diverse combinations of threats to ecosystems, at a probably unprecedented pace. The evolutionary history of life is marked with environmental challenges, in response to which local adaptations, dispersal, and phenotypic plasticity have evolved.

Do historical adaptations to previous environmental challenges help or hinder populations to respond adequately to current, multifaceted environmental changes? An answer to this question is far from obvious. For example, the alternation of glaciation and deglaciation episodes during the past million years caused repeated drastic changes in the distributions of most temperate-zone species (Dynesius and Jansson 2000). While today's loss and deterioration of habitats, which result from urban and agricultural development, might be envisaged as imposing similar challenges for the adaptation of species, the accompanying habitat fragmentation represents a novel impediment to range shifts and gene flow (see Chapters 11 to 14 for theoretical accounts of this issue, and Chapters 12 and 15 for an empirical perspective; see also Davis and Shaw 2001). A wealth of evidence from controlled experiments, artificial selection in plant and animal breeding and analyses of paleontological records underscores that adaptive evolution can proceed on short time scales (Chapters 5 and 6). On the other hand, however, it has also been demonstrated that, sometimes, genetic interdependence among traits (Chapter 7) can retard evolutionary responses to a point at which evolutionary rescue becomes unlikely (Davis and Shaw 2001; Etterson and Shaw 2001).

Human activities also impact greatly on the genetic and specific variation of communities upon which selective forces operate, often with deleterious consequences. The loss of genetic diversity is expected to hamper adaptation and trap populations in evolutionary dead ends (Chapters 1 and 5). By contrast, biotic exchanges, for which humans are effective agents in all regions of the globe, result in injuriously accelerated evolution (Vitousek *et al.* 1996; Mooney and Cleland 2001; Novacek and Cleland 2001; Chapters 17 and 18). Some of the more dramatic examples, such as the introduction of Nile perch into Lake Victoria and the resultant loss of at least 200 endemic cichlid species (Witte *et al.* 1992), offer sobering experimental evidence for the potentially catastrophic effects of invasive species – aggravated in this case by the further alteration of the food web that resulted from the lake's eutrophication in the 1980s (Verschuren *et al.* 2002). Invaders in general can be expected to affect community adaptation in a rapid manner, by matching local selection pressures and by inducing evolutionary responses in native species (Thompson 1998; Huey *et al.* 2000).

Biotechnology introduces more human-mediated mechanisms that generate evolutionary novelty. Some genetically modified organisms result from the insertion of exogenous genes into domestic plants and animals – effectively increasing the rate at which new traits and trait combinations become available, and thus acting akin to macromutations (Chapter 8). When modified traits cross from domestic into wild species, they can undergo rapid spread and thus add to the fuel of evolution in natural populations (Abbo and Rubbin 2000; Palumbi 2001). The introgressive hybridization of cultivars and their "wild" ancestors can eventually lead to the evolution of aggressive weeds, the disruption of ecological processes, and the loss of native species (Chapter 18). Macromutations with unknown genetic effects may also arise as a result of increases in background mutagen concentrations, from increases in the ultraviolet B (UVB) mediation of ozone depletion by nitrous oxide and chlorofluorocarbons, and from nuclear waste storage. Such potentially serious threats will require vigilance and careful assessment by evolutionary conservation biologists.

19.3 Evolutionary Conservation in Anthropogenic Landscapes

Evolutionary conservation biology must aim at practical and effective conservation strategies in a world in which human populations and wildlife communities are highly integrated. One of the most acute challenges is raised by changes in land use, ranked as the most intensive driver of terrestrial environmental change in the 21st century (Sala *et al.* 2000; Novacek and Cleland 2001). Projections for the expected impact of land-use change on the planet's biota are so stark that any conservation efforts must be geared realistically against a continual tide of human activities. There already are two major directions in the effort to constrain the rampant destruction of natural habitats, to which evolutionary conservation biology should contribute:

- To identify "biodiversity hotspots" at the local scale of preserved areas, and to establish management priorities accordingly (Myers *et al.* 2000);
- To define and implement sustainable practices and management programs at the larger scale of highly populated areas.

On which basis should biodiversity hotspots be identified and ranked for intensive study and conservation efforts? Realizing that the current composition and structure of ecosystems represent the "canopy" of a forest of evolutionary trees, evolutionary conservation biology raises the issue of whether and how we should account for evolutionary history in defining such conservation targets. It has been argued that even if we lose 90% of the species on the planet, we may lose only 20% of the phylogenetic diversity (because most genera have several species, and the survival of one might capture most of the genetic variability that exists within the whole clade; Nee and May 1997). So is one tuatara worth 200 species of skinks? The tuatara has been dubbed "the world's most unique reptile" for being the last surviving species in an order that stretches right back to the Mesozoic (other such "living fossils" include the coelacanth fish, the horseshoe crab, and the native frogs of New Zealand). As emphasized by Loreau *et al.* (Chapter 17), evolutionary conservation biology in natural sanctuaries does value the phylogenetic uniqueness of the tuatara, but perhaps most importantly stresses the value for long-term and global conservation of a web of ecological interactions, such as those in which highly diverse communities of skinks are embedded (Woodruff 2001). As

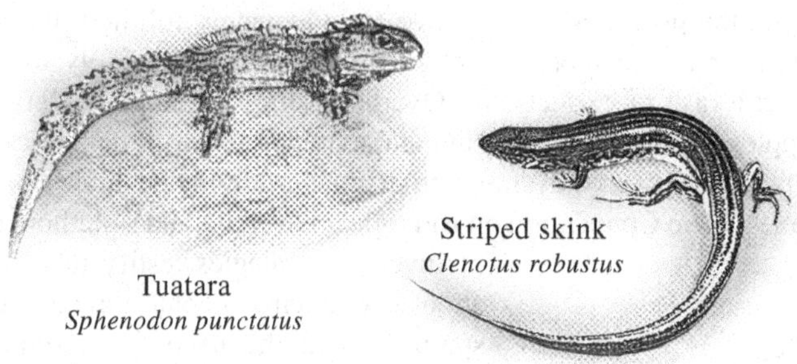

far as priorities are concerned, two lines of action should therefore be pursued and balanced under the constraints of limited financial and technical resources:

- Species-specific conservation effort, advocated not only as a matter of esthetics or biophilia, but most importantly whenever such species are critical to maintaining the basic ecological relationships and evolutionary processes within a community (Chapters 1–4 and 16–18);
- Conservation of groups of less charismatic and often poorly known organisms that may perform apparently redundant roles in an ecosystem, so-called "functional groups", to preserve the way that nature evolved to hedge its bets in the face of an uncertain future (Western 2001; Chapter 17).

Species-specific management in biodiversity hotspots raises several important issues for the genetic arm of evolutionary conservation biology (Hedrick 2001):

- *Detecting genetic erosion.* Genetic erosion, which is both a symptom and a cause of endangerment of small populations, can become a dominant concern in isolated wildlife reserves. The very detection of genetic erosion in small populations is problematic and requires integrated surveys of demography and genetics, and their interaction.
- *Linking inbreeding and adaptations.* As a consequence of genetic erosion, inbreeding is expected to impair adaptation primarily (Chapter 5); but, although the deleterious impact of inbreeding on population demography has been demonstrated clearly (Nieminen *et al.* 2001), the long-term consequences on and through the eco-evolutionary feedback loop remain poorly understood (Chapters 3 and 5). For example, habitat fragmentation has a direct effect on local levels of inbreeding, which may in turn alter selective pressures that act on dispersal, and thereby further modify rates of inbreeding (Chapter 12; Perrin and Mazalov 2000; Ebert *et al.* 2002).
- *Managing gene flow in the face of local maladaptation.* A fragmented habitat is also the substrate of local maladaptation (Chapters 13 and 15), which can be amplified by regional environmental change (as demonstrated in a demographic and physiological study of great tits, *Parus major*, by Thomas *et al.* 2001b). Thus, gene flow in fragmented landscapes subject to global change is not necessarily beneficial to population persistence and adaptability, and could be an important target of active management (Templeton *et al.* 2001).

Even when the priorities listed above are adequately fulfilled, the global network of biodiversity hotspots and other protected areas is likely to remain too small to avert a rash of extinctions. Overharvesting, resource depletion, and the growing ripples of by-products of human activities result in ecological homogenization, simplification, and dysfunction in human-dominated landscapes (Western 2001). The 1992 Rio Convention on Biological Diversity and a plethora of national biodiversity strategies testify to the consensus about the environmental threats of overconsumption and the need for sustainable practices at a global scale (Hempel 1996). Even those biodiversity hotspots that do or should receive the highest degree of official protection are highly vulnerable to threats from outside the system, including climate change, pollution, nitrogen deposition, and biological invasions (Dobson 1996). On the other hand, even in human-dominated landscapes not all species are losing ground to us. Some live with us and prosper – in German, these are known as *Kulturfolger*, culture followers. It is part of the research agenda of evolutionary conservation biologists to help discover how to share anthropogenic habitats with wild species to maintain and promote their diversity (Rosenzweig 2001, 2003). A growing number of studies pave the way in this respect. For example, "countryside biogeography" shows that some styles of land use are already compatible with the ecological and evolutionary needs of many species (Greenberg *et al.* 1997; Daily *et al.* 2001). "Reconciliation ecology" aims to combine controlled experiments and the analyses of large-scale ecological patterns to find how to preserve biodiversity in habitats that host high levels of human activity (Rosenzweig 2003).

Great tit
Parus major

To maintain and restore the evolutionary potential of ecosystems that persist in areas heavily impacted by human activities, evolutionary conservation biologists should seek ways to harness the forces of evolution to their advantage. Rarely has this been attempted so far (Ewald 1994, p. 215; Palumbi 2000), although encouraging examples come from virulence and pest management, on the basis of a fruitful dialogue between theory and practice (Dieckmann *et al.* 2002). A striking example is provided by the use of chemical control in which resistance includes a severe metabolic cost, and so makes resistant organisms less fit when the chemicals are removed (McKenzie 1996; Palumbi 2001). Methods currently used to achieve successful virulence management impact all three factors that drive evolutionary change: variation in fitness-related traits (e.g., in human immunodeficiency virus 1 by limiting the appearance of resistance mutations; Wainberg *et al.* 1996), directional selection (e.g., by varying the choice of antibiotics over time, Lipsitch *et al.* 2000), and heritability of fitness-related traits (e.g., by artificially increasing the proportion of individuals without resistance alleles; Mallet and Porter 1992). However, seldom have all three evolutionary factors been manipulated in the same system, and seldom has the engineering of the evolutionary process been attempted in a systematic fashion. In this vein, recent experimental work on selection at the level of the ecosystem (Swenson *et al.* 2000) opens interesting new directions of research, which may eventually lead to innovative practices in restoration ecology.

19.4 Culture's Role in the Eco-evolutionary Feedback Loop

The future of biodiversity will be shaped by our awareness of the global threats and our willingness to take suitable action. Our ability to do so is currently hampered by several factors, including the poor state of our biospheric and geospheric knowledge, the ignorance of human impact, and the lack of guidelines for sustainability. The paucity of good policies and the lack of incentives to adopt practices in daily life that would be compatible with sustainability are related to the presently still weak connection between biodiversity and human welfare (Western 2001). Put in a pointed manner, our systematic alteration of eco-evolutionary processes is indeed hard to fault based on our own evolutionary success to date.

Anthropogenic challenges to biodiversity take on a different complexion, however, when the growing problems of overconsumption, ecological and evolutionary side effects, and rising costs are considered (Western 2001). The cost of growing human consumption can be measured in falling yields, mounting pollution, and rising production costs (Botsford *et al.* 1997; Daily 1997; Myers and Kent 1998). Nearly half of the world's marine fish stocks are fully exploited and another quarter are overexploited (Botsford *et al.* 1997; FAO 1999), and a three-fold increase in the amount of pesticides used in agriculture is expected by 2050 (Tilman and Lehman 2001). Overall, the real costs of food, resource, energy, and materials production are disguised by subsidies and an oversight of negative externalities (Myers and Kent 1998).

The costs of the side effects of anthropogenic environmental change are more immediately visible, as they often have a direct bearing on human health. Ozone thinning and increased UV levels, toxic pollutants, endocrine-mimicking substances, immune suppression (Chivian 1997), and the emergence and spread of resistant diseases, including HIV, Ebola, and Marburg (Daszak *et al.* 2000), all cause grave concerns and mandate increasing health expenditures. Not even a money scale is required to assess the magnitude of the tragedy of "environmental refugees" – millions of people who can no longer gain a secure livelihood in their homelands because of drought, soil erosion, desertification, deforestation, and other human-induced environmental problems (Myers 2002). Thus, eco-evolutionary responses of ecosystems to human activities result in a global reduction of ecosystem services to humanity (Daily 1997; Ehrlich 2001). This occurs through the loss of species, genetic diversity, and ecological interactions (as with pollination; Chapter 16; Pimentel *et al.* 1997), through rising costs, and even through our sheer inability to access the remaining ecosystem services and avoid the side effects of our impact. This adds to Odum's (1971), McDonnell and Pickett's (1993), and O'Neill and Kahn's (2000) views that both ecology and socioeconomics, because of their limited paradigms, have artificially isolated *homo oeconomicus* from the ecosystems in which it functions.

That large-scale changes in ecosystem function can lead to dramatic societal changes – including population dislocations, urban abandonment, and state collapse (a process that, at a conceptual level, is perhaps akin to evolutionary suicide) – has been documented in several outstanding case studies drawn from New and Old World civilizations, including the classic Maya empire of Mesoamerica and the Akkadian empire of Mesopotamia (Weiss *et al.* 1993; Thompson *et al.* 1994; Hodell *et al.* 1995; Gill 2000; deMenocal 2001; Weiss and Bradley 2001). These examples show that, challenged by the unprecedented environmental stress of prolonged drought, whole empires collapsed and their people were diminished to much lower subsistence levels, whereas in other cases, populations migrated and adapted to new subsistence modes. In all these cases, the observed societal response reflects an interaction between human cultural elements (socioeconomic, political, and secular stresses) and persistent century-long shifts in climate. What makes these ancient events relevant to modern times is that they simultaneously document both the resilience and vulnerability of large, complex civilizations to ecosystem variability. Complex societies are neither powerless pawns nor infinitely plastic and adaptive to environmental change (deMenocal 2001).

Maya site of Tikal (ca. 800 AD)

The vast majority of humanity may currently see little reason to value most of biodiversity. However, the hazards, losses, and costs related to ecosystem degradation eventually impact our very survival, production, and reproduction – in short, our fitness (Western 2001); therefore, these processes can be expected to generate selective pressures on the evolution of our own culture and value systems, including the ethical obligation to preserve biodiversity (Ehrlich 2001). Thus, humans are not isolated from the eco-evolutionary feedback loop that has shaped the past and will continue to shape the future of biodiversity (Feldman and Laland 1996). The economically dominated cultural background against which the value of biodiversity is assessed will evolve under the selective pressures that economic activities generate themselves. Ethics can evolve at rates that easily surpass those of genetic evolution – for example, our circle of "caring" has widened rapidly, through the attribution of rights first to all human beings (as opposed to only some group of kin, color, or cast), then to domestic animals, then to charismatic animals, and eventually to all organisms and ecosystems (Ehrlich 2000). Evolutionary conservation biologists must contribute to and foster the evolution of new ethics that deal with various aspects of the human predicament, both by forging new paradigms in the form of sustainable alternatives, and by strengthening selective pressures through public education and interacting vigorously with researchers from other disciplines in the biological, earth, and atmospheric sciences, as well as in other walks of life (Woodruff 2001).

As principles lie at the basis of conservation and the development of operational policies, we should aim at the development of robust, yet relatively simple, models of interacting ecosystems and societies. Such models should help address the central question of identifying critical structures and thresholds for species, processes, and areas in terms of the sustainability of ecosystem services (Holling 1992; Costanza *et al.* 1997; Gatto and De Leo 2000). The integration of ecological and economic dynamics in simple models has been initiated in fisheries management (e.g., Walters 1986). Attempts have already been made to extend such a "systems analysis" approach to incorporate cultural variables that quantify the human valuation of some ecosystem services (Casagrandi and Rinaldi 2002). The perspective of adding an evolutionary dimension to such models – including important notions such as a geographic mosaic of coevolutionary processes (Thompson 1994), and evolutionary constraints that arise from a trade-off between the welfare of current and future generations (Costanza 1991) – opens exciting new directions for future research.

19.5 Concluding Comments

Until the past decade or so, despite large-scale questions and perspectives, conservation biology provided hardly more than reactive short-term and small-scale solutions to environmental threats (Western 2001). The necessity for a shift from saving things, the products of evolution, to saving the underlying process, evolution itself, has already been advocated strongly (e.g., Mace *et al.* 1998; Bowen 1999; Templeton *et al.* 2001; Woodruff 2001). Within a broadening scope and increasing depth of conservation efforts, evolutionary conservation biology has a natural and inevitable role – paving the way to go beyond the separation of humanity and nature that has been underlying conservation biology so far, ultimately to embrace the processes that shape human-dominated ecosystems as well as those that direct the evolution of human culture and ethical systems.

Evolutionary conservation biology is not in competition with the established fields of conservation research, and cannot progress on its own. Many of the individual points raised in this book have been made separately before, and a need for methodological pluralism remains. Evolutionary conservation biology should add a unifying perspective and an invigorated thrust. It is expected that disciplinary boundaries will be abandoned naturally when conservation researchers start to utilize all the tools available to tackle fundamental issues, including:

- Establishing closer links between individual behavior and population dynamics;
- Investigating the joint effects of phenotypic plasticity, local adaptation, and the evolution of dispersal on the viability of a population subject to environmental change;
- Analyzing the combined effects of different temporal and spatial scales of environmental change on the adaptive responses of multiple traits;

- Examining the role of frequency-dependent selection in the wild, and designing controlled experiments to evaluate its impact on population viability;
- Better understanding the ecological and genetic processes that can limit the speed of population responses to environmental threats – and likewise, those that can accelerate the evolution of undesirable adaptations that could prove deleterious to the population;
- Improving our grasp of the ecological and genetic mechanisms that underlie processes and patterns of community diversification, via endogenous speciation or exogenous invasions;
- Extending the empirical and theoretical scope of population genetics to the study of community genetics.

Ignoring evolutionary mechanisms and dynamics renders all our conservation efforts (and sometimes successes) as temporary only. To develop principles of sustainability that avoid evolutionary sclerosis or deleterious evolutionary acceleration may be the most important task ahead for ecologists (Western 2001). The ultimate test of evolutionary biology as a science will not be whether it solves the riddles of the past, but rather whether it enables us to manage the biosphere's future. In this sense, by turning around and facing forward in time, evolutionary biologists become conservation scientists (Woodruff 2001). In such a setting, the traditional dichotomy between one group doing fundamental research and the other doing applied work can be severely counterproductive. Conservation biology provides some of the most difficult problems ever tackled by evolutionary biology. If our greatest achievement in the past century was the collective understanding of what evolution meant to our own survival, the challenge of the present century is to develop a more predictive evolutionary conservation biology that can manage human-dominated ecosystems before it is too late to shape our environmental future in a desirable way.

References

Page numbers of reference citations in this volume are given in square brackets.

Abbo S & Rubbin B (2000). Transgenic crops: A cautionary tale. *Science* **287**:1927–1928 [*358*]

Abrams PA (1986). Adaptive responses of predators to prey and prey to predators: The failure of the arms-race analogy. *Evolution* **40**:1229–1247 [*339*]

Abrams PA & Matsuda H (1997). Fitness minimization and dynamic instability as a consequence of predator–prey coevolution. *Evolutionary Ecology Research* **11**:1–20 [*339*]

Abrams PA, Matsuda H & Harada Y (1993). Evolutionarily unstable fitness maxima and stable fitness minima of continuous traits. *Evolutionary Ecology* **7**:465–487 [*198, 310*]

Ahas R, Aasa A, Menzel A, Fedotova VG & Scheifinger H (2002). Changes in European spring phenology. *International Journal of Climatology* **22**:1727–1738 [*320*]

Aizen MA & Feinsinger P (1994). Forest fragmentation, pollination, and plant reproduction in a Chaco dry forest, Argentina. *Ecology* **75**:330–351 [*313*]

Allee WC (1931). *Animal Aggregations*. Chicago, IL, USA: University of Chicago Press [*24*]

Allee WC (1938). *The Social Life of Animals*. New York, NY, USA: WW Norton & Co. [*24*]

Allen JC, Schaffer WM & Rosko D (1993). Chaos reduces species extinction by amplifying local-population noise. *Nature* **364**:229–232 [*44, 197, 269*]

Allen-Wardell G, Bernhardt P, Bitner R, Burquez A, Buchmann S, Cane J, Cox PA, Dalton V, Feinsinger P, Ingram M, Inouye D, Jones CE, Kennedy K, Kevan P, Koopowitz H, Medellin R, Medellin-Morales S, Nabhan GP, Pavlik B, Tepedino V, Torchio P & Walker S (1998). The potential consequences of pollinator declines on the conservation of biodiversity and stability of food crop yields. *Conservation Biology* **12**:8–17 [*307*]

Alonso LE (1998). Spatial and temporal variation in the ant occupants of a facultative ant-plant. *Biotropica* **30**:201–213 [*320*]

Altmann J, Alberts SC, Haines SA, Dubach J, Muruthi P, Coote T, Geffen E, Cheesman DJ, Mututua RS & Saiyalel SN (1996). Behavior predicts genetic structure in a wild primate group. *Proceedings of the National Academy of Sciences of the USA* **93**:5797–5801 [*126*]

Anderson E (1949). *Introgressive Hybridization*. New York, NY, USA: John Wiley & Sons [*344*]

Anderson E & Stebbins GL (1954). Hybridization as an evolutionary stimulus. *Evolution* **8**:378–388 [*354*]

Andersson M (1994). *Sexual Selection*. Princeton, NJ, USA: Princeton University Press [*51*]

Andrén H (1994). Effects of habitat fragmentation on birds and mammals in landscapes with different proportions of suitable habitat: A review. *Oikos* **71**:355–366 [*285*]

Andrewartha HG & Birch LC (1954). *The Distribution and Abundance of Animals*. Chicago, IL, USA: University of Chicago Press [*88*]

Anstett MC, Michaloud G & Kjellberg F (1995). Critical population size for fig/wasp mutualism in a seasonal environment: Effect and evolution of the duration of receptivity. *Oecologia* **103**:453–461 [*316*]

Anstett MC, Hossaert-McKey M & McKey D (1997a). Modeling the persistence of small populations of strongly interdependent species: Figs and fig wasps. *Conservation Biology* **11**:204–213 [*316*]

Anstett MC, Hossaert-McKey M & Kjellberg F (1997b). Figs and fig pollinators: Evolutionary conflicts in a coevolved mutualism. *Trends in Ecology and Evolution* **12**:94–99 [*316, 338*]

Anstett MC, Gibernau M & Hossaert-McKey M (1998). Partial avoidance of female inflorescences of a dioecious fig by their mutualistic pollinating wasps. *Proceedings of the Royal Society of London B* **265**:45–50 [*311*]

Antilla CK, Dahler CC, Rank NE & Strong DR

(1998). Greater male fitness of a rare invader (*Spartina alternifolia*, Poaceae) threatens a common nature (*Spartina foliosa*) with hybridization. *American Journal of Botany* **85**:1597–1601 [*350*]

Antonovics J (1976). The nature of limits to natural selection. *Annals of the Missouri Botanical Garden* **63**:224–247 [*242*]

Antonovics J & Bradshaw AD (1970). Evolution in closely adjacent plant populations VIII: Clinal patterns at a mine boundary. *Heredity* **25**:349–362 [*111, 116*]

Antonovics J, Thrall P, Jarosz A & Stratton D (1994). Ecological genetics of metapopulations: The Silene–Ustilago plant–pathogen system. In *Ecological Genetics*, ed. Real LA, pp. 146–170. Princeton, NJ, USA: Princeton University Press [*293*]

Arano B, Lorente G, Garcia-Paris M & Herrero P (1995). Species translocation menaces Iberian waterfrogs. *Conservation Biology* **9**:196–198 [*346*]

Ariño A & Pimm SL (1995). On the nature of population extremes. *Evolutionary Ecology Research* **9**:429–443 [*36*]

Arnault C & Biemont C (1989). Heat shock does not mobilize mobile elements in the genome of *Drosophila melanogaster*. *Journal of Molecular Evolution* **33**:388–390 [*139*]

Arnault C, Heizmann A, Loevenbruck C & Biemont C (1991). Environmental stresses and mobilization of transposable elements in inbred lines of *Drosophila melanogaster*. *Mutation Research* **248**:51–60 [*139*]

Arnold ML (1993). *Iris nelsonii*: Origin and genetic composition of a homoploid hybrid species. *American Journal of Botany* **80**:577–583 [*352*]

Arnold SJ & Duvall D (1994). Animal mating systems: A synthesis based on selection theory. *The American Naturalist* **143**:317–348 [*47*]

Ashton PA & Abbott RJ (1992). Multiple origins and genetic diversity in the newly arisen allopolyploid species, *Senecio cambrensis* Rosser (Compositae). *Journal of Heredity* **68**:25–32 [*352*]

Asmussen S & Hering H (1983). *Branching Processes*. Boston, MA, USA: Birkhäuser [*50*]

Athreya KB & Ney PE (1972). *Branching Processes*. New York, NY, USA: Springer-Verlag [*32, 50*]

Avise JC (1994). *Molecular Markers, Natural History and Evolution*. New York, NY, USA: Chapman & Hall [*230*]

Avise JC, Pierce PC, Van Den Avyle MJ, Smith MH, Nelson WS & Asmussen MA (1997). Cytonuclear introgressive swamping and species turnover of bass after an introduction. *Journal of Heredity* **88**:14–20 [*350*]

Axelrod R & Hamilton WD (1981). The evolution of cooperation. *Science* **211**:1390–1396 [*307*]

Ballou JD (1997). Ancestral inbreeding only minimally affects inbreeding depression in mammalian populations. *Journal of Heredity* **88**:169–178 [*133*]

Balmford A (1996). Extinction filters and current resilience: The significance of past selection pressures for conservation biology. *Trends in Ecology and Evolution* **11**:193–196 [*9*]

Barker JSF & Krebs RA (1995). Genetic variation and plasticity of thorax length and wing length in *Drosophila aldrichi* and *Drosophila buzzatii*. *Journal of Evolutionary Biology* **8**:689–709 [*142, 147*]

Barrett SCH (1998). The evolution of mating strategies in flowering plants. *Trends in Plant Science* **3**:335–341 [*295*]

Barton NH & Keightley (2002). Understanding quantitative genetic variation. *Nature Reviews, Genetics* **3**:11–21 [*187*]

Barton NH & Turelli M (1989). Evolutionary quantitative genetics: How little do we know? *Annual Review of Genetics* **23**:337–370 [*124, 177*]

Bateman AJ (1948). Intra-sexual selection in *Drosophila*. *Heredity* **2**:349–368 [*52*]

Bauer ON & Hoffman GL (1976). Helminth range extension by translocation of fish. In *Wildlife Diseases*, ed. Page LA, pp. 163–172. New York, NY, USA: Plenum Press [*346*]

Beattie AJ (1985). *The Evolutionary Ecology of Ant–Plant Mutualisms*. Cambridge, UK: Cambridge University Press [*321*]

Belovsky GE, Mellison C, Larson C & Van Zandt PA (1999). Experimental studies of extinction dynamics. *Science* **286**:1175–1177 [*39*]

Belsky AJ (1986). Does herbivory benefit plants? A review of the evidence. *The American Naturalist* **127**:870–892 [*333*]

Belsky AJ, Carson WP, Jense CL & Fox GA (1993). Overcompensation by plants: Herbivore optimization or red herring? *Evolution-*

ary Ecology Research **7**:109–121 [*333, 335*]

Belyaev DK & Borodin PM (1982). The influence of stress on variation and its role in evolution. *Biologisches Zentralblatt* **100**:705–714 [*137*]

Bender EA, Case TJ & Gilpin ME (1984). Perturbation experiments in community ecology: Theory and practice. *Ecology* **65**:1–13 [*338*]

Bennett AF & Lenski RE (1993). Evolutionary adaptation to temperature. II. Thermal niches of experimental lines of *Escherichia coli*. *Evolution* **47**:1–12 [*91*]

Benton TG & Grant A (1999). Elasticity analysis as an important tool in evolutionary and population ecology. *Trends in Ecology and Evolution* **14**:467–471 [*58*]

Bergelson J & Crawley MJ (1992). Herbivory and *Ipomopsis aggregata*: The disadvantages of being eaten. *The American Naturalist* **139**:870–882 [*333*]

Bergelson J & Purrington CB (1996). Surveying patterns in the cost of resistance in plants. *The American Naturalist* **148**:536–558 [*335*]

Bergelson J, Juenger T & Crawley MJ (1996). Regrowth following herbivory in *Ipomopsis aggregata*: Compensation but not overcompensation. *The American Naturalist* **148**:744–755 [*333*]

Berger J (1990). Persistence of different-sized populations: An empirical assessment of rapid extinction in bighorn sheep. *Conservation Biology* **4**:91–98 [*57*]

Bergstrom CT & Lachmann M (2003). The Red King effect: When the slowest runner wins the coevolutionary race. *Proceedings of the National Academy of Sciences of the USA* **100**:593–598 [*325*]

Bernhard JM, Buck KR, Farmer MA & Bowser SS (2000). The Santa Barbara Basin is a symbiosis oasis. *Nature* **403**:77–80 [*307*]

Berry RJ (1990). Industrial melanism and peppered moths (*Biston betularia* (L)). *Biological Journal of the Linnean Society* **39**:301–322 [*111, 116*]

Blossey B & Nötzold R (1995). Evolution of increased competitive ability in invasive nonindigenous plants: A hypothesis. *Journal of Ecology* **83**:887–889 [*341*]

Blows MW & Sokolowski MB (1995). The expression of additive and nonadditive genetic variation under stress. *Genetica* **140**:1149–1159 [*147*]

Bolnick DI (2001). Intraspecific competition favours niche width expansion in *Drosophila melanogaster*. *Nature* **410**:463–466 [*5*]

Bond WJ (1994). Do mutualisms matter? Assessing the impact of pollinator and dispersal disruption on plant extinction. *Philosophical Transactions of the Royal Society of London B* **344**:83–90 [*307*]

Bond WJ (1995). Assessing the risk of plant extinction due to pollinator and disperser failure. In *Extinction Rates*, eds. Lawton JH & May RM, pp.174–191. Oxford, UK: Oxford University Press [*307, 317*]

Bond W & Slingsby P (1984). Collapse of an ant-plant mutualism: The Argentine ant (*Iridomyrmex humilis*) and myrmecochorous Proteaceae. *Ecology* **65**:1031–1037 [*313, 321*]

Bonnin I, Prosperi J-M & Olivieri I (1996). Genetic markers and quantitative genetic variation in *Medicago truncatula* (Leguminosae): A comparative analysis of population structure. *Genetics* **143**:1795–1805 [*296*]

Botsford LW, Castilla JC & Petersen CH (1997). The management of fisheries and marine ecosystems. *Science* **277**:509–515 [*361*]

Bowen BW (1999). Preserving genes, species, or ecosystems? Healing the fractured foundations of conservation policy. *Molecular Ecology* **8**:S5–S10 [*363*]

Boyce M (1992). Population viability analysis. *Annual Review oof Ecology and Systematics* **23**:481–506 [*38, 40*]

Brachet S, Olivieri I, Godelle B, Klein E, Frascaria-Lacoste N & Gouyon P-H (1999). Dispersal and metapopulation viability in a heterogeneous landscape. *Journal of Theoretical Biology* **198**:479–495 [*278–279*]

Bradshaw AD (1991). Genostasis and the limits to evolution. *Philosophical Transactions of the Royal Society of London B* **333**:289–305 [*244*]

Brakefield PM & Liebert TG (2000). Evolutionary dynamics of declining melanism in the peppered moth in The Netherlands. *Proceedings of the Royal Society of London B* **267**:1953–1957 [*94–95*]

Bramwell D (1990). Conserving biodiversity in the Canary Islands. *Annals of the Missouri Botanical Garden* **77**:28–37 [*349*]

Brauer F (1979). Harvesting strategies for population systems. *Rocky Mountain Journal of Mathematics* **9**:19–26 [*26*]

Bremer K (1994). *Asteraceae. Cladistics and Classification*. Portland, OR, USA: Timber Press [*297*]

Brewer BA, Lacy RC, Foster ML & Alaks G (1990). Inbreeding depression in insular and central populations of *Peromyscus* mice. *Journal of Heredity* **81**:257–266 [*133*]

Briggs D & Walters SM (1997). *Plant Variation and Evolution*, Third Edition. Cambridge, UK: Cambridge University Press [*94*]

Briscoe DA, Malpica JM, Robertson A, Smith GJ, Frankham R, Banks RG & Barker JSF (1992). Rapid loss of genetic variation in large captive populations of *Drosophila* flies: Implications for the genetic management of captive populations. *Conservation Biology* **6**:416–425 [*131*]

Brochmann C (1984). Hybridization and distribution of *Argyranthemum coronopifolium* (Asteraceae-Anthemideae) in the Canary Islands. *Nordic Journal of Botany* **4**:729–736 [*349*]

Bronstein JL (1992). Seed predators as mutualists: Ecology and evolution of the fig/pollinator interaction. In *Insect-Plant Interactions*, Volume IV, ed. Bernays E, pp. 1–44. Boca Raton, FL, USA: CRC Press [*316*]

Bronstein JL (1994). Our current understanding of mutualism. *Quarterly Review of Biology* **69**:31–51 [*305–306*]

Bronstein JL (1995). The plant/pollinator landscape. In *Mosaic Landscapes and Ecological Processes*, eds. Fahrig L, Hansson L & Merriam G, pp. 256–288. New York, NY, USA: Chapman and Hall [*318*]

Bronstein JL (2001a). Mutualisms. In *Evolutionary Ecology: Perspectives and Synthesis*, eds. Fox C, Fairbairn D & Roff D, pp. 315–330. Oxford, UK: Oxford University Press [*305–306, 325*]

Bronstein JL (2001b). The exploitation of mutualisms. *Ecology Letters* **4**:277–287 [*307*]

Bronstein JL & Hossaert-McKey M (1995). Hurricane Andrew and a Florida fig pollination mutualism: Resilience of an obligate interaction. *Biotropica* **27**:373–381 [*316*]

Bronstein JL, Gouyon PH, Gliddon C, Kjellberg F & Michaloud G (1990). Ecological consequences of flowering asynchrony in monoecious figs: A simulation study. *Ecology* **71**:2145–2156 [*316*]

Brooks JL & Dodson SI (1965). Predation, body size, and composition of plankton. *Science* **150**:28–35 [*310*]

Brooks WR & Rittschof D (1995). Chemical detection and host selection by the symbiotic crab *Porcellana sayana*. *Invertebrate Biology* **114**:180–185 [*318*]

Brown JH & Kodric-Brown A (1977). Turnover rates in insular biogeography: Effect of immigration on extinction. *Ecology* **58**:445–449 [*65*]

Brown JS & Vincent TL (1992). Organization of predator–prey communities as an evolutionary game. *Evolution* **46**:1269–1283 [*339*]

Brown BJ, Mitchell RJ & Graham SA (2002). Competition for pollination between an invasive species (purple loosestrife) and a native congener. *Ecology* **83**:2328–2336 [*321*]

Bryant EH, McCommas SA & Combs LM (1986). The effect of an experimental bottleneck upon quantitative genetic variation in the housefly. *Genetics* **114**:1191–1211 [*131*]

Bubli OA, Imasheva AG & Loeschcke V (1998). Selection for knockdown resistance to heat in *Drosophila melanogaster* at high and low larval densities. *Evolution* **52**:619–625 [*148*]

Bubli OA, Loeschcke V & Imasheva AG (2000). Effect of stressful and nonstressful growth temperatures on variation of sternopleural bristle number in *Drosophila melanogaster*. *Evolution* **54**:1444–1449 [*142–143, 148–149*]

Buchmann SL & Nabhan GP (1996). *The Forgotten Pollinators*. Washington, DC, USA: Island Press [*313–314, 326*]

Bull JJ & Rice WR (1991). Distinguishing mechanisms for the evolution of cooperation. *Journal of Theoretical Biology* **149**:63–74 [*307, 311*]

Bürger R (1998). Mathematical properties of mutation-selection models. *Genetica* **103**:279–298 [*201*]

Bürger R (1999). Evolution of genetic variability and the advantage of sex and recombination in changing environments. *Genetics* **153**:1055–1069 [*175, 181, 184–185, 187*]

Bürger R (2000). *The Mathematical Theory of Selection, Recombination, and Mutation*. Chichester, UK: John Wiley & Sons [*156, 173, 177, 184*]

Bürger R & Bomze IM (1996). Stationary distributions under mutation-selection balance: Structure and properties. *Advances in Applied Probability* **28**:227–251 [*201*]

Bürger R & Ewens WJ (1995). Fixation probabilities of additive alleles in diploid populations. *Journal of Mathematical Biology* **33**:557–575 [*158*]

Bürger R & Gimelfarb A (2002). Fluctuating environments and the mode of mutation in maintaining genetic variation. *Genetical Research* **80**:31–46 [*175*]

Bürger R & Hofbauer J (1994). Mutation load and mutation–selection balance in quantitative genetic traits. *Journal of Mathematical Biology* **32**:193–218 [*156*]

Bürger R & Lynch M (1995). Evolution and extinction in a changing environment: A quantitative-genetic analysis. *Evolution* **49**:151–163 [*175–176, 178–182, 186–187, 247, 250*]

Bürger R & Lynch M (1997). Adaptation and extinction in changing environments. In *Environmental Stress, Adaptation and Evolution*, eds. Bijlsma R & Loeschcke V, pp. 209–239. Basel, Switzerland: Birkhäuser [*34, 175, 179, 184*]

Bürger R, Wagner GP & Stettinger F (1989). How much heritable variation can be maintained in finite populations by mutation–selection balance? *Evolution* **43**:1748–1766 [*178*]

Bulmer MG (1980). *The Mathematical Theory of Quantitative Genetics*. Oxford, UK: Clarendon Press [*173*]

Burch CL & Chao L (1999). Evolution by small steps and rugged landscapes in the RNA virus ϕ6. *Genetics* **151**:921–927 [*166*]

Burgman MA, Ferson S & Akçakaya HR (1993). *Risk Assessment in Conservation Biology*. London, UK: Chapman & Hall [*38, 351*]

Burns JM (1968). Mating frequencies in natural populations of skippers and butterflies as determined by spermatophores counts. *Proceedings of the National Academy of Sciences of the USA* **61**:852–859 [*25*]

Burt A (1995). Perspective: The evolution of fitness. *Evolution* **49**:1–8 [*245*]

Bush GL (1969). Sympatric host race formation and speciation in frugivorous flies of the genus Rhagoletis (Diptera: Tephritidae). *Evolution* **23**:237–251 [*205*]

Butcher D (1995). Muller's ratchet, epistasis and mutation effects. *Genetics* **141**:431–437 [*164*]

Butlin RK & Tregenza T (1998). Levels of genetic polymorphism: Marker loci versus quantitative traits. *Philosophical Transactions of the Royal Society of London B* **353**:187–198 [*128*]

Byers DL & Meagher TR (1992). Mate availability in small populations of plant species with homomorphic incompatibility. *Journal of Heredity* **68**:353–359 [*354*]

Caballero A & Keightley PD (1998). Inferences on genome-wide deleterious mutation rates in inbred populations of *Drosophila* and mice. *Genetica* **102/103**:229–239 [*160*]

Cadet C (1998). *Dynamique adaptative de la dispersion dans une métapopulation: modèles stochastiques densité-dépendants*. MSc Thesis. Paris, France: University of Paris VI [*265*]

Cadet C, Ferrière R, Metz JAJ & Van Baalen M (2003). The evolution of dispersal under demographic stochasticity. *The American Naturalist* **162**:427–441 [*274*]

Cain AJ & Sheppard PM (1954). Natural selection in *Cepaea*. *Genetics* **39**:89–116 [*122*]

Callaway RM & Walker LR (1997). Competition and facilitation: A synthetic approach to interactions in plant communities. *Ecology* **78**:1958–1965 [*310*]

Calow P (1999). *Encyclopedia of Ecology and Environmental Management*. Oxford UK: Blackwell Publishing [*190*]

Carlquist S (1974). *Island Biology*. New York, NY, USA: Columbia University Press [*346*]

Carlton JT (1996). Biological invasions and cryogenic species. *Ecology* **77**:1653–1655 [*345*]

Carroll SP & Boyd C (1992). Host race radiation in the soapberry bug: Natural history with the history. *Evolution* **46**:1052–1069 [*111, 116*]

Carroll SP, Klassen SP & Dingle H (1998). Rapidly evolving adaptations to host ecology and nutrition in the soapberry bug. *Evolutionary Ecology* **12**:955–968 [*305*]

Casagrandi R & Rinaldi S (2002). A theoretical approach to tourism sustainability. *Conservation Ecology* **6**:13. [online] http://www.consecol.org/vol6/iss1/art13 [*363*]

Case TJ (1999). *An Illustrated Guide to Theoretical Ecology*. Oxford, UK: Oxford University Press [*190*]

Case TJ & Taper ML (2000). Interspecific competition, environmental gradients, gene flow, and the coevolution of species' borders. *The American Naturalist* **155**:583–605 [*259, 261*]

Castillo-Chavez C & Hsu Schmitz S-F (1997). The evolution of age-structured marriage functions: It takes two to tango. In *Structured-population Models in Marine, Terrestrial and Freshwater Systems*, eds. Tuljapurkar S & Caswell H, pp. 533–553. New York, NY, USA: Chapman & Hall [*46*]

Caswell H (1989). *Matrix Population Models.* Sunderland, MA, USA: Sinauer Associates Inc. [*42–43*]

Caswell H (2001). *Matrix Population Models: Construction, Analysis, and Interpretation*, Second Edition. Sunderland, MA, USA: Sinauer Associates Inc. [*32, 34, 42–43, 268*]

Caswell H & Cohen JE (1995). Red, white and blue: Environmental variance spectra and coexistence in metapopulations. *Journal of Theoretical Biology* **186**:301–316 [*23*]

Caswell H & Weeks DE (1986). Two-sex models: Chaos, extinction, and other dynamic consequences of sex. *The American Naturalist* **128**:707–735 [*46, 48*]

Charlesworth B (1980). *Evolution in Age Structured Populations*. Cambridge, UK: Cambridge University Press [*102*]

Charlesworth B (1984). The evolutionary genetics of life-histories. In *Evolutionary Ecology*, ed. Shorrocks B, pp. 117–133. Oxford, UK: Blackwell [*125*]

Charlesworth B (1987). The heritability of fitness. In *Sexual Selection: Testing the Alternatives*, eds. Bradbury JW & Andersson MB, pp. 21–40. Chichester, UK: John Wiley & Sons [*119*]

Charlesworth B (1993a). The evolution of sex and recombination in a varying environment. *Journal of Heredity* **84**:345–450 [*175, 187*]

Charlesworth B (1993b). Directional selection and the evolution of sex and recombination. *Genetical Research* **61**:205–224 [*175, 179, 185, 187*]

Charlesworth B (1994). *Evolution in Age-Structured Populations*. Cambridge, UK: Cambridge University Press [*119, 124–126, 130*]

Charlesworth B (1998). Measures of divergence between populations and the effect of forces that reduce variability. *Molecular Biology and Evolution* **15**:538–543 [*240*]

Charlesworth D & Charlesworth B (1987). Inbreeding depression and its evolutionary consequences. *Annual Review of Ecology and Systematics* **18**:237–268 [*96, 123, 131–132*]

Charlesworth B & Charlesworth D (1997). Rapid fixation of deleterious alleles can be caused by Muller's ratchet. *Genetical Research* **67**:27–41 [*165*]

Charlesworth B & Charlesworth D (1998). Some evolutionary consequences of deleterious mutations. *Genetica* **102/103**:3–19 [*171*]

Charlesworth B & Hughes KA (2000). The maintenance of genetic variation in life history traits. In *Evolutionary Genetics from Molecules to Morphology*, eds. Singh RS & CB Krimbas, pp. 369–391. Cambridge, UK: Cambridge University Press [*119, 122–123*]

Charlesworth D, Morgan MT & Charlesworth B (1992). The effect of linkage and population size on inbreeding depression due to mutational load. *Genetical Research* **59**:49–61 [*132*]

Charlesworth D, Morgan MT & Charlesworth B (1993a). Mutation accumulation in finite outbreeding and inbreeding populations. *Genetical Research* **61**:39–56 [*157, 165, 168*]

Charlesworth D, Morgan MT & Charlesworth B (1993b). The effect of deleterious mutations on neutral molecular variation. *Genetics* **134**:1289–1303 [*162*]

Charnov E (1993). *Life History Invariants: Some Explorations of Symmetry in Evolutionary Ecology*. Oxford, UK: Oxford University Press [*191*]

Chivian E (1997). Global environmental degradation and biodiversity loss. In *Biodiversity and Human Health*, eds. Grifo FC & Rosenthal J, pp. 7–38. Washington, DC, USA: Island Press [*361*]

Christensen KM, Whitham TG & Balda RP (1991). Discrimination among pinyon pine trees by Clark's nutcrackers: Effects of cone crop size and cone characters. *Oecologia* **86**:402–407 [*311*]

Christian CE (2001). Consequences of a biological invasion reveal the importance of mutualism for plant communities. *Nature* **413**:635–639 [*313, 321*]

Christiansen FB (1975). Hard and soft selection in a subdivided population. *The American Naturalist* **109**:11–16 [*123*]

Chung R (1994). Cycles in the two-sex problem: An investigation of a nonlinear demographic model. *Mathematical Population Studies* **5**:45–73 [*48*]

Clausen J, Keck DD & Hiesey WM (1940). *Experimental Studies on the Nature of Species.*

I. Effect of Varied Environments on Western North American Plants. Publication Number 520. Washington, DC, USA: Carnegie Institution of Washington [*348*]

Clobert J, Danchin E, Dhondt AA & Nichols JD (2001). *Dispersal*. Oxford, UK: Oxford University Press [*79*]

Clutton-Brock TH (1989). Mammalian mating systems. *Proceedings of the Royal Society of London B* **236**:339–372 [*47*]

Cody ML & Overton JM (1996). Short term evolution of reduced dispersal in island plant populations. *Journal of Ecology* **84**:53–61 [*290–291, 295*]

Cohen AS (1994). Extinction in ancient lakes: Biodiversity crises and conservation 40 years after J.L. Brooks. *Archiv für Hydrobiologie, Beiheft, Ergebnisse der Limnologie* **44**:451–479 [*305*]

Cohen D & Levin SA (1991). Dispersal in patchy environments: The effects of temporal and spatial structure. *Theoretical Population Biology* **39**:63–99 [*265*]

Colas B (1997). Centaurea corymbosa, *chronique d'une extinction annoncée*. PhD Thesis. Tours, France: Université de Tours [*298*]

Colas B, Riba M & Molina J (1996). Statut démographique de *Centaurea corymbosa* Pourret (Asteraceae), *Hormatophylla pyrenaica* (Lapeyr.) Cullen & Dudley (Brassicaceae), et *Marsilea strigosa* Willd. (Marsileaceae-Pteridophyta), trois plantes rares dans le sud de la France. *Acta Botanica Gallica* **143**:191–198 [*296*]

Colas B, Olivieri I & Riba M (1997). *Centaurea corymbosa*, a cliff-dwelling species tottering on the brink of extinction. A demographic and genetic study. *Proceedings of the National Academy of Sciences of the USA* **94**:3471–3476 [*9, 295–297*]

Colas B, Olivieri I & Riba M (2001). Spatio-temporal variation of reproductive success and conservation of the narrow-endemic *Centaurea corymbosa* (Asteraceae). *Biological Conservation* **99**:375–386 [*295–296*]

Colas B, Fréville H, Riba M, Mignot A, Imbert E, Olivieri I & Clobert J. Species viability analysis of a cliff-dwelling, narrow-endemic plant. Unpublished [*297–298*]

Cole FR, Madeiros AC, Loope LL & Zuehlke WW (1992). Effects of the Argentine ant on arthropod fauna of Hawaiian high-elevation shrubland. *Ecology* **73**:1313–1322 [*313*]

Coley PD, Bryant JP & Chapin FS III (1985). Resource availability and plant antiherbivore defense. *Science* **230**:895–899 [*329*]

Comins HN, Hamilton WD & May RM (1980). Evolutionarily stable dispersal strategies. *Journal of Theoretical Biology* **82**:205–230 [*265, 275*]

Compton SG & McCormack G (1999). The Pacific Banyan in the Cook Islands: Have its pollination and seed dispersal mutualisms been disrupted, and does it matter? *Biodiversity and Conservation* **8**:1707–1715 [*316*]

Costanza R (1991). *Ecological Economics: The Science and Management of Sustainability*. New York, NY, USA: Columbia University Press [*363*]

Costanza R, d'Arge R, deGroot R, Farber S, Grasso M, Hannon B, Limburg K, Naeem S, O'Neill RV, Paruelo J, Raskin RG, Sutton P & Van den Belt M (1997). The value of the world's ecosystem services and natural capital. *Nature* **387**:253–260 [*363*]

Courchamp F, Clutton-Brock T & Grenfell B (1999). Inverse density dependence and the Allee effect. *Trends in Ecology and Evolution* **14**:405–410 [*24, 46*]

Couvet D (2002). Deleterious effects of restricted gene flow in a metapopulation. *Conservation Biology* **16**:369–376 [*239*]

Cowling RM & Pressey RL (2001). Rapid plant diversification: Planning for an evolutionary future. *Proceedings of the National Academy of Sciences of the USA* **98**:5452–5457 [*11*]

Crandall KA, Bininda-Emonds ORP, Mace GM & Wayne RK (2000). Considering evolutionary processes in conservation biology. *Trends in Ecology and Evolution* **15**:290–295 [*10*]

Crow JF & Aoki K (1984). Group selection for a polygenic behavioral trait: Estimating the degree of population subdivision. *Proceedings of the National Academy of Sciences of the USA* **81**:6073–6077 [*231*]

Crow JF & Kimura M (1964). The theory of genetic loads. In *Proceedings of the XI International Congress of Genetics*, ed. Geerts S, pp. 495–505. Oxford, UK: Pergamon Press [*178*]

Crow JF & Kimura M (1970). *An Introduction to Population Genetics Theory*. New York, NY, USA: Harper & Row [*97, 121, 131, 135, 157–158, 163, 231*]

Crow JF & Simmons MJ (1983). The mutation load in *Drosophila*. In *The Genetics and Biology of* Drosophila, Vol. 3c, eds. Ashburner

M, Carson HL & Thompson JN Jr, pp. 1–35. New York, NY, USA: Academic Press [*160*]

Cruickshank I, Gurney WSC & Veitch AR (1999). The characteristics of epidemics and invasions with thresholds. *Theoretical Population Biology* **56**:279–292 [*201*]

Cuddington KM & Yodzis P (1999). Black noise and population persistence. *Proceedings of the Royal Society of London B* **266**:969–973 [*36–37*]

Czech B & Krausman PR (1997). Distribution and causation of species endangerment in the United States. *Science* **277**:1116–1117 [*305*]

Dahlgaard J, Loeschcke V, Michalak P & Justesen J (1998). Induced thermotolerance and associated expression of the heat-shock protein HSP70 in adult *Drosophila melanogaster*. *Functional Ecology* **12**:786–788 [*140*]

Daily GC, ed. (1997). *Nature's Services: Societal Dependence on Natural Ecosystems*. Washington, DC, USA: Island Press [*361*]

Daily GC, Ehrlich PR & Sanchez-Azofeifa A (2001). Countryside biogeography: Use of human-dominated habitats by the avifauna of southern Costa Rica. *Ecological Applications* **11**:1–13 [*360*]

Daley DJ, Hull DM & Taylor JM (1986). Bisexual Galton–Watson branching processes with superadditive mating functions. *Journal of Applied Probability* **23**:585–600 [*33–34*]

Darroch J & Seneta E (1965). On quasi-stationary distributions in absorbing discrete-time finite Markov chains. *Journal of Applied Probability* **2**:88–100 [*74*]

Darwin C (1859). *On the Origin of Species*. London, UK: John Murray [*188*]

Darwin C (1871). *The Descent of Man and Selection in Relation to Sex*. London, UK: J. Murray [*51*]

Daszak P, Cunningham AA & Hyatt AD (2000). Wildlife ecology – Emerging infectious diseases of wildlife – Threats to biodiversity and human health. *Science* **287**:443–449 [*361*]

David JR, Moreteau B, Gautier JP, Petavy G, Stockel A & Imasheva AG (1994). Reaction norms in size characters in relation to growth temperatures in *Drosophila melanogaster*: An isofemale line analysis. *Genetics Selection Evolution* **26**:229–251 [*142, 147*]

Davies NB & Lundberg A (1984). Food distribution and a variable mating system in the dunnock (*Prunella modularis*). *Journal of Animal Ecology* **53**:895–912 [*47*]

Davies EK, Peters AD & Keightley PD (1999). High frequency of cryptic deleterious mutations in *Caenorhabditis elegans*. *Science* **285**:1748–1751 [*162*]

Davis MB & Shaw RG (2001). Range shifts and adaptive responses to Quaternary climate change. *Science* **292**:673–679 [*357*]

Dawley RM & Bogart JP (1989). *Evolution and Ecology of Unisexual Vertebrates*. Albany, NY, USA: New York State Museum [*352*]

Dawson PS & Riddle RA (1983). Genetic variation, environmental heterogeneity, and evolutionary stability. In *Population Biology: Retrospect and Prospect*, eds. King CF & Dawson PS, pp. 147–170. New York, NY, USA: Columbia University Press [*123*]

Dayan T & Simberloff D (1994). Character displacement, sexual dimorphism, and morphological variation among British and Irish mustelids. *Ecology* **75**:1063–1073 [*305*]

De Jong G (1989). Phenotypically plastic characters in isolated populations. In *Evolutionary Biology of Transient Unstable Populations*, ed. Fontdevila A, pp. 3–18. Heidelberg, Germany: Springer-Verlag [*141*]

De Jong G (1990). Quantitative genetics of reaction norms. *Journal of Evolutionary Biology* **3**:447–468 [*141*]

De Jong TJ (1995). Why fast-growing plants do not bother about defense. *Oikos* **74**:545–548 [*329*]

DeJoode DR & Wendel JF (1992). Genetic diversity and origin of the Hawaiian Islands cotton, *Gossypium tomentosum*. *American Journal of Botany* **79**:1311–1319 [*350*]

Delpuech J-M, Moreteau B, Chiche J, Pla E, Vouidibio J & David JR (1995). Phenotypic plasticity and reaction norms in temperate and tropical populations of *Drosophila melanogaster*: Ovarian size and developmental temperatures. *Evolution* **49**:670–675 [*142, 150*]

DeMarais BD, Dowling TE, Douglas ME, Minckley WL & Marsh PC (1992). Origin of *Gila seminuda* (Teleostei: Cyprinidae) through introgressive hybridization: Implications for evolution and conservation. *Proceedings of the National Academy of Sciences of the USA* **89**:2747–2751 [*352*]

De Mauro M (1990). Relationship of breeding system to rarity in the lakeside daisy (*Hy-*

menoxys acaulis, var. *glabra*). *Conservation Biology* **7**:542–550 [*354*]

De Mazancourt C & Loreau M (2000a). Grazing optimization, nutrient cycling and spatial heterogeneity of plant–herbivore interactions: Should a palatable plant evolve? *Evolution* **54**:81–92 [*335*]

De Mazancourt C & Loreau M (2000b). Effect of herbivory and plant species replacement on primary production. *The American Naturalist* **155**:735–754 [*334*]

De Mazancourt C, Loreau M & Abbadie L (1998). Grazing optimization and nutrient cycling: When do herbivores enhance plant production? *Ecology* **79**:2242–2252 [*333*]

De Mazancourt C, Loreau M & Abbadie L (1999). Grazing optimization and nutrient cycling: Potential impact of large herbivores in a savanna system. *Ecological Applications* **9**:784–797 [*334*]

De Mazancourt C, Loreau M & Dieckmann U (2001). Can the evolution of plant defense lead to plant–herbivore mutualism? *The American Naturalist* **158**:109–123 [*335–337, 339*]

deMenocal PB (2001). Cultural responses to climate change during the late Holocene. *Science* **292**:667–673 [*362*]

De Moed GH, de Jong G & Scharloo W (1997). Environmental effects on body size variation in *Drosophila melanogaster* and its cellular basis. *Genetical Research* **70**:35–43 [*147*]

Dempster JP (1991). Fragmentation, isolation and mobility of insect populations. In *The Conservation of Insects and their Habitats*, eds. Collins NM & Thomas JA, pp. 143–154. London, UK: Academic Press [*287, 289, 291*]

Dempster JP, King ML & Lakhani KH (1976). Status of swallowtail butterfly in Britain. *Ecological Entomology* **1**:71–84 [*287, 289*]

Dennis B (1989). Allee effects: Population growth, critical density, and the chance of extinction. *Natural Resource Modeling* **3**:481–538 [*22, 24–25, 30, 210*]

Dennis RLH (1993). *Butterflies and Climate Change*. New York, NY, USA: Manchester University Press [*88–89*]

Dennis B, Desharnais RA, Cushing JM & Costantino RF (1995). Nonlinear demographic dynamics: Mathematical models, statistical methods, and biological experiments. *Ecological Monographs* **65**:261–281 [*21*]

Densmore LD III, Moritz CC, Wright JW & Brown WM (1989). Mitochondrial DNA analyses and the origin and relative age of parthenogenic lizards (genus *Cnemidophorus*). IV. Nine *sexlineatis* group unisexuals. *Evolution* **43**:969–983 [*352*]

Dercole F, Ferrière R & Rinaldi S (2002). Ecological bistability and evolutionary reversals under asymmetrical competition. *Evolution* **56**:1081–1090 [*209–210, 213*]

Després L & Jaeger N (1999). Evolution of oviposition strategies and speciation in the globeflower flies *Chiastocheta* spp. (Anthomyiidae). *Journal of Evolutionary Biology* **12**:822–831 [*307*]

Dessauer H & Cole CJ (1989). Diversity between and within nominal forms of unisexual lizards. In *Evolution and Ecology of Unisexual Vertebrates*, eds. Dawley RM & Bogart JP, pp. 49–71. Albany, NY, USA: New York State Museum [*352*]

De Visser JA, Hoekstra RF & Van den Ende H (1997). An experimental test for synergistic epistasis and its application in *Chlamydomonas*. *Genetics* **145**:815–819 [*163*]

Dewar RC & Watt AD (1992). Predicted changes in the synchrony of larval emergence and budburst under climatic warming. *Oecologia* **89**:557–559 [*90*]

Dhondt AA, Adriaensen F, Matthyssen E & Kempenaers B (1990). Nonadaptive clutch size in tits. *Nature* **348**:723–725 [*292*]

Dias PC (1996). Sources and sinks in population biology. *Trends in Ecology and Evolution* **11**:326–330 [*293*]

Diaz S (1996). Effects of elevated CO_2 at the community level mediated by root symbionts. *Plant and Soil* **187**:309–320 [*314*]

Dickerson GE (1955). Genetic slippage in response to selection. Cold Spring Harbor Symposium. *Quantitative Biology* **20**:213–224 [*125*]

Dieckmann U (1994). *Coevolutionary Dynamics of Stochastic Replicator Systems*. Jülich, Germany: Central Library of the Research Center Jülich [*188, 200*]

Dieckmann U (1997). Can adaptive dynamics invade? *Trends in Ecology and Evolution* **12**:128–131 [*329, 337*]

Dieckmann U (2002). Adaptive dynamics of pathogen–host interactions. In *Adaptive Dynamics of Infectious Diseases: In Pursuit of Virulence Management*, eds. Dieckmann U, Metz JAJ, Sabelis MW & Sigmund K, pp. 39–59. Cambridge, UK: Cambridge University Press [*188*]

Dieckmann U & Doebeli M (1999). On the origin of species by sympatric speciation. *Nature* **400**:354–357 [*205–206*]

Dieckmann U & Doebeli M (2004). Adaptive dynamics of speciation: Sexual populations. In *Adaptive Speciation*, eds. Dieckmann U, Doebeli M, Metz JAJ & Tautz D, pp. 76–111. Cambridge, UK: Cambridge University Press [*205*]

Dieckmann U & Heino M. The adaptive dynamics of function-valued traits. Unpublished [*201*]

Dieckmann U & Law R (1996). The dynamical theory of coevolution: A derivation from stochastic ecological processes. *Journal of Mathematical Biology* **34**:579–612 [*188, 197, 199–201, 308, 310, 324*]

Dieckmann U, Marrow P & Law R (1995). Evolutionary cycling in predator–prey interactions: Population dynamics and the red queen. *Journal of Theoretical Biology* **178**:91–102 [*200, 215, 339*]

Dieckmann U, Metz JAJ, Sabelis MW & Sigmund K (2002). *Adaptive Dynamics of Infectious Diseases: In Pursuit of Virulence Management*. Cambridge, UK: Cambridge University Press [*360*]

Dieckmann U, Doebeli M, Metz JAJ & Tautz D, eds. (2004). *Adaptive Speciation*. Cambridge, UK: Cambridge University Press [*204*]

Diekmann O & Heesterbeek JAP (1999). *Mathematical Epidemiology of Infectious Diseases: Model Building, Analysis and Interpretation*. New York, NY, USA: John Wiley & Sons [*62*]

Diekmann O, Metz JAJ & Sabelis MW (1988). Mathematical models of predator–prey–plant interactions in a patchy environment. *Experimental and Applied Acarology* **5**:319–342 [*65*]

Diekmann O, Metz JAJ & Sabelis MW (1989). Reflections and calculations on a prey–predator–patch problem. *Acta Applicandae Mathematicae* **14**:23–25 [*65*]

Diekmann O, Heesterbeek JAP & Metz JAJ (1990). On the definition of the basic reproduction ratio R_0 in models for infectious diseases in heterogeneous populations. *Journal of Mathematical Biology* **28**:365–382 [*62*]

Diekmann O, Gyllenberg M, Metz JAJ & Thieme HR (1998). On the formulation and analysis of general deterministic structured population models. I. Linear theory. *Journal of Mathematical Biology* **36**:349–388 [*66*]

Diekmann O, Gyllenberg M, Huang H, Kirkilionis M, Metz JAJ, & Thieme HR (2001). On the formulation and analysis of general deterministic structured population models. II. Nonlinear theory. *Journal of Mathematical Biology* **43**:157–189 [*63, 66–67*]

Dobson AP (1996). *Conservation and Biodiversity*. Scientific American Libraries. New York, NY, USA: WA Freeman & Company [*8, 360*]

Dodd ME, Silverton J & Chase MW (1999). Phylogenetic analysis of trait evolution and species diversity variation among angiosperm families. *Evolution* **53**:732–744 [*307*]

Doebeli M & Dieckmann U (2000). Evolutionary branching and sympatric speciation caused by different types of ecological interactions. *The American Naturalist* **156**:S77–S101 [*205*]

Doebeli M & Dieckmann U (2003). Speciation along environmental gradients. *Nature* **421**:259–264 [*205, 207*]

Doebeli M & Dieckmann U (2004). Adaptive dynamics of speciation: Spatial structure. In *Adaptive Speciation*, eds. Dieckmann U, Doebeli M, Metz JAJ & Tautz D, pp. 140–167. Cambridge, UK: Cambridge University Press [*205, 207*]

Doebeli M & Knowlton N (1998). The evolution of interspecific mutualisms. *Proceedings of the National Academy of Sciences of the USA* **95**:8676–8680 [*307*]

Doebeli M & Ruxton GD (1997). Evolution of dispersal rates in metapopulation models: Branching and cyclic dynamics in phenotype space. *Evolution* **51**:1730–1741 [*267*]

Douglas AE (1994). *Symbiotic Interactions*. New York, NY, USA: Oxford University Press [*306*]

Douglas AE & Smith DC (1989). Are endosymbioses mutualistic? *Trends in Ecology and Evolution* **4**:350–352 [*338*]

Drake JA, Mooney HA, di Castri F, Groves RH, Kruger FJ, Rejmanek M & Williamson M, eds. (1989). *Biological Invasions: A Global Perspective*. Brisbane, Australia: John Wiley & Sons [*201*]

Drechsler M & Wissel C (1997). Separability of local and regional dynamics in metapopulations. *Theoretical Population Biology* **51**:9–21 [*77*]

Dressler RL (1981). *The Orchids: Natural History and Classification*. Cambridge, MA, USA: Harvard University Press [*318*]

Dugatkin LA (1997). *Cooperation Among Animals: An Evolutionary Perspective*. New York, NY, USA: Oxford University Press [*306*]

Dunne JA, Harte J & Taylor KJ (2003). Subalpine meadow flowering phenology responses to climate change: Integrating experimental and gradient methods. *Ecological Monographs* **73**:69–86 [*320*]

Dwyer G, Levin SA & Buttel L (1990). A simulation model of the population dynamics and evolution of myxomatosis. *Ecological Monographs* **60**:423–447 [*305, 319*]

Dyer MI, DeAngelis DL & Post WM (1986). A model of herbivore feedback on plant productivity. *Mathematical Biosciences* **79**:171–184 [*333*]

Dynesius M & Jansson R (2000). Evolutionary consequences of changes in species' geographical distributions driven by Milankovitch climate oscillations. *Proceedings of the National Academy of Sciences of the USA* **97**:9115–9120 [*357*]

Ebert D, Haag C, Kirkpatrick M, Riek M, Hottinger JW & Pajunen VI (2002). A selective advantage to immigrant genes in a Daphnia metapopulation. *Science* **295**:485–488 [*359*]

Eckert CG, Manicacci D & Barrett SCH (1996). Genetic drift and founder effect in native versus introduced populations of an invading plant, *Lythrum salicaria* (Lythraceae). *Evolution* **50**:1512–1519 [*295*]

Ehrlich PR (2000). *Human Natures: Genes, Cultures, and the Human Prospect*. Washington, DC, USA: Island Press [*362*]

Ehrlich PR (2001). Intervening in evolution: Ethics and actions. *Proceedings of the National Academy of Sciences of the USA* **98**:5477–5480 [*361–362*]

Ehrlich PR, Dobkin DS & Wheye D (1988). *The Birder's Handbook*. New York, NY, USA: Simon and Schuster [*31*]

Ehrman L, White MM & Wallace B (1991). A long-term study involving *Drosophila melanogaster* and toxic media. *Evolutionary Biology* **25**:175–209 [*95*]

Eisen EJ (1975). Population size and selection intensity effects on long-term selection response in mice. *Genetics* **79**:305–323 [*87*]

Elena SF & Lenski RE (1997). Test of synergistic interactions among deleterious mutations in bacteria. *Nature* **390**:395–398 [*163*]

Elliott JK, Elliott JM & Mariscal RN (1995). Host selection, location, and association behaviors of anemonefishes in field settlement experiments. *Marine Biology* **122**:377–389 [*318*]

Ellstrand NC (1992). Gene flow by pollen: Implications for plant conservation genetics. *Oikos* **63**:77–86 [*344*]

Ellstrand NC, Prentice HC & Hancock FC (1999). Gene flow and introgression from domesticated plants into their wild relatives. *Annual Review of Ecology and Systematics* **30**:539–563 [*350–351*]

Emlen JM (1987). Evolutionary ecology and the optimality assumption. In *The Latest on the Best*, ed. Dupre J, pp. 163–177. Cambridge, MA, USA: MIT Press [*189*]

Endler JA (1980). Natural selection on color patterns in *Poecilia reticulata*. *Evolution* **34**:76–91 [*107*]

Engels WR (1983). The P family of transposable elements in *Drosophila*. *Annual Review of Genetics* **17**:315–344 [*347*]

Erhardt A & Rusterholz HP (1997). Effects of elevated CO_2 on flowering phenology and nectar production. *Acta Oecologica* **18**:249–253 [*320*]

Ernande B, Dieckmann U & Heino M (2002). Fisheries-induced changes in age and size at maturation and understanding the potential for selection-induced stock collapse. In *The Effects of Fishing on the Genetic Composition of Living Marine Resources*, ICES CM 2002/Y:06. Copenhagen, Denmark: International Council for the Exploration of the Seas [*214*]

Errington PL (1940). Natural restocking of muskrat-vacant habitats. *Journal of Wildlife Management* **4**:173–185 [*24*]

Eshel I (1983). Evolutionary and continuous stability. *Journal of Theoretical Biology* **103**:99–111 [*193, 197, 199, 335*]

Eshel I & Akin E (1983). Coevolutionary instability of mixed Nash solutions. *Journal of Mathematical Biology* **18**:123–133 [*340*]

Eshel I & Motro U (1981). Kin selection and strong stability of mutual help. *Theoretical Population Biology* **19**:420–433 [*193, 197, 199, 335*]

Etienne RS & Heesterbeek JAP (2001). Rules of thumb for conservation of metapopulations based on a stochastic winking-patch model. *The American Naturalist* **158**:389–407 [*61, 73, 75*]

Etterson JR & Shaw RG (2001). Constraint to adaptive evolution in response to global warming. *Science* **292**:151–154 [*357*]

Ewald P (1994). *Evolution of Infectious Disease.* Oxford, UK: Oxford University Press [*360*]

Ewens WJ (1989). The effective population sizes in the presence of catastrophes. In *Mathematical Evolutionary Theory*, ed. Feldman MW, pp. 9–25. Princeton, NJ, USA: Princeton University Press [*231*]

Eyre-Walker A & Keightley PD (1999). High genomic deleterious mutation rates in hominids. *Nature* **397**:344–347 [*160*]

Falconer DS (1960). *Introduction to Quantitative Genetics*. New York, NY, USA: Ronald Press Company [*107*]

Falconer DS (1989). *Introduction to Quantitative Genetics*, Third Edition. New York, NY, USA: John Wiley & Sons [*248*]

Falconer DS & Mackay TFC (1996). *Introduction to Quantitative Genetics*, Fourth Edition. Harlow, UK: Longman [*96, 120, 126–127, 144, 173–174, 177*]

Falk DA & Holsinger KE (1991). *Genetics and Conservation of Rare Plants.* Oxford, UK: Oxford University Press [*59*]

FAO (1999). *The State of World Fisheries and Aquaculture 1998.* Rome, Italy: FAO [*361*]

Feder ME & Hofmann G (1999). Heat-shock proteins, molecular chaperones and the stress response: Evolutionary and ecological physiology. *Annual Review of Physiology* **61**:243–282 [*140*]

Feder M & Krebs RA (1997). Ecological and evolutionary physiology of heat shock proteins and the stress response in *Drosophila*: Complementary insights from genetic engineering and natural variation. In *Environmental Stress, Evolution and Adaptation*, eds. Bijlsma R & Loeschcke V, pp. 155–173. Basel, Switzerland: Birkhäuser [*140*]

Feder JH, Rossi JM, Solomon J, Solomon N & Lindquist S (1992). The consequences of expressing HSP70 in *Drosophila* cells at normal temperatures. *Genes & Development* **6**:1402–1413 [*140*]

Feldman MW & Laland KN (1996). Gene-culture coevolutionary theory. *Trends in Ecology and Evolution* **11**:453–457 [*362*]

Felsenstein J (1974). The evolutionary advantage of recombination. *Genetics* **78**:737–756 [*165*]

Felsenstein J (1981). Skepticism towards Santa Rosalia, or why are there so few kinds of animals? *Evolution* **35**:124–238 [*205*]

Fenster CB & Galloway LF (2000). Inbreeding and outbreeding depression in natural populations of *Chamaechrita fasciculata* (Fabaceae). *Conservation Biology* **14**:1406–1412 [*284*]

Ferrière R (2000). Adaptive responses to environmental threats: Evolutionary suicide, insurance, and rescue. *Options* Spring, pp. 12–16. Laxenburg, Austria: International Institute for Applied Systems Analysis [*188, 208, 266, 278, 282*]

Ferrière R & Gatto M (1993). Chaotic population dynamics can result from natural selection. *Proceedings of the Royal Society of London B* **251**:33–38 [*44*]

Ferrière R & Gatto M (1995). Lyapunov exponents and the mathematics of invasion in oscillatory or chaotic populations. *Theoretical Population Biology* **48**:126–171 [*188*]

Ferrière R, Sarrazin F, Legendre S & Baron J-P (1996). Matrix population models applied to viability analysis and conservation: Theory and practice using the ULM software. *Acta Oecologica* **17**:629–656 [*43*]

Ferrière R, Belthoff JR, Olivieri I & Krackow S (2000). Evolving dispersal: Where to go next? *Trends in Ecology and Evolution* **15**:5–7 [*79, 276*]

Ferrière R, Bronstein JL, Rinaldi S, Law R & Gauduchon M (2002). Cheating and the evolutionary stability of mutualism. *Proceedings of the Royal Society of London B* **269**:773–780 [*309–311*]

Festa-Bianchet M, Jorgenson JT, Lucherini M & Wishart WD (1995). Life history consequences of variation in age of primiparity in bighorn ewes. *Ecology* **76**:871–881 [*53*]

Fieberg J & Ellner SP (2001). Stochastic matrix models for conservation and management: A comparative review of methods. *Ecology Letters* **4**:244–266 [*43*]

Fischer M & Matthies D (1998). Effects of population size on performance in the rare plant *Gentianella germanica. Journal of Ecology* **86**:195–204 [*286*]

Fisher RA (1918). The correlation between relatives on the supposition of Mendelian inheritance. *Transactions of the Royal Society of Edinburgh* **52**:399–433 [*173*]

Fisher RA (1930). *The Genetical Theory of Natural Selection.* Oxford, UK: Clarendon Press [*119, 166, 174, 189–190*]

Fisher RA (1941). Average excess and average effect of a gene substitution. *Annals of Eugenics* **11**:53–63 [*174*]

Fisher RA (1958). *The Genetical Theory of*

Natural Selection, Second Revised Edition. New York, NY, USA: Dover Publications [*51, 245*]

Fitter AH & Fitter RSR (2002). Rapid changes in flowering time in British plants. *Science* **296**:1689–1691 [*320*]

Fitzsimmons NN, Buskirk SW & Smith MH (1997). Genetic changes in reintroduced rocky mountain bighorn sheep populations. *Journal of Wildlife Management* **61**:863–872 [*55*]

Foley P (1994). Predicting extinction times from environmental stochasticity and carrying capacity. *Conservation Biology* **8**:124–137 [*36*]

Forbes VE & Calow P (1997). Responses of aquatic organisms to pollutant stress: Theoretical and practical implications. In *Environmental Stress, Evolution and Adaptation*, eds. Bijlsma R & Loeschcke V, pp. 25–32. Basel, Switzerland: Birkhäuser [*136*]

Frank SA (1986). Dispersal polymorphisms in subdivided populations. *Journal of Theoretical Biology* **122**:303–309 [*265*]

Frank SA (1993). Evolution of host–parasite diversity. *Evolution* **47**:1721–1732 [*340*]

Frank SA & Slatkin M (1992). Fisher's fundamental theorem of natural selection. *Trends in Ecology and Evolution* **7**:92–95 [*190*]

Frank K & Wissel C (1994). Ein Modell über den Einfluß räumlicher Aspekte auf das Überleben von Metapopulationen. *Verhandlungen der Gesellschaft für Ökologie* **23**:303–310 [*73*]

Frank K & Wissel C (1998). Spatial aspects of metapopulation survival: From model results to rules of thumb for landscape management. *Landscape Ecology* **13**:363–379 [*73*]

Frank K & Wissel C (2002). A formula for the mean lifetime of metapopulations in heterogeneous landscapes. *The American Naturalist* **159**:530–552 [*73, 75, 78*]

Frankham R (1983). Origin of genetic variation in selection lines. *Proceedings of the Thirty-Second Annual Breeders' Roundtable* **32**:1–18 [*87*]

Frankham R (1995a). Conservation genetics. *Annual Review of Genetics* **29**:305–327 [*96, 99, 133*]

Frankham R (1995b). Inbreeding and extinction: A threshold effect. *Conservation Biology* **9**:792–799 [*99*]

Frankham R (1995c). Effective population size/adult population size ratios in wildlife: A review. *Genetical Research* **66**:95–107. [*162*]

Frankham R (1996). Relationship of genetic variation to populations size in wildlife. *Conservation Biology* **10**:1500–1508 [*97, 130*]

Frankham R & Ralls K (1998). Conservation biology: Inbreeding leads to extinction. *Nature* **392**:441–442 [*96*]

Frankham R, Lees K, Montgomery ME, England PR, Lowe E & Briscoe DA (1999). Do population size bottlenecks reduce evolutionary potential? *Animal Conservation* **2**:255–260 [*98*]

Frankham R, Ballou JD & Briscoe DA (2002). *Introduction to Conservation Genetics*. Cambridge, UK: Cambridge University Press [*97*]

Frankham R, Gilligan DM, Lowe E, Woodworth L, Montgomery ME & Briscoe DA. Unpublished data [*99*]

Freidlin MI & Wentzell AD (1998). *Random Perturbations of Dynamical Systems*, Second Edition. New York, NY, USA: Springer-Verlag [*77–78*]

Fréville H, Colas B, Ronfort J, Riba M & Olivieri I (1998). Predicting endemism from population structure of widespread species: Case study in *Centaurea maculosa* Lam. (Asteraceae). *Conservation Biology* **12**:1269–1278 [*297*]

Fréville H, Justy F & Olivieri I (2001). Comparative allozyme and microsatellite population structure in a narrow endemic plant species, *Centaurea corymbosa* Pourret (Asteraceae). *Molecular Ecology* **10**:879–889 [*296*]

Fritts TH & Rodda GH (1998). The role of introduced species in the degradation of island ecosystems: A case history of Guam. *Annual Review of Ecology and Systematics* **29**:113–140 [*305*]

Fritz RS (1999). Resistance of hybrid plants to herbivores: Genes, environment, or both? *Ecology* **80**:382–391 [*348*]

Fritz RS, Nichols-Orians CM & Brunsfeld SJ (1994). Interspecific hybridization of plants and resistance to herbivores: Hypotheses, genetics and variable responses in a diverse herbivore community. *Oecologia* **97**:106–117 [*346*]

Fry JD, Keightley PD, Heinsohn SL & Nuzhdin SV (1999). New estimates of the rates and effects of mildly deleterious mutation in *Drosophila melanogaster*. *Proceed-*

ings of the National Academy of Sciences of the USA **96**:574–579 [*160*]

Fu YB (1999). Patterns of the purging of deleterious genes with synergistic interactions in different breeding schemes. *Theoretical and Applied Genetics* **98**:337–346 [*132*]

Fu Y-B, Namkoong G & Carlson JE (1998). Comparison of breeding strategies for purging inbreeding depression via simulation. *Conservation Biology* **12**:856–864 [*132*]

Gabriel W & Bürger R (1992). Survival of small populations under demographic stochasticity. *Theoretical Population Biology* **41**:44–71 [*27, 31, 38, 50*]

Gabriel W & Bürger R (1994). Extinction risk by mutational meltdown: Synergistic effects between population regulation and genetic drift. In *Conservation Genetics*, eds. Loeschcke V, Tomiuk J & Jain SK, pp. 70–84. Basel, Switzerland: Birkhäuser [*160*]

Gabriel W, Bürger R & Lynch M (1991). Population extinction by mutational load and demographic stochasticity. In *Population Biology and Conservation*, eds. Seitz A & Loeschcke V, pp. 49–59. Basel, Switzerland: Birkhäuser [*31*]

Gadgil M & Bossert PW (1970). Life historical consequences of natural selection. *The American Naturalist* **104**:1–24 [*102*]

Gaggiotti O (1996). Population genetic models of source–sink metapopulations. *Theoretical Population Biology* **50**:178–208 [*241–242*]

Gaggiotti O. Stochastic migration facilitates the spread of beneficial mutations in sink populations. Unpublished [*243*]

Gaggiotti O & Smouse P (1996). Stochastic migration and maintenance of genetic variation in sink populations. *The American Naturalist* **147**:919–945 [*241–242*]

Galen C & Gregory T (1989). Interspecific pollen transfer as a mechanism of competition: Consequences of foreign pollen contamination for seed set in the alpine wildflower *Polemonium viscosum*. *Oecologia* **81**:120–123 [*347*]

Gallagher R & Carpenter B (1997). Human-dominated ecosystems. *Science* **277**:485 [*356*]

Galton F (1889). *Natural Inheritance*. London, UK: MacMillan [*173*]

Gandon S (1999). Kin competition, the cost of inbreeding and the evolution of dispersal. *Journal of Theoretical Biology* **200**:245–364 [*265*]

Gandon S & Michalakis Y (1999). Evolutionarily stable dispersal rate in a metapopulation with extinctions and kin competition. *Journal of Theoretical Biology* **199**:275–290 [*265*]

Gandon S, Capowiez Y, Dubois Y, Michalakis Y & Olivieri I (1996). Local adaptation and gene-for-gene coevolution in a metapopulation model. *Proceedings of the Royal Society of London B* **263**:1003–1009 [*339–340*]

García-Dorado A & Caballero A (2001). On the average coefficient of dominance of deleterious spontaneous mutations. *Genetics* **155**:1991–2001 [*163*]

García-Dorado A, López-Fanjul C & Caballero A (1999). Properties of spontaneous mutations affecting quantitative traits. *Genetical Research* **74**:341–350 [*160*]

Gardiner CW (1983). *Handbook of Stochastic Methods for Physics, Chemistry and the Natural Sciences*. Berlin, Germany: Springer-Verlag [*77–78*]

Garrett L (1994). *The Coming Plague: Newly Emerging Diseases in a World Out of Balance*. London, UK: Virago Press [*96*]

Gaston KJ (1994). *Rarity*. London, UK: Chapman & Hall [*285, 345*]

Gaston KJ & Kunin WE (1997). Rare–common differences: An overview. In *The Biology of Rarity*, eds. Kunin WE & Gaston KJ, pp. 12–29. London, UK: Chapman & Hall [*345*]

Gatto M (1993). The evolutionary optimality of oscillatory and chaotic dynamics in simple population-models. *Theoretical Population Biology* **43**:310–336 [*197*]

Gatto M & De Leo G (2000). Pricing biodiversity and ecosystem services: The never-ending story. *BioScience* **50**:347–355 [*363*]

Gebhardt MD & Stearns SC (1988). Reaction norms for developmental time and weight at eclosion in *Drosophila mercatorum*. *Journal of Evolutionary Biology* **1**:335–354 [*146*]

Gebhardt MD & Stearns SC (1992). Phenotypic plasticity for life-history characters in *Drosophila melanogaster*: Effect of the environment on genetic parameters. *Genetical Research* **60**:87–101 [*146*]

Geist V (1971). *Mountain Sheep*. Chicago, IL, USA: University of Chicago Press [*53*]

Geritz SAH & Kisdi É (2000). Adaptive dynamics in diploid sexual populations and the evolution of reproductive isolation. *Proceedings of the Royal Society of London B* **267**:1671–1678 [*205*]

Geritz SAH, Metz JAJ, Kisdi E & Meszéna G

(1997). Dynamics of adaptation and evolutionary branching. *Physical Review Letters* **78**:2024–2027 [*196, 199, 309*]

Geritz SAH, Kisdi É, Meszéna G & Metz JAJ (1998). Evolutionarily singular strategies and the adaptive growth and branching of the evolutionary tree. *Evolutionary Ecology Research* **12**:35–57 [*329, 337*]

Geritz SAH, Kisdi É, Meszéna G, Metz JAJ (2004). Adaptive dynamics of speciation: Ecological underpinnings. In *Adaptive Speciation*, eds. Dieckmann U, Doebeli M, Metz JAJ & Tautz D, pp. 54–75. Cambridge, UK: Cambridge University Press [*205*]

Gerritsen J (1980). Sex and parthenogenesis in sparse populations. *The American Naturalist* **115**:718–742 [*24, 46*]

Gething MJ, ed. (1997). *Guidebook to Molecular Chaperones and Protein-Folding Catalysts*. Oxford, UK: Sambrook and Tooze Publication at Oxford University Press [*140*]

Gibernau M & Hossaert-McKey M (1998). Are olfactory signals sufficient to attract fig pollinators? *Ecoscience* **5**:306–311 [*316*]

Gibbs HL & Grant PR (1987). Oscillating selection on Darwin's finches. *Nature* **327**:511–513 [*110, 116*]

Gilissen LJ & Linskens HF (1975). Pollen tube growth in styles of self-incompatible *Petunia* pollinated with radiated pollen and foreign pollen mixtures. In *Gamete Competition in Plants and Animals*, ed. Mulcahy DL, pp. 201–205. Amsterdam, Netherlands: North Holland [*347*]

Gill FB (1987). Allozymes and genetic similarity of blue-winged and golden-winged warblers. *Auk* **104**:444–449 [*345*]

Gill RB (2000). *The Great Maya Droughts: Water, Life, and Death*. Albuquerque, NM, USA: University of New Mexico Press [*362*]

Gillespie JH (1991). *The Causes of Molecular Evolution*. Oxford, UK: Oxford University Press [*123*]

Gillespie JH & Turelli M (1989). Genotype–environment interactions and the maintenance of polygenic variation. *Genetics* **121**:129–138 [*123, 148*]

Gilligan DM, Woodworth LM, Montgomery ME, Briscoe DA & Frankham R (1997). Is mutation accumulation a threat to the survival of endangered populations? *Conservation Biology* **11**:1235–1241 [*97*]

Gilmer K & Kadereit JW (1989). The biology and affinities of *Senecio teneriffae* Schultz Bip., an annual endemic from the Canary Islands. *Botanische Jahrbücher* **11**:263–273 [*350*]

Gimelfarb A (1988). Processes of pair formation leading to assortative mating in biological populations: Encounter-mating model. *The American Naturalist* **131**:865–884 [*46*]

Gingerich PD (1983). Rates of evolution: Effects of time and temporal scaling. *Science* **222**:159–161 [*110*]

Godø OR (2000). *Maturation Dynamics of Arcto-Norwegian Cod*. IIASA Interim Report IR-00-024. Laxenburg, Austria: International Institute for Applied Systems Analysis [*5*]

Gomulkiewicz R & Holt RD (1995). When does evolution by natural selection prevent extinction? *Evolution* **49**:201–207 [*117, 186, 220, 247–248, 278*]

Gomulkiewicz R, Holt RD & Barfield M (1999). The effects of density dependence and immigration on local adaptation and niche evolution in a black-hole sink environment. *Theoretical Population Biology* **55**:283–296 [*243, 247, 253, 255, 257–258*]

Gonzalez M & Molina M (1996). On the limit behavior of a superadditive bisexual Galton–Watson branching process. *Journal of Applied Probability* **33**:960–967 [*33*]

Goodman D (1987a). Considerations of stochastic demography in the design and management of natural reserves. *Natural Resource Modeling* **1**:205–234 [*21, 27*]

Goodman D (1987b). The demography of chance extinction. In *Viable Populations for Conservation*. ed. Soulé ME, pp. 11–43. New York, NY, USA: Cambridge University Press [*21, 27, 38, 77*]

Goodnight CJ (1988). Epistasis and the effect of founder events on the additive genetic variance. *Evolution* **42**:441–454 [*148*]

Gosselin F & Lebreton JD (2000). Potential of branching processes as a modeling tool for conservation biology. In *Quantitative Methods for Conservation Biology*, eds. Ferson S & Burgman M, pp. 199–225. New York, NY, USA: Springer-Verlag [*32–33, 45*]

Gottelli D, Sillero-Zubiri C, Applebaum GD, Roy MS, Girman D, Garcia-Moreno J, Ostrander EA & Wayne RK (1994). Molecular genetics of the most endangered canid: The Ethiopian wolf, *Canis simensis*. *Molecular Ecology* **3**:301–312 [*350*]

Gould SJ (2000). Deconstructing the "science

wars" by reconstructing an old mold. *Science* **287**:253–261 [*356*]

Gould SJ & Lewontin RC (1979). The spandrels of San Marco and the Panglossian paradigm: A critique of the adaptationist programme. *Proceedings of the Royal Society of London B* **205**:581–598 [*190*]

Goverde M, Schweizer K, Baur B & Erhardt A (2002). Small-scale habitat fragmentation effects on pollinator behaviour: Experimental evidence from the bumblebee *Bombus veteranus* on calcareous grasslands. *Biological Conservation* **104**:293–299 [*313*]

Graham JH, Freeman DC & Emlen JM (1993a). Developmental stability: A sensitive indicator of populations under stress. In *Environmental Toxicology and Risk Assessment*, eds. Landis WG, Hughes JS & Lewis MA, pp. 136–158. Philadelphia, PA, USA: American Society for Testing and Materials [*142*]

Graham JH, Roe KE & West TB (1993b). Effects of lead and benzene on developmental stability of *Drosophila melanogaster*. *Ecotoxicology* **2**:185–195 [*143*]

Grant PR & Grant BR (1992). Hybridization of bird species. *Science* **256**:193–197 [*353*]

Grant BR & Grant PR (1996). High survival of Darwin's finch hybrids: Effects of beak morphology and diets. *Ecology* **77**:500–509 [*353*]

Grasman J & HilleRisLambers R (1997). On local extinction in a metapopulation. *Ecological Modelling* **103**:71–80 [*77*]

Grasman J & Van Herwaarden OA (1999). *Asymptotic Methods for the Fokker–Planck Equation and the Exit Problem in Applications*. Berlin, Germany: Springer-Verlag [*77–78*]

Greenberg R, Bichier P & Sterling J (1997). Acacia, cattle and migratory birds in southeastern Mexico. *Biological Conservation* **80**:235–247 [*360*]

Greene C, Umbanhowar J, Mangel M & Caro T (1998). Animal breeding systems, hunter selectivity, and consumptive use in wildlife conservation. In *Behavioral Ecology and Conservation Biology*, ed. Caro T, pp. 271–305. New York, NY, USA: Oxford University Press [*58*]

Griffis MR & Jaeger RG (1998). Competition leads to an extinction-prone species of salamander: Interspecific territoriality in a metapopulation. *Ecology* **79**:2494–2502 [*346*]

Griffith B, Scott MJ, Carpenter JW & Reed C (1989). Translocation as a species conservation tool: Status and strategy. *Science* **245**:477–480 [*45*]

Groom MJ (1998). Allee effects limit population viability of an annual plant. *The American Naturalist* **151**:487–496 [*321*]

Gronemeyer PA, Dilger BJ, Bouzat JL & Paige KN (1997). The effects of herbivory on paternal fitness in scarlet gilia: Better moms also make better pops. *The American Naturalist* **150**:592–602 [*333*]

Guckenheimer J & Holmes P (1997). *Nonlinear Oscillations, Dynamical Systems, and Bifurcations of Vector Fields*. Third Edition. New York, NY, USA: Springer-Verlag [*212, 219*]

Gyllenberg M. Metapopulations. In *Branching Processes in Biology: The Ramifications of Life and Death*, eds. Jagers P & Haccou P. Cambridge, UK: Cambridge University Press. In press [*61, 73*]

Gyllenberg M & Hanski I (1992). Single-species metapopulation dynamics: A structured model. *Theoretical Population Biology* **42**:35–62 [*64–65*]

Gyllenberg M & Hanski I (1997). Habitat deterioration, habitat destruction and metapopulation persistence in a heterogeneous landscape. *Theoretical Population Biology* **52**:198–215 [*64, 72*]

Gyllenberg M & Jagers P. Branching processes and structured population dynamics. In *Branching Processes in Biology: The Ramifications of Life and Death*, eds. Jagers P & Haccou P. Cambridge, UK: Cambridge University Press. In press [*66*]

Gyllenberg M & Metz JAJ (2001). On fitness in structured metapopulations. *Journal of Mathematical Biology* **43**:545–560 [*64, 66, 79, 213, 265, 272–273, 277*]

Gyllenberg M & Parvinen K (2001). Necessary and sufficient conditions for evolutionary suicide. *Bulletin of Mathematical Biology* **63**:981–993 [*209, 211, 266, 276–277, 282*]

Gyllenberg M & Silvestrov DS (1994). Quasi-stationary distributions of a stochastic metapopulation model. *Journal of Mathematical Biology* **33**:35–70 [*61*]

Gyllenberg M & Silvestrov DS (1999). Quasi-stationary phenomena for semi-Markov processes. In *Semi-Markov Models and Applications*, eds. Janssen J & Limnios N, pp. 33–60. Dordrecht, Netherlands: Kluwer Academic Publishers [*61, 73*]

Gyllenberg M & Silvestrov DS (2000). Nonlinearly perturbed regenerative processes and pseudo-stationary phenomena for stochastic systems. *Stochastic Processes and Their Applications* **86**:1–27 [*61, 73*]

Gyllenberg M, Söderbacka G & Ericsson S (1993). Does migration stabilize local population dynamics? Analysis of a discrete metapopulation model. *Mathematical Biosciences* **118**:25–49 [*61*]

Gyllenberg M, Högnas G & Koski T (1994). Population models with environmental stochasticity. *Journal of Mathematical Biology* **32**:92–108 [*35*]

Gyllenberg M, Osipov AV & Söderbacka G (1996). Bifurcation analysis of a metapopulation model with sources and sinks. *Journal of Nonlinear Science* **6**:329–366 [*61*]

Gyllenberg M, Hanski I & Hastings A (1997). Structured metapopulation models. In *Metapopulation Biology: Ecology, Genetics and Evolution*, eds. Hanski I & Gilpin ME, pp. 93–122. London, UK: Academic Press [*63–67, 71*]

Gyllenberg M, Parvinen K & Dieckmann U (2002). Evolutionary suicide and evolution of dispersal in structured metapopulations. *Journal of Mathematical Biology* **45**:79–105 [*64, 68, 71, 188, 211, 213, 266, 273, 276–277, 280–282*]

Gyllenberg M, Metz JAJ & Parvinen K. A basic reproduction ratio for structured metapopulations. Unpublished [*68*]

Hackney EE & McGraw JB (2001). Experimental demonstration of an Allee effect in American ginseng. *Conservation Biology* **15**:129–136 [*321*]

Haig SM, Belthoff JR & Allen DH (1993). Population viability analysis for a small population of red-cockaded woodpeckers and and evaluation of enhancement strategies. *Conservation Biology* **7**:289–301 [*100*]

Haikola S, Fortelius W, O'Hara RB, Kuussaari M, Wahlberg N, Saccheri IJ, Singer MC & Hanski I (2001). Inbreeding depression and the maintenance of genetic load in *Melitaea cinxia* metapopulations. *Conservation Genetics* **2**:325–335 [*287*]

Haldane JBS (1930). A mathematical theory of natural and artificial selection. VI. Isolation. *Proceedings of the Cambridge Philosophical Society* **26**:220–230 [*252*]

Haldane JBS (1932). *The Causes of Evolution*. London, UK: Harper [*6, 153, 208*]

Haldane JBS (1937). The effect of variation on fitness. *The American Naturalist* **71**:337–349 [*156*]

Haldane JBS (1949a). Suggestions as to quantitative measurement of rates of evolution. *Evolution* **3**:51–56 [*110*]

Haldane JBS (1949b). Parental and fraternal correlations in fitness. *Annals of Eugenics* **14**:288–292 [*123*]

Haldane JBS (1953). Animal populations and their regulation. *New Biology* **15**:9–24 [*24*]

Haldane JBS (1962). Conditions for stable polymorphism at an autosomal locus. *Nature* **193**:1108 [*124*]

Halley JM (1996). Ecology, evolution and $1/f$-noise. *Trends in Ecology and Evolution* **11**:33–37 [*36*]

Halley JM & Iwasa Y (1998). Extinction rate of a population under both demographic and environmental stochasticity. *Theoretical Population Biology* **53**:1–15 [*24, 35*]

Hamilton WD (1967). Extraordinary sex ratios. *Science* **156**:477–488 [*191, 199*]

Hamilton WD & May RM (1977). Dispersal in stable habitats. *Nature* **269**:578–581 [*265*]

Hammond A (1998). *Which World? Scenarios for the 21st Century*. London, UK: Earthscan [*356*]

Hamrick JL & Godt MJW (1989). Allozyme diversity in plant species. In *Plant Population Genetics, Breeding, and Genetic Resources*, eds. Brown AHD, Clegg MT, Kahler AL & Weir BS, pp. 43–63. Sunderland, MA, USA: Sinauer Associates Inc. [*97*]

Hanski I (1985). Single-species spatial dynamics may contribute to long-term rarity and commonness. *Ecology* **66**:335–343 [*65, 71*]

Hanski I (1989). Metapopulation dynamics: Does it help to have more of the same? *Trends in Ecology and Evolution* **4**:113–114 [*243, 285*]

Hanski I (1994). A practical model of metapopulation dynamics. *Journal of Animal Ecology* **63**:151–162 [*72, 288*]

Hanski I (1998). Metapopulation dynamics. *Nature* **396**:41–49 [*65*]

Hanski I (1999). *Metapopulation Ecology*. Oxford, UK: Oxford University Press [*59–60, 65, 72, 79, 265, 284–285, 287*]

Hanski I (2001). Spatially realistic theory of metapopulation ecology. *Naturwissenschaften* **88**:372–381 [*59*]

Hanski IA & Gilpin ME, eds. (1997).

Metapopulation Biology: Ecology, Genetics, and Evolution. New York, NY, USA: Academic Press [*59, 265, 284*]

Hanski I & Gyllenberg M (1993). Two general metapopulation models and the core–satellite species hypothesis. *The American Naturalist* **142**:17–41 [*64, 71*]

Hanski I & Heino M. Evolution of insect plant preference driven by extinction-colonization dynamics. Unpublished [*293*]

Hanski I & Ovaskainen O (2000). The metapopulation capacity of a fragmented landscape. *Nature* **404**:756–758 [*72*]

Hanski I & Simberloff D (1997). The metapopulation approach, its history, conceptual domain, and application to conservation. In *Metapopulation Biology: Ecology, Genetics and Evolution*, eds. Hanski I & Gilpin ME, pp. 5–26. London, UK: Academic Press [*59, 286*]

Hanski I & Singer M (2001). Spacial dynamics and host choice. *The American Naturalist* **158**:341–353 [*292–293*]

Hanski I, Alho J & Moilanen A (2000). Estimating the parameters of survival and migration of individuals in metapopulations. *Ecology* **81**:239–251 [*288*]

Hanson FB & Tuckwell HC (1978). Persistence times of populations with large random fluctuations. *Theoretical Population Biology* **14**:46–61 [*28*]

Harder LD, Cruzan MB & Thomson JD (1993). Unilateral incompatibility and the effects of interspecific pollination for *Erythronium americanum* and *Erythronium albidum* (Liliaceae). *Canadian Journal of Botany* **71**:353–358 [*347*]

Hardy O, González-Martinez SC, Fréville H, Bocquien G, Mignot A, Colas B & Olivieri I. Fine-scale mapping of gene dispersal in a narrow-endemic, self-incompatible plant species. Unpublished [*296*]

Harlan JR (1976). Diseases as a factor in evolution. *Annual Review of Phytopathology* **14**:31–51 [*346*]

Harlan JR (1992). *Crops and Man*. Madison, WI, USA: American Society of Agronomy [*353*]

Harper JL, Clatworthy JN, McNaughton IH & Sager GR (1961). The evolution and ecology of species living in the same area. *Evolution* **15**:209–227 [*344*]

Harrington R, Woiwod I & Sparks T (1999). Climate change and trophic interactions. *Trends in Ecology and Evolution* **14**:146–150 [*90*]

Harris TE (1963). *The Theory of Branching Processes*. Berlin, Germany: Springer-Verlag [*32, 50*]

Harris PD (1986). Species of *Gyrodactylus* von Nordmann 1932 (Monogenea Gyrodactylidae) from poeciliid fishes, with a description of *G. turnbulli* sp. Nov. from the guppy, *Poecilia reticulata* Peters. *Journal of Natural History* **20**:183–191 [*346*]

Harrison RD (2000). Repercussions of El Niño: Drought causes extinction and the breakdown of mutualism in Borneo. *Proceedings of the Royal Society of London B* **267**:911–915 [*316*]

Harrison JF & Fewell JH (1995). Thermal effects on feeding behavior and net energy-intake in a grasshopper experiencing large diurnal fluctuations in body temperature. *Physiological Zoology* **68**:453–473 [*93*]

Harrison S & Hastings A (1996). Genetic and evolutionary consequences of metapopulation structure. *Trends in Ecology and Evolution* **11**:180–183 [*236, 242, 284*]

Hassell MP (1975). Density-dependence in single-species populations. *Journal of Animal Ecology* **44**:283–295 [*23*]

Hassell MP, Lawton JH & May RM (1976). Patterns of dynamical behaviour in single-species populations. *Journal of Animal Ecology* **45**:471–486 [*23*]

Hastings A (1980). Disturbance, coexistence, history, and competition for space. *Theoretical Population Biology* **18**:363–373 [*202*]

Hastings A (1983). Can spatial variation alone lead to selection for dispersal. *Theoretical Population Biology* **24**:244–251 [*265, 267*]

Hastings A & Harrison S (1994). Metapopulation dynamics and genetics. *Annual Review of Ecology and Systematics* **25**:167–188 [*236*]

Hedrick PW (1986). Genetic polymorphism in heterogeneous environments: A decade later. *Annual Review of Ecology and Systematics* **17**:535–566 [*123*]

Hedrick PW (1990). Genotype-specific habitat selection: A new model. *Heredity* **65**:145–149 [*123*]

Hedrick PW (1994). Purging inbreeding depression and the probability of extinction: Full-sib mating. *Heredity* **73**:363–372 [*132*]

Hedrick PW (1995). Gene flow and genetic restoration: The Florida Panther as a case study. *Conservation Biology* **9**:996–1007 [*10*]

Hedrick PW (2000). *Genetics of Populations*, Second Edition. Sudbury, MA, USA: Jones and Bartlett [*252*]

Hedrick PW (2001). Conservation genetics: Where are we now? *Trends in Ecology and Evolution* **16**:629–636 [*359*]

Hedrick PW & Gilpin ME (1997). Genetic effective size of a metapopulation. In *Metapopulation Biology: Ecology, Genetics and Evolution*, eds. Hanski I & Gilpin ME, pp. 165–181. London, UK: Academic Press [*238*]

Hedrick PW & Kalinowski ST (2000). Inbreeding depression in conservation biology. *Annual Review of Ecology and Systematics* **31**:139–162 [*131–132*]

Hedrick PW, Ginevan ME & Ewing EP (1976). Genetic polymorphism in heterogenous environments. *Annual Review of Ecology and Systematics* **7**:1–32 [*123*]

Hedrick PW, Hedgecock D & Hamelberg S (1995). Effective population size in winter-run chinook salmon. *Conservation Biology* **9**:615–624 [*132–133*]

Heeney JL, Evermann JF, McKeirnan AJ, Marker-Kraus L, Roelke ME, Bush M, Wildt DE, Meltzer L, Colly D, Lucas GLJ, Manton VJ, Caro T & O'Brien SJ (1990). Prevalence and implications of feline coronavirus infections of captive and freeranging cheetahs (*Acinonyx jubatus*). *Journal of Virology* **64**:1964–1972 [*346*]

Heino M & Hanski I (2001). Evolution of migration rate in a spatially realistic metapopulation model. *The American Naturalist* **157**:495–511 [*9, 72, 265, 278, 288, 291*]

Heino M, Kaitala V, Ranta E & Lindström J (1997a). Synchronous dynamics and rates of extinction in spatially structured populations. *Proceedings of the Royal Society of London B* **264**:481–486 [*269*]

Heino M, Metz JAJ & Kaitala V (1997b). Evolution of mixed maturation strategies in semelparous life histories: The crucial role of dimensionality of feedback environment. *Philosophical Transactions of the Royal Society of London B* **352**:1647–1655 [*192, 194–195*]

Heino M, Metz JAJ & Kaitala V (1998). The enigma of frequency-dependence. *Trends in Ecology and Evolution* **13**:367–370 [*7, 46, 192*]

Heino M, Dieckmann U & Godø OR (2000). Shrinking cod: Fishery-induced change in an oceanic stock. *Options*, Spring, pp. 6–8. Laxenburg, Austria: International Institute for Applied Systems Analysis [*5*]

Heino M, Dieckmann U & Godø OR (2002). Reaction norm analysis of fisheries-induced adaptive change and the case of the Northeast Arctic cod. *ICES CM 2002/Y*:14 [*5*]

Helander ML, Neuvonen S & Ranta H (1996). Natural variation and effects of anthropogenic environmental changes on endophytic fungi in trees. In *Endophytic Fungi in Grasses and Woody Plants*, eds. Redin SC & Carris ML, pp. 197–207. Saint Paul, MN, USA: APS Press [*314*]

Hempel LC (1996). *Environmental Governance*. Washington, DC, USA: Island Press [*360*]

Hendry AP & Kinnison MT (1999). The pace of modern life: Measuring rates of contemporary microevolution. *Evolution* **53**:1637–1653 [*4, 84*]

Hendry AP, Hensleigh JE & Reisenbichler RR (1998). Incubation temperature, developmental biology, and the divergence of sockeye salmon (*Oncorhynchus nerka*) within Lake Washington. *Canadian Journal of Fisheries and Aquatic Sciences* **55**:1387–1394 [*91*]

Herre EA, Knowlton N, Mueller UG & Rehner SA (1999). The evolution of mutualisms: Exploring the paths between conflict and cooperation. *Trends in Ecology and Evolution* **14**:49–53 [*306*]

Heschel MS & Paige KN (1995). Inbreeding depression, environmental stress, and population size variation in scarlet gilia (*Ipomopsis aggregata*). *Conservation Biology* **9**:126–133 [*286*]

Heywood V (1979). The future of island floras. In *Plants and Islands*, ed. Bramwell D, pp. 431–441. New York, NY, USA: Academic Press [*345*]

Hilbert DW, Swift DM, Detling JK & Dyer MI (1981). Relative growth rates and the grazing optimization hypothesis. *Oecologia* **51**:14–18 [*333*]

Hill WG (1982). Predictions of response to artificial selection from new mutations. *Genetical Research* **40**:255–278 [*87*]

Hill WG & Rasbash J (1986). Models of long-term artificial selection in finite populations. *Genetical Research* **48**:41–50 [*87*]

Hill WG & Robertson A (1966). The effect of linkage on limits to artificial selection. *Genetical Research* **8**:269–294 [*162*]

Hill JK, Thomas CD & Lewis OT (1999a).

Flight morphology in fragmented populations of a rare British butterfly, *Hesperia comma*. *Biological Conservation* **87**:277–284 [*289, 291*]

Hill JK, Thomas CD & Huntley B (1999b). Climate and habitat availability determine 20th century changes in a butterfly's range margins. *Proceedings of the Royal Society of London B* **266**:1197–1206 [*289, 291*]

Hill JK, Thomas CD & Blakeley DS (1999c). Evolution of flight morphology in a butterfly that has recently expanded its geographic range. *Oecologia* **121**:165–170 [*291*]

Hill JK, Lewis OT, Blakeley D & Thomas CD. Flight-related morphologies of swallowtail butterflies (*Papilio machaon*) from populations that differ in dispersal rates. Unpublished [*289*]

Hitchings SP & Beebee TJC (1998). Loss of genetic diversity and fitness in Common Toad (*Bufo bufo*) populations isolated by inimical habitat. *Journal of Evolutionary Biology* **11**:269–283 [*286*]

Hochberg ME & Holt RD (1995). Refuge evolution and the population dynamics of coupled host–parasitoid associations. *Evolutionary Ecology Research* **9**:633–661 [*339*]

Hochberg ME & Van Baalen M (1998). Antagonistic coevolution over productivity gradients. *The American Naturalist* **152**:620–634 [*340*]

Hodell DA, Curtis JH & Brenner M (1995). Possible role of climate in the collapse of classic Maya civilization. *Nature* **375**:391–394 [*362*]

Höglund J (1996). Can mating systems affect local extinction risks? Two examples of lek-breeding waders. *Oikos* **77**:184–188 [*47*]

Hoekstra RF, Bijlsma R & Dolman AJ (1985). Polymorphism from environmental heterogeneity: Models are only robust if the heterozygote is close in fitness to the favoured homozygote in each environment. *Genetical Research* **45**:299–314 [*123*]

Hofbauer J & Sigmund K (1988). *The Theory of Evolution and Dynamical Systems*. Cambridge, UK: Cambridge University Press [*61*]

Hofbauer J & Sigmund K (1990). Adaptive dynamics and evolutionary stability. *Applied Mathematics Letters* **3**:75–79 [*310*]

Hoffmann AA & Merilä J (1999). Heritable variation under favorable and unfavorable conditions. *Trends in Ecology and Evolution* **14**:96–101 [*138*]

Hoffmann AA & Parsons PA (1988). The analysis of quantitative variation in natural populations with isofemale strains. *Genetics Selection Evolution* **20**:87–98 [*146*]

Hoffmann AA & Parsons PA (1989). An integrated approach to environmental stress tolerance and life-history variation: Desiccation tolerance in *Drosophila*. *Biological Journal of the Linnean Society* **37**:117–136 [*92*]

Hoffmann AA & Parsons PA (1991). *Evolutionary Genetics and Environmental Stress*. Oxford, UK: Oxford University Press [*137, 139–140*]

Hoffmann AA & Schiffer M (1998). Changes in heritability of five morphological traits under combined environmental stresses in *Drosophila melanogaster*. *Evolution* **52**:1207–1208 [*147*]

Holbrook S, Schmitt RJ & Stephens JS (1997). Changes in an assemblage of temperate reef fishes associated with a climate shift. *Ecological Applications* **7**:1299–1310 [*2*]

Holland JN & DeAngelis DL (2001). Population dynamics and the ecological stability of obligate pollination mutualisms. *Oecologia* **126**:575–586 [*307*]

Holland JN, DeAngelis DL & Bronstein JL (2002). Population dynamics of mutualism: Functional responses of benefits and costs. *The American Naturalist* **159**:231–244 [*307*]

Holling CS (1992). Cross-scale morphology, geometry, and dynamics of ecosystems. *Ecological Monographs* **62**:447–502 [*363*]

Holt RD (1990). The microevolutionary consequences of climate change. *Trends in Ecology and Evolution* **5**:311–315 [*244*]

Holt RD (1995). Linking species and ecosystems: Where's Darwin? In *Linking Species and Ecosystems*, eds. Jones CG & Lawton JH, pp. 273–279. New York, NY, USA: Chapman & Hall [*327, 343*]

Holt RD (1996). Demographic constraints in evolution: Towards unifying the evolutionary theories of senescence and niche conservatism. *Evolutionary Ecology Research* **10**:1–11 [*258, 293*]

Holt RD & Gaines MS (1992). The analysis of adaptation in heterogeneous landscapes: Implications for the evolution of fundamental niches. *Evolutionary Ecology Research* **6**:433–447 [*244*]

Holt RD & Gomulkiewicz R (1997a). How does immigration influence local adaptation? A reexamination of a familiar paradigm. *The American Naturalist* **149**:563–572 [*242–243, 252, 258*]

Holt RD & Gomulkiewicz R (1997b). The evolution of species' niches: A population dynamic perspective. In *Case Studies in Mathematical Biology*, eds. Othmar HG, Adler FR, Lewis MA & Dallon JC, pp. 25–50. Englewood Cliffs, NJ, USA: Prentice-Hall [*248, 252, 258*]

Holt RD & McPeek M (1996). Chaotic population dynamics favors the evolution of dispersal. *The American Naturalist* **148**:709–718 [*267*]

Holt RD, Grover J & Tilman D (1994). Simple rules for interspecific dominance in systems with exploitative and apparent competition. *The American Naturalist* **144**:741–771 [*336*]

Holt RD, Gomulkiewicz R & Barfield M. The phenomenology of niche evolution via quantitative traits in a "black-hole" sink: A mechanism for punctuated evolution? Unpublished [*247, 250, 255*]

Holway DA (1998). Effect of Argentine ant invasions on ground-dwelling arthropods in northern California riparian woodlands. *Oecologia* **116**:252–258 [*313*]

Horvitz CC & Schemske DW (1990). Spatiotemporal variation in insect mutualists of a neotropical herb. *Ecology* **71**:1085–1097 [*320*]

Houghton J (1997). *Global Warming: The Complete Briefing*, Second Edition. Cambridge, UK: Cambridge University Press [*88*]

Houle D (1989a). Allozyme associated heterosis in *Drosophila melanogaster*. *Genetics* **123**:788–901 [*128*]

Houle D (1989b). The maintenance of polygenic variation in finite populations. *Evolution* **43**:1767–1789 [*131*]

Houle D (1992). Comparing evolvability and variability of quantitative traits. *Genetics* **130**:185–204 [*120, 126, 129, 145*]

Houle D, Morikawa B & Lynch V (1996). Comparing mutational variabilities. *Genetics* **143**:1467–1483 [*139*]

Houle D, Hughes KA, Assimacopoulos S & Charlesworth B (1997). The effects of spontaneous mutation on quantitative traits. 2. Dominance of mutations with effects on LHTs. *Genetical Research* **138**:773–785 [*121*]

Houllier F & Lebreton J-D (1986). A renewal equation approach to the dynamics of stage grouped populations. *Mathematical Biosciences* **79**:185–197 [*44*]

Hubbard AL, McOrist S, Jones TW, Boid R, Scott R & Easterbee N (1992). Is survival of European wildcats *Felis sylvestris* in Britain threatened by interbreeding with domestic cats? *Biological Conservation* **61**:203–208 [*350*]

Huberman G (1978). Qualitative behavior of a fishery system. *Mathematical Biosciences* **42**:1–14 [*26*]

Huey RB & Kingsolver JG (1989). Evolution of thermal sensitivity of ectotherm performance. *Trends in Ecology and Evolution* **4**:131–135 [*91*]

Huey RB & Kingsolver JG (1993). Evolutionary responses to extreme temperatures in ectotherms. *The American Naturalist* **143**:S21–S46 [*86, 91, 182*]

Huey RB, Gilchrist GW, Carlson ML, Berrigan D & Serra L (2000). Rapid evolution of a geographic cline in size in an introduced fly. *Science* **287**:308–309 [*90, 357*]

Hughes KA (1995a). The evolutionary genetics of male life-history traits in *Drosophila melanogaster*. *Evolution* **49**:521–537 [*120*]

Hughes KA (1995b). The inbreeding decline and average dominance of genes affecting male life-history characters in *Drosophila melanogaster*. *Genetical Research* **65**:41–52 [*130, 132*]

Hughes KA (1997). Quantitative genetics of sperm precedence in *Drosophila melanogaster*. *Genetics* **145**:139–151 [*130*]

Hughes L (2000). Biological consequences of global warming: Is the signal already apparent? *Trends in Ecology and Evolution* **15**:56–61 [*2*]

Hughes KA, Du L, Rod FH & Reznick DN (1999). A test of frequency-dependent mate preference in guppies (*Poecilia reticulata*). *Animal Behaviour* **58**:907–916 [*122*]

Hull DM (1998). A reconsideration of Galton's problem (using a two-sex population). *Theoretical Population Biology* **54**:105–116 [*33, 50*]

Humphries CJ (1976). Evolution and endemism in *Argyranthemum* Webb ex Schultz Bip. (Compositae-Anthemidae). *Botanica Macaronesia* **1**:25–50 [*349*]

Hurtado L, Castrezana S, Mateos M, McLaurin D, Tello MK, Campoy J & Markow T (1997). Developmental stability and environmental stress in natural populations of

Drosophila pachea. *Ecotoxicology* **6**:233–238 [*143*]

Huston MA & DeAngelis DL (1994). Competition and coexistence: The effects of resource transport and supply rates. *The American Naturalist* **144**:954–977 [*331, 336*]

Huxel GR (1999). Rapid displacement of native species by invasive species: Effects of hybridization. *Biological Conservation* **89**:143–152 [*351*]

Imasheva AG, Loeschcke V, Zhivotovsky LA & Lazenby OE (1997). Effects of extreme temperatures on phenotypic variation and developmental stability in *Drosophila melanogaster* and *Drosophila buzzatii*. *Biological Journal of the Linnean Society* **61**:117–126 [*142–144, 150*]

Imasheva AG, Loeschcke V, Zhivotovsky LA & Lazenby OE (1998). Stress temperatures and quantitative variation in *Drosophila melanogaster*. *Heredity* **81**:246–253 [*142, 146–147, 150*]

Imasheva AG, Bosenko DV & Bubli OA (1999). Variation in morphological traits of *Drosophila melanogaster* (fruit fly) under nutritional stress. *Heredity* **82**:187–192 [*142–143, 146–147*]

Ingvarsson PK (1997). The effect of delayed population growth on the genetic differentiation of local populations subject to frequent extinctions and recolonizations. *Evolution* **51**:29–35 [*237*]

Inouye DW, Morales MA & Dodge GJ (2002). Variation in timing and abundance of flowering by *Delphinium barbeyi* Huth (Ranunculaceae): The roles of snowpack, frost, and La Niña, in the context of climate change. *Oecologia* **130**:543–550 [*320*]

IUCN (1994). *IUCN Red List Categories*. Gland, Switzerland: International Union for Conservation of Nature and Natural Resources [*99*]

Iwasa Y & Harada Y (1998). Female mate preference to maximize paternal care. II. Female competition leads to monogamy. *The American Naturalist* **151**:367–382 [*52*]

Iwasa Y, Pomiankowski A & Nee S (1991). The evolution of costly mate preferences. II. The "handicap" principle. *Evolution* **45**:1431–1442 [*52*]

Jablonski D (1993). The tropics as a source of evolutionary novelty through geological time. *Nature* **364**:142–144 [*202*]

Jacobs J (1984). Cooperation, optimal density and low density thresholds: Yet another modification of the logistic model. *Oecologia* **64**:389–395 [*26*]

Jansen VAA & Mulder GSEE (1999). Evolving biodiversity. *Ecology Letters* **2**:379–386 [*204*]

Janzen DH & Martin P (1982). Neotropical anachronisms: What the gomphotheres ate. *Science* **215**:19–27 [*314*]

Jimenez JA, Hughes K, Alaks G, Graham L & Lacy RC (1994). An experimental study of inbreeding depression in a natural habitat. *Science* **266**:271–273 [*132*]

Joffe A & Spitzer F (1967). On multitype branching processes with $\rho \leq 1$. *Journal of Mathematical Analysis and Applications* **19**:409–430 [*33*]

Johnson PA, Hoppensteadt FC, Smith JJ & Bush GL (1996). Conditions for sympatric speciation: A diploid model incorporating habitat fidelity and non-habitat assortative mating. *Evolutionary Ecology* **10**:187–205 [*205*]

Johnston IA & Bennett AF (1996). *Animals and Temperature: Phenotypic and Evolutionary Adaptation*. Cambridge, UK: Cambridge University Press [*88*]

Johst K, Doebeli M & Brandl R (1999). Evolution of complex dynamics in spatially structured populations. *Proceedings of the Royal Society of London B* **266**:1147–1154 [*267*]

Jones DA & Wilkins DA (1971). *Variation and Adaptation in Plant Species*. London, UK: Heinemann [*94*]

Jones LP, Frankham R & Barker JSF (1968). The effects of population size and selection intensity in selection for a quantitative character in *Drosophila*. II. Long-term response to selection. *Genetical Research* **12**:249–266 [*87*]

Jordano P (1994). Spatial and temporal variation in the avian-frugivore assemblage of *Prunus mahaleb*: Patterns and consequences. *Oikos* **71**:479–491 [*320*]

Jørgensen T (1990). Long-term changes in age at sexual maturity of Northeast Arctic cod (*Gadus morhua* L.). *Journal du Conseil International pour l'Exploration de la Mer* **46**:235–248 [*5*]

Jorgenson JT, Festa-Bianchet M, Gaillard J-M & Wishart WD (1997). Effects of age, sex, disease, and density on survival of bighorn sheep. *Ecology* **78**:1019–1032 [*53*]

Jules ES & Rathcke BJ (1999). Mechanisms of reduced *Trillium* recruitment along edges of old-growth forest remnants. *Conservation*

Biology **13**:784–793 [*313*]

Junakovic N, di Franco C, Barsanti P & Palumbo G (1986). Transpositions of copia-like elements can be induced by heat shock. *Journal of Molecular Evolution* **24**:89–93 [*139*]

Kalinowski ST (1999). *Conservation Genetics of Endangered Species*. Tempe, AZ, USA: Arizona State University Department of Biology [*134*]

Kalinowski ST & Hedrick PW (1998). An improved method for estimating inbreeding depression in pedigrees. *Zoo Biology* **17**:481–497 [*131*]

Kalinowski ST, Hedrick PW & Miller PS (2000). Inbreeding depression in the Speke's gazelle captive breeding program. *Conservation Biology* **14**:1375–1384 [*133–134*]

Kallman KD (1989). Genetic control of size at maturity in Xiphophorus. In *Ecology and Evolution of Livebearing Fishes (Poeciliidae)*, eds. Meffe GK & Meffe FFS, pp. 163–184. Englewood Cliffs, NJ, USA: Prentice-Hall [*107*]

Karban R (1986). Interspecific competition between folivorous insects on *Erigeron glaucus*. *Ecology* **67**:1063–1072 [*310*]

Karban R (1989). Fine-scale adaptation of herbivorous thrips in individual host plants. *Nature* **340**:60–61 [*293*]

Kareiva PM, Kingsolver JG & Huey RB, eds. (1993). *Biotic Interactions and Global Change*. Sunderland, MA, USA: Sinauer Associates Inc. [*100, 314*]

Karhu A, Hurme P, Karjalainen M, Karvonen P, Karkkainen K, Neale D & Savolainen O (1996). Do molecular markers reflect patterns of differentiation in adaptive traits of conifers? *Theoretical and Applied Genetics* **93**:215–221 [*128–129*]

Karlin S & Taylor HM (1981). *A Second Course in Stochastic Processes*. New York, NY, USA: Academic Press [*28*]

Kawecki TJ (1995). Demography of source–sink populations and the evolution of ecological niches. *Evolutionary Ecology Research* **7**:155–174 [*258*]

Kawecki TJ (2000). Adaptation to marginal habitats: Contrasting influence of dispersal on the fate of rare alleles with small and large effects. *Proceedings of the Royal Society of London B* **267**:1315–1320 [*258*]

Kawecki TJ & Holt RD (2002). Evolutionary consequences of asymmetric dispersal rates. *The American Naturalist* **160**:333–347 [*258*]

Kearns CA, Inouye DW & Waser NM (1998). Endangered mutualisms: The conservation of plant-pollinator interactions. *Annual Review of Ecology and Systematics* **29**:83–112 [*307, 313*]

Keightley P (1994). The distribution of mutation effects on viability in *Drosophila melanogaster*. *Genetics* **138**:1315–1322 [*162*]

Keller LF, Arcese P, Smith JNM, Jochacjla WM & Stearns SC (1994). Selection against inbred song sparrows during a natural population bottleneck. *Nature* **372**:356–357 [*132*]

Kettlewell HBD (1973). *The Evolution of Melanism*. Oxford, UK: Clarendon Press [*93*]

Khadari B, Gibernau M, Anstett MC, Kjellberg F & Hossaert-McKey M (1995). When figs wait for pollinators: The length of fig receptivity. *American Journal of Botany* **82**:992–999 [*316, 318*]

Kiesecker JM, Blaustein AR & Belden LK (2001). Complex causes of amphibian population declines. *Nature* **410**:681–684 [*2*]

Kim JH & Yang SW (1993). Systematic studies of the genus *Cobitis* (Pisces: Cobotidae) in Korea. IV. Introgressive hybridization between two spined loach subspecies of the genus *Cobitis*. *Korean Journal of Zoology* **36**:535–544 [*345*]

Kimura M (1953). "Stepping stone" model of population. *Annual Report of the National Institute of Genetics* **3**:62–63 [*234*]

Kimura M (1957). Some problems of stochastic processes in genetics. *Annals of Mathematical Statistics* **28**:882–901 [*158, 163*]

Kimura M (1965). A stochastic model concerning maintenance of genetic variability in quantitative characters. *Proceedings of the National Academy of Sciences of the USA* **54**:731–735 [*201*]

Kimura M & Maruyama T (1971). Pattern of neutral polymorphism in a geographically structured population. *Genetical Research* **18**:125–131 [*235*]

Kimura M & Ohta T (1971). *Theoretical Aspects of Population Genetics*. Monographs in Population Biology. Princeton, NJ, USA: Princeton University Press [*233*]

Kimura M & Weiss WH (1964). The stepping stone model of genetic structure and the decrease of genetic correlation with distance. *Genetics* **49**:561–576 [*234–235*]

Kimura M, Maruyama T & Crow JF (1963). The mutation load in small populations. *Genetics* **48**:1303–1312 [*156*]

Kingsolver JG (2000). Feeding, growth and the thermal environment of Cabbage White caterpillars, *Pieris rapae* L. *Physiological and Biochemical Zoology* **73**:621–628 [*94*]

Kingsolver JG & Woods HA (1997). Thermal sensitivity of feeding and digestion in *Manduca* caterpillars. *Physiological Zoology* **70**:631–638 [*93–94*]

Kinzie RAI (1992). Predation by the introduced carnivorous snail *Euglandina rosea* (Ferussac) on endemic aquatic lymnaeid snails in Hawaii. *Biological Conservation* **60**:149–155 [*305*]

Kirchner F, Ferdy J-B, Andalo C, Colas B & Moret J (2003). Role of corridors in plant dispersal: An example with the endangered *Ranunculus nodiflorus* L. *Conservation Biology* **17**:401–410 [*299*]

Kirkpatrick M (1982). Sexual selection and the evolution of female choice. *Evolution* **36**:1–12 [*52, 214*]

Kirkpatrick M (1996). Genes and adaptation: A pocket guide to theory. In *Adaptation*, eds. Rose MR & Lauder GV, pp. 125–128. San Diego, CA, USA: Academic Press [*192, 214*]

Kirkpatrick M & Barton NH (1997). Evolution of a species' range. *The American Naturalist* **150**:1–23 [*242, 259–261*]

Kirkpatrick M & Jarne P (2000). The effects of a bottleneck on inbreeding depression and the genetic load. *The American Naturalist* **155**:154–167 [*157*]

Kisdi É (1999). Evolutionary branching under asymmetric competition. *Journal of Theoretical Biology* **197**:149–162 [*309*]

Kisdi É (2002). Dispersal: Risk spreading versus local adaptation. *The American Naturalist* **159**:579–596 [*265, 267, 283*]

Kisdi É & Meszéna G (1993). Density-dependent life-history evolution in fluctuating environments. In *Adaptation in a Stochastic Environment*, eds. Yoshimura J & Clark C, p. 26–62, Lecture Notes in Biomathematics Vol. 98. Berlin, Germany: Springer-Verlag [*196, 199*]

Kjellberg F & Valdeyron G (1990). Species-specific pollination: A help or a limitation to range extension? In *Biological Invasions in Europe and the Mediterranean Basin*, eds. diCastri F & Hansen AJ, pp. 371–378. The Hague, Netherlands: Dr W Junk Publishers [*316*]

Knowlton N (2003). Microbial mutualisms on coral reefs: The host as habitat. *The American Naturalist* **S161**. In press [*326*]

Kokko H & Ebenhard T (1996). Measuring the strength of demographic stochasticity. *Journal of Theoretical Biology* **183**:169–178 [*50*]

Kondrashov AS & Yampolsky LY (1996a). High genetic variability under the balance between symmetric mutation and fluctuating stabilizing selection. *Genetical Research* **68**:157–164 [*175*]

Kondrashov AS & Yampolsky LY (1996b). Evolution of amphimixis and recombination under fluctuating selection in one and many traits. *Genetical Research* **68**:165–173 [*175*]

Kostitzin VA (1940). Sur la loi logistique et ses généralisations. *Acta Biotheoretica* **5**:155–159 [*26*]

Kozlowski J (1993). Measuring fitness in life history studies. *Trends in Ecology and Evolution* **8**:84–85 [*191*]

Kozlowski J & Wiegert RG (1986). Optimal allocation of energy to growth and reproduction. *Theoretical Population Biology* **29**:16–37 [*191*]

Krebs CJ (1985). *Ecology: The Experimental Analysis of Distribution and Abundance*. New York, NY, USA: Harper & Row [*338*]

Krebs RA & Feder ME (1997). Natural variation in the expression of the heat-shock protein HSP70 in a population of *Drosophila melanogaster* and its correlation with tolerance of ecologically relevant thermal stress. *Evolution* **51**:173–179 [*140*]

Krebs RA & Loeschcke V (1994). Costs and benefits of activation of the heat-shock response in *Drosophila melanogaster*. *Functional Ecology* **8**:730–737 [*140*]

Kulhavy DL, Hooper RG & Costa R, eds. (1995). *Red-cockaded Woodpecker: Recovery, Ecology and Management*. Nacogdoches, TX, USA: Centre for Applied Studies, Stephen F. Austin State University [*100*]

Kuussaari M, Singer M & Hanski I (2000). Local specialization and landscape-level influence of host use in a herbivorous insect. *Ecology* **81**:2177–2187 [*292*]

Kuznetsov YA (1995). *Elements of Applied Bifurcation Theory*. Berlin, Germany: Springer-Verlag [*219*]

Lacy RC (1993). Impacts of inbreeding in natural and captive populations of vertebrates: Implications for conservation. *Perspectives in Biology and Medicine* **36**:480–495 [*132*]

Lacy RC (1997). Importance of genetic variation to the viability of mammalian populations. *Journal of Mammology* **78**:320–355 [*133*]

Lacy RC & Ballou JD (1998). Effectiveness of selection in reducing the genetic load in populations of *Peromyscus polionotus* during generations of inbreeding. *Evolution* **52**:900–909 [*133*]

Lacy RC & Horner BE (1996). Effects of inbreeding on skeletal development of *Rattus villosissimus*. *Journal of Heredity* **87**:277–287 [*133*]

Lande R (1975). The maintenance of genetic variation by mutation in a polygenic character with linked loci. *Genetical Research* **26**:221–235 [*175, 178*]

Lande R (1976). Natural selection and random genetic drift in phenotypic evolution. *Evolution* **30**:314–334 [*173, 176–177*]

Lande R (1979). Quantitative genetic analysis of multivariate evolution, applied to brain:body size allometry. *Evolution* **33**:402–416 [*107–108, 173, 177*]

Lande R (1980a). Sexual dimorphism, sexual selection, and adaptation in polygenic characters. *Evolution* **34**:292–305 [*52*]

Lande R (1980b). The genetic covariance between characters maintained by pleiotropic mutations. *Genetics* **94**:203–215 [*182–183*]

Lande R (1982). A quantitative genetic theory of life-history evolution. *Ecology* **63**:607–615 [*58*]

Lande R (1987). Extinction thresholds in demographic models of territorial populations. *The American Naturalist* **130**:624–635 [*62*]

Lande R (1988). Genetics and demography in biological conservation. *Science* **241**:1455–1460 [*25*]

Lande R (1993). Risks of population extinction from demographic and environmental stochasticity and random catastrophes. *The American Naturalist* **142**:911–927 [*27–29, 45*]

Lande R (1994). Risk of population extinction from fixation of new deleterious mutations. *Evolution* **48**:1460–1469 [*53, 157, 159–160, 162–163, 168–169*]

Lande R (1995). Mutation and conservation. *Conservation Biology* **9**:782–791 [*53, 96, 121, 126, 155*]

Lande R (1998a). Demographic stochasticity and Allee effect on a scale with isotropic noise. *Oikos* **83**:353–358 [*30*]

Lande R (1998b). Risk of population extinction from fixation of deleterious and reverse mutations. *Genetica* **102/103**:21–27 [*165–166, 169*]

Lande R & Barrowclough GF (1987). Effective population size, genetic variation, and their use in population management. In *Viable Populations for Conservation*, ed. Soulé ME, pp. 87–123. New York, NY, USA: Cambridge University Press [*229–230*]

Lande R & Orzack S (1988). Extinction dynamics of age-structured populations in a fluctuating environment. *Proceedings of the National Academy of Sciences of the USA* **85**:7418–7421 [*58*]

Lande R & Shannon S (1996). The role of genetic variation in adaptation and population persistence in a changing environment. *Evolution* **50**:434–437 [*175, 179, 184–185, 187*]

Lank DB, Smith CM, Hanotte O, Burke TA & Cooke F (1995). Genetic polymorphism for alternative mating behavior in lekking male ruff, *Philomachus pugnax*. *Nature* **378**:59–62 [*123*]

Lappalainen JH, Koricheva J, Helander ML & Haukioja E (1999). Densities of endophytic fungi and performance of leafminers (Lepidoptera: Eriocraniidae) on birch along a pollution gradient. *Environmental Pollution* **104**:99–105 [*314*]

Latter BDH (1970). Selection in finite population with multiple alleles. II: Centripetal selection, mutation, and isoallelic variation. *Genetics* **66**:165–186 [*176*]

Latter BDH, Mulley JC, Reid D & Pascoe L (1995). Reduced genetic load revealed by slow inbreeding in *Drosophila melanogaster*. *Genetics* **139**:287–297 [*99, 133*]

Latour B (1999). *Pandora's Hope*. Cambridge, MA, USA: Harvard University Press [*356*]

Law R (1979). Optimal life histories under age-specific predation. *The American Naturalist* **114**:399–417 [*102, 202*]

Law R & Dieckmann U (1998). Symbiosis through exploitation and the merger of lineages in evolution. *Proceedings of the Royal Society of London B* **265**:1245–1253 [*338*]

Law R & Grey DR (1989). Evolution of yields from populations with age-specific cropping. *Evolutionary Ecology Research*

3:343–359 [*125*]

Law R, Marrow P & Dieckmann U (1997). On evolution under asymmetric competition. *Evolutionary Ecology* **11**:485–501 [*309*]

Law R, Bronstein JL & Ferrière RG (2001). On mutualists and exploiters: Plant–insect coevolution in pollinating seed-parasite systems. *Journal of Theoretical Biology* **212**:373–389 [*307*]

Lawrey JD & Hale JME (1979). Lichen growth responses to stress induced by automobile exhaust pollution. *Science* **204**:423–424 [*314*]

Lawton JH (1981). Asymmetrical competition in insects. *Nature* **289**:793–795 [*310*]

Leary RF & Allendorf FW (1989). Fluctuating asymmetry as an indicator of stress in conservation biology. *Trends in Ecology and Evolution* **4**:214–217 [*142*]

Lee TN & Stone AA (1998). Pollinator preferences and the persistence of crop genes in wild radish populations (*Raphanus raphanistrum*, Brassicaceae). *American Journal of Botany* **85**:333–339 [*314*]

Le Galliard JF, Ferrière R & Dieckmann U (2003). The adaptive dynamics of altruism in spatially heterogeneous populations. *Evolution* **57**:1–17 [*213*]

Legendre S, Clobert J, Møller AP & Sorci G (1999). Demographic stochasticity and social mating system in the process of extinction of small populations: The case of passerines introduced to New Zealand. *The American Naturalist* **153**:449–463 [*46, 49, 51*]

Lebreton JD (1981). *Contribution à la dynamique des populations d'oiseaux: Modèles mathématiques en temps discret*. PhD thesis. Villeurbanne, France: Université Lyon I [*32*]

Lebreton JD, Burnham KP, Clobert J & Anderson DR (1992). Modeling survival and testing biological hypotheses using marked animals: A unified approach with case studies. *Ecological Monographs* **62**:67–118 [*21*]

Leigh EG Jr (1981). The average lifetime of a population in a varying environment. *Journal of Theoretical Biology* **90**:213–239 [*21, 27, 38*]

Leimar O & Nordberg U (1997). Metapopulation extinction and genetic variation in dispersal-related traits. *Oikos* **80**:448–458 [*288*]

Leirs H, Stenseth NC, Nichols JD, Hines JE, Verhagen R & Verheyen W (1997). Stochastic seasonality and nonlinear density-dependent factors regulate population size in an African rodent. *Nature* **389**:176–180 [*21, 45*]

Lennartsson T, Tuomi J & Nilsson P (1997). Evidence for an evolutionary history of overcompensation in the grassland biennal *Gentianella campestris* (Gentianaceae). *The American Naturalist* **149**:1147–1155 [*333*]

Lennartsson T, Nilsson P & Tuomi J (1998). Induction of overcompensation in the field gentian, *Gentianella campestris*. *Ecology* **79**:1061–1072 [*333*]

Levin DA (1978). The origin of isolating mechanisms in flowering plants. *Evolutionary Biology* **11**:185–317 [*347*]

Levin DA (1995). Plant outliers: An ecogenetic perspective. *The American Naturalist* **145**:109–118 [*299*]

Levin DA (2000). *The Origin, Expansion, and Demise of Plant Species*. Oxford, UK: Oxford University Press [*349*]

Levin SA, Cohen D & Hastings A (1984). Dispersal strategies in patchy environments. *Theoretical Population Biology* **26**:165–191 [*265, 267*]

Levin DA, Francisco-Ortega J & Jansen RK (1996). Hybridization and the extinction of rare plant species. *Conservation Biology* **10**:10–16 [*344, 349, 355*]

Levins R (1962a). Theory of fitness in a heterogeneous environment. I. The fitness set and adaptive function. *The American Naturalist* **96**:361–373 [*190*]

Levins R (1962b). Theory of fitness in a heterogeneous environment. II. Developmental flexibility and niche selection. *The American Naturalist* **97**:74–90 [*190*]

Levins R (1968). *Evolution in Changing Environments*. Princeton, NJ, USA: Princeton University Press [*190*]

Levins R (1969). Some demographic and genetic consequences of environmental heterogeneity for biological control. *Bulletin of the Entomological Society of America* **15**:237–240 [*59, 265*]

Levins R (1970). Extinction. In *Some Mathematical Problems in Biology*, ed. Gerstenhaber E, pp. 77–107. Providence, RI, USA: American Mathematical Society [*265*]

Lewis H (1961). Experimental sympatric populations of *Clarkia*. *The American Naturalist* **95**:155–168 [*347*]

Lewis OT, Thomas CD, Hill JK, Brookes MI, Robin Crane TP, Graneau YA, Mallet JL &

Rose OC (1997). Three ways of assessing the metapopulation structure in the butterfly *Plebejus argus*. *Ecological Entomology* **22**:283–293 [*290*]

Lewontin RC (1974). *The Genetic Basis of Evolutionary Change*. New York, NY, USA: Columbia University Press [*97, 119*]

Lewontin RC (1979). Fitness, survival, and optimality. In *Analysis of Ecological Systems*, eds. Horn DJ, Stairs GR & Mitchell RD, pp. 3–22. Columbus, OH, USA: Ohio State University Press [*189*]

Lewontin RC (1987) The shape of optimality. In *The Latest on the Best*, ed. Dupre J, pp. 151–159. Cambridge, MA, USA: MIT Press [*189*]

Lewontin RC & Birch LC (1966). Hybridization as a source of variation for adaptation to new environments. *Evolution* **20**:315–336 [*354*]

Lewontin RC & Cohen D (1969). On population growth in a randomly varying environment. *Proceedings of the National Academy of Sciences of the USA* **62**:1056–1060 [*29, 34*]

Lindström J & Kokko H (1998). Sexual reproduction and population dynamics: The role of polygyny and demographic sex differences. *Proceedings of the Royal Society of London B* **265**:483–488 [*48*]

Lipsitch M, Bergstrom CT & Levin BR (2000). The epidemiology of antibiotic resistance in hospitals: Paradoxes and prescriptions. *Proceedings of the National Academy of Sciences of the USA* **97**:1938–1943 [*360*]

Liu F, Zhang L & Charlesworth D (1998). Genetic diversity in *Leavenworthia* population with different inbreeding levels. *Proceedings of the Royal Society of London B* **265**:293–301 [*241*]

Livingstone FB (1992). Polymorphism and differential selection for the sexes. *Human Biology* **64**:649–657 [*124*]

Loeschcke V, Krebs RA & Barker JSF (1994). Genetic variation for resistance and acclimation to high temperature stress in *Drosophila buzzatii*. *Biological Journal of the Linnean Society* **52**:83–92 [*140*]

Loeschcke V, Krebs RA, Dahlgaard J & Michalak P (1997). High temperature stress and the evolution of thermal resistance. In *Environmental Stress, Adaptation and Evolution*, eds. Bijlsma R & Loeschcke V, pp. 175–191. Basel, Switzerland: Birkhäuser [*150*]

Loeschcke V, Bundgaard J & Barker JSF (1999). Reaction norms across and genetic parameters at different temperatures for thorax and wing size traits in *Drosophila aldrichi* and *Drosophila buzzatii*. *Journal of Evolutionary Biology* **12**:605–623 [*142–143, 147, 150*]

Lomnicki A (1988). *Population Ecology of Individuals*. Princeton, NJ, USA: Princeton University Press [*23*]

Lopez-Fanjul C & Villaverde A (1989). Inbreeding increases genetic variance for viability in *Drosophila melanogaster*. *Evolution* **43**:1800–1804 [*130*]

Lopez-Vaamonde C, Rasplus JY, Weiblen GD & Cook JM (2001). Molecular phylogenies of fig wasps: Partial cocladogenesis of pollinators and parasites. *Molecular Phylogenetics and Evolution* **21**:55–71 [*307*]

Lord JM (1991). Pollination and seed dispersal in *Freycinetia baueriana*, a dioecious liane that has lost its bat pollinator. *New Zealand Journal of Botany* **29**:83–86 [*314*]

Loreau M (1995). Consumers as maximizers of matter and energy flow in ecosystems. *The American Naturalist* **145**:22–42 [*333*]

Loreau M (1996). Coexistence of multiple food chains in a heterogeneous environment: Interactions among community structure, ecosystem functioning, and nutrient dynamics. *Mathematical Biosciences* **134**:153–188 [*331, 336*]

Loreau M (1998a). Ecosystem development explained by competition within and between material cycles. *Proceedings of the Royal Society of London B* **265**:33–38 [*329, 331, 333*]

Loreau M (1998b). Biodiversity and ecosystem functioning: A mechanistic model. *Proceedings of the National Academy of Sciences of the USA* **95**:5632–5636 [*331, 336*]

Loreau M (2002). Evolutionary processes in ecosystems. In *Encyclopedia of Global Environmental Change*, Volume 2, eds. Mooney HA & Canadell J, pp. 292–297. Chichester, UK: John Wiley & Sons [*328*]

Loreau M & de Mazancourt C (1999). Should plants in resource-poor environments invest more in antiherbivore defense? *Oikos* **87**:195–200 [*329–330*]

Loreau M, Barbault R, Kawanabe H, Higashi M, Alvarez-Buylla E & Renaud F (1995). Dynamics of biodiversity at the community and ecosystem level. In *Global Biodiversity Assessment*, ed. United Nations Environment Program, pp. 245–274. Cambridge, UK: Cambridge University Press [*327*]

Lorimer G (1997). Folding with a two stroke motor. *Nature* **388**:720–723 [*140*]

Losos JB (1996). Ecological and evolutionary determinants of the species–area relation in Caribbean anoline lizards. *Philosophical Transactions of the Royal Society of London B* **351**:847–854 [*202*]

Losos JB & Schluter D (2000). Analysis of an evolutionary species–area relationship. *Nature* **408**:847–850 [*202, 207*]

Losos JB, Warhelt KI & Schoener TW (1997). Adaptive differentiation following experimental island colonization in *Anolis* lizards. *Nature* **387**:70–73 [*5, 116*]

Louda SM, Kendall D, Connor J & Simberloff D (1997). Ecological effects of an insect introduced for the biological control of weeds. *Science* **277**:1088–1090 [*305*]

Luckinbill LS & Clare MJ (1985). Selection for life span in *Drosophila melanogaster*. *Heredity* **55**:9–18 [*148*]

Ludwig D (1996). The distribution of population survival times. *The American Naturalist* **147**:506–526 [*38*]

Ludwig D, Jones DD & Holling CS (1978). Qualitative analysis of insect outbreak systems: The spruce budworm and forest. *Journal of Animal Ecology* **47**:315–332 [*26*]

Lundberg P, Ranta E, Ripa J & Kaitala V (2000). Population variability in space and time. *Trends in Ecology and Evolution* **15**:460–464 [*270*]

Lush JL (1937). *Animal Breeding Plans*. Ames, IA, USA: Iowa State University Press [*173*]

Lynch CB (1977). Inbreeding effects upon animals derived from a wild population of *Mus musculus*. *Evolution* **31**:526–537 [*133*]

Lynch M & Gabriel W (1990). Mutation load and the survival of small populations. *Evolution* **44**:1725–1737 [*121, 159, 164*]

Lynch M & Lande R (1993). Evolution and extinction in response to environmental change. In *Biotic Interactions and Global Change*, eds. Kareiva PM, Kingsolver JG & Huey RB, pp. 234–250. Sunderland, MA, USA: Sinauer Associates Inc. [*86, 175, 179, 186*]

Lynch M & Walsh B (1998). *Genetics and Analysis of Quantitative Traits*. Sunderland, MA, USA: Sinauer Associates Inc. [*127–128, 160, 173, 178*]

Lynch M, Gabriel W & Wood AM (1991). Adaptive demographic responses of plankton populations to environmental change. *Limnology and Oceanography* **36**:1301–1312 [*175, 179*]

Lynch M, Conery J & Bürger R (1995a). Mutation accumulation and the extinction of small populations. *The American Naturalist* **146**:489–518 [*53, 96, 133, 156–157, 159, 162, 168, 245*]

Lynch M, Conery J & Bürger R (1995b). Mutational meltdowns in sexual populations. *Evolution* **49**:1067–1080 [*96, 121, 156–157, 168*]

Lynch M, Latta L, Hicks J & Giorgianni M (1998). Mutation, selection, and the maintenance of life-history variation in a natural population. *Evolution* **52**:727–733 [*121*]

Lynch M, Blanchard J, Houle D, Kibota T, Schultz S, Vassilieva L & Willis J (1999). Perspective: Spontaneous deleterious mutation. *Evolution* **53**:645–663 [*160*]

MacArthur RH (1972). *Geographical Ecology: Patterns in the Distribution of Species*. New York, NY, USA: Harper & Row [*77*]

MacArthur RH & Wilson EO (1967). *The Theory of Island Biogeography*. Princeton, NJ, USA: Princeton University Press [*21, 207*]

Mace GM, Balmford A & Ginsberg JR, eds. (1998). *Conservation in a Changing World*. Cambridge, UK: Cambridge University Press [*363*]

Mackay TFC (1980). Genetic variance, fitness, and homeostasis in varying environments: An experimental check of the theory. *Evolution* **34**:1219–1222 [*123*]

Mackay TFC (1981). Genetic variation in varying environments. *Genetical Research* **37**:79–93 [*123*]

Mackay TFC & Fry JD (1996). Polygenic mutation in *Drosophila melanogaster*: Genetic interactions between selection lines and candidate quantitative trait loci. *Genetics* **144**:671–688 [*124*]

Mackay TFC, Lyman RF, Jackson MX, Terzian C & Hill WG (1992a). Polygenic mutation in *Drosophila melanogaster*: Estimates from divergence among inbred strains. *Evolution* **46**:300–316 [*124*]

Mackay TFC, Lyman RF & Jackson MS (1992b). Effects of P-element insertions on quantitative traits in *Drosophila melanogaster*. *Genetics* **130**:315–332 [*162*]

Mackay TFC, Fry JD, Lyman RF & Nuzhdin SV (1994). Polygenic mutation in *Drosophila melanogaster*: Estimates from response to selection of inbred strains. *Genetics* **136**:937–951 [*87, 124*]

Mackay TFC, Lyman RF & Hill WG (1995). Polygenic mutation in *Drosophila melanogaster*: Non-linear divergence among unselected strains. *Genetics* **139**:849–859 [*124*]

Madsen T, Shine R, Loman J & Håkansson T (1992). Why do female adders copulate so frequently? *Nature* **355**:440–441 [*25*]

Madsen T, Shine R, Olsson M & Wittzell H (1999). Restoration of an inbred adder population. *Nature* **402**:34–35 [*3*]

Majerus MEN (1998). *Melanism: Evolution in Action.* Oxford, UK: Oxford University Press [*82, 93–95*]

Malécot G (1952). Les processes stochastiques et la méthode des fonctions génératrices ou caracteréstiques. *Publications de l'Institut de Statistique de l'Université de Paris 1* **3**:1–16 [*158*]

Malécot G (1968). *The Mathematics of Heredity.* New York, NY, USA: WA Freeman & Company [*234*]

Mallet J (1989). The evolution of insecticide resistance: Have the insects won? *Trends in Ecology and Evolution* **4**:336–340 [*111, 116*]

Mallet J & Porter P (1992). Preventing insect adaptation to insect-resistant crops – Are seed mixtures or refugia the best strategy? *Proceedings of the Royal Society of London B* **250**:165–169 [*360*]

Margulis S (1998a). Differential effects of inbreeding at juvenile and adult life-history stages in *Peromyscus polionotus*. *Journal of Mammology* **79**:326–336 [*132*]

Margulis S (1998b). Relationships among parental inbreeding, parental behaviour and offspring viability in oldfield mice. *Animal Behaviour* **55**:427–438 [*132*]

Martcheva M (1999). Exponential growth in age-structured two-sex populations. *Mathematical Biosciences* **157**:1–22 [*46, 48*]

Martin PS & Klein RG (1984). *Quaternary Extinctions: A Prehistoric Revolution.* Tucson, AZ, USA: University of Arizona Press [*314*]

Maruyama T (1970). Effective number of alleles in a subdivided population. *Theoretical Population Biology* **1**:273–306 [*235*]

Massot M, Clobert J, Pilorge T, Lecomte J & Barbault R (1992). Density dependence in the common lizard: Demographic consequences of a density manipulation. *Ecology* **73**:1742–1756 [*21*]

Matessi C, Gimelfarb A & Gavrilets S (2001). Long term buildup of reproductive isolation promoted by disruptive selection: How far does it go? *Selection* **2**:41–64 [*206*]

Mather K (1943). Polygenic inheritance and natural selection. *Biological Reviews* **18**:32–64 [*137*]

Mathews JNA (1994). The benefits of overcompensation and herbivory: The difference between coping with herbivores and liking them. *The American Naturalist* **144**:528–533 [*333*]

Mathias A & Kisdi É (2002). Adaptive diversification of germination strategies. *Proceedings of the Royal Society of London B* **269**:151–155 [*209*]

Mathias A, Kisdi É & Olivieri I (2001). Divergent evolution of dispersal in a heterogeneous landscape. *Evolution* **55**:246–259 [*267*]

Matsuda H (1985). Evolutionarily stable strategies for predator switching. *Journal of Theoretical Biology* **115**:351–366 [*196*]

Matsuda H & Abrams PA (1994a). Runaway evolution to self-extinction under asymmetrical competition. *Evolution* **48**:1764–1772 [*209, 211, 214*]

Matsuda H & Abrams PA (1994b). Timid consumers – self-extinction due to adaptive change in foraging and anti-predator effort. *Theoretical Population Biology* **45**:76–91 [*188, 209, 214*]

Matsuda H & Abrams PA (1994c). Plant-herbivore interactions and theory of coevolution. *Plant Species Biology* **9**:155–161 [*309*]

May RM (1974). *Stability and Complexity in Model Ecosystems.* Princeton, NJ, USA: Princeton University Press [*20*]

May RM (1976). Models for two interacting populations. In *Theoretical Ecology: Principles and Applications*, ed. May RM, pp. 49–70. Philadelphia, PA, USA: Saunders [*307*]

May RM (1977). Thresholds and breakpoints in ecosystems with a multiplicity of steady states. *Nature* **269**:471–477 [*26*]

May RM & Nowak MA (1994). Superinfection, metapopulation dynamics, and the evolution of diversity. *Journal of Theoretical Biology* **170**:95–114 [*202*]

May RM & Oster GF (1976). Bifurcations and dynamic complexity in simple ecological models. *The American Naturalist* **110**:573–599 [*23, 44, 197*]

Maynard Smith J (1966). Sympatric speciation. *The American Naturalist* **100**:637–650 [*205*]

Maynard Smith J (1982). *Evolution and the Theory of Games*. Cambridge, UK: Cambridge University Press [*52, 191, 199*]

Maynard Smith J (1998). *Evolutionary Genetics*, Second Edition. Oxford, UK: Oxford University Press [*186*]

Maynard Smith J & Haigh J (1974). The hitchhiking effect of a favorable gene. *Genetical Research* **23**:23–35 [*162*]

Maynard Smith J & Hoekstra RF (1980). Polymorphism in a varied environment: How robust are the models? *Genetical Research* **35**:45–57 [*123*]

Maynard Smith J & Price GR (1973). Logic of animal conflict. *Nature* **246**:15–18 [*191, 199*]

Maynard Smith J & Slatkin M (1973). The stability of predator–prey systems. *Ecology* **54**:384–391 [*23*]

Maynard Smith J, Burian R, Kauffman S, Alberch P, Campbell J, Goodwin B, Lande R, Raup D & Wolpert L (1985). Developmental constraints and evolution. *Quarterly Review of Biology* **60**:265–287 [*190*]

Mayo O (1987). *The Theory of Plant Breeding*, Second Edition. Oxford, UK: Clarendon Press [*173–174*]

Mayr E (1963). *Animal Species and Evolution*. Cambridge, MA, USA: The Belknap Press of Harvard University Press [*205, 242*]

Mayr E (1982). *The Growth of Biological Thought: Diversity, Evolution, and Inheritance*. Cambridge, MA, USA: The Belknap Press of Harvard University Press [*205*]

McCarthy MA (1997). The Allee effect, finding mates and theoretical models. *Ecological Modelling* **103**:99–102 [*25*]

McColl G, Hoffmann AA & McKechnie SW (1996). Response of two heat shock genes to selection for knockdown heat resistance in *Drosophila melanogaster*. *Genetics* **143**:1615–1627 [*140*]

McCullough DR, ed. (1996). *Metapopulations and Wildlife Conservation*. Washington, DC, USA: Island Press [*59*]

McDonald JF (1995). Transposable elements: Possible catalysts of organismic evolution. *Trends in Ecology and Evolution* **10**:123–126 [*139*]

McDonnell MJ & Pickett STA (1993). *Humans as Components of Ecosystems*. New York, NY, USA: Springer-Verlag [*361*]

McKenzie JA (1996). *Ecological and Evolutionary Aspects of Insecticide Resistance*. Austin, TX, USA: Academic Press [*360*]

McKenzie JA & Batterham P (1994). The genetic, molecular and phenotypic consequences of selection for insecticide resistance. *Trends in Ecology and Evolution* **9**:166–169 [*86, 96*]

McKey D (1989). Population biology of figs: Applications for conservation. *Experientia* **45**:661–673 [*316, 318*]

McLain DK, Moulton MP & Redfearn TP (1995). Sexual selection and the risk of extinction of introduced birds on oceanic islands. *Oikos* **74**:27–34 [*52*]

McNaughton SJ (1979). Grazing as an optimization process: Grass–ungulate relationships in the Serengeti. *The American Naturalist* **113**:691–703 [*333–334*]

McNaughton SJ (1983). Compensatory plant growth as a response to herbivory. *Oikos* **40**:329–336 [*333*]

McPeek M & Holt RD (1992). The evolution of dispersal in spatially and temporally varying environments. *The American Naturalist* **140**:1010–1027 [*267*]

Menge BA (1995). Indirect effects in marine rocky intertidal interaction webs: Patterns and importance. *Ecological Monographs* **65**:21–74 [*333*]

Mertz DB (1971). The mathematical demography of the California condor population. *The American Naturalist* **105**:437–453 [*24*]

Meszéna G & Metz JAJ (1999). *Species Diversity and Population Regulation: The Importance of Environmental Feedback Dimensionality*. IIASA Interim Report IR-99-045. Laxenburg, Austria: International Institute for Applied Systems Analysis [*7*]

Meszéna G, Kisdi É, Dieckmann U, Geritz SAH & Metz JAJ (2001). Evolutionary optimisation models and matrix games in the unified perspective of adaptive dynamics. *Selection* **2**:193–210 [*188, 193*]

Metz JAJ & de Roos AM (1992). The role of physiologically structured population models within a general individual-based modeling perspective. In *Individual-based Models and Approaches in Ecology*, eds. DeAngelis DL & Gross LJ, pp. 88–111. London, UK: Chapman & Hall [*63*]

Metz JAJ & Diekmann O (1986). *The Dynamics of Physiologically Structured Populations*, Lecture Notes in Biomathematics, Vol. 68. Berlin, Germany: Springer-Verlag

[*65*]

Metz JAJ & Gyllenberg M (2001). How should we define fitness in structured metapopulation models? Including an application for the calculation of evolutionarily stable dispersal strategies. *Proceedings of the Royal Society of London B* **268**:499–508 [*64, 66, 79, 213, 265, 273–274, 277*]

Metz JAJ, Nisbet RM & Geritz SAH (1992). How should we define "fitness" for general ecological scenarios. *Trends in Ecology and Evolution* **7**:198–202 [*188, 191, 196–200, 308, 329*]

Metz JAJ, Geritz SAH, Meszéna G, Jacobs FJA & Van Heerwaarden JS (1996a). Adaptive dynamics, a geometrical study of the consequences of nearly faithful reproduction. In *Stochastic and Spatial Structures of Dynamical Systems*, eds. Van Strien SJ & Verduyn Lunel SM, pp. 183–231. Amsterdam, Netherlands: North-Holland [*7, 188, 196–197, 199*]

Metz JAJ, Mylius SD & Diekmann O (1996b). *When Does Evolution Optimize? On the Relation between Types of Density Dependence and Evolutionarily Stable Life History Parameters*. IIASA Working Paper WP-96-004. Laxenburg, Austria: International Institute for Applied Systems Analysis [*188, 192, 194–195*]

Meyer A (1993). Phylogenetic relationships and evolutionary processes in east African cichlid fishes. *Trends in Ecology and Evolution* **8**:279–284 [*205*]

Michod RE (1979). Evolution of life histories in response to age-specific mortality factors. *The American Naturalist* **113**:531–550 [*102*]

Middleton DAJ & Nisbet RM (1997). Population persistence times: Estimates, models, and mechanisms. *Ecological Applications* **7**:107–117 [*38*]

Miller PS & Hedrick P (1993). Inbreeding and fitness in captive populations: Lessons from *Drosophila*. *Zoo Biology* **12**:333–351 [*132*]

Miller PS & Lacy RC (1999). *VORTEX Version 8 Users Manual. A Stochastic Simulation of the Simulation Process*. Apple Valley, MN, USA: IUCN/SSC Conservation Breeding Specialist Group [*130*]

Mills LS & Allendorf FW (1996). The one-migrant-per-generation rule in conservation and management. *Conservation Biology* **10**:1509–1518 [*233, 286*]

Mills LS & Smouse PE (1994). Demographic consequences of inbreeding in remnant populations. *The American Naturalist* **144**:412–431 [*53*]

Mills LS, Hayes SG, Baldwin C, Wisdon MJ, Citta J, Mattson DJ & Murphy K (1996). Factors leading to different viability predictions for a grizzly bear data set. *Conservation Biology* **10**:863–873 [*43*]

Mills LS, Doak DF & Wisdom MJ (1999). Reliability of conservation actions based on elasticity analysis of matrix models. *Conservation Biology* **13**:815–829 [*44*]

Mitchell RJ (1994). Effects of floral traits, pollinator visitation, and plant size on *Ipomopsis aggregata* fruit production. *The American Naturalist* **143**:870–889 [*311*]

Mode CJ & Pickens GT (1986). Demographic stochasticity and uncertainty in population projections: A study by computer simulation. *Mathematical Biosciences* **79**:55–72 [*32*]

Moilanen A (1999). Patch occupancy models of metapopulation dynamics: Efficient parameter estimation using implicit statistical inference. *Ecology* **80**:1031–1043 [72]

Moilanen A (2000). The equilibrium assumption in estimating the parameters of metapopulation models. *Journal of Animal Ecology* **69**:143–153 [*72*]

Mole S (1994). Trade-offs and constraints in plant–herbivore defense theory: A life-history perspective. *Oikos* **71**:3–12 [*334–335*]

Møller AP & Birkhead TR (1994). The evolution of plumage brightness in birds is related to extrapair paternity. *Evolution* **48**:1089–1100 [*47*]

Møller AP & Legendre S (2001). Allee effect, sexual selection and demographic stochasticity. *Oikos* **92**:27–34 [*46*]

Møller AP & Swaddle JP (1992). *Asymmetry, Developmental Stability and Evolution*. Oxford, UK: Oxford University Press [*143*]

Mollison D (1991). Dependence of epidemic and population velocities on basic parameters. *Mathematical Biosciences* **107**:255–287 [*201*]

Mongold JA, Bennett AF & Lenski RE (1996). Experimental investigations of evolutionary adaptation to temperature. In *Animals and Temperature: Phenotypic and Evolutionary Adaptation*, eds. Johnston IA & Bennett AF, pp. 239–264. Cambridge, UK: Cambridge University Press [*92*]

Mooney HA & Cleland EE (2001). The evolutionary impact of invasive species. *Proceedings of the National Academy of Sciences of the USA* **98**:5446–5451 [*201, 357*]

Mooney HA & Hobbs RJ, eds. (2000). *Invasive Species in a Changing World*. Washington, DC, USA: Island Press [*201*]

Mönkkönen M & Reunanen P (1999). On critical thresholds in landscape connectivity: A management perspective. *Oikos* **84**:302–305 [*299*]

Monson G & Sumner L (1981). *The Desert Bighorn*. Tucson, AZ, USA: The University of Arizona Press [*53*]

Morin JP, Moreteau B, Petavy G, Imasheva AG & David JR (1996). Body size and developmental temperature in *Drosophila simulans*: Comparison of reaction norms with sympatric *Drosophila melanogaster*. *Genetics Selection Evolution* **28**:415–436 [*147*]

Morris PJ, Ivany LC, Schopf KM & Brett CE (1995). The challenge of paleontological stasis: Reassessing sources of evolutionary stability. *Proceedings of the National Academy of Sciences of the USA* **92**:11269–11273 [*4*]

Morris WF, Bronstein JL & Wilson WG (2003). Three-way coexistence in obligate mutualist-exploiter interactions: The potential role of competition. *The American Naturalist* **161**:860–875 [*307*]

Morton NE, Crow JF & Muller H (1956). An estimate of the mutational damage in man from data on consanguineous marriages. *Proceedings of the National Academy of Sciences of the USA* **42**:855–863 [*131*]

Motro U (1982a). Optimal rates of dispersal. 1. Haploid populations. *Theoretical Population Biology* **21**:394–411 [*265, 275*]

Motro U (1982b). Optimal rates of dispersal. 2. Diploid populations. *Theoretical Population Biology* **21**:412–429 [*265*]

Moulton MP (1993). The all-or-none pattern in introduced Hawaiian passeriformes: The role of competition sustained. *The American Naturalist* **141**:105–119 [*305*]

Mouquet N, Mulder GSEE, Jansen VAA & Loreau M (2001). The properties of competitive communities with coupled local and regional dynamics. In *Dispersal*, eds. Clobert J, Danchin E, Dhondt AA & Nichols JD, pp. 311–326. Oxford, UK: Oxford University Press [*203*]

Mousseau TA & Roff DA (1987). Natural selection and the heritability of fitness components. *Heredity* **59**:181–197 [*129*]

Muir WM & Howard RD (1999). Possible ecological risks of transgenic organism release when transgenes affect mating success: Sexual selection and the Trojan gene hypothesis. *Proceedings of the National Academy of Sciences of the USA* **96**:13853–13856 [*208*]

Mukai T (1964). The genetic structure of natural populations of *Drosophila melanogaster* I. Spontaneous mutation rate of polygenes controlling viability. *Genetics* **50**:1–19 [*162–163*]

Muller HJ (1950). Our load of mutations. *American Journal of Human Genetics* **2**:111–176 [*156*]

Muller HJ (1964). The relation of recombination to mutational advance. *Mutation Research* **1**:2–9 [*165*]

Myers N (2002). Environmental refugees: A growing phenomenon of the 21st century. *Philosophical Transactions of the Royal Society of London B* **357**:609–613 [*361*]

Myers N & Kent J (1998). *Perverse Subsidies*. Winnipeg, Canada: International Institute for Sustainable Development [*361*]

Myers N & Knoll AH (2001). The biotic crisis and the future of evolution. *Proceedings of the National Academy of Sciences of the USA* **98**:5389–5392 [*12, 202*]

Myers N, Mittermeier RA, Mittermeier CG, da Fonsca GAB & Kent J (2000). Biodiversity hotspots for conservation priorities. *Nature* **403**:853–858 [*358*]

Mylius SD & Diekmann O (1995). On evolutionarily stable life histories, optimization and the need to be specific about density dependence. *Oikos* **74**:218–224 [*188, 194*]

Nadel H, Frank JH & Knight JRJ (1992). Escapees and accomplices: The naturalization of exotic *Ficus* and their associated faunas in Florida. *Florida Entomologist* **75**:29–38 [*318*]

Nagy J. Adaptive dynamics of dispersal in a vertebrate metapopulation: A case study. In *Elements of Adaptive Dynamics*, eds. Dieckmann U & Metz JAJ. Cambridge, UK: Cambridge University Press. In press [*265*]

Nason JD, Herre EA & Hamrick JL (1998). The breeding structure of a tropical keystone plant resource. *Nature* **391**:685–687 [*316*]

Nee S & May RM (1997). Extinction and the loss of evolutionary history. *Science* **278**:692–694 [*358*]

Nei M (1973). Analysis of gene diversity in subdivided populations. *Proceedings of the National Academy of Sciences of the USA*

70:3321–3323 [*230*]

Nei M & Li WH (1979). Mathematical model for studying genetic variation in terms of restriction endonucleases. *Proceedings of the National Academy of Sciences of the USA* **76**:5269–5273 [*230*]

Neifakh AA & Hartl DL (1993). Genetic control of the rate of embryonic development: Selection for faster development at elevated temperatures. *Evolution* **47**:1625–1631 [*148*]

Nieminen M, Singer MC, Fortelius W, Schöps K & Hanski I (2001). Experimental confirmation of inbreeding depression increasing extinction risk in butterfly populations. *The American Naturalist* **157**:237–244 [*287, 359*]

Nilsson LA (1992). Orchid pollination biology. *Trends in Ecology and Evolution* **7**:255–259 [*306*]

Noach EJK, de Jong G & Scharloo W (1996). Phenotypic plasticity in morphological traits in two populations of *Drosophila melanogaster*. *Journal of Evolutionary Biology* **9**:831–834 [*147*]

Noë R & Hammerstein P (1995). Biological markets. *Trends in Ecology and Evolution* **10**:336–339 [*306*]

Noss RF (1996). Ecosystem as conservation targets. *Trends in Ecology and Evolution* **11**:351 [*12*]

Novacek MJ & Cleland EE (2001). The current biodiversity extinction event: Scenarios for mitigation and recovery. *Proceedings of the National Academy of Sciences of the USA* **98**:5466–5470 [*356–358*]

Nowak MA & May RM (1994). Superinfection and the evolution of parasite virulence. *Proceedings of the Royal Society of London B* **255**:81–89 [*202*]

Nowak M & Sigmund K (1989). Oscillations in the evolution of reciprocity. *Journal of Theoretical Biology* **137**:21–26 [*197, 199*]

O'Brien SJ, Roelke ME, Marker L, Newman A, Winkler C, Meltzer A, Colly D, Evermann L, Bush JFM & Wildt DE (1985). Genetic basis for species vulnerability in the cheetah. *Science* **227**:1428–1434 [*346*]

Odum HT (1971). *Environment, Power, and Society.* New York, NY, USA: John Wiley & Sons [*361*]

Olivieri I & Gouyon P-H (1997). Evolution of migration rate and other traits: The metapopulation effect. In *Metapopulation Biology: Ecology, Genetics and Evolution*, eds. Hanski I & Gilpin ME, pp. 293–323. London, UK: Academic Press [*275–276, 278, 288*]

Olivieri I, Couvet D & Gouyon P-H (1990). The genetics of transient populations: Research at the metapopulation level. *Trends in Ecology and Evolution* **5**:207–210 [*284, 291*]

Olivieri I, Michalakis Y & Gouyon P-H (1995). Metapopulation genetics and the evolution of dispersal. *The American Naturalist* **146**:202–228 [*265, 284, 288, 291, 295*]

O'Neill RV & Kahn JR (2000). Homo economus as a keystone species. *BioScience* **50**:333–337 [*361*]

Orians GH (1969). On the evolution of mating systems in birds and mammals. *The American Naturalist* **103**:589–603 [*47*]

Orians GH (1997). Evolved consequences of rarity. In *The Biology of Rarity*, eds. Kunin WE & Gaston KJ, pp. 190–208. London, UK: Chapman & Hall [*345*]

Orr HA (1998). The population genetics of adaptation: The distribution of factors fixed during adaptive evolution. *Evolution* **52**:935–949 [*253*]

Otto SP & Whitlock MC (1997). Fixation of beneficial mutations in a population of changing size. *Genetics* **146**:723–733 [*167*]

Ouborg NJ, Van Treuren R & Van Damme JMM (1991). The significance of genetic erosion in the process of extinction. II. Morphological variation and fitness components in populations of varying size of *Salvia pratensis* L. and *Scabiosa columbaria* L. *Oecologia* **86**:359–367 [*286*]

Ovaskainen O & Hanski I (2001). Spatially structured metapopulation models: Global and local assessment of metapopulation capacity. *Theoretical Population Biology* **60**:281–302 [*72*]

Ovaskainen O & Hanski I (2002). Transient dynamics in metapopulation response to perturbation. *Theoretical Population Biology* **61**:285–295 [*72*]

Ovaskainen O & Hanski I. How much does an individual habitat fragment contribute to metapopulation dynamics and persistence? Unpublished [*72*]

Owen DF & Wiegert RG (1976). Do consumers maximize plant fitness? *Oikos* **27**:488–492 [*333*]

Owen DF & Wiegert RG (1981). Mutualism between grasses and grazers: An evolutionary hypothesis. *Oikos* **36**:376–378 [*333*]

Ownbey M (1950). Natural hybridization and amphiploidy in the genus *Tragopogon*. *American Journal of Botany* **37**:487–499 [*352*]

Paige KN (1992). Overcompensation in response to mammalian herbivory: From mutualistic to antagonistic interactions. *Ecology* **73**:2076–2085 [*333*]

Paige KN & Whitham TG (1987). Overcompensation in response to mammalian herbivory: The advantage of being eaten. *The American Naturalist* **129**:407–416 [*333*]

Palmer AR (1994). Fluctuating asymmetry analyses: A primer. In *Developmental Stability: Its Origins and Evolutionary Implications*. ed. Markow TA, pp. 335–364. Dordrecht, Netherlands: Kluwer [*142*]

Palmer AR (1996). Waltzing with asymmetry. *Bioscience* **46**:518–532 [*142*]

Palmer AR & Strobeck C (1986). Fluctuating asymmetry: Measurement, analysis, patterns. *Annual Review of Ecology and Systematics* **17**:391–421 [*142*]

Palumbi SR (2000). *The Evolution Explosion.* New York, NY, USA: WW Norton & Co. [*360*]

Palumbi SR (2001). Humans as the world's greatest evolutionary force. *Science* **293**:1786–1790 [*356–358, 360*]

Pamilo P, Nei M & Li W-H (1987). Accumulation of mutations in sexual and asexual populations. *Genetical Research* **49**:135–146 [*165*]

Panetsos C & Baker HG (1968). The origin of variation in "wild" *Raphanus sativus* (Cruciferae) in California. *Genetica* **38**:243–278 [*353*]

Park T (1933). Studies in population physiology. II. Factors regulating initial growth of *Tribolium confusum* populations. *Journal of Experimental Zoology* **65**:17–42 [*24*]

Parsell DA & Lindquist S (1994). Heat shock proteins and stress tolerance. In *The Biology of Heat Shock Proteins and Molecular Chaperones*, eds. Morimoto RI, Tissières A & Georgopoulos C, pp. 457–494. Cold Spring Harbor, NY, USA: Cold Spring Harbor Laboratory Press [*140*]

Parsons PA (1961). Fly size, emergence time and sternopleural chaeta number in *Drosophila*. *Heredity* **16**:445–473 [*137, 142*]

Parsons PA (1987). Evolutionary rates under environmental stress. *Evolutionary Biology* **21**:311–347 [*137*]

Partridge L & Barton NH (1993). Optimality, mutation and the evolution of ageing. *Nature* **362**:305–311 [*126*]

Partridge L, Barrie B, Barton NH, Fowler K & French V (1995). Rapid laboratory evolution of adult life-history traits in *Drosophila melanogaster* in response to temperature. *Evolution* **49**:538–544 [*92–93*]

Parvinen K (1999). Evolution of migration in a metapopulation. *Bulletin of Mathematical Biology* **61**:531–550 [*265, 267–269, 271–272*]

Parvinen K (2001a). *Evolutionary Branching of Dispersal Strategies in Structured Metapopulations*. TUCS Technical Report 399. Turku, Finland: Turku Centre for Computer Science [*64*]

Parvinen K (2001b). *Adaptive Metapopulation Dynamics*. PhD Thesis, Turku, Finland: University of Turku [*64, 79, 265*]

Parvinen K (2002). Evolutionary branching of dispersal strategies in structured metapopulations. *Journal of Mathematical Biology* **45**:106–124 [*273, 278, 280*]

Parvinen K & Dieckmann U. Evolutionary suicide. In *Elements of Adaptive Dynamics*, eds. Dieckmann U & Metz JAJ. Cambridge, UK: Cambridge University Press. In press [*213*]

Parvinen K & Dieckmann U. Even frequency-independent selection can cause self-extinction. Unpublished [*189*]

Parvinen K, Dieckmann U, Gyllenberg M & Metz JAJ (2003). Evolution of dispersal in metapopulations with local density dependence and demographic stochasticity. *Journal of Evolutionary Biology* **16**:143–153 [*273–275, 282*]

Pascual MA, Kareiva P & Hilborn R (1997). The influence of model structure on conclusions about the viability and harvesting of Serengeti wildebeest. *Conservation Biology* **11**:966–976 [*38*]

Pásztor L, Meszéna G & Kisdi É (1996). R_0 or r: A matter of taste? *Journal of Evolutionary Biology* **9**:511–518 [*188*]

Pearson K (1903). Mathematical contributions to the theory of evolution. XI. On the influence of natural selection on the variability and correlation of organs. *Philosophical Transactions of the Royal Society of London A* **200**:1–66 [*173*]

Pease CM, Lande R & Bull JJ (1989). A model of population growth, dispersal, and evolution in a changing environment. *Ecology*

70:1657–1664 [*244, 258–260, 263, 265*]

Peitgen H-O & Saupe D (1988). *The Science of Fractal Images*. New York, NY, USA: Springer-Verlag [*36*]

Pellmyr O & Leebens-Mack J (1999). Forty million years of mutualism: Evidence for Eocene origin of the yucca-yucca moth association. *Proceedings of the National Academy of Sciences of the USA* **96**:9178–9183 [*307*]

Pemberton RW & Irving DW (1990). Elaiosomes on weed seeds and the potential for myrmecochory in naturalized plants. *Weed Science* **38**:615–619 [*321*]

Perrin N & Mazalov V (2000). Local competition, inbreeding, and the evolution of sex-biased dispersal. *The American Naturalist* **155**:116–127 [*359*]

Perrings C, Williamson M, Barbier EB, Delfino D, Dalmazzone S, Shogren J, Simmons P & Watkinson A (2002). Biological invasion risks and the public good: An economic perspective. *Conservation Ecology* **6**:1 [*201*]

Petchey OL, Gonzalez A & Wilson HB (1997). Effects on population persistence: The interaction between environmental noise colour, intraspecific competition and space. *Proceedings of the Royal Society of London B* **264**:1841–1847 [*36*]

Petit C, Fréville H, Mignot A, Colas B, Riba M, Imbert E, Hurtrez-Boussès S, Virevaire M & Olivieri I (2001). Gene flow and local adaptation in two endemic plant species. *Biological Conservation* **100**:21–34 [*296*]

Phillips PC, Otto SP & Whitlock MC (2000). Beyond the average: The evolutionary importance of epistasis and the variability of epistatic effects. In *Epistasis and the Evolutionary Process*, eds. Wolf J, Brodie III ED & Wade MJ, pp. 20–38. Oxford, UK: Oxford Press [*163*]

Pimentel D, Wilson C, McCullum C, Huang R, Dwen P, Flack J, Tran Q, Salman T & Cliff B (1997). Economic and environmental benefits of biodiversity. *BioScience* **47**:747–757 [*361*]

Pimm SL & Redfearn A (1988). The variability of animal populations. *Nature* **334**:613–614 [*36*]

Pinxten R & Eens M (1990). Polygyny in the European starling: Effect of female reproductive success. *Animal Behaviour* **40**:1035–1047 [*47*]

Poggiale JC (1998). From behavioural to population level: Growth and competition. *Mathematical Computer Modeling* **27**:41–49 [*25*]

Pogson GH & Zouros E (1994). Allozyme and RFLP heterozygosities as correlates of growth rate in the scallop *Placopecten magellanicus*: A test of the associative overdominance hypothesis. *Genetics* **137**:221–231 [*128*]

Pogson GH, Mesa DA & Boutilier RG (1995). Genetic population structure and gene flow in the Atlantic cod *Gadus morhua*: A comparison of allozyme and nuclear RFLP loci. *Genetics* **139**:375–385 [*239*]

Pollard E (1979). Population ecology and change in range of the white admiral butterfly *Ladoga camilla* L. in England. *Ecological Entomology* **4**:61–74 [*89*]

Pollard E (1988). Temperature, rainfall and butterfly numbers. *Journal of Applied Ecology* **25**:819–828 [*90*]

Pomiankowski A, Isawa Y & Nee S (1991). The evolution of costly mate preferences. I. Fisher and biased mutation. *Evolution* **45**:1422–1430 [*52*]

Poon A & Otto SP (2000). Compensating for our load of mutations: Freezing the meltdown of small populations. *Evolution* **54**:1467–1479 [*167*]

Pounds JA, Fogden MPL, Campbell JH & Savage JM (1999). Biological responses to climate change on a tropical mountain. *Nature* **398**:611–615 [*2*]

Prevosti A, Ribo G, Serra L, Aguade M, Balana J, Monclus M & Mestres F (1988). Colonization of America by *Drosophila subobscura*: Experiment in natural populations that supports the adaptive role of chromosomal-inversion polymorphism. *Proceedings of the National Academy of Sciences USA* **85**:5597–5600 [*90*]

Primack RB (1985). Patterns of flowering phenology in communities, populations, individuals, and single flowers. In *Population Structure of Vegetation*, ed. White J, pp. 571–594. Dordrecht, Netherlands: Kluwer Academic Publishers [*318*]

Puccia CJ & Levins R (1985). *Qualitative Modeling of Complex Systems*. Cambridge, MA, USA: Harvard University Press [*333*]

Pulliam HR (1988). Sources, sinks, and population regulation. *The American Naturalist* **132**:652–661 [*241*]

Pulliam HR (1996). Sources and sinks: Empirical evidence and population consequences.

In *Population Dynamics in Space and Time*, eds. Rhodes OE, Chesser RK & Smith MH, pp. 45–70. Chicago, IL, USA: University of Chicago Press [*251, 293*]

Pullin AS, ed. (1995). *Ecology and Conservation of Butterflies*. London, UK: Chapman & Hall [*60*]

Quattro JM, Avise JC & Vrijenhoek RC (1991). Molecular evidence for multiple origins of hybridogenetic fish clones (Poeciliidae: *Poeciliopsis*). *Genetics* **127**:391–398 [*352*]

Quilichini A, Debussche M & Thompson JD (2001). Evidence for local outbreeding depression in the Mediterranean island endemic *Anchusa crispa* Viv. (Boraginaceae). *Heredity* **87**:190–197 [*284*]

Rabinowitz D (1981). Seven forms of rarity. In *The Biological Aspects of Biological Conservation*, ed. Synge H, pp. 205–217. Chichester, UK: John Wiley & Sons [*345*]

Ralls K, Ballou JD & Templeton A (1988). Estimates of lethal equivalents and the cost of inbreeding in mammals. *Conservation Biology* **2**:185–193 [*96, 132*]

Rasplus J-Y (1996). The one-to-one species-specificity of the Ficus–Agaonine mutualism: How casual? In *The Biodiversity of African Plants*, eds. van der Maesen LJG, van der Burgt XM & van Medenbach de Rooy JM, pp. 639–649. Dordrecht, Netherlands: Kluwer Academic Publishers [*316*]

Rassi P, Alanen A, Kanerva T & Mannerkoski I (2001). Suomen lajien uhanalaisuus 2000 [The red-listed species of Finland in 2000]. Helsinki, Finland: Ympäristöministeriö & Suomen ympäristökeskus [*59*]

Ratner VF, Zabanov SA, Kolesnikova OV & Vasilyeva LA (1992). Induction of mobile genetic element *Dm412* induced by a severe heat shock in *Drosophila*. *Proceedings of the National Academy of Sciences of the USA* **89**:5650–5654 [*139*]

Reinartz JA & Les DH (1994). Bottleneck-induced dissolution of self-incompatibility and breeding system consequences in *Aster furcatus* (Asteraceae). *American Journal of Botany* **81**:446–455 [*295, 354*]

Renault O & Ferrière R. Population size and persistence under demographic stochasticity and overcompensatory density dependence. Unpublished [*189, 197*]

Reznick DN (1982). The impact of predation on life history evolution in Trinidadian guppies: The genetic components of observed life history differences. *Evolution* **36**:1236–1250 [*103, 106*]

Reznick DN & Bryga H (1987). Life-history evolution in guppies. 1. Phenotypic and genotypic changes in an introduction experiment. *Evolution* **41**:1370–1385 [*106–107*]

Reznick DN & Bryga H (1996). Life-history evolution in guppies (*Poecilia reticulata*: Poeciliidae). V. Genetic basis of parallelism in life histories. *The American Naturalist* **147**:339–359 [*103, 112*]

Reznick DN & Endler JA (1982). The impact of predation on life history evolution in Trinidadian guppies (*Poecilia reticulata*). *Evolution* **36**:160–177 [*103, 105*]

Reznick DN, Bryga H & Endler JA (1990). Experimentally induced life-history evolution in a natural population. *Nature* **346**:357–359 [*106–107, 125*]

Reznick DN, Butler MJ, Rodd FH & Ross P (1996). Life history evolution in guppies (*Poecilia reticulata*). 6. Differential mortality as a mechanism for natural selection, *Evolution* **50**:1651–1660 [*102–105*]

Reznick DN, Shaw FH, Rodd FH & Shaw RG (1997). Evaluation of the rate of evolution in natural populations of guppies (*Poecilia reticulata*). *Science* **275**:1934–1937 [*107–108*]

Rhymer JM & Simberloff D (1996). Extinction by hybridization and introgression. *Annual Review of Ecology and Systematics* **27**:83–109 [*344*]

Rhymer JM, Williams MJ & Braun MJ (1994). Mitochondrial analysis of gene flow between New Zealand mallards (*A. platyrhynchos*) and grey ducks (*A. superciliosa*). *Auk* **111**:970–978 [*346*]

Rice WR (1992). Sexually antagonistic genes: Experimental evidence. *Science* **256**:1436–1439 [*124*]

Rice WR (1996). Sexually antagonistic male adaptation triggered by arrested female evolution. *Nature* **381**:232–234 [*124*]

Richardson DM, Allsopp N, D'Antonio C, Milton SJ & Rejmánek M (2000). Plant invasions: The role of mutualisms. *Biological Reviews* **75**:65–93 [*306, 314*]

Ricker WE (1954). Stock and recruitment. *Journal of the Fisheries Research Board Canada* **11**:559–623 [*270*]

Riddle RA, Dawson PS & Zirkle DF (1986). An experimental test of the relationship between genetic variation and environmental variation in *Tribolium* flour beetles. *Genet-*

ics **113**:391–404 [*123*]

Riedl H & Croft B (1978). The effects of photoperiod and effective temperatures on the seasonal phenology of the codling moth *Laspeyresia pomonella* (Lepidoptera: Tortricidae). *Canadian Entomologist* **110**:455–470 [*91*]

Rieseberg LH (1991). Hybridization in rare plants: Insights from case studies in *Cerocarpus* and *Helianthus*. In *Genetics and Conservation of Rare Plants*, eds. Falk DA & Holsinger KE, pp. 171–181. Oxford, UK: Oxford University Press [*344*]

Rieseberg LH & Gerber D (1995). Hybridization in the Catalina mountain mahogany (*Cerocarpus traskiae*): RAPD evidence. *Conservation Biology* **9**:199–203 [*348*]

Rieseberg LH, Zona S, Aberbom L & Martin TD (1989). Hybridization in the island endemic, Catalina mahogany. *Conservation Biology* **3**:52–58 [*348*]

Rieseberg LH, Sinervo B, Linder CR, Ungerer M & Arias DM (1996). Role of gene interactions in hybrid speciation: Evidence from ancient and experimental hybrids. *Science* **272**:741–745 [*351–352*]

Ripa J & Lundberg P (1996). Noise colour and the risk of population extinctions. *Proceedings of the Royal Society of London B* **263**:1751–1753 [*36*]

Riska B, Prout TT & Turelli M (1989). Laboratory estimates of heritabilities and genetic correlations in nature. *Genetics* **123**:865–872 [*128*]

Ritland K (1996). A marker-based method for inferences about quantitative inheritance in natural populations. *Evolution* **50**:1062–1073 [*126*]

Roberts G & Sherratt TN (1998). Development of cooperative relationships through increasing investment. *Nature* **394**:175–179 [*307*]

Robertson A (1960). A theory of limits in artificial selection. *Proceedings of the Royal Society of London B* **153**:234–249 [*87*]

Rodd FH & Reznick DN (1997). Variation in the demography of natural populations of guppies: The importance of predation and life histories. *Ecology* **78**:405–418 [*112*]

Roff DA (1974). The analysis of a population model demonstrating the importance of dispersal in a heterogeneous environment. *Oecologia* **5**:259–275 [*265*]

Roff DA (1992). *The Evolution of Life Histories: Theory and Analysis*. London, UK: Chapman & Hall [*124–126, 188, 191*]

Roff DA (2002). *Life History Evolution*. Sunderland, MA, USA: Sinauer Associates Inc. [*126, 190*]

Roff DA & Mousseau TA (1987). Quantitative genetics and fitness: Lessons from *Drosophila*. *Heredity* **58**:103–118 [*120, 129*]

Ronce O & Kirkpatrick M (2001). When sources become sinks: Migrational meltdown in heterogeneous habitats. *Evolution* **55**:1520–1531 [*258*]

Ronce O, Perret F & Olivieri I (2000a). Evolutionarily stable dispersal rates do not always increase with local extinction rates. *The American Naturalist* **155**:485–496 [*265, 273, 282, 288*]

Ronce O, Perret F & Olivieri I (2000b). Landscape dynamics and evolution of colonizer syndromes: Interactions between reproductive effort and dispersal in a metapopulation. *Evolutionary Ecology Research* **14**:233–260 [*283*]

Root TL (1988). Energy constraints on avian distributions and abundances. *Ecology* **69**:330–339 [*90*]

Rose MR (1982). Antagonistic pleiotropy, dominance, and genetic variation. *Heredity* **48**:63–78 [*125*]

Rosen DE & Bailey RM (1963). The poeciliid fishes (Cyprindontiformes), their structure, zoogeography, and systematics. *Bulletin of the American Museum of Natural History* **126**:1–176 [*101*]

Rosenzweig ML (1978). Competitive speciation. *Biological Journal of the Linnean Society* **10**:275–289 [*205*]

Rosenzweig ML (1995). *Species Diversity in Space and Time*. Cambridge, UK: Cambridge University Press [*202, 204, 207*]

Rosenzweig ML (2001). Loss of speciation rate will impoverish future diversity. *Proceedings of the National Academy of Sciences of the USA* **98**:5404–5410 [*202, 360*]

Rosenzweig ML (2003). *Win-Win Ecology*. Oxford, UK: Oxford University Press [*360*]

Roughgarden J (1979). *Theory of Population Genetics and Evolutionary Ecology: An Introduction*. New York, NY, USA: Macmillan [*190, 194, 215*]

Roughgarden J (1995). *Anolis Lizards of the Caribbean*. Oxford, UK: Oxford University Press [*207*]

Roush W (1997). Hybrids consummate species invasion. *Science* **277**:316–317 [*350*]

Roush RT & McKenzie JA (1987). Ecological genetics of insecticide and acaricide resistance. *Annual Review of Entomology* **32**:361–380 [*96*]

Rowe G, Beebee TJC & Burke T (1999). Microsatellite heterozygosity, fitness and demography in natterjack toads *Bufo calamita*. *Animal Conservation* **2**:85–92 [*286*]

Roy MS, Geffen E, Smith D, Ostrander EA & Wayne RK (1994). Patterns of differentiation and hybridization in North American wolf-like canids, revealed by analysis of microsatellite loci. *Molecular Biology and Evolution* **11**:553–570 [*352*]

Royama T (1992). *Analytical Population Dynamics*. London, UK: Chapman & Hall [*24*]

Rozhnov VV (1993). Extinction of the European mink: Ecological catastrophe or a natural process? *Lutreola* **1**:10–16 [*347*]

Ruelle D (1989). *Chaotic Evolution and Strange Attractor*. Cambridge, UK: Cambridge University Press [*44*]

Rutherford SL & Lindquist S (1998). HSP90 as a capacitor for morphological evolution. *Nature* **396**:336–342 [*140–141*]

Ryckman RE & Ueshima N (1964). Biosystematics of the *Hesperocimex* complex (Hemiptera: Cimicidae) and avian hosts (Piciformes: Picidae; Passeriformes: Hirudinidae). *Annals of the Entomological Society of America* **57**:624–638 [*347*]

Saccheri I, Kuussaari M, Kankare M, Vikman P, Fortelius W & Hanski I (1998). Inbreeding and extinction in a butterfly metapopulation. *Nature* **392**:491–494 [*2, 96, 287*]

Saether BE, Engen S & Lande R (1999). Finite metapopulation models with density-dependent migration and stochastic local dynamics. *Proceedings of the Royal Society of London B* **266**:113–118 [*237*]

Sala OE, Chapin FS III, Armesto JJ, Berlow E, Bloomfield J, Dirzo R, Huber-Sanwald E, Henneke LF, Jackson RB, Kinzig A, Leemans R, Lodge DM, Mooney HA, Oesterheld M, Poff NL, Sykes MT, Walker BH, Walker M & Wall DH (2000). Global biodiversity scenarios for the year 2100. *Science* **287**:1770–1774 [*358*]

Salas-Pascual M, Acebes-Ginoves JR & Del Acro-Aguilar M (1993). *Arbutus* X *androsterilis*, a new interspecific hybrid between *Arbutus canariensis* and *A. unedo* from the Canary Islands. *Taxon* **42**:789–792 [*350*]

Sarrazin F & Legendre S (2000). Demographic approach to releasing adults versus young in reintroduction. *Conservation Biology* **14**:474–487 [*58*]

Savidge JA (1987). Extinction of an island forest avifauna by an introduced snake. *Ecology* **68**:660–668 [*305*]

Savolainen O (1994). Genetic variation and fitness: Conservation lessons from pines. In *Conservation Genetics*, eds. Loeschcke V, Tomiuk J & Jain SK, pp. 27–36. Basel, Switzerland: Birkhäuser [*129*]

Savolainen O & Hedrick P (1995). Heterozygosity and fitness: No association in Scots pine. *Genetics* **140**:755–766 [*128*]

Schei PJ, Sandlund OT & Strand R, eds. (1999). *Proceedings of the Norway/UN Conference on the Ecosystem Approach for Sustainable Use of Biological Diversity*. Trondheim, Norway: Norwegian Institute for Nature Research [*327*]

Schemske DW, Husband BC, Ruckelshaus MH, Goodwillie C, Parker IM & Bishop JG (1994). Evaluating approaches to the conservation of rare and endangered plants. *Ecology* **75**:584–606 [*344*]

Schliewen UK, Tautz D & Pääbo S (1994). Sympatric speciation suggested by monophyly of crater lake cichlids. *Nature* **368**:629–632 [*205*]

Schluter D (2000). *The Ecology of Adaptive Radiation*. Oxford, UK: Oxford University Press [*205*]

Schmalhausen II (1949). *Factors of Evolution*. Philadelphia, PA, USA: Blakiston [*139*]

Schnell FW & Cockerham CC (1992). Multiplicative vs. arbitrary gene action in heterosis. *Genetics* **131**:461–469 [*123*]

Schoener TW (1983). Field experiments on interspecific competition. *The American Naturalist* **122**:240–285 [*338*]

Schoener TW (1989). The ecological niche. In *Ecological Concepts*, ed. Cherrett JM, pp. 79–114. Oxford, UK: Blackwell [*244*]

Schofield EK (1989). Effects of introduced plants and animals on island vegetation: Examples from the Galapagos archipelago. *Conservation Biology* **3**:227–238 [*305*]

Schroeder M (1991). *Fractals, Chaos, Power Laws: Minutes from an Infinite Paradise*. New York, NY, USA: WH Freeman and Co. [*36*]

Schultz ST & Lynch M (1997). Mutation and extinction: The role of variable mutational effects, synergistic epistasis, beneficial mutations, and degree of outcrossing. *Evolution* **51**:1363–1371 [*162–163, 165, 167*]

Schultz ST & Willis JH (1995). Individual variation in inbreeding depression: The roles of inbreeding history and mutation. *Genetics* **141**:1209–1223 [*133*]

Schuss Z (1980). *Theory and Applications of Stochastic Differential Equations*. New York, NY, USA: John Wiley & Sons [*77–78*]

Searle SR (1971). Topics in variance component estimation. *Biometrics* **27**:1–76 [*128*]

Seger J (1992). Evolution of exploiter–victim relationships. In *Natural Enemies: The Population Biology of Predators, Parasites, and Diseases*, ed. Crawley MJ, pp. 3–25. Oxford, UK: Blackwell [*339–340*]

Selye H (1955). Stress and disease. *Science* **122**:625–631 [*136*]

Sgrò CM & Hoffmann AA (1998a). Effects of temperature extremes on genetic variances for life-history traits in *Drosophila melanogaster* as determined from parent–offspring regression. *Journal of Evolutionary Biology* **11**:1–20 [*147*]

Sgrò CM & Hoffmann AA (1998b). Effects of stress combinations on the expression of additive genetic variation for fecundity in *Drosophila melanogaster*. *Genetical Research* **72**:13–18 [*142, 147*]

Shaffer ML (1981). Minimum population size for conservation. *BioScience* **31**:131–134 [*20*]

Shaffer ML (1987). Minimum viable populations: Coping with uncertainty. In *Viable Populations for Conservation*, ed. Soulé ME, pp. 69–86. New York, NY, USA: Cambridge University Press [*20, 45*]

Shanahan M, So S, Compton S & Corlett R (2001a). Fig-eating by vertebrate frugivores: A global review. *Biological Reviews* **76**:529–572 [*316*]

Shanahan M, Harrison RD, Yamuna R, Boen W & Thornton IWB (2001b). Colonization of an island volcano, Long Island, Papua New Guinea, and an emergent island, Motmot, in its caldera lake. V. Colonization by figs (*Ficus* spp.), their dispersers and pollinators. *Journal of Biogeography* **28**:1365–1377 [*316*]

Sharp PM (1984). The effect of inbreeding on competitive male-mating ability in *Drosophila melanogaster*. *Genetics* **106**:601–612 [*133*]

Shaw RG (1987). Maximum likelihood approaches to quantitative genetics of natural populations. *Evolution* **41**:812–826 [*127*]

Shenk TM, White GC & Burnham KP (1998). Sampling-variance effects on detecting density dependence from temporal trends in natural populations. *Ecological Monographs* **68**:445–463 [*21*]

Sibly R & Calow P (1989). A life-cycle theory of responses to stress. *Biological Journal of the Linnean Society* **37**:101–116 [*137*]

Sih A, Jonsson BG & Luikart G (2000). Habitat loss: Ecological, evolutionary and genetic consequences. *Trends in Ecology and Evolution* **15**:132–134 [*265, 278*]

Silvela L, Rodgers R, Barrera A & Alexander DE (1989). Effect of selection intensity and population size on percent oil in maize, *Zea mays* L. *Theoretical and Applied Genetics* **78**:298–304 [*87*]

Silvertown JW (1982). No evolved mutualism between grasses and grazers. *Oikos* **38**:253–259 [*333*]

Simberloff D (1981). Community effects of introduced species. In *Biotic Crises in Ecological and Evolutionary Time*, ed. Nitecki M, pp. 53–81. New York, NY, USA: Academic Press [*305*]

Simberloff D (1988). Metapopulation dynamics: Does it help to have more of the same? *Trends in Ecology and Evolution* **4**:113–114 [*243*]

Simberloff D & Boecklen W (1991). Patterns of extinction in the introduced Hawaiian avifauna: A reexamination of the role of competition. *The American Naturalist* **138**:300–327 [*346*]

Simberloff D & von Holle B (1999). Positive interactions of nonindigenous species: Invasional meltdown? *Biological Invasions* **1**:21–32 [*306*]

Simmons MJ & Crow JF (1977). Mutations affecting fitness in *Drosophila* populations. *Annual Review of Genetics* **11**:49–78 [*130, 132*]

Simms EL & Rausher MD (1987). Costs and benefits of plant resistance to herbivory. *The American Naturalist* **130**:570–581 [*335*]

Simpson GG (1944). *Tempo and Mode in Evolution*. New York, NY, USA: Columbia University Press [*6, 82*]

Sinervo B & Lively CM (1996). The rock–paper–scissors game and the evolution of alternative male strategies. *Nature* **380**:240–243 [*193*]

Sinervo B, Svensson E & Comendant T (2000). Density cycles and an offspring quantity and quality game driven by natural selection.

Nature **406**:985–988 [*193*]

Singer MC (1983). Determinants of multiple host use by a phytophagous insect population. *Evolution* **37**:389–403 [*294*]

Singer MC & Thomas CD (1996). Evolutionary responses of a butterfly metapopulation to human and climate-caused environmental variation. *The American Naturalist* **148**:9–39 [*294, 305*]

Singer MC, Ng D & Thomas CD (1988). Heritability of oviposition preference and its relationship to offspring performance within a single insect population. *Evolution* **42**:977–985 [*293*]

Singer MC, Thomas CD & Parmesan C (1993). Rapid human-induced evolution of insect diet. *Nature* **366**:681–683 [*293–294, 314, 319*]

Slatkin M (1977). Gene flow and genetic drift in a species subject to frequent local extinctions. *Theoretical Population Biology* **12**:253–262 [*236*]

Slatkin M (1985). Gene flow in natural populations. *Annual Review of Ecology and Systematics* **16**:393–430 [*231*]

Slatkin M (1987). Gene flow and the geographic structure of natural populations. *Science* **236**:787–792 [*231*]

Slatkin M (1991). Inbreeding coefficients and coalescence times. *Genetical Research* **58**:167–175 [*235*]

Slatkin M (1993). Isolation by distance in equilibrium and non-equilibrium populations. *Evolution* **47**:264–279 [*235*]

Slatkin M & Lande R (1976). Niche width in a fluctuating environment-density dependent model. *The American Naturalist* **110**:31–55 [*187*]

Slatkin M & Voelm L (1991). F_{ST} in a hierarchical island model. *Genetics* **127**:627–629 [*233*]

Smith HF (1936). A discriminant function for plant selection. *Annals of Eugenics* **7**:240–250 [*173*]

Smith JF (2001). High species diversity in fleshy-fruited tropical understory plants. *The American Naturalist* **157**:646–653 [*307*]

Smith SV & Buddemeier RW (1992). Global change and coral reef ecosystems. *Annual Review of Ecology and Systematics* **23**:89–118 [*326*]

Smith TB, Freed LA, Lepson JK & Carothers JH (1995). Evolutionary consequences of extinctions in populations of a Hawaiian honeycreeper. *Conservation Biology* **9**:107–113 [*319*]

Smouse PE & Long JC (1992). Matrix correlation analysis in anthropology and genetics. *Yearbook of Physical Anthropology* **35**:187–213 [*235*]

Snow AA & Palma PM (1997). Commercialization of transgenic plants: Potential ecological risks. *Bioscience* **47**:86–96 [*314*]

Soberon Mainero J & Martinez del Rio C (1985). Cheating and taking advantage in mutualistic associations. In *The Biology of Mutualism*, ed. Boucher DH, pp. 192–216. New York, NY, USA: Oxford University Press [*307*]

Soltis DE & Soltis PS (1993). Molecular data and the dynamic nature of polyploidy. *Critical Reviews in Plant Science* **12**:243–273 [*352*]

Soltis PG, Plunkett G, Novak S & Soltis D (1995). Genetic variation in *Tragopogon* species: Additional origins of the allotetraploids *T. mirus* and *T. miscellus*. *American Journal of Botany* **82**:1329–1341 [*352*]

Sorci G, Møller AP & Clobert J (1998). Plumage dichromatism predicts introduction success in New Zealand. *Journal of Animal Ecology* **67**:263–269 [*52*]

Soulé ME (1976). Allozyme variation, its determinants in space and time. In *Molecular Evolution*, ed. Ayala FJ, pp. 60–77. Sunderland, MA, USA: Sinauer Associates Inc. [*97*]

Soulé ME (1980). Thresholds for survival: Maintaining fitness and evolutionary potential. In *Conservation Biology: An Evolutionary-Ecological Perspective*, eds. Soulé ME & Wilcox BA, pp. 151–170. Sunderland, MA, USA: Sinauer Associates Inc. [*202*]

Soulé ME (1987). *Viable Populations for Conservation*. New York, NY, USA: Cambridge University Press [*26, 59*]

Stace CA (1975). *Hybridization and the Flora of the British Isles*. New York, NY, USA: Academic Press [*350*]

Stacey PB & Taper M (1992). Environmental variation and the persistence of small populations. *Ecological Applications* **2**:18–29 [*284*]

Stacey PB, Johnson VA & Taper ML (1997). Migration within metapopulations: The impact upon local population dynamics. In *Metapopulation Biology: Ecology, Genetics and Evolution*, eds. Hanski I & Gilpin ME, pp. 267–292. London, UK: Academic Press

[*65*]

Staddon PL & Fitter AH (1998). Does elevated atmospheric carbon dioxide affect arbuscular mycorrhizas? *Trends in Ecology and Evolution* **13**:455–458 [*314*]

Stanton ML (2003). Interacting guilds: Moving beyond the pairwise perspective on mutualisms. *The American Naturalist* **162**:S10–S23 [*321*]

Steadman DW (1995). Prehistoric extinctions of Pacific Island birds: Biodiversity meets zooarcheology. *Science* **267**:1123–1131 [*305*]

Stearns SC (1992). *The Evolution of Life Histories*. Oxford, UK: Oxford University Press [*42, 110, 188, 190–191*]

Stearns SC & Koella JC (1986). The evolution of phenotypic plasticity in life-history traits – predictions of reaction norms for age and size at maturity. *Evolution* **40**:893–913 [*191*]

Stebbins GL (1958). The inviability, weakness, and sterility of interspecific hybrids. *Advances in Genetics* **9**:147–215 [*347*]

Steinberg PD, Estes JA & Winter FC (1995). Evolutionary consequences of food chain length in kelp forest communities. *Proceedings of the National Academy of Sciences of the USA* **92**:8145–8148 [*332*]

Stephens PA & Sutherland WJ (1999). Consequences of the Allee effect for behaviour, ecology and conservation. *Trends in Ecology and Evolution* **14**:401–405 [*24, 46*]

Stephens PA, Sutherland WJ & Freckleton RP (1999). What is the Allee effect? *Oikos* **87**:185–190 [*46*]

Storfer A (1999). Gene flow and endangered species translocations: A topic revisited. *Biological Conservation* **87**:173–180 [*284, 292*]

Strauss SY & Agrawal AA (1999). The ecology and evolution of plant tolerance to herbivory. *Trends in Ecology and Evolution* **14**:179–185 [*335*]

Stuart C & Stuart T (1991).The feral cat problem in southern Africa. *African Wildlife* **45**:13–15 [*350*]

Surtees G & Wright JB (1960). Insect populations at low densities. *Pest Infestation Research* **1960**:48–49 [*25*]

Swenson W, Wilson DS & Elias R (2000). Artificial ecosystem selection. *Proceedings of the National Academy of Sciences of the USA* **97**:9110–9114 [*361*]

Tabachnick WJ & Powell JR (1977). Adaptive flexibility of "marginal" and "central" populations of *Drosophila willistoni*. *Evolution* **31**:692–694 [*95*]

Takabayashi J & Dicke M (1996). Plant-carnivore mutualism through herbivore-induced carnivore attractants. *Trends in Plant Sciences* **1**:109–113 [*318*]

Tantawy AO & Mallah GS (1961). Studies on natural populations of *Drosophila*: Heat resistance and geographical variation in *Drosophila melanogaster* and *Drosophila simulans*. *Evolution* **15**:1–14 [*142*]

Tauber MJ, Tauber CA & Masaki S (1986). *Seasonal Adaptations of Insects*. Oxford, UK: Oxford University Press [*91*]

Taylor PD (1989). Evolutionary stability in one-parameter models under weak selection. *Theoretical Population Biology* **36**:125–143 [*196*]

Teesdale C (1940). Fertilization in the tsetse fly, *Glossina palpalis*, in a population of low density. *Journal of Animal Ecology* **9**:24–26 [*25*]

Temple SA (1977). Plant–animal mutualism: Coevolution with dodo leads to near extinction of plant. *Science* **197**:885–886 [*315*]

Templeton AR (1986). Coadaptation and outbreeding depression. In *Conservation Biology: The Science of Scarcity and Diversity*, ed. Soulé ME, pp. 105–116. Sunderland, MA, USA: Sinauer Associates Inc. [*347*]

Templeton AR & Read B (1983). The elimination of inbreeding depression in a captive herd of Speke's gazelle. In *Genetics and Conservation: A Reference for Managing Wild Animal and Plant Populations*, eds. Schonewald-Cox CM, Chambers SM, MacBride B & Thomas WL, pp. 241–261. Menlo Park, CA, USA: Benjamin/Cummings [*132–133*]

Templeton AR & Read B (1984). Factors eliminating inbreeding depression in a captive herd of Speke's gazelle (*Gazella spekei*). *Zoo Biology* **3**:177–199 [*132–133*]

Templeton AR, Robertson RJ, Brisson J & Strasburg J (2001). Disrupting evolutionary processes: The effect of habitat fragmentation on collared lizards in the Missouri Ozarks. *Proceedings of the National Academy of Sciences of the USA* **98**:5426–5432 [*359, 363*]

ter Braak JF, Hanski IA & Verboom J (1998). The incidence function approach to modeling of metapopulation dynamics. In *Modeling Spatiotemporal Dynamics in Ecology*, eds. Bascompte J & Solé RV, pp. 167–188. New York, NY, USA: Springer-Verlag [*72*]

Thévenon S & Couvet D (2002). The impact of inbreeding depression on population survival depending on demographic parameters. *Animal Conservation* **5**:53–60 [*53*]

Thomas CD (2000). Dispersal and extinction in fragmented landscapes. *Proceedings of the Royal Society of London B* **267**:139–145 [*292*]

Thomas CD & Hanski I (1997). Butterfly metapopulations. In *Metapopulation Biology: Ecology, Genetics and Evolution*, eds. Hanski I & Gilpin ME, pp. 359–386. London, UK: Academic Press [*60, 65*]

Thomas CD & Jones TM (1993). Partial recovery of a skipper butterfly (*Hesperia comma*) from population refuges: Lessons for conservation in a fragmented landscape. *Journal of Animal Ecology* **62**:472–481 [*291*]

Thomas CD, Ng D, Singer MC, Mallet JLB, Parmesan C & Billington HL (1987). Incorporation of a European weed into the diet of a North-American herbivore. *Evolution* **41**:892–901 [*293*]

Thomas RB, Richter DD, Ye H, Heine PR & Strain BR (1991). Nitrogen dynamics and growth of seedlings of an N-fixing tree (*Gliricidia sepium* (Jacq.) Walp.) exposed to elevated atmospheric carbon dioxide. *Oecologia* **88**:415–421 [*314*]

Thomas CD, Singer MC & Boughton DA (1996). Catastrophic extinction of population sources in a butterfly metapopulation. *The American Naturalist* **148**:957–975 [*294*]

Thomas CD, Hill JK & Lewis OT (1998). Evolutionary consequences of habitat fragmentation in a localized butterfly. *Journal of Animal Ecology* **67**:485–497 [*289–290*]

Thomas CD, Bodsworth EJ, Wilson RJ, Simmons AD, Davies ZG, Musche M & Conradt L (2001a). Ecological and evolutionary processes at expanding range margins. *Nature* **411**:577–581 [*291–292*]

Thomas DW, Blondel J, Perret P, Lambrechts MM & Speakman JR (2001b). Energetic and fitness costs of mismatching resource supply and demand in seasonally breeding birds. *Science* **291**:2598–2600 [*359*]

Thomas-Orillard M & Legendre S (1996). *Drosophila* C virus and host-population dynamics. *Comptes Rendus de l'Académie des Sciences de Paris* **319**:615–621 [*58*]

Thompson JN (1994). *The Coevolutionary Process*. Chicago, IL, USA: University of Chicago Press [*306, 363*]

Thompson JN (1996). Evolutionary ecology and the conservation of biodiversity. *Trends in Ecology and Evolution* **11**:300–303 [*305*]

Thompson JN (1997). Evaluating the dynamics of coevolution among geographically structured populations. *Ecology* **78**:1619–1623 [*340*]

Thompson JN (1998). Rapid evolution as an ecological process. *Trends in Ecology and Evolution* **13**:329–332 [*12, 357*]

Thompson JD (1999a). Population differentiation in Mediterranean plants: Insights into colonization history and the evolution of endemic species. *Heredity* **82**:229–236 [*298*]

Thompson JN (1999b). The evolution of species interactions. *Science* **284**:2116–2118 [*12*]

Thompson LG, Davis ME & Mosley-Thompson E (1994). Glacial records of global climate – A 1,500-year tropical ice core record of climate. *Human Ecology* **22**:83–95 [*362*]

Thomson JD, Plowright RC & Thaler GR (1985). Matacil insecticide spraying, pollinator mortality, and plant fecundity in New Brunswick forests. *Canadian Journal of Botany* **63**:2056–2061 [*321*]

Thornhill NW, ed. (1993). *The Natural History of Inbreeding and Outbreeding: Theoretical and Empirical Perspectives*. Chicago, IL, USA: University of Chicago Press [*96*]

Tilman D (1982). *Resource Competition and Community Structure*. Princeton, NJ, USA: Princeton University Press [*329*]

Tilman D (1994). Competition and biodiversity in spatially structured habitats. *Ecology* **75**:2–16 [*202*]

Tilman D & Lehman C (2001). Human-caused environmental change: Impacts on plant diversity and evolution. *Proceedings of the National Academy of Sciences of the USA* **98a**:5433–5440 [*203, 361*]

Tilman D, May RM, Lehman CL & Nowak MA (1994). Habitat destruction and the extinction debt. *Nature* **371**:65–66 [*202*]

Travis JA (1984). Anuran size at metamorphosis: Experimental test of a model based on intraspecific competition. *Ecology* **65**:1155–1160 [*19*]

Travis JMJ & Dytham C (1999). Habitat persistence, habitat availability and the evolution of dispersal. *Proceedings of the Royal Society of London B* **266**:723–728 [*265*]

Travis JMJ, Murrell DJ & Dytham C (1999). The evolution of density-dependent dispersal. *Proceedings of the Royal Society of London B* **266**:1837–1842 [*265*]

Tuljapurkar S (1990). *Population Dynamics in Variable Environments*. Berlin, Germany: Springer-Verlag [*43, 45*]

Turelli M (1984). Heritable genetic variation via mutation–selection balance: Lerch's zeta meets the abdominal bristle. *Theoretical Population Biology* **25**:138–193 [*177–178*]

Turelli M (1985). Effects of pleiotropy on predictions concerning mutation–selection balance for polygenic traits. *Genetics* **111**:165–195 [*183*]

Turelli M (1988). Population genetic models for polygenic variation and evolution. In *Proceedings of the Second International Conference on Quantitative Genetics*, eds. Weir BS, Eisen EJ, Goodman NM & Namkoong G, pp. 601–618. Sunderland, MA, USA: Sinauer Associates Inc. [*177*]

Turelli M & Barton NH (1994). Genetic and statistical analyses of strong selection on polygenic traits: What, me normal? *Genetics* **138**:913–941 [*177*]

Tutin YG, Heywood VH, Burges NA, Moore DM, Valentine DH, Walters SM & Webb DA (1976). *Flora Europaea*, Vol. 4. Cambridge, UK: Cambridge University Press [*297*]

Van der Meijden E, de Jong TJ, Klinkhamer PGL & Kooi RE (1985). Temporal and spatial dynamics in populations of biennials plants. In *Structure and Functioning of Plant Populations. 2. Phenotypic and Genotypic Variation in Plant Populations*, eds. Haeck J & Woldendorp JW, pp. 91–103. Amsterdam, Holland: North-Holland Publishing Company [*284*]

Vanderplank FL (1948). Experiments in cross breeding tsetse flies (*Glossina* species). *Annals of Tropical Medicinal Parasitology* **42**:131–152 [*347*]

Van Dyck H, Matthyssen E & Dhondt AA (1997). Mate-locating strategies are related to relative body length and wing colour in the speckled wood butterfly *Pararge aegeria*. *Ecological Entomology* **22**:116–120 [*291*]

Van Tienderen PH & de Jong G (1986). Sex-ratio under the haystack model – polymorphism may occur. *Journal of Theoretical Biology* **122**:69–81 [*196, 199*]

Van Valen L (1971). Group selection and the evolution of dispersal. *Evolution* **25**:591–598 [*271*]

Vasquez CG & Bohren BB (1982). Population size as a factor in response to selection for 8-week body weight in White-Leghorns. *Poultry Science* **61**:1273–1278 [*87*]

Verboom J, Lankester K & Metz JAJ (1991). Linking local and regional dynamics in stochastic metapopulation models. *Biological Journal of the Linnean Society* **42**:39–55 [*77*]

Verschuren D, Johnson TC, Kling HJ, Edgington DN, Leavitt PR, Brown ET, Talbot MR & Hecky RE (2002). History and timing of human impact on Lake Victoria, East Africa. *Proceedings of the Royal Society of London B* **269**:289–294 [*357*]

Vigouroux Y & Couvet D (2000). The hierarchical island model revisited. *Genetics Selection Evolution* **32**:395–402 [*233*]

Vitousek PM, D'Antonio CM, Loope LL & Westbrooks R (1996). Biological invasions as global environmental change. *American Scientist* **84**:468–478 [*357*]

Vitousek PM, Mooney HA, Lubchenco J & Melillo JM (1997). Human domination of Earth's ecosystems. *Science* **277**:494 [*356*]

Vrijenhoek RC (1994). Unisexual fish: Model systems for studying ecology and evolution. *Annual Review of Ecology and Systematics* **25**:71–96 [*352*]

Waddington CH (1961). Genetic assimilation. *Advanced Genetics* **10**:257–293 [*139*]

Wade MJ (1990). Genotype–environment interaction for climate and competition in a natural population of flour beetles, *Tribolium castaneum*. *Evolution* **44**:2004–2011 [*293*]

Wade MJ & Arnold SJ (1980). The intensity of sexual selection in relation to male sexual behaviour, female choice, and sperm precedence. *Animal Behaviour* **28**:446–461 [*47, 52*]

Wade MJ & McCauley DE (1988). Extinction and recolonization: Their effects on the genetic differentiation of local populations. *Evolution* **42**:995–1005 [*236–237*]

Wagner GP (1989). Multivariate mutation–selection balance with constrained pleiotropic effects. *Genetics* **122**:223–234 [*183*]

Wagner GP & Gabriel W (1990). Quantitative variation in finite parthenogenetic populations: What stops Muller's ratchet in the absence of recombination? *Evolution* **44**:715–731 [*167*]

Wainberg MA, Drosopoulos WC, Salomon H, Hsu M, Borkow G, Parniak MA, Gu ZX, Song QB, Manne J, Islam S, Castriota G & Prasad VR (1996). Enhanced fidelity of 3TC-selected mutant HIV-1 reverse transcriptase. *Science* **271**:1282–1285 [*360*]

Wall DH & Moore JC (1999). Interactions underground: Soil biodiversity, mutualism, and ecosystem processes. *BioScience* **49**:109–117 [*307*]

Walters C (1986). *Adaptive Management of Renewable Resources*. New York, NY, USA: Macmillan [*363*]

Wang J, Hill WG, Charlesworth D & Charlesworth B (1999). Dynamics of inbreeding depression due to deleterious mutations in small populations: Mutation parameters and inbreeding rate. *Genetical Research* **74**:165–178 [*132–133*]

Ward RD, Skibinski DOF & Woodwark M (1992). Protein heterozygosity, protein structure, and taxonomic differentiation. *Evolutionary Biology* **26**:73–159 [*97*]

Ware AB, Kaye PT, Compton SG & Noort SV (1993). Fig volatiles: Their role in attracting pollinators and maintaining pollinator specificity. *Plant Systematics and Evolution* **186**:147–156 [*318*]

Waser NM & Price MV (1989). Optimal outcrossing in *Ipomopsis aggregata*: Seed set and offspring fitness. *Evolution* **43**:1097–1109 [*284*]

Waser NM, Chittka L, Price MV, Williams NM & Ollerton J (1996). Generalization in pollination systems, and why it matters. *Ecology* **77**:1043–1060 [*306*]

Washitani I (1996). Predicted genetic consequences of strong fertility selection due to pollinator loss in an isolated population of *Primula sieboldii*. *Conservation Biology* **10**:59–64 [*313*]

Watkinson AR & Sutherland WJ (1995). Sources, sinks and pseudo-sinks. *Journal of Animal Ecology* **64**:126–130 [*251*]

Watmough SA & Hutchinson TC (1997). Metal resistance in red maple (*Acer rubrum*) callus cultures from mine and smelter sites in Canada. *Canadian Journal of Forest Research* **27**:693–700 [*95*]

Watt KEF (1968). *Ecology and Resource Management*. New York, NY, USA: McGraw-Hill [*24*]

Wayne RK & Jenks SM (1991). Mitochondrial DNA analysis implying extensive hybridization of the endangered red wolf *Canus rufus*. *Nature* **351**:565–568 [*352*]

Wcislo WT & Cane JH (1996). Floral resource utilization by solitary bees (Hymenoptera: Apoidea) and exploitation of their stored food by natural enemies. *Annual Review of Entomology* **41**:257–286 [*318*]

Weber KE (1990). Increased selection response in larger populations. I. Selection for wing-tip height in *Drosophila melanogaster* at three population sizes. *Genetics* **125**:579–584 [*87*]

Weber KE & Diggins LT (1990). Increased selection response in larger populations. II. Selection for ethanol vapor resistance in *Drosophila melanogaster*, at two population sizes. *Genetics* **125**:585–597 [*87*]

Weiblen GD (2002). How to be a fig wasp. *Annual Review of Entomology* **47**:299–330 [*316*]

Weiss H & Bradley RS (2001). What drives societal collapse? *Science* **291**:609–610 [*362*]

Weiss H, Courty MA, Wetterstrom W, Senior L, Meadow R, Guichard F & Curnow A (1993). The genesis and collapse of 3rd millennium North Mesopotamian civilization. *Science* **261**:995–1004 [*362*]

Wells H, Strauss E, Rutter M & Wells P (1998). Mate location, population growth and species extinction. *Biological Conservation* **86**:317–324 [*24*]

Wendel JF & Percy RG (1990). Allozyme diversity and introgression in the Galapagos Islands endemic *Gossypium darwinii* and its relationship to continental *G. barbadense*. *Biochemical Systematics and Ecology* **18**:517–528 [*350*]

Westerman JM & Parsons PA (1973). Variation in genetic architecture at different doses of gamma-radiation as measured by longevity in *Drosophila melanogaster*. *Canadian Journal of Genetical Cytology* **15**:289–298 [*137, 146*]

Western D (2001). Human-modified ecosystems and future evolution. *Proceedings of the National Academy of Sciences of the USA* **98**:5458–5465 [*356, 359–364*]

Western D & Pearl M (1989). *Conservation Biology in the 21st Century*. Oxford, UK: Oxford University Press [*59*]

White GC (2000). Population viability analysis: Data requirements and essential analysis. In *Research Techniques in Animal Ecology*, eds. Boitani L & Fuller TK, pp. 288–331. New York, NY, USA: Columbia University Press [*40*]

White EB, DeBach P & Garber MJ (1970). Artificial selection for genetic adaptation to temperature extremes in *Aphytis lingnanensis. Hilgardiaa* **40**:161–192 [*92*]

Whitlock MC (2000). Fixation of new alleles and the extinction of small populations: Drift load, beneficial alleles, and sexual selection. *Evolution* **54**:1855–1861 [*164, 167, 169*]

Whitlock MC & Barton NH (1997). The effective size of a subdivided population. *Genetics* **146**:427–441 [*233, 238*]

Whitlock MC & Bourguet D (2000). Factors affecting the genetic load in *Drosophila*: Synergistic epistasis and correlations among fitness components. *Evolution* **54**:1654–1660 [*163–164*]

Whitlock MC & McCauley DE (1990). Some population genetic consequences of colony formation and extinction: Genetic correlation within founding groups. *Evolution* **44**:1717-1724 [*237–238*]

Wielgus RB, Sarrazin F, Ferrière R, Clobert J (2001). Estimating effects of adult male mortality on grizzly bear population growth and peristence using matrix models. *Biological Conservation* **98**:293–303 [*58*]

Wilbur HM (1984). Complex life cycles and community organization in amphibians. In *A New Ecology: Novel Approaches to Interactive Systems*, ed. Price PW, pp. 114–131. New York, NY, USA: John Wiley & Sons [*19*]

Wilcox BA & Murphy DD (1985). Conservation strategy: The effects of fragmentation on extinction. *The American Naturalist* **125**:879–887 [*285*]

Williamson M (1996). *Biological Invasions*. New York, NY, USA: Chapman & Hall [*201, 305, 346*]

Williamson M & Brown KC (1986). The analysis and modelling of British invasions. *Philosophical Transactions of the Royal Society of London B* **314**:505–522 [*313*]

Willis JH & Orr HA (1993). Increased heritable variation following population bottlenecks: The role of dominance. *Evolution* **47**:949–957 [*131, 148*]

Willis K & Wiese RJ (1997). Elimination of inbreeding depression from captive populations: Speke's gazelle revisited. *Zoo Biology* **16**:9–16 [*133*]

Wilson DS (1980). *The Natural Selection of Populations and Communities*. Menlo Park, CA, USA: Benjamin/Cummings [*335*]

Wilson EO (1992). *The Diversity of Life*. New York, NY, USA: WW Norton & Company [*315*]

Wilson CA, Beckman DW & Dean JM (1987). Calcein as a fluorescent marker of otoliths of larval and juvenile fish. *Transactions of the American Fisheries Society* **116**:668–670 [*104*]

Wilson WG, Morris WF & Bronstein JL (2003). Coexistence of mutualists and exploiters on spatial landscapes. *Ecological Monographs* **73**:397–413 [*307*]

Witmer MC & Cheke AS (1991). The dodo and the tambalacoque tree: An obligate mutualism reconsidered. *Oikos* **61**:133–137 [*315*]

Witte R, Goldschmied T, Wanink J, Van Oijen M, Goudswaard K, Witte Maas E & Bouton N (1992). *Environmental Biology of Fishes* **34**:1–28 [*357*]

Wolin CL (1985). The population dynamics of mutualistic systems. In *The Biology of Mutualism*, ed. Boucher DH, pp. 248–269. New York, NY, USA: Oxford University Press [*308*]

Woodruff DS (2001). Declines of biomes and biotas and the future of evolution. *Proceedings of the National Academy of Sciences of the USA* **98**:5471–5476 [*358, 363–364*]

Woods RE, Sgrò CM, Hercus M & Hoffmann AA (1999). The association between fluctuating asymmetry, trait variability, trait heritability, and stress: A multiply replicated experiment on combined stresses in *Drosophila melanogaster*. *Evolution* **53**:493–505 [*142–143, 146–147*]

Woodworth LM (1996). *Population Size in Captive Breeding Programs*. PhD Dissertation. Sydney, Australia: Macquarie University [*97*]

Wootton JT (1994). The nature and consequences of indirect effects in ecological communities. *Annual Review of Ecology and Systematics* **25**:443–466 [*333*]

Wright S (1921). Systems of mating. *Genetics* **6**:111–178 [*173*]

Wright S (1931). Evolution in Mendelian populations. *Genetics* **16**:97–159 [*6, 193*]

Wright S (1932). The roles of mutation, inbreeding, crossbreeding and selection in evolution. *Proceedings of the 6th International Congress of Genetics* **1**:356–366 [*189, 193*]

Wright S (1940). Breeding structure of populations in relation to speciation. *The American Naturalist* **74**:232–248 [*236*]

Wright S (1943). Isolation by distance. *Genetics* **28**:114–138 [*235*]

Wright S (1951). The genetical structure of populations. *Annals of Eugenics* **15**:323–354 [*232*]

Wright S (1967). Surfaces of selective value. *Proceedings of the National Academy of Sciences of the USA* **102**:81–84 [*193*]

Wright S (1969). *Evolution and the Genetics of Populations*, Volume 2. Chicago, IL, USA: University of Chicago Press [*189*]

Wright S (1977). *Evolution and the Genetics of Populations. Volume 3. Experimental Results and Evolutionary Deductions.* Chicago, IL, USA: University of Chicago Press [*96*]

Wright S (1988). Surfaces of selective value revisited. *The American Naturalist* **131**:115–123 [*193*]

Yamamura N & Tsuji N (1995). Optimal strategy of plant antiherbivore defense: Implications for apparency and resource-availability theories. *Ecological Research* **10**:19–30 [*329*]

Yodzis P (1988). The indeterminacy of ecological interactions as perceived through perturbation experiments. *Ecology* **69**:508–515 [*333*]

Yodzis P (1989). *Introduction to Theoretical Ecology*. New York, NY, USA: Harper & Row [*190, 194*]

Young AG & Clarke GM, eds. (2000). *Genetics, Demography and Viability of Fragmented Populations*. Cambridge, UK: Cambridge University Press [*59*]

Young TP, Stubblefield CH & Isbell LA (1997). Ants on swollen-thorn acacias: Species coexistence in a simple system. *Oecologia* **109**:98–107 [*320*]

Yu DW (2001). Parasites of mutualisms. *Biological Journal of the Linnean Society* **72**:S29–S46 [*307*]

Zhivotovsky LA (1997). Environmental stress and evolution: A theoretical study. In *Environmental Stress, Adaptation and Evolution*, eds. Bijlsma R & Loeschcke V, pp. 241–255. Basel, Switzerland: Birkhäuser [*141*]

Zhivotovsky LA, Feldman MW & Bergman A (1996). On the evolution of phenotypic plasticity in a spatially heterogeneous environment. *Evolution* **50**:547–588 [*141*]

Zirkle DF & Riddle RA (1983). Quantitative genetic response to environmental heterogeneity in *Tribolium confusum*. *Evolution* **37**:637–639 [*123*]

Zouros E, Romero-Dorey M & Mallet AL (1988). Heterozygosity and growth in marine bivalves: Further data and possible explanations. *Evolution* **42**:1332–1341 [*128*]

Index

The International Institute for Applied Systems Analysis

is an interdisciplinary, nongovernmental research institution founded in 1972 by leading scientific organizations in 12 countries. Situated near Vienna, in the center of Europe, IIASA has been producing valuable scientific research on economic, technological, and environmental issues for nearly three decades.

IIASA was one of the first international institutes to systematically study global issues of environment, technology, and development. IIASA's Governing Council states that the Institute's goal is: *to conduct international and interdisciplinary scientific studies to provide timely and relevant information and options, addressing critical issues of global environmental, economic, and social change, for the benefit of the public, the scientific community, and national and international institutions.* Research is organized around three central themes:

- Energy and Technology;
- Environment and Natural Resources;
- Population and Society.

The Institute now has National Member Organizations in the following countries:

Austria
The Austrian Academy of Sciences

China
National Natural Science Foundation of China

Czech Republic
The Academy of Sciences of the Czech Republic

Egypt
Academy of Scientific Research and Technology (ASRT)

Estonia
Estonian Association for Systems Analysis

Finland
The Finnish Committee for IIASA

Germany
The Association for the Advancement of IIASA

Hungary
The Hungarian Committee for Applied Systems Analysis

Japan
The Japan Committee for IIASA

Netherlands
The Netherlands Organization for Scientific Research (NWO)

Norway
The Research Council of Norway

Poland
The Polish Academy of Sciences

Russian Federation
The Russian Academy of Sciences

Sweden
The Swedish Research Council for Environment, Agricultural Sciences and Spatial Planning (FORMAS)

Ukraine
The Ukrainian Academy of Sciences

United States of America
The National Academy of Sciences

The International Institute for Applied Systems Analysis